An Anthology of Visual Double Stars

Modern telescopes of even modest aperture can show thousands of double stars. Many are faint and unremarkable but hundreds are worth searching out. Veteran double star observer Bob Argyle and his co-authors take a close-up look at their selection of 175 of the night sky's most interesting double and multiple stars. The history of each system is laid out from the original discovery to what we know at the present time about the stars. Wide-field finder charts are presented for each system, along with plots of the apparent orbits and predicted future positions for the orbital systems. Recent measurements of each system are included, which will help you to decide whether they can be seen in your telescope, as well as giving advice on the aperture needed. Double star observers of all levels of experience will treasure the level of detail given in this guide to these jewels of the night sky.

Bob Argyle has observed double stars since 1966. He has been Director of the Webb Deep-Sky Society's Double Star Section since 1970. He edited *Observing and Measuring Visual Double Stars* (Springer 2012) and writes monthly columns on double stars for *Astronomy Now* and the Webb Society. He is a Fellow of the Royal Astronomical Society, a Member of the International Astronomical Union, and Editor of *Observatory* magazine.

Mike Swan worked for the Ordnance Survey in England. He has extensive experience in computer graphics and uranography and was solely responsible for the *Webb Society Star Atlas*. He is currently completing a two-part *Atlas of Galactic Clusters*. For the present volume he has produced the finder charts, the all-sky charts, and the orbital plots.

Andrew James has been interested in double stars since the late 1970s. He is a long-term member, and past President, of the Astronomical Society of New South Wales (ASNSW), and formed its Double Star Section in 1979. His interests include the historical backgrounds and works of various discoverers of southern double stars.

An Anthology of Visual Double Stars

Bob Argyle
University of Cambridge

Mike Swan
Webb Deep-Sky Society

Andrew James
Astronomical Society of New South Wales

CAMBRIDGE
UNIVERSITY PRESS

University Printing House, Cambridge CB2 8BS, United Kingdom

One Liberty Plaza, 20th Floor, New York, NY 10006, USA

477 Williamstown Road, Port Melbourne, VIC 3207, Australia

314–321, 3rd Floor, Plot 3, Splendor Forum, Jasola District Centre, New Delhi – 110025, India

79 Anson Road, #06–04/06, Singapore 079906

Cambridge University Press is part of the University of Cambridge.

It furthers the University's mission by disseminating knowledge in the pursuit of education, learning, and research at the highest international levels of excellence.

www.cambridge.org
Information on this title: www.cambridge.org/9781316629253
DOI: 10.1017/9781316823163

© Robert W. Argyle, Mike Swan, and Andrew James 2019

This publication is in copyright. Subject to statutory exception and to the provisions of relevant collective licensing agreements, no reproduction of any part may take place without the written permission of Cambridge University Press.

First published 2019

Printed in the United Kingdom by TJ International Ltd. Padstow Cornwall

A catalogue record for this publication is available from the British Library.

Library of Congress Cataloging-in-Publication Data
Names: Argyle, Robert W., author. | Swan, Mike, 1952– author. | James, Andrew (Astronomer), author.
Title: An anthology of visual double stars / Robert W. Argyle (University of Cambridge),
    Mike Swan (Webb Deep-Sky Society), Andrew James (Astronomical Society of New South Wales).
Description: Cambridge ; New York, NY : Cambridge University Press, 2019. |
    Chiefly a catalog of double stars. | Includes bibliographical references and index.
Identifiers: LCCN 2018061697 | ISBN 9781316629253 (pbk. : alk. paper)
Subjects: LCSH: Double stars–Catalogs. | Double stars–Observers' manuals. |
    Astronomy–Observers' manuals. | Amateur astronomy.
Classification: LCC QB821 .A74 2019 | DDC 523.8/41–dc23
LC record available at https://lccn.loc.gov/2018061697

ISBN 978-1-316-62925-3 Paperback

Chapter openings image credit: 1412/Moment/Getty Images

Cambridge University Press has no responsibility for the persistence or accuracy of URLs for external or third-party internet websites referred to in this publication and does not guarantee that any content on such websites is, or will remain, accurate or appropriate.

Dedicated to Angela Argyle, Angela Kelly, and the memory of
Catherine McMahon (1964–2001)

CONTENTS

ABOUT THE AUTHORS

Bob Argyle

I have been observing double stars ever since I acquired a telescope in 1966. To this day I do not know what prompted me to give them special attention but I do know that using a copy of *Norton's Star Atlas* which was given to me by an early mentor, Frank Acfield of Newcastle-upon-Tyne, had an effect. I suspect it was the lack of up-to-date information about the separation and position angles included in the lists which accompanied each of the star maps that prompted me to start making observations, by eye at first. In my copy (the 15th edition of 1964), for example, the date of the given position angle and separation for the bright binaries was no later than 1938 and many of the wider pairs were 20 or 30 years older than that. This gave me the distinct impression that here was something useful that could be done, and I remember thinking that I must get a micrometer in order to do some of this work properly. At that time, however, micrometers were rare and expensive objects and I never did come across one to use on my 10-inch reflector.

Around this time I bought a copy of Webb's *Celestial Objects for Common Telescopes*, Volume 2, which was available then in the paperback reproduction issued by Dover. It is a little treasure chest of double stars and considerably expanded the number available to a small telescope compared to the lists in *Norton*. As a subscriber to *The Astronomer* magazine I followed the columns of 'From the Night Sky' written by John Larard, which described his observations of double stars made at Mill Hill using an 8-inch Cooke refractor.

John spent much, perhaps too much, of his energy bringing the Webb Society into being as he felt there was a distinct lack of direction in deep-sky and double star observation amongst the amateur community.

I joined the Webb Society in 1968 and was soon sending in observations of double stars to John, who was then Director of the Double Star Section. After a reorganisation in 1970 John became Director of the Nebulae and Clusters Section, and I was asked to direct the Double Star Section. I'm still doing it.

Figure 0.1 The 8-inch (20-cm) Thorrowgood Cooke refractor at the Observatories, Cambridge. It was delivered to Dawes in 1865 and was left to the Royal Astronomical Society in 1927. It has been at the Institute of Astronomy, Cambridge, since 1929 (R. W. Argyle)

The 8-inch refractor at Cambridge Observatories is an historic instrument. Measurements started in 1990 and the programme continues to this day – 29 years on. This volume contains some results of these observations.

Mike Swan

I was employed as a cartographer with the Ordnance Survey (OS) for over 25 years, working in Southampton and Birmingham. Nearly 20 years of that time was spent working in all aspects of digital mapping. I retired from the OS early and moved to the dark skies of western Ireland.

Combining cartography with an enthusiasm for astronomy, it was only natural that I became a uranographer. I started creating maps of the stars from the mid 1970s, producing them professionally for various books and publications, most notably the 18th and 19th editions of Norton's *Star Atlas*. With the advent of home computing, and easy access to databases and software programs, it became easier and quicker to create maps of the stars. I have produced the *Webb Star Atlas* for the Webb Deep-Sky Society and I am currently working on Volume 2 of my two-part publication, *Atlas of Open Star Clusters*, the first part of which is now available.

For this present volume I produced the all-sky charts, the finder charts, and the orbital plots.

Andrew James

I have been interested in double stars since the late 1970s. I am a long-term member, and past President, of the Astronomical Society of New South Wales (ASNSW) and formed its Double Star Section in 1979. I presented many papers on double stars to the National Australian Convention of Amateur Astronomy (NACAA) between 1980 and 2014. In recent times I have focussed on southern double stars and the historical backgrounds and works of various discoverers in Australia, including James Dunlop and Charles Rümker, who made the first southern double star catalogues.

Another associated interest is the astronomer Henry Chamberlain Russell, and various observational assistants, who found and measured many new doubles within the Sydney Observatory Double Star Programme between 1870 and 1900.

Further current investigations have recently extended to a new examination of Sir Thomas Brisbane's Paramatta Star Catalogue, created during the 1820s at Sydney, and its important connection to the discovery of double stars and deep-sky objects.

General southern historical accounts and information on double stars and some selected double stars also appear on my website, Southern Astronomical Delights www.southastrodel.com. I am still active in the local astronomical scene in Sydney, and between 2013 and 2015 I acted as a consultant to the design revamp of Sydney Observatory's new East Dome, used by the public – especially aimed for access by the disabled and seniors. My astronomical experience also extends to lecturing on the subject to Evening Colleges, and I have presented many talks over a large range of subjects. I am presently the Planetary Nebulae and Deep-Sky Section Leader of the ASNSW.

ACKNOWLEDGEMENTS

The data in this book relies heavily on several sources. The *WDS Catalog*, maintained by the United States Naval Observatory, is the clearing house for all observations of visual double stars. This database has been managed for a number of years by Dr Brian Mason and Dr Bill Hartkopf, whose prompt and generous cooperation with requests for data has been much appreciated.

When it was released the Hipparcos catalogue contained high-quality astrometric data for about 118,000 bright stars down to $V = 11$ or so. In particular the trigonometrical parallaxes were most welcome, especially in the case of binary stars where estimates of the total stellar mass in the system could be appreciably improved. In addition, photometry in two colours was also taken and that data has been used in this catalogue.

In April 2018, the Gaia project [682] released its second interim catalogue (*DR2*). This unprecedented compilation lists positions, proper motions, and parallaxes for about 1.6 billion stars, and, for a subset of these, radial velocities are also available. The astrometry of many bright, close, visual binaries will not be available until *DR3*, which is expected to be issued in 2020. The catalogue in this book uses *DR2* data wherever possible.

This work has made use of data from the European Space Agency (ESA) mission Gaia, `www.cosmos.esa.int/web/gaia`, processed by the Gaia Data Processing and Analysis Consortium (DPAC), `www.cosmos.esa.int/web/gaia/dpac/consortium`. Funding for the DPAC has been provided by national institutions, in particular the institutions participating in the Gaia Multilateral Agreement.

The authors are also indebted to the compilers of the SIMBAD catalogue, which has been invaluable, especially for checking references to papers on each system discussed.

RWA also would like to thank Christopher Taylor for permission to reproduce entries from his observing logbook, Michael Greaney for the conversion table in the Appendix, and Drs Henry Zirm and Jack Drummond for high-resolution images for inclusion in the text.

Throughout the course of this project, Mark Hurn, Librarian at the Institute of Astronomy, has been a reliable and sympathetic source of help, especially in locating the most obscure references.

Continuous help and encouragement from staff at CUP, notably Esther Miguéliz Obanos and Vince Higgs, is gratefully acknowledged.

CHAPTER

1

Introduction

Christian Mayer published the first proper double star catalogue in 1784. However, the study of visual double stars effectively started with William Herschel. He not only systematically searched for them, initially to try to determine stellar parallax, but ended up by proving that some pairs of stars that appeared close together really were attracted by a common gravitational pull. By 1830 we knew of more than 3000 double stars but this figure did not exceed 10,000 until 1892. The current number is 146,000 and we stand on the brink of a revolution in terms of discovery which will come when the Gaia satellite ends its current mission. It is expected that millions of new pairs will be catalogued; however, Gaia will not only find new systems, but will also give us more detail on existing systems, in particular distances, proper motions, and multiplicity.

This volume aims to fill out the story on some of the brighter and more interesting visual double stars which can be seen in small and moderate apertures. Here we may define small as 20 cm or less, medium as 20 to 40 cm and large as 40 to 60 cm. Many double stars have contrasting colours; some are very unequal, some are at or beyond the resolving limit. To make a selection of pairs which can all be seen in a small aperture risks discouraging observers with larger telescopes who wish to push them optically. Its also fair to say that views of many pairs improve with aperture. Some systems have three or more components. None are the same.

It is, of course, impossible to definitively choose the finest visual double stars in the sky. Such a distinction depends too heavily on personal taste. The catalogue of objects that is discussed here does not necessarily contain the brightest, the most spectacularly coloured, the closest or the most-difficult-to-see pairs that exist in the heavens. Rather it is a summary of all these properties: visual pairs are presented that have something which attracts the observer to observe them. What can be said is that the vast majority have been observed (or attempted) by the authors, and hopefully this fact will recommend them to the reader.

Visual double stars come in two main forms. *Optical* pairs consist of two stars which appear to be angularly close on the sky but which are in reality totally unrelated because one star is much further away than the other. *Binary* stars appear close together on the sky because they are, in fact, physically connected and are rotating around a common centre of gravity. For the common telescope user the periods in which they do this range from about 25 years to tens of thousands of years. Binary stars are important to stellar astronomers. Measurements of the relative positions of the two components give good information on stellar masses and can also indicate how far away such stars are.

There has been a historical imbalance in the professional approach to research into visual double stars. This is clear from the numbers of observations made of southern binaries as opposed to those visible from the northern hemisphere. As an example, γ Virginis has had 1720 measures whereas its southern equivalent, γ Cen, has only 181, yet the latter has half the orbital period of the former.

The whole sky is covered in this volume. Not to do so would be to leave out some of the best objects in this class. In fact the very finest double star in the sky, α Centauri, is only practically visible to anyone south of about +25° – say, the latitude of Hawaii. Not every object in the catalogue will be resolvable, even with a significant aperture, but, because of the changing nature of binary stars, that will not always be the case. That well-known test object, the companion to the star

γ And B, is now way beyond the resolution of most telescopes and will remain so for ten years or so; in the case of ζ Boo it might be 20 years before it can be seen again. As apertures and techniques used by amateurs are constantly getting larger, these bland statements of fact might be seen as challenges to the observer wishing to test his or her telescope optics. When it comes to deciding whether stars of a particular separation can be divided by a given aperture there are a number of formulae can be used to check this possibility ([685]). We have the Rayleigh limit and the Dawes limit; whilst other more complex predictors exist, in my experience the Dawes limit is as good (and as simple) as any. The atmosphere and state of the telescope are just as important in obtaining a resolution. Go and find out for yourself.

If you are new to double star observing, it will take a little time to get used to the look of the images but, eventually, even though you cannot see two separate images, the fact that a close pair is double will usually be apparent given enough time. Even on nights when the atmosphere is very unsteady there will be the odd moment of calm when the images sharpen. This is the basis of the lucky imaging technique, which some observers are using to good effect with CCD cameras. How far can you go visually? With careful collimation of telescope optics and some experience of examining images from close pairs of stars, it is possible to see duplicity in stars at the $0''.2$ level with a 32-cm telescope whilst the Dawes limit is formally accepted as $0''.36$.

Other bright binaries move sufficiently quickly that the angular motion can be seen in the course of a year. Eyepieces contain a fine illuminated mesh of grid lines, called graticule eyepieces, can be used to make more precise measurements of the relative angle and separation between two stars but are not, in general, suitable for binary stars, which tend to be too close. Other instruments, such as micrometers for the visual observer or a CCD camera for those who prefer their stars served electronically, can be employed. It is outside the scope of this book to describe these techniques, but see the volume edited by Argyle [85].

2 Observing Double Stars

Visual double stars are usually observed for one or more of three main reasons. Firstly, they offer a range of colour often missing in other deep-sky objects. Secondly, they can be used to test the performance of an objective lens or mirror as they offer a whole range of separations and magnitudes. Finally, measuring visual double stars makes a contribution to the determination of stellar masses. There are many pairs, some of them quite bright, that have periods of centuries, and regular and continuous measurement of these helps to pin down the orbits.

Visual double stars can be observed in the worst of sky conditions or in the periods near Full Moon, and even the smallest aperture can show some impressive sights. In fact a good pair of stabilised binoculars can reveal some of the prettiest double stars in the sky – i.e. Albireo, θ Serpentis, and Mizar, to name but a few, but an aperture of 10 to 15 cm will open up a vista of hundreds of double stars. When I first started in astronomy I spent several years just noting down my impressions of about a thousand systems, occasionally making a sketch if the field was particularly interesting, estimating colours, position angles, and separations. I started off with a homemade 6-inch (15-cm reflector) on a concrete pillar, quickly followed by an 8.3-inch (21-cm), observing from a garden in North Shields with a very bright sky and fairly restricted horizons. The southern aspect, for instance, was over the house and down towards the River Tyne, and in those days it was still a hive of shipbuilding industry. Needless to say, the heat from the houses meant that the seeing was never very good, so actually dividing pairs closer than 2 arcseconds I regarded as a bit of an achievement. Occasionally the sky relented and pairs closer than 1 arcsecond could be seen. Towards the end of this period I installed a 10-inch (25-cm) f/6 Newtonian in an observatory.

This book is intended for those who observe double stars visually but, for the sake of completeness, some brief descriptions of other methods will be given below. The books by Argyle [85], MacEvoy & Tirion [86], and Benavides *et al.* (in Spanish) [100] contain substantially more detail.

Methods of Observation

Naked Eye

The resolution of the human eye depends on many factors not the least of which is age; also important is the brightness of the two stars. The naked-eye pair that most people can see is Mizar and Alcor in Ursa Major. These stars, of magnitudes 2.0 and 4.0, are more than 11 arcminutes apart. The dark-adapted diameter of the pupil can reach 8 mm, which implies that double stars of separation 17″ should be resolvable without optical aid, but in practice this value is much larger and also depends on other factors. Jerry Lodigruss [92] considers that no-one can resolve 1 arcminute with the naked eye and thinks that the limit is nearer 2 arcminutes for keen eyes and 3 arcminutes for average eyes. A good test for the keen eye is ϵ Lyrae, but on his website Lodigruss gives a range of naked-eye pairs, of which the following make a good starting point:

θ Tau, magnitudes 3.4, 3.8, sep. 5′.6; α Cap, magnitudes 3.6, 4.3, sep. 6′.3; ϵ Lyr, magnitudes 4.6, 4.7, sep. 3′.5; and μ Sco, magnitudes 3.0, 3.6, sep 5′.8.

The first visual double noted without any optical assistance was the two stars which form ν Sgr.

Binoculars

Whilst having the advantages of relative cheapness and lightness compared with telescopes, binoculars also suffer from lack of aperture and lack of magnification. A typical pair will be described as 8 × 30 or 10 × 50. This means the magnification is ×8 (or ×10) and each objective lens is 30 mm (or 50 mm) in diameter. In practice 8 × 50 will limit you to only the brightest and widest pairs (i.e. brighter than magnitude 5 or 6 and wider than about 30″). A good list of stars within range can be found on the Astronomical League website – look for Binocular Double Stars. A significant improvement can be made if a tripod can be acquired on which to mount the binoculars, or stabilised binoculars are used. These contain an active compensation system which damps out to a large degree the vibration imparted to the binoculars when an instrument is hand-held. With a pair of 15 × 50 stabilised binoculars the resolution increases significantly, and pairs nearer 10″ might be seen. For more information take at a look at 'Double Stars for Binoculars' [93].

Telescopes

Although a number of attractive pairs can be reached with binoculars, a small telescope is the best way of making inroads into the catalogue of the more interesting visual double stars, in particular the visual binaries. Such instruments will often reach the diffraction limit, and it does not need any advanced equipment except a good telescope and a range of eyepieces to allow the observer to take advantage of good seeing. For an 8-inch refractor, up to ×600 can be used on special objects (e.g. γ Vir at periastron). Detection of duplicity need not stop at the resolution limit – structure is still visible beyond that – Christopher Taylor has pushed his 12.5-inch (32-cm) Calver to better than 0″.3. For those who do not want to draw,

Figure 2.1 The double star ι Leonis (STF 1536) observed in 2017, by Christopher Taylor with a 4-inch refractor.

verbal descriptions can be recorded. Here is an example of an observation with a 4-inch refractor by Christopher Taylor (see also Figure 2.1):

Tuesday 23rd May 2017 21h05-22h00

At the 4″ OG as usual: the seeing I–II on α, δ and θ Leonis, all beautiful, sharp, round, rock-steady star disks continuously at $\sim$ 200 and rings only light flickering, taking this first double star more as a mere 'pretty' initially, rather than as one of the evening's serious targets. The following turned out quite unexpectedly to be, in fact one of JCT's most memorable double star observations ever visually and one of the most important in its implications. An exquisite image of ι Leonis = Σ 1536 at $\sim$ 200, with the companion seen almost instantly like the most infinitesimal needle-prick of light having no perceptible size, easily and continuously visible on, or just outside of A's 1st ring – or rather, where that ring would be if its light were not confined almost entirely to the opposite preceding semicircle – thus:

Having adjusted the illuminated crosswire accurately preceding – following the tiny star was seen to be just below the parallel from 'A' in position angle estimated at $93°$–$95°$ by D. W. – one of the most exquisite double star sights ever seen, like a minuter and more delicate version of ϵ Bootis, perhaps indeed, the most perfect and beautiful view ever of such a delicate double star, in my telescope – in a 4″ glass!

Drawing

Jeremy Perez has years of experience in drawing the fields of double stars (see his publications [87], for instance, or his website [89]). This is quite a challenging exercise as star colours can be rather subtle and the scale needs to be chosen so that both components of the target double are clearly shown and the surrounding field is also given, to include any other components and to aid identification.

Objective Gratings

First used in the nineteenth century were objective gratings with metal bars to diffract the light from very bright stars and distribute it into a series of images equally spaced but getting rapidly fainter, so that faint companion stars could also be measured. Gratings were used by amateur observers in the last century but the main disadvantage is that up to half the incident light is stopped by the bars of the grating.

In addition, the separation between the diffracted orders is slightly dependent on the wavelength of the starlight. On the plus side, such gratings are easy to make and both position angle and separation can be derived from measurements of grating angles (see [85]).

Lunar Occultations

Watching stars disappear or reappear behind the limb of the Moon, it is perfectly possible to see a double event as the limb covers first one and then the other component of a double star, so discoveries may still be made with relatively small apertures. In order to estimate position angle and separation, a second occultation (or reappearance), where the lunar limb is at a different orientation to the line between the stars, is needed. The best events are those where the star is occulted by the dark limb of the Moon. The most difficult event is to observe a reappearance at the bright limb of the Moon. The main body for this work is IOTA (the International Occultation Timing Association), who have been active since 1983. Their website is `https://occultations.org`.

Coronography

By placing a circular obstruction in the focal plane of a telescope, relatively long exposure images around very bright stars can be made, looking for close-in companions. The name 'coronography' derives from a similar method in which a circular mask is used to cut out the bright disk of the Sun in order to examine the solar corona. Stellar images are much smaller than the solar disk at the focal plane so any obstructions need to be quite small. One method is to insert a bar made from silver paper, which gives very sharp edges, through the centre of the eyepiece field (see Daley [5]).

Shaped Apertures

The use of aperture masks of various types – hexagonal etc. – to see faint companions to bright stars, e.g. Sirius, should be considered. The apexes of the hexagonal mask should be at the full aperture of the telescope and the mask should be rotatable.

Graticule Eyepieces

For basic measurements consider a graticule eyepiece. Several of these devices are commercially available (from Celestron and Meade, for instance). Their use is described by Tom Teague in the volume edited by Argyle [95]. They are capable of significant accuracy, given that enough care is taken with measurements – see for instance the work of Neil Webster [96], and Dave Arnold [99].

Micrometry

Invented by William Gascoigne in the late 1630s [90] the astronomical micrometer is used for measuring small angles in the optical paths of telescopes. It is essentially simple in principle but requires a high standard of manufacture, particularly in the construction of the screw threads and the separation and position angle readouts.

In the 1980s and 1990s commercially available micrometers were available from several sources, all of which appear to have dried up now. The RETEL micrometer used by one of the authors (RWA) has 12 micron wires, but care needs to be taken to match the apparent diameter of the threads with the resolution of the telescope – usually a $\times 2$ or $\times 3$ Barlow lens can be used to achieve this. To use a micrometer effectively requires an equatorial telescope with a good right ascension (RA) clock, and though it has traditionally been done with refractors, can be used with reflectors as well.

CCD Imaging

There are many skilled observers using CCD cameras to make direct images of the wider double star systems, where 'wider' means about $5''$, but this relatively large value can be compensated for to some degree by the limiting magnitude V reached – often beyond $V = 15$. In these cases the exposure times used run into tens of seconds or a few minutes and so the images formed are integrated over the time of exposure. If very short exposure times are used (corresponding to, say, 10 to 50 mas) the quality of the images obtained is often variable – some are sharp and others diffuse.

A technique called *lucky imaging* takes advantage of the fact that perhaps one in ten or one in 20 of short-exposure images will produce imaging down to the resolution limit of the telescope, and two sharp nuclei will be visible. If this is repeated for 1000 exposures, the number of sharp images may exceed 50. These can then be co-added in software to produce an integrated sharp image of a close pair which can be measured for position angle and separation. Large apertures tend to be more affected by atmospheric seeing, so benefit from longer runs of images, but in any case this will only take a few minutes per object. The advantage is that the data can be post-processed. Lucky imaging goes down to the diffraction limit provided that the telescope has good optics and is well collimated, the pixel size is correctly chosen, and the exposure time can be reduced far enough to freeze the images. See, for instance, the recent work of Rainer Anton [97]; more details of the technique can be found in Anton [88].

Speckle Interferometry

This technique is beyond the scope of this book but is being used by a number of non-professional observers. It can be used with medium apertures, and software packages such as REDUC [91] are freely available to handle the data reduction for those who want to try it.

CHAPTER

3 History of Measurement Techniques

The measurements given in the Catalogue section were made using a variety of methods, which were described in more detail in Chapter 2. All the earliest measurements were obtained using filar micrometers, whilst a significant number of the most recent results were made with the techniques that are now current. The following is a brief summary of the various methods and how they were developed and used.

Whilst double stars were first seen telescopically in 1617 it was to be another 150 years before an attempt was made to measure them, even though the idea of using a micrometer fitted to the eyepiece of a telescope to measure the separation of two stars and the orientation of the line between them had already been presented by William Gascoigne in England around the time of the English Civil War (1640).

It was William Herschel who developed the idea to such a pitch that even today Herschel's measures are taken seriously, and not just because he extended the time range of measures another 40 years or so beyond the measures of F. G. W. Struve at Dorpat. Herschel had two micrometers. One consisted of two wires, one of which could be rotated 360° whilst the other was fixed. This allowed Herschel to measure the position angle directly whilst the 'wedge' formed by the two wires placed over the two components gave a measurement of their separation, using some trigonometry. The second device was placed at some distance from the eyepiece and essentially produced two star-like point images, which could be separated and rotated by a complex mechanism of ropes, pulleys, and wooden blocks, with the illumination produced by a candle. This was an ingenious forerunner of the comparison-image micrometer which was developed in England, but hardly used, in the early twentieth century.

The modern wire micrometer first saw the light of day around 1825 when Fraunhofer made one for the Dorpat refractor used by the elder Struve. Here, there was a fixed wire but two other wires were set at right angles to it, both of which could be moved independently by a micrometer screw and whose position could be read in linear measure, which was then converted into angular measure by the use of an equatorial transit star. This is known as a bifilar micrometer; a variant called the filar micrometer is sometimes used. In this case only one of the two parallel wires is controlled by a micrometer screw. These micrometers dominated double star astronomy for well over a century.

Towards the end of the nineteenth century Karl Schwarzschild developed a device that was essentially a coarse diffraction grating, which fitted over the objective of a telescope. It was hinged in the centre and could be opened or closed so that the apparent width of the grating bars and spaces could be varied at will. What the observer saw in the eyepiece was a bright star accompanied by a series of equally spaced but diminishingly fainter images, which spread out in a direction perpendicular to the direction of the grating bars. Every star in the field produced a similar pattern, the separation of which depended almost entirely on the width of the grating bars and spaces but also slightly on the wavelength of the starlight. By rotating the grating, certain patterns could be obtained, which allowed the separation and the position angle of the two stars to be estimated quite accurately without the need for wires or field illumination.

This multi-slit device, using the constructive and destructive interference of light, was followed by a variant of the Michelson interferometer applied to a large telescope. In the

1920s Pease and Anderson fitted a series of flat mirrors to the end of the 100-inch reflector at Mount Wilson in California. This increased the effective aperture of the telescope to 20 feet. Light from a bright star from each set of mirrors was brought to a focus in the telescope, where two slits were placed. By moving the mirrors apart, a series of dark and bright fringes was obtained from which the separation and position angle of close binary stars or the diameter of nearby giant stars could be directly measured. Pease and Anderson were able to show that the bright star Capella was actually a very close pair of equally bright stars around $0''.05$ apart. Their measurements confirmed that the two stars rotated around each other in 104 days, which had already been determined spectroscopically. This technique worked only on bright objects which had some geometrical structure and thus was not used very much.

Photographic astrometry assumed a much greater importance during the twentieth century, especially for the measurement of proper motions and parallaxes. Large programmes using refractors of up to 30 inches were in place around the world, but attempts to measure binary stars were severely limited by the atmosphere and the resolution of photographic emulsions. The visual method using a filar micrometer and a large refractor remained supreme until well into the 1970s, when observers such as van den Bos and van Biesbroeck were making use of reflectors in the 2-metre aperture class for the micrometry of close pairs down to separations of $0''.1$ or even below. However, this was little better than what Burnham and Aitken could achieve with the Lick 36-inch refractor around 1900.

In the early 1970s Antoine Labeyrie developed speckle interferometry. The name derives from the 'speckles' which observers using large refractors regularly saw in the images of double stars which they were observing. These speckles were actually diffraction-limited images of the star being viewed, and if that star was a close double then the speckles faithfully reproduced the true image of the star. They were formed by interference of the light travelling through different parts of the atmosphere and were in constant motion, so that trying to place the wire of a micrometer on them was very difficult. Labeyrie realised that if he could isolate the speckles then this would effectively cancel out the effect of the atmosphere and allow telescopes to work to the full resolution implied by their aperture. Labeyrie used short exposures and a narrow-band filter in large telescopes to freeze speckle patterns. Then, by adding together a number of exposures and illuminating with a laser, a series of fringes is formed, the spacing of which is directly related to the separation of the stars and the alignment of which is $90°$ to the position angle. Modern detectors use CCD cameras, which avoids the need for time-consuming photographic processes.

Amateur interest in double star measurement was at a low level during much of the twentieth century but interest began to pick up again around 1980, and one or two commercially made filar micrometers appeared on the market. They did not really catch on, however; it was the advent of the video camera followed by the CCD camera that inspired a number of observers to take up the observation of double stars. The spread of astronomical data on the internet also meant that an individual could do research on many double star systems via facilities such as SIMBAD, where astrometry could be done on epochal survey plates and checks for common proper motion made. The observation of close binary stars, however, cannot be done in this way because the resolution is not good enough. A number of skilled amateurs are now using speckle interferometry to produce good and reliable position angles and separations for binary stars at well below the 0.5 arcsecond level.

Whilst speckle interferometry requires more complex reduction methods, an increasingly popular technique called lucky imaging has been gaining traction amongst suitably equipped amateur astronomers. The basic principle is very simple. By taking a short enough exposure of a double star, the distortions of the star images induced by the atmosphere can be essentially frozen. Much of the time the image will appear as a formless mess, but every now and again the atmosphere relents and a diffraction-limited image appears briefly. With the modern processing power of laptops, it is an easy matter to take several thousand short exposure images and then extract the 5% or 10% of which contain useful information, i.e. clearly show both components of a pair. These select images can then be shifted and added so that a better signal-to-noise image is built up, which can then be analysed using available software techniques, and a position angle and separation obtained. The skilled practitioners of this art can produce results comparable with professional measurements using speckle interferometers.

Lunar occultations afford a way for observers with small telescopes to detect pairs of stars which are much closer than the telescope resolution limit. There are a number of groups across the world dedicated to this activity, the most prominent of which is IOTA [106] (the International Occultation Timing Association).

Large telescopes also use adaptive optics to help reduce the effect of the atmosphere on direct imaging. Lasers fired near to the optical axis can sample the turbulence in the upper atmosphere at an altitude of about 70 km and feed back information to the optics in the telescope, which can then be 'altered' to correct for the changes in the atmosphere above the telescope aperture.

Surveys for exoplanets and for low-mass, faint companions to bright stars using both very sensitive spectrographs which can measure radial velocities to a few metres per second and coronagraphs to essentially mask out the very bright primary star, allowing the space nearby to be examined, have also added to our knowledge of duplicity in the solar neighbourhood. Apparently stellar multiplicity is not a barrier to the formation of planetary systems.

The current state of the art in terms of resolution remains on the ground with the construction of large multi-aperture array telescopes such as the CHARA array at Mount Wilson in California, where several pairs of 1-metre telescopes can be combined over baselines of 340 metres to produce resolutions at the level of a few milliarcseconds. This is at least ten times better than can be achieved with speckle on the biggest telescopes. The drawback is that it is a complex and time-consuming procedure but has given us remarkable glimpses into stars such as the eclipsing system β Persei, where all three components have been imaged.

Above the atmosphere the future of double star astronomy will be carried by Gaia. It has a relatively small aperture (there are two telescopes, each with a rectangular mirror of 1.45×0.50 metre^2 aperture, inclined to each other at an angle of $106°.5$), and the focal plane is filled with 106 CCDs occupying an area of 1×0.5 metre^2. It will revolutionise what we know about the frequencies of double stars and will give us unprecedentedly accurate values for the parallaxes, which, in turn, with a knowledge of the orbital parameters, will give the masses of the stars in each binary system. Not only will positions and proper motions also be obtained to unprecedented accuracy and depth but radial velocities will also be available. Here the drawback is the length of time for the mission. It is anticipated that the observational phase will last five years, so many binaries with periods longer than that will need to be observed by alternative methods to extract more information from them, but Gaia will create a catalogue of millions of new pairs. Although it cannot match the angular resolution from the ground, being limited to $0''.1$ or $0''.2$, it has a tremendous multiplex advantage in that many objects can be observed simultaneously.

4 Observational Double Star Groups

At the time of writing there are a number of active groups of double star observers, most of which produce an online bulletin, freely available; these report on the observational activities of double star observers around the world. The list of bulletins below is not expected to be comprehensive.

Webb Society Double Star Section

Formed in 1967, observations consisted of visual estimates of colours and positions until around 1979 when, thanks to a correspondence with Maurice Duruy in the south of France, experiments were made by several members using the grating micrometer, a simpler form of the objective device developed by Karl Schwarzchild and later taken up in the 1930s by Lawrence Richardson in England. M. Duruy had extensive experience of micrometry using his suite of telescopes at Le Rouret in the Alpes Maritimes. With the availability of the commercial micrometers made by Ron Darbinian and the RETEL company in the 1980s, measurements of binary stars were commenced. Contributions are currently made by observers using a variety of techniques – filar micrometers, graticule eyepieces, CCD cameras, and internet astrometry. The results are published in the Double Star Circulars of the Webb Society Double Star Section (DSSC) and appear approximately annually, edited by Bob Argyle. Go to www.webbdeepsky.com/dssc/.

El Observador de Estrellas Dobles

The first edition of this Spanish journal appeared on-line in April 2009 and to date 21 editions have been published, new editions being issued every January and July. The journal is edited by Rafael Benavides, Juan-Luis González, and Edgardo Masa. The website is www.elobservadord eestrellasdobles.wordpress.com.

Société Astronomique de France Étoiles Doubles Section

This group has been established for about 40 years. There are meetings twice yearly – usually in Paris in the spring and in the autumn at a regional centre. Membership includes both amateur and professional astronomers, some of whom use for double star work the large refractors at the Observatory of Nice. Observations are published in *Observations et Travaux*, https://saf-astronomie.fr/etoiles-doubles/.

Il Bollettino delle Stelle Doppie

Published in Italian and edited by Antonio Adigrat, Giuseppe Micello, and Gianluca Sordiglioni, the Bollettino was first issued in April 2012. It has so far run to 25 editions and it is issued twice per year. Go to https://sites.google.com/site/ilbollettinodellestelledoppie/.

Journal of Double Star Observations

In the USA in 1990 Ronald Tanguay formed the Association of Binary Star Observers. The magazine was called the *Double Star Observer*. Both the association and the magazine have since ceased to exist but in spring 2005 a successor magazine, the *Journal of Double Star Observations*, appeared. It is issued every quarter and is available at www.jdso.org. There is an editorial committee consisting of R. Kent Clark, Russell Genet, Richard Harshaw, Jo Johnson, and Rod Mollise which is able to able to call on acknowledged experts in the field such as Brian Mason of USNO to help referee contributions. The 50th issue was published in 2018.

Pro-Am Meetings

Since 2002, an occasional series of conferences has been arranged to bring together non-professional and professional astronomers with an interest in double star observation. The first meeting was held in Castelldefels, near Barcelona in Spain, in 2002 and the fourth (in 2014) was held nearby in Vilanova i la Geltrú. The list of talks at the fourth meeting is available on the Garraf Observatory website (www.oagarraf.net) and a report appears in *El Observador de Estrellas Dobles*, no. 16, 2016.

Discussion and Help Groups

Below is a selection of the most useful internet sites for the double star observer.

Star Splitters

Star Splitters, www.bestdoubles.wordpress.com, is an internet discussion forum which has been established for about eight years. The main contributors are John Nanson and Greg Stone. They use a combination of catalogue data and plotting programs to support a lot of observation using readily available apertures and to share experiences of observing and make more widely known some of the obscure pairs and multiple systems in the catalogues.

Cloudy Nights

Despite the name of the group there is a substantial amount of observational experience on display at www.cloudynights.com. Go to Forums, then to Observing, and then Double Star Observing. At last count there were 2800 topics covering all aspects of observing and discussing many different systems.

The Astronomical League

The AL covers a large range of astronomical topics, amongst which are programmes dedicated to the observation of double stars. There are two at present: The Binocular Double Star Observing programme (coordinator: Robert Kerr). The basic observing list has about 120 wide pairs, which range in separation from about $20''$ to $15'$ and whose faintest stars reach magnitude 9.4. For those with more powerful instruments, the Advanced Binocular Double Star list has 110 systems which push the closest separations down to $10''$, and the faintest components are near $V \sim 10$.

The Double Star programme is coordinated by Cliff Mygatt and the observing list currently has 100 stars on it, many of which are in common with the pairs in this volume.

The Astronomical League website is www.astroleague.org.

Observational Centre in Namibia

The International Amateur Sternwarte (IAS) is located at a site at an altitude of 1834 metres at Hakos Farm, which is 135 km southwest of Windhoek, in Namibia. Nearby is the Gamsberg plateau where, since 1970, the Max-Planck Institute for Astronomy has maintained a presence. At the present time amateur astronomers are allowed access to a 70-cm photographic and visual telescope on the plateau but most work is done at the farm. In the last ten years, Rainer Anton and Karl-Ludwig Bath have made a number of observing trips and have accumulated some high-resolution images of visual double stars, obtained by means of lucky imaging and speckle interferometry. The results can be found in the *Journal of Double Star Observations* (see above).

CHAPTER

5 Double Star Resources

Online Resources

There are many useful sites on the internet which are of interest to the double star observer. This is especially convenient if catalogue information is required, because of the size of the data files involved. The Gaia DR2 catalogue, for instance is 550 GByte even as a compressed download file. The WDS catalogues can be downloaded quickly, enabling them to be updated regularly.

United States Naval Observatory Catalogues

Currently there are more than 146,340 double star systems in the *Washington Double Star* (*WDS*) *Catalog*, which remains the primary source of reference for the double star observer. The size of the WDS precludes a paper format and it can be found online at `http://ad.usno.navy.mil/wds`. It is maintained by the United States Naval Observatory (USNO, in particular by Dr Brian Mason) and is updated nightly. The current size is 20 Mbyte. There are three different formats. One contains, in addition to the basic data on each component of each system, more accurate positions of each star. Another lists both the first and last recorded measures and adds the Bayer and Flamsteed designations and the constellation in which the pair sits, and the third is a structured query language (SQL) database.

In addition, the *Sixth Orbital Catalog* from the same institution is the primary source reference for orbital elements of visual binaries. It is not necessary for a user to be mathematically able since the predicted positions at annual increments are given for each system, along with the apparent orbit. Superimposed on the orbit are the available historical measures of the system. This gives a good indication of how good the orbit description is and whether a particular system is beginning to stray from its predicted path. The *Observations Catalog* is not publicly available, but the measurement history of particular double stars can be requested from USNO. It currently contains 1.715 million entries, of which five basic observational techniques have contributed 87% of the total number of observations. Still the most productive area is micrometric work but in the frequencies considered it is rapidly becoming a less significant proportion of the database. The next two categories are almost as productive but they have only been in use for about 20 years. These are wide-field CCD imaging, such as is used in all-sky surveys, and the use of space techniques such as Hipparcos and HST. The remaining two categories are photographic imaging and photometry. The complete set of measures for each system may be requested from the USNO. All reasonable requests for data in this way are usually met promptly and courteously.

Also of use is the *Fourth Catalog of Interferometric Measures* (*INT4*). This draws together all the extant measures of visual double stars made with interferometric techniques, i.e. speckle, ground-based arrays, and the results from the Hipparcos satellite and, no doubt, from Gaia when the final double star data is available. Many of the systems are substantially below the resolving limit of small telescopes, but there are pairs included here with separations of around an arcsecond or two and the accuracy and contemporaneity of the measurements make them excellent tests for telescopes of all apertures. The full catalogue is about 42 MByte but you can download a gzipped version (8 MByte) or select a particular hour of right ascension (RA).

A recent addition is the *Washington Double Star Supplemental (WDSS) Catalog*, which is designed to include the very large numbers of faint pairs found in all-sky surveys such as 2MASS and Gaia. At the time of writing there were 590,895 entries on 113,538 objects listed under a single WDSS ID number. To distinguish between moving groups and clusters, and double or multiple stars, the number of components in any one system in the WDSS has been limited to eight. The catalogue has also been introduced to test out a new format, the need for which has been under discussion for several decades. This new format is required in order to tackle the increasing number of exoplanetary bodies around double star systems and to incorporate spectroscopic sub-systems, for instance, into the scheme. The catalogue is available on the WDS website, split into four equal bands of RA in order to make download faster; the current size of the whole catalogue is 135 MByte.

Using the Washington Double Star Catalog

This is the most useful reference for the visual double star observer. It is maintained by the Astrometry Department of USNO in Washington, DC, and can be found on the USNO website. There are three main files.

(a) The main catalogue contains one line for each pair of stars. Triple stars will therefore have two entries, quadruple stars three entries, and so on. Go to `www.ad.usno.navy.mil/wds` and click on one of the three options:

  1 WDS Catalog with precise last observation only. This includes more precise RA and Dec for each system, but only gives the latest measurement.

  2 WDS Catalog as an SQL database.

  3 WDS with constellation and Bayer/Flamsteed designation (when applicable) appended. This also gives both the first and last recorded measures.

  The guide to what each column represents can be found below. The WDS website gives more extensive explanatory notes.

  The catalogue is currently 19.6 MByte in size so can be downloaded fairly quickly. It is updated frequently.

(b) The notes catalogue is called `wdsnewnotes_main .txt`. This contains extra useful information for each entry, for instance: alternative catalogue names; details of orbital motion, and where an orbit can be found, if any; expressions of doubt if the pair has not been confirmed to exist or has been lost; the existence of any spectroscopic

WDS SUMMARY CATALOG KEY

Column	Format	Data
1–10	A10	WDS designation (J2000)
11–17	A7	Discoverer and number
18–22	A5	Components
24–27	I4	Date (first)
29–32	I4	Date (last)
34–37	I4	Number of observations
39–41	I3	Position angle (first measure)
43–45	I3	Position angle (last measure)
47–51	F5.1	Separation (first measure)
53–57	F5.1	Separation (last measure)
59–63	F5.2	Magnitude of star A
65–69	F5.2	Magnitude of star B
71–79	A9	Spectral type (A/B)
81–84	I4	Proper motion of A (RA)
85–88	I4	Proper motion of A (Dec)
90–93	I4	Proper motion of B (RA)
94–97	I4	Proper motion of B (Dec)
99–106	A8	Durchmusterung number
108–111	A4	Notes
113–130	A18	2000 arcsecond coordinates

companions or distant common proper motion stars and other literature references if relevant. The current size is 2 MByte.

(c) The third file is the list of references (`wdsnewref.txt`). It is the key to finding more information in the astronomical literature including where the discovery of a star was announced and where the current orbits were published. It also explains the three-letter abbreviation given to each observer that now tends to replace the older system, which contained some Greek letters such as Σ for F. G. W. Struve (now STF), β for S. W. Burnham (now BU), and ϕ for W. S. Finsen (now FIN). Currently this is 2.2 MByte in size.

The WDS team also maintains the *Observations Catalog*. This is a comprehensive compilation of all published measurements. It is extremely useful, for instance, when one is considering the calculation of a new orbit or checking on the history of a system, to get all the available observations to date.

There are 1.6 million measures currently included in this file, which is not generally available, but requests for data from this resource can be submitted to the USNO team. The USNO site also has a 'Frequently Asked Questions' section which covers most of the queries that observers might have.

Fourth Catalog of Interferometric Measurements of Binary Stars

The *Fourth Catalog of Interferometric Measurements of Binary Stars* (INT4) began in 1982 as an internal database at the Georgia State University Center for High Angular Resolution Astronomy (CHARA), tabulating binary star observations made using the technique of speckle interferometry by that group's speckle camera. The *Speckle Catalog* soon grew to encompass other published speckle efforts, then all published astrometric and photometric data for binary stars (and single stars observed in duplicity surveys) obtained by other high-angular-resolution techniques (lunar occultations, adaptive optics, eyepiece interferometry, Hipparcos, etc.) as well. This extended the catalogue baseline of observations back by nearly a century, to the efforts of Schwarzschild & Villiger in 1896 [105]. Surveys based on various infrared speckle or imaging techniques were later added, even though some of these don't really qualify as 'high resolution'.

The complete catalogue (as of July 2016) contains 228,437 observations of 88,426 stars, 62,926 of which were resolved, and it occupies 44 Mbyte although an 8 MByte gzipped version is available. The catalogue is also available in bands of one hour of RA. The great majority of measures are of systems well below $1''$ but recent measures of the brighter and wider binaries are useful for checking the resolution of telescopes. Such measures can be given greater weight than orbital ephemerides for this purpose.

Orbital Elements

P   period of revolution in years or days

T   epoch of periastron

e   numerical eccentricity of true orbit

a   semi-major axis in arcseconds

I   inclination in degrees of true orbital plane to plane of sky

Ω line of nodes or position angle in degrees on the apparent orbit, at which the true orbital plane intersects the plane of the sky

ω longitude of periastron in the true orbit in degrees, i.e. the angle between the nodes and the periastron in the direction of motion

Using the USNO Orbital Catalog

Also on the USNO website is the *Sixth Orbital Catalog* (ORB6) which is a compilation of the published elements of visual binary stars, which have been assessed and graded on a scale of 1 to 5, as follows.

1 *Definitive*

Well-distributed coverage exceeding one revolution; no revisions expected except for minor adjustments.

2 *Good*

Most of a revolution, well observed, with sufficient curvature to give considerable confidence in the derived elements. No major changes in the elements likely.

3 *Reliable*

At least half the orbit is defined, but the lesser coverage (in number or distribution) or data consistency leaves the possibility of larger errors than in Grade 2.

4 *Preliminary*

Individual elements entitled to little weight, and they may be subject to substantial revisions. The quantity $3 \log(a) - 2 \log(P)$ should not be grossly erroneous. This class contains: orbits with less than half the ellipse defined; orbits with weak or inconsistent data; orbits showing deteriorating representations of recent data.

5 *Indeterminate*

The elements may not be even approximately correct. The observed arc is usually too short, with little curvature, and frequently there are large residuals associated with the computations.

Other grades exist but refer to astrometric or spectroscopic binaries (which cannot be seen visually). Only a few per cent of all the objects are graded Category 1. Many more have been observed over only a small arc of their apparent orbits and need continuing observations; the published elements represent the motion so far observed, as well as can be judged.

For each orbit, an ephemeris of predicted future position angle and separation with time is given, and there is a plot of the apparent orbit on which is given the observations which have been compiled so far.

The Multiple Star Catalog

This resource has been created and is maintained by Andrei Tokovinin, one of the most active observers of double and multiple stars with large apertures. It is extremely useful, since it not only details the orbital parameters of each system and sub-system but also displays them on a diagram showing how all the stars fit into the system hierarchy. It was updated in 2017 at www.ctio.noao .edu/atokovinin/stars/.

Gaia DR2 Catalog

The Gaia DR2 Catalogue (issued 25 April 2018) was added to the SIMBAD catalogue at the end of June 2018. The DR2 catalogue can be quizzed directly using any one of several mirror sites such as https://gea.esac.esa.int/archive/. Selecting SEARCH offers the user the choice of inputting a recognised star catalogue ID such as HIP or HD or of inputting a field centre and then choosing a search radius in arcseconds. The maximum number of records per search is currently 2000 (default 500), which can easily be reached at low galactic latitude even with a relatively small search radius. A default number of parameters is displayed including RA, Dec, parallax and G magnitude (G is the instrumental equivalent of the V band). Click on Basic to go back to the previous page and then by clicking on Display Columns you will be presented with a large array of parameters, which can be switched on or off at will. The mean epoch and equinox of the positions is J2015.5.

The Stelle Doppie Website

Maintained by Gianluca Sordiglioni, this facility (available by registering at stelledoppie.goaction.it) allows for searches of the WDS using up to 35 different variables, including positions, magnitudes, separations, catalogue names, and so on. It therefore allows you to create a precisely defined search list of stars that would match your own criteria. Lists of stars based on a wide range of properties can be made – popularity, binocular doubles, triples, multiples, or neglected pairs, for instance. By selecting SHOW at the beginning of each data line, an extensive range of data is displayed along with a list of nearby double stars, and if the pair is binary, the apparent orbit and ephemeris are shown.

James Kaler's 'Stars' Website

James Kaler has been writing a description of a particular star, some of which are double or binary, each week for about 20 years. It is particularly useful for details of astrophysical interest and also includes linear diameters and separations, for instance. The library of these contributions now amounts to about 950 objects and is invaluable for obtaining both information and interpretation in one visit. (stars.astro.illinois/sow/sowlist.html).

Mayer's Catalog

As mentioned in Chapter 1, the first proper double star catalogue was compiled by Christian Mayer and published in 1784. A recent compilation has been made by Juerg Schlimmer [29] and can be found in the archives of the *Journal of Double Star Observations*.

Herschel's Double Stars

Bruce MacEvoy has recompiled the William Herschel double stars and on his webpage (www.handprint.com/ ASTRO/herschel.html) can be found a complete catalogue of 805 objects in both .html and .xlxs format. A subset of the catalogue, called the Herschel 500, has been selected from the catalogues of 1782 and 1785 and thus provides an excellent list for the beginner as it contains many of the finest pairs in the sky down to $-25°$ or so. An additional benefit of this work is that the WDS catalogue now contains a comprehensive cross reference file that links the H stars with other double star catalogues and with both Bayer and Flamsteed numbers. The same author has also compiled a comprehensive treatise an double star astronomy, which is available at [685].

Star Splitters

Star Splitters, https://bestdoubles.wordpress.com is an internet discussion forum which has been established for about eight years. The main contributors are John Nanson and Greg Stone. They use a combination of catalogue data and plotting programs to support a lot of observations using readily available apertures.

Double Star of the Month

Author Robert Argyle (RWA) writes a monthly column highlighting two double stars visible in the sky at that particular time, one in each hemisphere. This commenced in November 2006 and all the stars featured since then are freely available from the following address: www .webbdeepsky.com/double-stars/double//- star-of-the-month/.

Software

There are many software packages which contain lists of double stars and can be used to display star fields allowing each pair to be located. These are extremely useful when looking for fainter pairs beyond the range of the commonly available star atlas. For more information see the chapter by Owen Brazell in Argyle [85]. I found CARTES DU CIEL by Patrick Chevalley particularly useful on a recent observing trip. This is freely available and can be downloaded from `www.sourceforge.net/projects/skychart`.

The following notes describe those books and atlases which are most helpful to the observer, in the estimation of the present authors. In any case, not everyone has a laptop or other online terminal available at the eyepiece.

Atlases and Observing Guides

The Cambridge Double Star Atlas

The Cambridge Double Star Atlas [7] (2nd edition) by Bruce McEvoy and Wil Tirion is the best of the paper atlases which can be purchased. Sadly, Norton's *Star Atlas* cannot be recommended here. Until the 20th edition double stars were labelled on the star charts but this practice was then dropped (ISBN 978-1107534209).

Burnham's Celestial Handbook

Burnham's *Celestial Handbook* [9] remains a firm source of useful data and targets although the position angles and separations are now 50 years out of date. It is a tour-de-force for the keen observer and consists of three volumes totalling 2138 pages. The lack of up-to-date information is compensated by the sheer number of double stars included. At the time of writing, copies were easily available through internet booksellers for extremely reasonable prices (ISBN 048623567X, Volume 1; ISBN 0486235688, Volume 2; ISBN 0486236730, Volume 3).

Double Stars for Small Telescopes

Sissy Haas's 2006 publication [13] is indispensable for the double star observer who prefers to wander around the sky looking for the best pairs to observe. Many of the double and multiple stars contained within the pages (and there are more than 2100 systems described) were observed by Mrs Haas

using a 60-mm refractor. A suitable star atlas such as that by MacEvoy and Tirion would be a useful adjunct with this volume (ISBN 1931559325).

Webb Society Double Star cards

As an adjunct at the telescope, the cards produced by Mike Swan may be found useful. These are in encapsulated A3 format and available from the author.

Toshimi Taki's Atlas of Double Stars

This internet atlas is the most comprehensive available. It features 36 A4 pages in Modified Transverse Mercator projection at a scale of 3.9 mm per degree and is drawn to magnitude 7.0. There are 2053 double stars plotted. (See `www.geocities.jp/toshimi_taki`).

Millennium Star Atlas – ESA

A by-product of the Hipparcos project, this three-volume set contains 1548 maps plotted at a typical scale of 31 mm per degree ($112''$ per mm) and includes more than one million stars down to $V = 11$. It is too bulky to take to the eyepiece but fields of interest can be traced beforehand. Double stars are not labelled but they are indicated by means of a line, emanating from the star symbol, whose length is proportional to the separation of the stars (but on a logarithmic scale) and whose angle corresponds to the position angle. The cost of these books hardly justifies acquiring them – it is more advisable to use one of the software packages listed below (ISBN 978-0933346840).

Hartung's Astronomical Objects for Southern Telescopes

This is a revised and expanded version of the original book written by Ernst Hartung in 1968 and published by Melbourne University Press in 1995. It contains visual descriptions of double stars and deep-sky objects observed by the author with his 30-cm reflector in Victoria, Australia. The observations of double stars include estimates of the colours of each component in many cases. The number of objects (which can be found between $+50°$ and $-90°$ declination) was increased from 1017 to 1129. The authors of the revision are D. Malin and D. J. Frew (ISBN 0522845533).

Observing and Measuring Visual Double Stars

This guide to the practicalities of measuring visual pairs, contains chapters on CCD (including lucky) imaging, DSLR imaging, internet astrometry, filar micrometers, and speckle interferometry, written by independent experts in their fields. Edited by R. W. Argyle (ISBN 978-1461439448).

Observation of double stars

Recently published in Spanish under the title *Observación de Estrellas Dobles* by Marcombo, this profusely illustrated guide to all aspects of observing double stars can be appreciated without an extensive knowledge of the language. The authors are Rafael Benavides, Juan-Luis González, and Edgardo Masa (ISBN 978-8426723826).

Historical guidebooks

The Bedford Catalogue – Smyth

This book written by Admiral W. H. Smyth [10] is the second part of *The Cycle of Celestial Objects*, which, after it appeared in 1844, resulted in Smyth being awarded the Gold Medal of the RAS and then being appointed President. It was printed privately but a copy found its way into the hands of the Reverend T. W. Webb, who based his observing on its contents (see below). Original copies are rare but *The Bedford Catalogue* was issued again as a paperback in 1986 by Willman-Bell. Descriptions of 850 objects are included, of which 519 are double or multiple stars. The descriptions of star colours are perhaps the attraction here, and there is much enjoyable browsing to be had (ISBN 0943396107).

In 1881 G. F. Chambers, having obtained the material which Smyth had accumulated for a proposed second edition of the *Bedford Catalogue*, edited the book and saw it published. Chambers removed some of the more prolix passages and, more importantly for double star observers, added many more recent double star measures, which were checked by S. W. Burnham. The number of objects was increased from 850 to 1604 and includes the far southern sky. The Chambers version has been scanned by Google and the text is available online.

Celestial Objects for Common Telescopes – Webb

In 1859 the Reverend T. W. Webb [8] wrote the first easily available guide to double stars and the deep sky. He based his observation lists on those of Admiral W. H. Smyth, whose *Cycle of Celestial Objects* is a classic book, but its initial cost placed it out of the reach of many small telescope users when it was published in 1844. Webb used a 3.7-inch refractor at first and then progressed to a 9.4-inch reflector.

After Webb's death the book was expanded by the Reverend T. H. Espin and divided into two volumes, the second one covering the stars. Webb's *Celestial Objects for Common Telescopes* in two volumes was last re-issued in paperback by Dover in 1962. Combridge University Press have reproduced the first edition of 1859, which is also available on CDs from the Webb Society. This is of interest, but the Dover reprint contains significantly more double stars. The writing retains the charm with which its Victorian author was associated (ISBN 978-1108014076).

A Field Guide to the Heavens – Olcott & Puttnam

William Tyler Olcott (1871–1936) was an American lawyer who had an abiding interest in astronomy. The last of his five books [11] (written in collaboration with G. Puttnam) is a compact guide to the deep-sky and double stars.

The Binary Stars – Aitken

The classic book on the physics and astronomy of binary stars, by R. G. Aitken [1], this volume also contains some useful tips for the serious observer. The Dover paperback edition can sometimes be picked up in bookstores and on-line.

6 Biographies of Visual Double Star Observers

Brief History of Double Star Observing

Whilst the first naked-eye double star was noted by Ptolemy and appears in the Almagest (ν^1, ν^2 Sgr) – subtends $13'$ on the sky – it was not until the invention of the telescope that interest in double stars was reborn. Using a telescope of Galileo's manufacture, his friend Benedetto Castelli noted that the well-known naked-eye double star Mizar and Alcor was actually triple: Mizar has a companion of magnitude 4 some $14''$ away. Not long after, the pairs β Mon (by Castelli) and three components in θ Orionis (by Galileo) were first seen.

Towards the end of the seventeenth century, telescopes were becoming much better optically, and the availability of higher magnifications more widespread. In 1664 Robert Hooke, in observing the comet Hevelius, saw γ Arietis in his telescope's field of view. This was the closest pair ($8''$) that had yet been seen, and the near equality of the components impressed Hooke, who had not come across anything like it in his years of observing.

Over the course of the next century a dozen or so double stars were discovered in a casual way, and in the late 1770s a Czech priest called Christian Mayer was the first to systematically catalogue double stars using a telescope. At the same time, in England, William Herschel was beginning to start his sweeps of the night sky, which would lead to the discovery of thousands of nebulae and double stars. Herschel presented his double star discoveries in three catalogues, published in 1782, 1784, and 1821. He was succeeded by his son John who worked in collaboration with James South in the 1820s. At the same time F. G. W. Struve was beginning an all-sky survey using a new 9.3-inch Fraunhofer refractor at Dorpat Observatory in modern Estonia. In two years Struve recorded 3100 pairs, of which about 2650 were new discoveries. His son Otto used the new 15-inch refractor at Pulkovo to produce a smaller catalogue of 547 pairs over 35 years, some of which were very close and difficult and included δ Equ, with a period of 5.7 years.

It was assumed by some that Struve had discovered all the double stars that were likely to be found, until Sherburne Wesley Burnham came on the scene. Using a refracting telescope by Alvan Clark with an exquisite 6-inch lens he started to survey the sky anew and, thanks to remarkable visual acuity, he began to find hundreds of new double stars, often of great difficulty, and his skill and endeavour gained him access to some of the biggest and best American refractors. Burnham did not start out as a professional astronomer but his pioneering work earned him a post at Lick Observatory in California, where he used the 36-inch refractor to add to his list of discoveries. It is notable, however, that most were made with smaller apertures. In the 1890s other American astronomers such as Aitken and Hussey started to plan a systematic survey with the Lick telescope. The result of that work, completed by about 1910, was another 5000 new pairs, mostly close and some very faint.

Aitken and his colleagues surveyed the northern sky and it was left to Hussey to organise a similar project in the south. He persuaded a wealthy old school friend to finance the construction of a large refractor to be placed in the southern hemisphere. A 27-inch refractor was duly established at Bloemfontein in South Africa but Hussey did not live to see it. His assistant Robert Rossiter was left to supervise the erection of the telescope and played a major part in the discovery programme, which added another 7000 pairs to the catalogue. At the same time Robert Innes had set up a 26.5-inch refractor at Johannesburg and, along with Willem van den Bos and William Finsen, carried out a parallel survey programme, which was initiated in 1925.

In France significant efforts were also being made. Robert Jonckheere, during a long career, compiled in 1962 a catalogue of 3350 pairs discovered by him with a number of refracting, and latterly reflecting, telescopes. These pairs were largely optical in nature. After Jonckheere retired, a more important contribution came from Paul Couteau and Paul Muller, who between them organised a search of the northern sky to somewhat fainter magnitudes than that carried out by Aitken and Hussey. Couteau found 2750 new pairs, mostly with the large refractors at the Observatoire de Nice, whilst Muller logged another 700 or so. This work constituted the last organised visual survey work in this field. The next major contributions would be made by space-based missions.

Hipparcos was a satellite built by the European Space Agency and which flew into orbit in 1989 carrying out a 30-month mission to catalogue stars down to about $V = 11$. The aim was to determine positions, proper motions, and parallaxes at the milliarcsecond level. With the satellite spinning slowly, star fields passed in front of a series of slits and detectors recorded star profiles, the positions and timing of which could be related to position on the sky. The slits also allowed the measurement of composite profiles of double or multiple stars, and a substantial catalogue of new pairs was built up – almost 15,000 in all.

The successor to Hipparcos, Gaia, is now operating in orbit and sending back huge amounts of data. Its limiting magnitude $G = 21$ (G is the instrumental equivalent of the V band) and angular resolution are significantly better than that of Hipparcos and, for stars around $G = 10$, positions at the microarcsecond level are expected. As this volume goes to press, the second of three catalogues of Gaia data has been issued. Catalogue DR1 came out in 2017 and contained information including positions, proper motion, parallax, and photometry in the satellite's visual band for about 2 million single stars, and positions only for a further 1.14 billion. DR2 was issued in late April 2018 and extends the full astrometric coverage to 1.3 billion stars with G magnitudes between 3 and 21. DR3, due out in late 2020, will contain improved results plus radial velocities and the non-single-star solutions. The final catalogue is currently expected around late 2022, but there is a proposal in train to extend the mission by another five years. Gaia is expected to have a resolution of approximately 0.1–0.2 arcseconds, given a sufficient number of transits, and will increase the number of double stars by several millions. It will therefore represent a seismic shift in our knowledge of these star systems. For the most part the short lifetime of the Gaia mission will be insufficient to trace out the orbits of long-period systems, and so ground-based observations will continue to be needed.

Biographies

Giovanni Batista Hodierna

Born in Ragusa, Sicily, on 13 April 1597. Died in Palma de Montechiaro on 6 April 1660.

Probably self-educated in science, Giovanni Battista Hodierna observed the three comets of 1618–1619 from Ragusa with a telescope of Galilean type and fixed magnification 20. After ordination he served as a priest in Ragusa and taught mathematics and astronomy there. Hodierna was an enthusiastic follower of Galileo and was particular impressed by Galileo's observation that the Milky Way could be resolved into stars and 'nebulae' such as Praesepe. He continued to be interested in nebulae although most of his work concerned the Solar System.

In 1637 the Dukes of Montechiaro gave him a house and a piece of land to live on and funded his publications, and he served them first as a chaplain and parish priest. In 1644, he earned a doctorate in theology, becoming court mathematician in 1655. Besides his duties as a priest, Hodierna practised astronomy, as well as natural philosophy, physics, botany, and other sciences. He studied light passing through a prism and formulated a vague explanation of the rainbow.

Hodierna's contributions to astronomy, though interesting and remarkable, are largely unknown because his publications had only a small circulation. Perhaps his most interesting work is his 1654 *De Systemate Orbis Cometici; deque Admirandis Coeli Characteribus [Of the Systematics of the World of Comets, and on the Admirable Objects of the Sky]*. The first part of the book deals with comets but in the second, more interesting, part he describes and lists 40 nebulae he had observed, with finder charts and some sketches, of which 25 could be identified with real deep-sky objects (mostly open clusters, and some predating Messier); the others are either asterisms (visual collections of stars) or insufficiently described for identification. This publication also contains a section discussing telescopic double stars. There appear to be 13 pairs of stars described but these have not yet been fully identified. He was compiling a map of the sky but this was never completed.

Christian Mayer

Born in Velke Mezirici, Czech Republic, on 20 August 1719. Died in Heidelberg, Germany, on 17 April 1783.

In 1745 he entered the Jesuit order in Mannheim and taught mathematics at a Jesuit school before becoming interested in observational astronomy. In 1752 he became professor of mathematics and physics at Heidelberg but started to concentrate his attention on astronomy. Mayer was elected state astronomer by the Elector of the Palatinate, and a small observatory at the Elector's summer residence was set up. In 1771 a large observatory at Mannheim was constructed and equipped with up-to-date instruments from the finest London makers – Bird, Troughton, Dollond, and Ramsden. Mayer himself records 'My residence is now at Manheim, in a new observatory, fitted for every astronomical purpose, and well finished with the most precious and accurate instruments made at London; amongst which the chief is a brass mural quadrant of eight feet radius, the workmanship of that celebrated artist Mr. Bird, finished in the year 1775, fitted with an achromatic telescope, and fixed to a solid wall towards the meridian. With this instrument I have made many observations of the heavens, when the weather will permit, and two years ago I distinctly discovered, amongst many of the *fixed* stars (from the *first* to the *sixth* magnitude) other *concomitant* or *attendant little stars* … But what surprised me most was, that none of these *attendant* little stars, a few perhaps excepted, has ever been noted in any catalogue which I have seen.'

In 1779 he published his first catalogue of double stars, based on observations made in 1776–1777 with a 2.5-inch achromatic telescope by Dollond mounted on an 8-foot mural quadrant made by John Bird. The difference in RA between the two components of each double star was determined with reference to a pendulum clock, whereas the difference in declination could be measured directly from the quadrant circle. Mayer was criticised by the Austrian astronomer Maximilian Hell and others who had understood Mayer to be saying that he had discovered stars with planets; the pressure from Hell pushed Mayer into publishing his catalogue, but Mayer clarified that he had been talking about the coupling of stars.

Frederick William Herschel

Born in Hanover, Germany, on 15 November 1738. Died in Slough, Berkshire, England, on 25 August 1822.

William Herschel was the son of a music professor, the fourth child in a family of ten. By the age of 14 he was already a competent performer on the oboe and the viol. He was a member of the court orchestra in Hanover and also of the band of the Hanoverian Guard, but when the French invaded Hanover he took his chance and fled to England.

Herschel was then 19 and had to work hard to establish himself as a musician but by the age of 28 he was the organist of the Octagon Chapel in Bath, and he became an accomplished composer. He studied the theory of music, which brought him into contact with mathematics and then optics; eventually he became familiar with the workings of telescopes and an interest in astronomy followed naturally.

By 1776 Herschel possessed a telescope of his own construction and his skill in both optical fabrication and observing developed rapidly. Following a suggestion made originally by Galileo Galilei, he reasoned that pairs of stars which appeared close together on the sky might be expected, especially if they were unequally bright, to be at different distances from us. By observing the position and orientation of the fainter star from the brighter star in a pair, he expected to measure the parallactic shift as the nearer star moved back and forth due to the Earth's rotation around the Sun. In 1779 he started to look for pairs of stars where he could test this theory [387]. 'I resolved to examine every star in the heavens with the utmost attention, and a very high power, that I might collect materials for this research as would enable me to fix my observations on those that would best answer my end. The subject has already proved so extensive, and still promises so rich a harvest to those who are inclined to be diligent in the pursuit, that I cannot help inviting every lover of astronomy to join with me in observations that must inevitably lead to new discoveries.'

His first list of 269 stars was published in 1782 [33]. Of these nine were known before Mayer's time, 33 were found by Mayer, and the remaining 227 were pairs new to science. Herschel divided them into six classes (see the Appendix for the full list) depending on the separation in arcseconds, ranging from Class-I close pairs requiring 'indeed a very superior telescope, the utmost clearness of air' (essentially closer than 2″) to Class VI with pairs separated between 1′ and 2′. A second list followed in 1784 [34], which contained 484 pairs, after which his observational attention was directed elsewhere. A third list of 145 pairs was published in 1822 [103].

The nature of the motion in the six double stars which Herschel found convinced him that this motion was not due to parallactic effects but it was in fact orbital motion in which the stars moved around a common centre of gravity. This was announced in his paper of 1803: 'Account of the Changes that Have Happened, during the Last Twenty Five Years, in the Relative Situation of Double Stars with an Investigation of the Cause to which They are Owing' [363].

Herschel used two micrometers to make measurements of position angle and separation. One was a compact eyepiece in which one wire could rotate through a measurable angle and, in conjunction with a fixed wire in the same focal plane, measurements of position angle and separation could be made. The other was an ingenious comparison-image micrometer in which artificial double stars were set up at a distance from the telescope and could be viewed from the telescope at the same time as the real pairs were being observed with the telescope eyepiece. Through a series of wires, Herschel could adjust the artificial pair to resemble the real pair and then derive the apparent position angle and separation.

James Dunlop

Born in Dalry, Ayrshire, on 31 October 1793. Died in Boora Boora, New South Wales, Australia, on 22 September 1848.

James Dunlop was the son of John Dunlop, a weaver, and his wife Janet. He became interested in astronomy at an early age and was constructing telescopes in 1810. By good fortune, in 1820 he made the acquaintance of the astronomically inclined Sir Thomas Brisbane. In the same year, Brisbane was appointed as the new Governor of New South Wales and decided to set up an astronomical observatory in the new colony. Prior to leaving Britain, Dunlop was then appointed as his second scientific assistant, and both travelled to Sydney in 1821.

After arriving, Brisbane almost immediately started building his observatory at Parramatta and it was Dunlop who was employed to do the astrometric observations for a new accurate southern star catalogue. Also employed was the German-born Carl Ludwig Christian Rümker, who had been recruited by Brisbane as first astronomical assistant. Rümker soon left the observatory in protest at his treatment during 1823, leaving Dunlop in charge of the astrometric measures and general maintenance of the astronomical instruments and the observatory.

Between June 1823 and February 1826, Dunlop made 40,000 observations and catalogued some 7385 stars, which included 166 double stars and references to several bright deep-sky objects near the bright stars he catalogued. By the beginning of March 1826 he had left Parramatta Observatory and continued working at home. There, over the next 18 months, he constructed telescopes and other equipment and organised his own southern sky survey for the discovery of double stars and deep-sky objects.

Sir Thomas Brisbane, before finally departing Sydney for the last time in December 1825, arranged to sell all his instruments to the government so that the observatory could continue to function. Some of the equipment he gave to Dunlop, which he used at his home, especially the useful small equatorial-mounted 8-centimetre (3.1-inch) refracting tele-scope that Rümker, and later Dunlop, both used for obtaining measures of the important double stars as their own personal projects.

Dunlop made several noteworthy discoveries in the southern hemisphere sky and in 1828 published *A Catalogue of Nebulae and Clusters of Stars in the Southern Hemisphere observed in New South Wales*, which contains 629 objects. Dunlop's other major observational work was the observation of 253 southern double stars or 'pairs' below a declination of about 30° S. Many of these pairs were new discoveries, though the most northerly of them had been earlier discoveries made by other observers. These double star observations were all made roughly between December 1825 and December 1826, either using Dunlop's homemade 9-foot 23-cm (9-inch) speculum Newtonian reflector or by measuring the separated distances and position angles of selected double stars using the small 8-centimetre (3.1-inch) equatorially mounted refracting telescope.

Dunlop left Sydney for Scotland in February 1827 and was employed for four years at the observatory of Sir Thomas Brisbane. He was awarded the Gold Medal of the Royal Astronomical Society of London on 8 February 1828. Sir John Herschel, when making the presentation, spoke in the highest terms of the value of the work done by Dunlop in New South Wales. On arrival, Dunlop prepared his southern double star prepared and deep-sky observations for publication.

These two detailed astronomical papers were received with many accolades from his peers, which lasted until his observations were scrutinised by John Herschel and Thomas Maclear in South Africa. It seems that the double star observations were made when the atmospheric conditions were quite unsuitable for looking at deep-sky objects; they were made either under unsteady astronomical seeing or when the sky was illuminated by the bright Moon. John Herschel, immediately on his arrival in South Africa in 1834 and 1835, re-observed all of Dunlop's double stars. Re-observing the double stars found at Paramatta caused Herschel problems. In many cases, either the double star he found did not resemble its catalogue entry or the catalogue position was very discordant with his own observations.

In April 1831, Dunlop was appointed superintendent of the Government observatory at Parramatta. He arrived at Sydney and found the observatory in a deplorable condition. Dunlop succeeded in getting the building repaired and started on his work with energy, but around 1835 his health began to fail; he had no assistant, and the building fell gradually into decay. In August 1847, he resigned his position, and went to live on his farm in Brisbane Waters, where he died on 22 September 1848. In 1816 Dunlop had married his cousin Jean Service, who survived him.

Friedrich Georg Wilhelm von Struve

Born in Altona, Hamburg, Germany on 15 April 1793. Died in St Petersburg, Russia, on 25 November 1864.

Altona was, and still is, in the westernmost borough of Hamburg. Friedrich (or Wilhelm) was one of 14 children born to Jacob and Maria Struve. In 1794 Jacob became Rector of Altona but he was also adept at mathematics and wrote a textbook. With Altona under the threat of war, Wilhelm was sent to Dorpat in Estonia to be with other members of the family for safety. He came to know George Parot, who was instrumental in establishing the study of astronomy at Dorpat University. Wilhelm enrolled at Dorpat and was encouraged to work in the observatory, starting by using the Dollond transit circle to determine the latitude and longitude of Dorpat. At the end of 1821 he became Professor of Mathematics, a post he would hold until 1839. In 1819 he married and in that year his son Otto was born.

Having heard that Fraunhofer was building a 9-inch achromatic refractor, in 1820 Struve ordered such a telescope for Dorpat. It was delivered in 1824. On 129 nights between 11 February 1825 and 11 February 1827 he used the telescope to make a survey of new double stars and ended up with a list of 3112 systems, of which 2343 (Wright [683]) were his own discoveries and the remainder were actually found by Bessel, South, and, in particular, William Herschel. He allowed the sky to drift through the eyepiece field whilst he moved an arc of 7.5° in declination. Using this technique he could examine 400 stars an hour and found that on average one in every 35 was a double. The results were published as *Catalogus Novus Stellarum Duplicium and Multiplicium* and for which he received the Gold Medal of the Royal Astronomical Society in 1826. Struve called the telescope The Great Refractor partly because he suffered a broken leg whilst travelling to escort it to Dorpat and again whilst using it. Struve followed the *Catalogus Novus* up with the *Mensurae Micrometricae* in 1837, in which he presented measures of all the pairs in the *Catalogus Novus*.

John Frederick William Herschel

Born in Slough, Buckinghamshire, England, on 7 March 1792. Died in Collingwood, Kent, England, on 11 May 1871.

John Frederick William Herschel was the son of William and Mary Herschel. He went to Eton College for a few months before being educated at home, where his mathematical talent was nurtured, later going up to St John's College, Cambridge. He graduated as Senior Wrangler in 1814. He left Cambridge in 1816 and returned to Slough to help his father in his astronomical activities. He made an 18-inch aperture reflector of 20 feet focal length, and between 1821 to 1823 he re-observed his father's double star discoveries with Sir James South. In 1833, having spent some years studying the apparent orbits of double stars and how to convert them into true orbits, especially for the pair γ Virginis, Herschel was awarded the Royal Medal of the Royal Society.

The following year, on 15 January, Herschel and his family set out by ship from England to South Africa. Arriving in 1834 he built an observatory at Feldhausen, now called Claremont, which is on the southwest slope of Table Mountain. He set up an 18.5-inch reflector (the 20-foot) with three mirrors available for his use, along with a 5-inch refractor (the 7-foot). Over the next four years or so he carried out a systematic search using the sweep method developed by his father. The results of this work were presented in a privately printed volume in 1847 [142], in which another 2103 new pairs were included (HJ 3347 to HJ 5549). In later years Herschel compiled a catalogue of 10,400 double stars.

Otto Wilhelm von Struve

Born in Dorpat (now Tartu), Estonia, 7 May 1819. Died in Karlsruhe, Germany, on 14 April 1905.

The third of 19 children born to Wilhelm von Struve and his wife Emilie, Otto started to help his father in the Observatory at Dorpat in 1837.

In 1830 Tsar Nicholas I, aware of the parlous state of the observatory at St Petersburg, suggested that a new observatory be established at Pulkovo, in the suburb about 20 km south of the city centre. Wilhelm visited Germany and Austria in order to purchase instruments and took his son Otto along with him. During this period Otto met such men as Bessel, Encke, and von Humboldt, which is likely to have had an influence on his choice of career. Wilhelm ordered a 15-inch lens, tube, and mounting from the German firm of Merz and Mahler. The telescope, when completed in 1839, was the largest in the world. Otto finished his gymnasium education in 1835 and was allowed to sit in on lectures at Dorpat University, where his father taught.

In March 1839 the Struves left Dorpat and moved to Pulkovo. Otto was entrusted with the use of the 15-inch telescope to search for new double stars which his father might have missed, as Wilhelm gradually began to withdraw from active observing. In this work, Otto found more than 500 pairs, identified with the code 0Σ. In 1845 Otto became Deputy Director and took over much of the administrative work from his father, eventually succeeding to the Directorship in 1862 and remaining in post until 1889, when he retired and moved to Karlsruhe.

The initial catalogue of new pairs numbered 514 but, in the 30 years following this, Otto added a few new objects and it eventually contained 547 objects. Included in this list are δ Equulei, with a period of 5.7 years, and the companion to γ And B.

Otto was the second generation of the Struve dynasty in astronomy; his son Ludwig was the third, and Ludwig's nephew, the well-known American astrophysicist Otto Struve (1897–1963), who also specialized in binary stars, was to extend it into a fourth. Both Ottos, Ludwig, and Wilhelm received the Gold Medal of the Royal Astronomical Society.

William Rutter Dawes

Born in London, England, on 19 March 1799. Died in Haddenham, Buckinghamshire, England, on 15 February 1868.

Trained as a physician, Dawes practiced in Haddenham in Buckinghamshire and (from 1826) in Liverpool; subsequently he became a Nonconformist clergyman. In 1829 he set up a private observatory at Ormskirk, Lancashire, where he measured more than 200 double stars before taking charge of George Bishop's observatory at South Villa, Regents Park, London, in 1839.

He was extremely skilled at micrometric measurement. His eyesight was legendary and he was referred to as 'the eagle-eyed'. He is most well known for his formulation of the Dawes limit, an empirical value for the resolution of a telescope as a function of the aperture. Whilst not strictly applicable to all apertures, it does represent well the performance of small apertures on equally bright double stars, of magnitude 6, say. Simply put, the resolution of a telescope of aperture d inches is $4''.56/d$, but this breaks down when the stars are unequally bright or are both faint.

He later set up private observatories at Cranbrook, Kent (1844), Wateringbury, Kent (1850), and Haddenham, Buckinghamshire (1857). Dawes received the Gold Medal of the Royal Astronomical Society in 1855 and was elected a Fellow of the Royal Society in 1865. A crater on the Moon is named after him. His last telescope, an 8-inch Cooke refractor, is now housed at the Observatories of the University of Cambridge.

Ercole Dembowski

Born in Milan, Italy, on 12 January 1812. Died in Solbiate Arno, Lombardy, Italy, on 19 December 1881.

At the age of 13 the young Ercole was left an orphan, and he entered the Austrian Naval College in Venice. He became a midshipman and later saw action in the eastern Mediterranean, especially in Syria, where in 1833 and 1840 he took part in naval actions, the latter an attack on the fortifications around St John of Acre.

In 1842, duty took him to England but he retired from the Navy in 1843, possibly because he suffered frequent attacks of gout. He moved to Naples in the hope that the pleasant climate there would be beneficial for his health and, with the aid of a small family inheritance, he set up home and became increasingly interested in science and literature. It was in Naples that he met Antonio Nobile, an astronomer who worked at Capodimonte Observatory, at that time under the directorship of Ernesto Copocci. It seems likely that it was Nobile who suggested that Dembowski should purchase a 5-inch Plössl dialyte, with which he made his first double star observations. In 1851 at San Giorgio a Cremano, Dembowski began the series of double star measurements that would extend until 1878 and which forms his life's work. In Naples, Dembowski married Enrichetta (Henrietta) Bellelli, a lady of noble birth, and they had three children, Matilde, born in 1851, Francesca (1856), and Filippo (1859).

By 1858 his health had improved to the extent that he felt able to move to the north of Italy and settled again in the town of Gallerate, which lies between Milan and Varese. Here he rented the Villa Calderara, purchased a larger Merz refractor, and from 1862 continued his measurements. However, his attacks of gout were becoming a daily occurrence and were so bad that he was forced to spend months in bed, unable to use his hands or feet. His last observation in Gallarate was made in October 1878 and shortly afterwards he bought a house in the nearby hamlet of Monte in the town of Solbiate Arno, by Lake Maggiore. Although he resurrected the telescope there he was unable to make any more measures and died on 19 January 1881.

Fortunately his children appreciated their father's achievements and presented the manuscripts of his observations to the Brera Observatory in Milan. The double star measures were eventually published in two large volumes by the Reale Accademia dei Lincei in 1883 and 1884.

Towards the end of 1851, Dembowski was established at Borgo di San Giorgio a Cremano, some 6 km southeast of Naples and located in the foothills of Vesuvius. In the next six years he made about 2000 sets of measures of mostly Dorpat stars with the 5-inch Plössl. The high quality of the measures made with this telescope are all the more remarkable when it is realised that the telescope had no driving clock (overcome by Dembowski using 'practice and patience'), the micrometer had no position angle circle, and there was no way to adjust the parallelism of the fixed and mobile threads. In fact Dembowski recognised that this instrument was really not suitable for double star work, but he came up with a scheme using various sets of parallel vertical and horizontal wires in the micrometer field to make the observations.

When he moved to Gallarate, Dembowski started upon the most fruitful period of his observational career. He acquired a Merz refractor of 187.2-mm aperture (7.5-inches), complete with equatorial mounting, driving clock, and micrometer.

In the years to 1878, Dembowski made some 18,000 measures, of which 13,000 are of the pairs in the Dorpat catalogue. The Pulkovo catalogue takes up another 3000 measures and some 1700 measures were made of other pairs, principally those stars found as double by Burnham with his 6-inch Clark refractor but which he was unable to measure for lack of a micrometer. With the 2000 sets of measures made at Naples and a further 700 sets of measures taken as a special investigation of observational errors, the sum total of the work comes to 21,000 measures. This is equivalent to all the work done by both Struves at Dorpat and Pulkovo and many pairs were measured on more than three nights. Indeed, in cases such as γ Virginis or ξ UMa, it was not unusual for Dembowski to make measures of each system on 15 or more nights a year. Taking into account the care with which each measure was made, it is easy to appreciate just how reliable these mean measures were. They were also carried out during a time where there was little parallel activity. In the USA Burnham was beginning his career but he was more concerned to find new pairs than to measure those already known. Dembowski's contribution was therefore unique and irreplaceable, but such was his modesty that it was only after his death that the true magnitude of his work was apparent.

According to Dembowski, the dialyte which he used in Naples was made by Simon Plössl in Vienna. It has an aperture of 5.3 inches (13.54 cm) and a focal length of 66 inches (167.5 cm). He states that 'Notwithstanding the small aperture, the telescope gives images of extraordinary clarity and without any aberrations. In good seeing, bright stars separated by one arcsecond are sharply defined. In fact I was surprised that I was able to measure the position angles of several *Lucidae* (bright stars) of the first two classes of F. G. W. Struve. The achromatism can be said to be perfect'.

The Merz, with a clear aperture of 7.4 inches (187 mm), obtained in 1858, was equipped with an equatorial mount, a driving clock, and a micrometer. The final and most significant part of Dembowski's work was carried out with this telescope in Gallarate. The telescope currently resides in the University of Padua and during the International Year of Astronomy in 2009 it was reassembled and placed on display in the Vatican Museum in Rome.

The 21,319 measures accumulated by Dembowski are summed up in the two great volumes published after his death by the Reale Accademia dei Lincei in 1883 and 1884. Volume I begins with a biographical note by Otto Struve and Giovanni Schiaparelli.

Dembowski received the Gold Medal of the Royal Astronomical Society in 1878 and later that year was elected as an Associate of the Society.

When S. W. Burnham was making the first discoveries with his 6-inch Clark refractor, he turned to Dembowski both to confirm the new pairs and to make measures of them, as Burnham's telescope was not then equipped with a micrometer. The unstinting help received from Dembowski obviously impressed Burnham very much, so much so that he writes the following notes in the introduction to his *General Catalogue of 1290 Double Stars*, published in 1900: 'As an observer with the micrometer he had no superior and few, if any, equals. His work is of the highest degree of accuracy. He made no mistakes, and wasted no time in idle speculations. He left a record of honest, thorough and consistent work which will be an honor to his memory for all time. Baron Dembowski was to me an example so inspiring, a critic so genial and frank, a friend so warm-hearted and disinterested that simple justice as well as friendship impels me to inscribe this volume to his memory'.

Sherburne Wesley Burnham

Born in Thetford, Vermont, USA, on 12 December 1838. Died in Chicago, Illinois, USA, on 11 March 1921.

Sherburne Wesley Burnham was educated locally and his first job was as a court recorder. In 1859 he was in New York but during the Civil War he followed his occupation with the Union Army at New Orleans. It was during this time that he became interested in astronomy by the chance purchase of a book by Burritt called *The Geography of the Heavens*. Burnham acquired his first telescope whilst still in New Orleans but was constantly changing telescopes to get something better. After the Civil War he settled down in Chicago near Dearborn Observatory, which had just been equipped with an Alvan Clark refractor of 18.5-inch aperture, and this may have persuaded Burnham that his current telescope, a 3.75-inch refractor with which he had been drawn to the observation of double stars, was not good enough for this purpose.

It was in 1869 that a chance meeting with Alvan Clark in Chicago spurred Burnham into ordering a 6-inch refractor, stipulating that 'it should do on double stars all that it was possible for any instrument of that aperture to do'. Burnham's obituarist John Jackson, writing in *Monthly Notices of the Royal Astronomical Society* (*MNRAS*), considered that his interest in double stars may have been decided by the fact that one of the few books that Burnham owned was Webb's *Celestial Objects for Common Telescopes*. When Burnham began discovering new and difficult double stars with the 6-inch, Webb wrote to congratulate him but warned him that he would not be able to maintain this rate of discovery. In fact, Burnham showed that there were many more undiscovered pairs but they were difficult to see – either the stars were very close or very unequal. Burnham had remarkably acute vision, which allowed him to see pairs with the 6-inch that other astronomers found difficult to measure with larger apertures. His telescope was not equipped with a micrometer, so he asked Baron Ercole Dembowski to measure the new pairs for him.

Burnham's reputation gave him access to some of the biggest refractors in the eastern USA, but he remained an unpaid observer until 1888 when he accepted an appointment with Lick Observatory and moved for four years with his family to California, allowing him the use of the 36-inch on Mount Hamilton. In 1892 he returned to Chicago and used the 40-inch refractor on two nights per week whilst beginning to compile a catalogue of his own discoveries, which was published in 1900. He followed this with work on a general catalogue of all known double stars from the North Pole down to $-31°$. This meant spending a lot of time searching out and confirming double stars, discovered by others, which might have been misidentified or had positions wrongly recorded. In 1906 the *General Catalogue of Double Stars within 121° of the North Pole* was published in two volumes by the Carnegie Institution of Washington. It contained 13,665 entries.

The remaining six years of Burnham's active career were spent in making a long series of observations of widely spaced pairs that displayed significant proper motion.

Thomas Lewis

Born in London, England, on 12 June 1856. Died in Wivenhoe, Essex, England, on 5 June 1927.

At the age of 11, Lewis entered the Royal Hospital School at Greenwich; three years later he became a pupil teacher and by December 1874 had enrolled as a student at the College for training naval teachers. In 1879 he became a qualified naval schoolmaster and was appointed Assistant Teacher at Devonport Dockyard School; this was achieved through a Civil Service examination. Shortly after he took another examination for appointment to the Royal Observatory at

Greenwich and was successful. He started in 1881 and his first job was to take charge of the Meridian Circle reductions. Later that year he became Superintendent of the Time Department and supervised the Navy's chronometers until he retired in 1917. His interest in double stars began in 1882 and he started to collect old measures, particularly of double stars in the F. G. W. Struve catalogue the *Mensurae Micrometricae*. The Astronomer Royal at this time was F. Dyson, who was very encouraging to Lewis and was in charge when the 28-inch refractor arrived in 1893 and a proper double star measurement programme was initiated. The following year Lewis made a few measures with the 12.8-inch refractor but was the principal observer when the 28-inch was fully operational and indeed did most of the observing himself for the first few years. In 1906 Lewis published his accumulation of the measures of the Struve double stars in a volume which is notable for its thoroughness and, because it brought to a wider astronomical world the catalogue of Struve, Lewis was awarded the Lalande Prize of the Paris Academy of Sciences for this achievement. He also published a paper in *The Observatory* [14], in which he assessed the magnification most likely to be used by double star observers. Using thousands of values taken from 36 prominent observers with a range of apertures between 3.8-inch and 36-inch, Lewis determined that the optimum magnification was $140\sqrt{D}$, where D is the aperture in inches.

Thomas Henry Espinell Compton Espin

Born in Birmingham, England, on 28 May 1858. Died in Tow Law, County Durham, England, on 2 December 1934.

Thomas Espin was educated at Haileybury College and whilst there acquired an interest in astronomy which was aroused by the appearance of Coggia's Comet in 1874 and by some weekly talks on astronomy by his form master. Going to Exeter College, Oxford, his interest deepened and he was allowed the use of the 13-inch refractor on condition that he helped students in their practical telescopic work. He took Holy Orders when he left Oxford and in 1888 became Vicar of Tow Law, County Durham, where he remained until his death 46 years later. He established a correspondence relation with T. W. Webb in 1876 and assisted him with revised editions of *Celestial Objects for Common Telescopes*. Between 1885 and 1900 Espin was interested in red stars (as was Webb), instituted a spectroscopic search for them, and in 1890 published a new version of John Birmingham's *Catalogue of Red Stars* and also contributed annual lists of stars with peculiar spectra to *AN* from 1897 to 1996.

In 1900 he turned his observational attention to double stars and, thanks to a legacy from Webb, he was able to buy a $17\frac{1}{4}$-inch Calver reflector to carry out a systematic search for double stars. In 1914 he added a 24-inch reflector and turned over the $17\frac{1}{4}$-inch to William Milburn (1896–1982), the son of one of Espin's friends and who remained Espin's assistant for the rest of his career, discovering 1170 pairs himself. Espin did not aim to find the closest pairs that his telescope could show but tended to seek out fainter stars separated by several arcseconds; his last list published in *MNRAS* contains the running number 2575, discovered by the age of 72 in 1932. S. W. Burnham, another correspondent, expressed surprise that Espin was able to detect stars that Burnham had found not easy even using the Yerkes 40-inch telescope. Milburn ended up with 673 new pairs. The *Sixth Orbit Catalog* contains only one of Espin's systems (ES 608, $P = 280$ years) but MLB 377 (also known as VYS 2) is an interesting binary, each component of which has an astrometric perturbation.

Espin took over the production of the fifth and sixth editions of Webb's *Celestial Objects* and split the original book into two volumes.

Robert Thorburn Ayton Innes

Born in Edinburgh, Scotland, on 10 November 1861. Died in Surbiton, London, England, on 13 March 1933.

Innes developed an early interest in astronomy and was elected Fellow of the Royal Astronomical Society when he was 17 years old. He emigrated to Australia, where he set up in business as a wine merchant in Sydney and made the acquaintance of Walter Gale – an active observer who loaned Innes a 6.3-inch Cooke refractor with a power of 360 which he used to survey naked-eye stars, looking for new double stars. The first list of Innes' discoveries (26 in all) appeared in *MNRAS* in December 1894, although Innes noted that three were subsequently found to have been discovered already. Number I10 is δ Argûs, now δ Vel, a complex and interesting multiple star containing the brightest eclipsing binary in the sky. I12 must have been separated by not much more than $0''.3$, whilst I22 consisted of stars of magnitudes 7.2 and 8.9 at $0''.6$ – remarkable feats given the aperture employed.

In 1896 he was appointed as a clerical assistant on the staff of the Royal Observatory at the Cape; David Gill could see the nascent skill as an observer which Innes possessed, and so he allowed him to use the 7-inch Merz refractor for double star sweeps and measurement.

In 1903 a meteorology service was installed in the Transvaal and Innes became Government Meteorologist on the recommendation of Gill. He carried out his duties assiduously but his mind was on something grander – a large telescope with which to survey the southern sky for new double stars, in the way that Aitken and Hussey had done in the northern hemisphere. The new Meteorological Office was located in Johannesburg on a kopje, or hill, in an area which is itself 6000 feet above sea level and from where only a few houses could be seen. Innes was also acutely aware of the quality of the sky at Johannesburg and felt it would be an ideal place for astronomical observation.

The telescope was ordered from Sir Howard Grubb in 1909 and Innes expected that it would be two years until it was operational. In the event the telescope became operational in early 1925 and Innes retired in 1927. He was able to balance organisational skill with observational acumen, and his visual double star discoveries number more than 1600. He was also one of the first to appreciate the power of the blink microscope and used the technique to discover Proxima Centauri in 1915.

Robert Grant Aitken

Born in Jackson, California, USA, on 31 December 1864. Died in Berkeley, California, USA, on 29 October 1951.

Robert Grant Aitken entered Williams College in Massachusetts in 1883 with the intention of studying for the ministry. After graduating in 1887 he married and got a job at Livermore University in California, moving on to the University of the Pacific in 1891, where he taught classics but also supervised the modest observatory and 6-inch refractor on site. At this time he started a correspondence with Edward Holden, the Director of Lick Observatory, and by 1895 Aitken was appointed to a one-year position on the staff. His energy and ability were such that after a few weeks he was made an Assistant Astronomer and his first project was to measure a list of double stars selected by E. E. Barnard.

By 1899 Aitken was convinced that measuring known pairs was not as important as a systematic survey for new double stars made with a modern telescope and to a given magnitude limit. In collaboration with W. J. Hussey, who joined him at Lick, the survey started in 1899 and they divided the sky from $+90°$ to $-22°$ into zones which they shared between them. By 1915 Aitken had found about 3100 pairs and Hussey 1300, almost all of which were closer than $5''$. Aitken found that at least one star in every 18 as faint as magnitude 9 was double in the 36-inch. In addition to observing, Aitken was a regular computer of orbits and in 1918 his classic textbook *The Binary Stars* was issued. This book, whilst dealing with the technical aspects of visual and spectroscopic binary orbits, also contains a wealth of commonsense advice for the observer. Dover Publications reproduced the book in 1963 and copies may be found on eBay. From 1920 Aitken had inherited Burnham's double star catalogue, which had, since 1912, been maintained by E. Doolittle, and in 1932 produced the *New General Catalogue of Double Stars within 120° of the North Pole*, now commonly called the ADS. In 1930 Aitken became Director of Lick Observatory; he retired in 1935 but continued to maintain the card catalogue of double star observations that was intended to form a supplement to the ADS.

Richard Alfred Rossiter

Born in Oswego, New York State, USA, on 19 December 1886. Died in Bloemfontein, South Africa, on 26 January 1977.

Richard Rossiter studied at Wesleyan University and then taught mathematics for five years before joining the University of Michigan astronomy department, whose head at the time was William J. Hussey. Hussey was a visual double star specialist who had participated in the search for new northern double stars at Lick Observatory. He was convinced that the search should be extended to the southern hemisphere and, thanks to his long-standing friendship with the wealthy industrialist Robert Patterson Lamont, he was able to realise this ambition. In 1910 Lamont authorised the funds to establish an observatory in South Africa which would be equipped with a new 24-inch refractor. After the war, Lamont purchased two 27.5-inch optical disks from Zeiss, which were then figured by McDowells of Pittsburgh to provide an achromatic doublet of 27.25-inches clear aperture.

In 1923, Rossiter had completed his PhD, in which he proved the existence of stellar rotation for the first time, in a study of the brighter component of the ellipsoidal variable β Lyrae. Hussey wanted Rossiter to help him establish the new observatory and, in 1926, Hussey and his wife and Rossiter with his wife and children travelled to London en-route to South Africa. Unfortunately Hussey died suddenly in London and Rossiter was left in charge. Hussey, however, had already established that the telescope should be sited near Bloemfontein in the Orange Free State. On arrival in South Africa Rossiter established that a good site for the Lamont–Hussey Observatory (LHO) would be on Naval Hill, just outside Bloemfontein.

By May 1928 Rossiter had overseen the clearing of the site, the construction of the dome and the erection of the telescope, and he began a systematic search of the southern skies aided by two Michigan graduates, Henry Donner and Morris Jessup. Each observer was allocated several 4-degree declination bands to cover. Jessup stayed three years, discovering 803 new

pairs, and Donner stayed for six years and found 1031, but after 1933 Rossiter worked virtually unaided until the end of 1952, when he took retirement. The observers at LHO used charts prepared from the Cordoba Photographic Durchmusterung (CPD) and included stars down to magnitude 9.5; each chart covered 4° in declination and 12 minutes in RA, and the survey covered the declination zones +1° to −90° (apart from −70° to −73°). An iris photometer on the 4-inch finder was used to measure the brightness of each star surveyed. By 1952, Rossiter had found 5534 new pairs, many close and faint, and which Rossiter himself re-observed in the search for significant orbital motion. After retirement Rossiter produced a catalogue of southern double stars enshrining all the measures and discoveries made by the Bloemfontein observers, which he dedicated to the memory of Hussey.

He retired to live in Pietermaritzburg, but moved back to Bloemfontein two years before his death. His impressive achievement of 23,000+ measures needs to be seen in context with the circumstances in which he found himself. After 1933 he was alone at Bloemfontein, and when the University of Michigan withdrew funding in 1937 he had to make up the shortfall by appealing to the Bloemfontein Town Council. He also manned the telescope for public observing nights, which took place once a week, showing almost 20,000 citizens the southern heavens.

The telescope was briefly used again for double star work by Frank Holden in the 1970s but eventually closed for good in 1972. The objective, along with the micrometer, is in Ann Arbor, Michigan, whilst some parts of the telescope and mounting are being kept at the fire station in Bloemfontein. The dome is now the first digital planetarium in southern Africa.

Georges-Achille van Biesbroeck

Born in Ghent, Belgium, on 21 January 1880. Died in Tucson, Arizona, USA, on 23 February 1974.

More often referred to as van B by his colleagues in the USA, Georges van Biesbroeck had an early interest in astronomy but attended the University of Ghent to study engineering. This led him to work in Brussels on the city roads between 1902 and 1908. Not long after he arrived in Brussels he volunteered as an observer at the Royal Observatory, Uccle, in the south of the city. In 1908 he took up a position as adjunct observer at Uccle, by which time he had been measuring double stars for five years. The pages of the sixth edition of Webb's *Celestial Objects for Common Telescopes* contain some of his measures from as early as 1903.

In 1914, S. W. Burnham had just retired and E. E. Barnard was becoming seriously ill, so Edwin Frost, the Director of Yerkes Observatory, was looking for replacement observers and he invited van Biesbroeck to come to Yerkes, which he did for a 10-month stay in 1915. Afterwards he returned to Europe but in 1917, after the offer of a permanent post at Yerkes, he and his family made a perilous journey back to the USA.

Although he had an abiding interest in comets, asteroids, variable stars, and solar eclipses, he continued to make visual micrometric measurements all his life, observing with the 90-inch reflector at Steward Observatory on his 90th birthday.

Willem Hendrik van den Bos

Born in Rotterdam, Holland, on 25 September 1896. Died in Johannesburg, South Africa, on 30 March 1974.

Willem van den Bos entered the University of Leiden in 1913 but his studies were cut short by the outbreak of World War I. He served as a lieutenant in the Netherlands Coast Artillery and at the end of hostilities resumed his place at Leiden. At the end of 1920, at the suggestion of Ejnar Hertzsprung, he started a programme of micrometric measurements of visual double stars using the 10.5-inch Clark–Repsold refractor, which finished at the end of June 1925.

An invitation from R. T. A. Innes, the Director of the Union Observatory in Johannesburg, saw van den Bos travel to South Africa almost immediately to take up a temporary research assistant post. He never returned. He threw himself into observations with the newly installed 26.5-inch refractor, which would continue for over 40 years.

One of his first jobs was to help Innes compile the *Southern Double Star Catalogue* in 1927. After Innes' successor, Wood, retired in 1941, van den Bos became Director, remaining in post until he retired in 1956.

He was President of the International Astronomical Union (IAU) Commission 26 (Double Stars) from 1938 to 1952 and maintained a card catalogue of all pairs in the zones $-19°$ to $-90°$ south. On a visit to Lick Observatory he struck up a collaboration with Hamilton Jeffers, which resulted, with the help of Frances M. Greeby, in the publication of the *Lick Index Catalogue of Double Stars* in 1961. This was the first all-sky double star catalogue and contained 64,247 entries. It was also the last time that an all-sky double star catalogue appeared in paper form.

van den Bos was, by far, the most prolific measurer of visual double stars. In a career lasting about 50 years he made more than 73,000 individual measures, almost twice as many as the next person on the list, and calculated several hundred orbits. His discoveries number more than 3100 and include α Doradûs, with a period of 12.11 years. Most of his measures were made at Johannesburg but he also paid visits to American observatories such as Yerkes and Lick.

William Stephen Finsen

Born in Johannesburg, South Africa, on 28 July 1905. Died in Johannesburg on 16 May 1979.

Bill Finsen was the nephew of Nils R. Finsen, winner of the 1903 Nobel Prize in Physiology or Medicine and founder of the Finsen Institute in Copenhagen. Finsen spent his pre-school years in Denmark and returned to South Africa in 1912. He attended the King Edward VII School in Johannesburg between 1920 and 1923 and in the following year became a volunteer observer at the Union Observatory, then under the Directorship of R. T. A. Innes.

Whilst he took part in most of the activities of the observatory, including taking charge of the Time Department for a number of years, his speciality was in observational and theoretical work in visual double stars. He took part with van de Bos in the Union Observatory's systematic search for new double stars south of $-19°$ with the 26.5-inch refractor. He also computed orbits, calculated dynamical parallaxes, and helped in the preparation and maintenance of the comprehensive card catalogue of double stars south of $10°$. After World War II he experimented with improvements to the Anderson double star interferometer (following earlier promising tests in 1933) and this led him to design and construct an eyepiece interferometer in 1950. With this instrument he made more than 13,000 examinations of 8117 stars between $+20°$ and $-75°$, yielding 73 new pairs, 11 of which were found to have periods ranging from 2.65 to 21 years. In addition, nearly 6000 measures of pairs too close for the micrometer were made.

Finsen also took colour photographs of Mars with the big refractor, during the oppositions of 1954 and 1956, which were reproduced in many magazines of the time such as *Life* magazine. During his career he published 135 papers, including three editions of a catalogue of visual binary star orbits in 1934, 1939, and, latterly, with C. E. Worley, in 1970.

Paul Achille-Ariel Baize

Born in Paris, France, on 11 March 1901. Died in Laval, Mayenne, France, on 6 October 1995.

Paul Baize was the son of a doctor, who encouraged his son to take up medicine. This he did and became a very successful paediatrician at the Hospital for Sick Children (l'Hôpital des Enfants Malades) in Paris.

At the age of 20 he acquired a 4.25-inch (10.8-cm) refractor and used it to observe double stars. The first results appeared in the *Aitken Double Star Catalogue*, which was published in 1932. From 1930 onwards he was able to use the refractors at the Paris Observatory for double star measures, firstly a 12-inch (30.5-cm) and then the 15-inch (38-cm) equatorial in the East Tower from 1947.

He retired in 1971 and returned to Normandy, but not until he had made 25,000 micrometric measures of high quality and demonstrated the potential of instruments of moderate resolving power. He was not just an observer, however, and also wrote more than 150 papers and calculated 471 orbits, the last of which appeared when he was 93 years old, just two years before his death.

His achievements are all the more noteworthy because he remained an amateur observer all his life. His astronomical activities had to be dovetailed in with his full-time work for the medical profession.

Charles Edmund Worley

Born in Iowa City, Iowa, USA, on 22 May 1935. Died in Washington, DC, USA, on 31 December 1997.

Charles Worley grew up in Des Moines where his father was a doctor. He became interested in astronomy at the age of nine. His first observational work as an amateur astronomer was the plotting and recording of more than 10,000 meteors for the American Meteor Society. Continuing his love for astronomy he attended Swarthmore College, where he took part in the parallax programme. He also met the other love of his life, his wife, Jane. He obtained a BA in mathematics from San Jose State College in 1959. He worked for the Lick Observatory in California (1959–1961) as a research astronomer under a Naval Research Grant to observe double stars. After arriving at the US Naval Observatory (UNSO) in 1961 he was the motive force behind an extensive programme of double star observation (being, himself, a prolific observer, having the second largest number of double star measurements ever achieved by one person), instrumental innovation, and double star cataloguing. He quickly gained recognition as one of the world's leading experts in the field of double star astronomy.

In 1965 Worley arranged for the database of double star data, the *Index Catalogue of Visual Double Stars* (*IDS*), to be transferred from the Lick Observatory to the USNO. This database became a truly comprehensive resource under his guidance and is formally recognised as the international source of double star data by the IAU. He updated the database on a continuing basis, adding 290,400 observational records to the original 179,000 and increasing the original 64,000 systems by an additional 17,100 through careful literature searches and extensive communication with other double star observers throughout the world. He extended the scope and utility of the database, now known as the *Washington Double Star Catalog* (*WDS*), by adding accurate photometric data, improved spectral types, and identification information. The project was completed in 1996, and the revised WDS is available on the world wide web. He oversaw the addition of 15,000 Hipparcos Catalogue double stars to the WDS. Requests for information from the WDS database arrive daily from astronomers all over the world.

In collaboration with William Finsen and, later, Wulff Heintz, Charles produced two *Catalogues of Orbits of Visual Binary Stars*, the more recent being published in 1983. At the time of his death he was preparing what would have been a new version.

In recent years an accurate knowledge of double and multiple star separations, position angles, and orbital motions has become increasingly important to astronomy. It is now realised that not only must double stars be identified and calibrated in order to produce the best astrometric catalogues of stellar positions, but also the varying centres of emission at different wavelength bands must be taken into account to meet modern high-precision astrometric needs. For Charles Worley's contribution to this aspect of astrometry, he received the 1994 US Naval Observatory Simon Newcomb Award for Scientific Research Achievement.

In 1991 he was elected vice-president of Commission 26 of the IAU (Multiple and Double Stars) and became president of that commission at the IAU General Assembly in 1994. He was a member of IAU Commission 5, the American Astronomical Society, including the AAS Historical Astronomy Division, and the Royal Astronomical Society. He was also an active supporter of the amateur community, and published a series of articles in *Sky and Telescope* and produced the double star section of the *Observer's Handbook*.

During his career Charles Worley made over 40,000 measures of double and multiple stars using the USNO filar micrometer on telescopes in the northern and southern hemispheres. In 1990 he obtained a speckle interferometer in order to improve the accuracy of double star measurements. He oversaw improvements in both instrumentation and software implementation that resulted in the USNO becoming the world's second largest producer of double star observations using a speckle interferometer. Under his direction more than 9200 observations were made with the speckle interferometer on 1100 systems down to separations of one-fifth of an arcsecond, the theoretical limit of the 26-inch refractor. More recently the speckle interferometer has been used to observe Hipparcos' problem stars on the McDonald 2.1-metre Otto Struve telescope. His special interest in nearby stars led to the discovery of 39 new, cool, stellar companions. These companions, which are faint and difficult to observe, provide critical census information on the solar neighbourhood. From 1954 to 1997 he published some 75 professional papers primarily on double star astronomy and gave numerous invited presentations at meetings. He was known for his exacting standards and high quality, best typified by his paper challenging all other double star observers: 'Is This Orbit Really Necessary [686]?'.

(Acknowledgements to Geoffrey G. Douglass (deceased), Thomas E. Corbin, and Brian D. Mason (US Naval Observatory) for permission to reproduce this text).

Paul Muller

Born in Lorquin, Moselle, France, on 17 November 1910. Died in Nice, France, on 9 July 2000.

Paul Muller joined the Strasbourg Observatory in 1931 when André Danjon asked him to investigate the astronomical applications of birefringent quartz prisms. This led, in 1936, to the development of a novel double-image micrometer for the measurement of visual double stars. Unfortunately his work was cut short by World War II and he was imprisoned. In 1948 he was able to defend his doctoral thesis on the novel double-image micrometer. This instrument had a precision and convenience (no field illumination was necessary, for instance) which made it superior to the standard wire micrometer. In 1956, Muller left Strasbourg for Paris; he regularly used the 83-cm refractor at Meudon and the 76-cm refractor at Nice for double star measurement. He initiated a survey of stars from +52° to the North Celestial Pole to continue Aitken's earlier survey of the sky, which was incomplete. In this endeavour he was joined by Paul Couteau at Nice, who tackled the declination zones from +17° to +52°. In all, Muller found 705 new pairs and made 13,000 measurements.

Wulff Dieter Heintz

Born in Würzburg, Germany, on 3 June 1930. Died in Swarthmore, Pennsylvania, USA, on 10 June 2006.

As a teenager living in Germany during World War II, Wulff Heintz would listen to his family radio for any news from the outside world. He used to say that he loved the blackouts during the bombing runs because it made it much easier to see the stars. When the allied troops invaded Germany in 1945, he volunteered to be a translator between the American and British soldiers and the local villagers.

Shortly after the war ended, he enrolled at Würzburg University, eventually completing his studies in 1950 with two majors, mathematics and chemistry. In 1950 he enrolled for graduate studies at Munich University. Most of the university buildings had been destroyed during the war, but the buildings and domes of the Munich-Bogenhausen Observatory, which housed the meridian circles and the telescopes, suffered only minor damage. Lectures in astronomy were given in one of the small surviving buildings on a tiny blackboard. Deplorable circumstances notwithstanding, he received a thorough instruction in astronomy and also gained practical training in meridian circles and position micrometers, learning to make binary star observations with the old Fraunhofer refractor (from 1835) of the Munich Observatory. One of his tutors was Wilhelm Rabe (1893–1958), who had a long career at Munich, making more than 37,000 individual measures, one of the most prolific visual observers of all time. It was here that his passion for binary stars was born.

In 1960, Wulff Heintz published an early but substantial paper, 'Die Doppelsterne im FK4', which was very important in the construction of the FK4 star catalogue and was still used in 1988 for the FK5. Subsequently, in 1961, he was invited to attend the IAU Symposium on Visual Double Stars at the University of California, Berkeley. The experience was inspirational and solidified his devotion to double star research.

In 1967 Wulff Heintz received an invitation from Professor Peter van de Kamp to come to the USA as a visiting astronomer at Swarthmore College, located outside Philadelphia. He joined the Department of Astronomy permanently as an Associate Professor in 1969 and was a full-time faculty member at Swarthmore until his retirement in 1998.

Over his long and distinguished career, Wulff Heintz pursued numerous research interests, including fundamental astrometry, stellar statistics, planetary studies, radial velocities, and, in his last years, the monitoring of slow variable stars using a CCD detector. Together with the committed staff of the Sproul Observatory, he determined about 800 precise trigonometric parallaxes of mostly faint, high-proper-motion stars. The lion's share of his attention over the period 1954–1997 was devoted to double and multiple stars, orbit theory, and relative astrometry. An assiduous observer, he logged many hours at the 24-inch Sproul refractor, striving to equal or better the record for total number of observations by a single observer set by William Herschel at the beginning of the nineteenth century. Over several decades, he made a total of 54,000 micrometer measures of double stars and discovered over 900 new pairs.

Some of his resolutions of new binaries have been confirmed only with speckle interferometry or by the Hipparcos satellite. In fact, in the latter case, several of the 'new' binaries resolved by Hipparcos had actually been previously resolved by Wulff Heintz years earlier.

In 1983, Wulff Heintz and Charles Worley of USNO collaborated on the *Fourth Catalog of Orbits of Visual Binary Stars*. He was the author of some 150 research papers, and author, co-author, or editor of nine books. His early monograph *Doppelsterne* (Goldmann, 1971) was recrafted and translated into English to become *Double Stars* (D. Reidel, 1978). (This summary is based on the obituary by Augener and Geyer [31] and is published by kind permission of the American Astronomical Society).

Paul Couteau

Born in La-Roche-sur-Yon, Vendee, France, on 31 December 1923. Died in Nice, France, on 27 August 2014.

Paul Couteau developed an interest in astronomy as a youngster, firstly after his mother had talked to him about it; his interest was reinforced after reading works by Baldet (on Mars), Flammarion, and Quenet. He made a telescope from Meccano, and by the age of 11 he was convinced that he wanted to be an astronomer.

After obtaining a degree in mathematics, he entered the Institut d'Astrophysique in Paris in 1948 and studied under Evry Schatzmann. His doctoral thesis (Sorbonne, 1956) was concerned with white dwarfs, but it is for his work on visual double star observation, discovery, and orbital calculation that he is best remembered. He started to make micrometric observations, encouraged by Robert Jonckheere. In 1958 a job had become vacant at Nice Observatory. André Danjon told Couteau 'Do you want this post? You can have it – but, a word of warning. Conditions are hard at Nice; you will be almost alone, certainly isolated, and I expect you to use the small equatorial to observe double stars. There is need for someone to do that'. With the encouragement of Jean-Claude Pecker he set about refurbishing and modernising the two large refractors of 50-cm and 76-cm aperture, no mean feat as the large one had not been used for half a century.

Once he had done that he resumed his double star measurement programme, which he had started in 1951 and which eventually spanned more than 45 years. His main work consisted of a systematic search for new pairs between northern declinations +17° to +52° in a collaborative programme with Paul Muller (1910–2000), who concentrated on the zones between +52° and the North Pole. As a result, Couteau found more than 2700 new double stars, made 25,600 measurements

(putting him in the top ten all-time observers list), and computed more than 50 orbits. The observing alone occupied, according to one estimate, 70,000 hours. He was the last major representative of a select band of visual double star observers who spent a lifetime dedicated to the subject, starting with S. W. Burnham in the 1860s and reaching a zenith with the work of R. G. Aitken in the northern hemisphere and W. H. van den Bos in the south. He was President of the Double Star Commission of the IAU (Commission 26) from 1967 to 1970 and from 1983 until 1993 he edited the *Circulars of the Commission*, and, with Paul Muller, maintained a catalogue of the elements of visual binaries. He was a continual source of encouragement to the amateur astronomical community in France and was Honorary President of the Commission d'Etoiles Doubles of the Société Astronomique de France.

Author RWA had the pleasure of meeting him only once – at a double star meeting in Santiago de Compostela in 1996. He stood ramrod straight with a serious countenance, speaking only French but with a distinct twinkle in his eye. He regarded himself as an astronomer monk. 'Do not forget that an astronomer who observes perfect images visually is a wild beast who devours his prey', he would say, 'Do not disturb him under any pretext. Let nature take its course.'

He received the Janssen Prize in 2007 but his greatest honour was to be made a Chevalier of the Legion d'Honneur in 2009.

RWA is grateful to Jean-Claude Thorel and Pierre Durand for additional information.

Modern Period

In the last 15 years interest in double star observing has certainly grown rapidly and there are active observers in a number of countries worldwide, but most predominantly in Spain, Italy, France, and the USA. The activities of organised groups have been discussed elsewhere in this volume, but the following notes give brief biographical details of some of the professional and amateur observers who are now involved in observing visual double stars.

The Professionals

Jose-Angel Docobo Durante – University of Santiago de Compostela, Spain

Jose Docobo was born in Galicia, Spain, in 1951 and graduated in mathematics at the University of Santiago de Compostela. Later, he obtained a doctoral degree in astronomy (1977) at

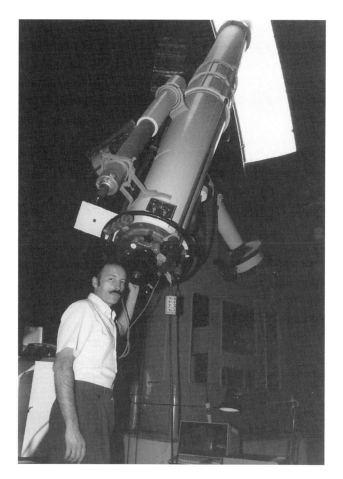

Figure 6.1 Jose Docobo (J. Docobo).

the University of Zaragoza with a dissertation concerning the theory of perturbations applied to triple star systems. Since then he has dedicated his efforts to research, teaching, and the diffusion of astronomy. He is a Full Professor in this discipline at the University of Santiago de Compostela. His passion for double stars is due to the influence of his uncle Angel, who, along with the director of his doctoral dissertation (Professor Rafael Cid), was a disciple of Dr Ramon Maria Aller, the person who introduced the study of double stars to Spain in the first half of the last century.

He began to carry out micrometric observations of double stars with various types of equipment. He used refractors of 0.12-metre (Ramon Maria Aller Observatory, Santiago de Compostela), 0.38-metre (Fabra Observatory, Barcelona), and 0.50-metre and 0.76-metre (Nice Observatory). Later, he used telescopes of 0.35-metre (Santiago de Compostela), metre (Calar Alto Observatory), and 2.0-metre (Pic du Midi Observatory). In the 1990s, keeping in mind the new technique of speckle interferometry, he was responsible for the acquisition of two cameras (ICCD and EMCCD) in order to obtain precise measurements utilising the 3.5-metre telescope at Calar

Alto and the 6.0-metre telescope at the Special Astrophysical Observatory, Russia.

Previously, he had used the 1.22-metre telescope (Calar Alto) to observe binaries using a CCD camera and, recently, Professor Docobo's research team performed speckle observations using the 2.6-metre telescope at BAO (Armenia) and the 4.2-metre SOAR telescope (Chile).

Professor Docobo was the President of Commission 26 (Double and Multiple Stars) of the IAU from 2009 to 2012. Moreover, among many other contributions in the field of binaries, he is the author of a very versatile analytical method for the calculation of double star orbits which has been used to obtain more than 350 orbits in the past few years.

William I. Hartkopf – United States Naval Observatory, Washington, DC, USA

William (Bill) I. Hartkopf was born in Syracuse, New York State, in 1951, and was the first in his family to attend college. A high school physics class sparked his interest, so he applied to Rensselaer Polytechnic Institute (RPI), receiving BS and MS degrees in that field in 1973 and 1975. An introductory astronomy course at RPI gave an additional spark, leading to a Master's project studying quasars and then further graduate work at the University of Illinois. He received MS and PhD degrees in Astronomy in 1977 and 1981, respectively, his doctoral thesis on the structure and evolution of our Galaxy being earned under advisor Kenneth Yoss.

Georgia State University Professor Harold McAlister gave a fortuitously timed colloquium at the University of Illinois in the spring of 1981, describing his work on the speckle interferometry of binary stars. Yet another spark! Bill applied for a temporary research assistant position working for McAlister, which evolved into an 18-year career at GSU as research astronomer, associate professor, and assistant director of the newly formed Center for High Angular Resolution Astronomy (CHARA).

Bill left CHARA in 1999 to take a position as research astronomer at the US Naval Observatory in Washington, DC, where he worked with GSU graduate and former CHARA colleague Brian Mason. He continued making speckle interferometric observations of binary stars using the USNO's historic 26-inch refractor, and (with Mason) maintained the Washington Double Star database and related double star catalogues. He received the observatory's Simon Newcomb Research Award in 2009 and its James Gillis Service Award in 2016.

Bill has long been involved with the double star commission of the International Astronomical Union, maintaining that commission's website for many years and serving terms as its vice-president and president. He has authored over 200 papers and edited two books of IAU colloquium proceedings. He retired in 2017.

Elliott D. Horch – Southern Connecticut State University, USA

Elliott Horch is an observer and instrument builder whose research interests include binary stars, star formation, stellar structure and evolution, exoplanets, high-resolution imaging, and interferometry. A graduate of the College of the University of Chicago (BA 1987), he obtained his PhD in Applied Physics from Stanford University in 1994 under the direction of J. Gethyn Timothy. After postdoctoral stints at Yale and the Rochester Institute of Technology, where he worked with William van Altena and Zoran Ninkov respectively, he held faculty appointments at the Rochester Institute of Technology (1999–2002) and at UMass Dartmouth (2002–2007). He went to Southern Connecticut State University (SCSU) in 2007 and is currently Professor of Physics. During his career, Professor Horch has used some of the world's largest telescopes in his research and has been the recipient of Hubble Space Telescope

time on multiple occasions. In the area of binary star research, Professor Horch has been the leader of the long programme of speckle imaging observations at the WIYN 3.5-metre telescope at Kitt Peak. His dual-channel speckle-imaging system, the Differential Speckle Survey Instrument (DSSI), was commissioned at WIYN in 2009 and was granted official visiting instrument status at the Gemini Observatory in 2012. Closer to home, he was awarded the Connecticut State University System Research Prize in 2011 and was named SCSU Faculty Scholar of the Year in 2012. His current projects include a speckle survey of K and M dwarfs in collaboration with Todd Henry and Gerard van Belle, exoplanet-related research in collaboration with Steve Howell, and the development of a new stellar intensity interferometer at SCSU using single-photon avalanche diode (SPAD) detectors.

Harold A. McAlister – Mount Wilson Observatory, California, USA

Harold (Hal) A. McAlister is Regents' Professor Emeritus of Astronomy at Georgia State University and founder and Director Emeritus of the Center for High Angular Resolution Astronomy (CHARA) at GSU. He received his PhD in astronomy from the University of Virginia in 1975 and spent the next two years as a postdoctoral fellow at Kitt Peak National Observatory, pioneering and establishing the technique of speckle interferometry as a powerful means of measuring the orbital motions of binary stars. He then spent the next 38 years as a faculty member at GSU where he played key roles in establishing his department's doctoral programme in astronomy and its Hard Labor Creek Observatory, located 50 miles east of Atlanta. He retired from GSU in 2015.

Figure 6.2 Elliott Horch (Southern Connecticut State University).

Figure 6.3 (left to right) Bill Hartkopf, Harold McAlister, and Brian Mason (Susan McAlister).

While continuing the speckle interferometry programme, in collaboration with a small team, Professor McAlister's primary focus was CHARA, which designed, funded, and now operates an optical interferometric telescope array that produces the highest-resolution images of stars ever made (see `http://www.chara.gsu.edu` for more information about this facility). The CHARA array is located in the grounds of the historic Mount Wilson Observatory (MWO) in the Angeles National Forest of southern California. From 2002 until 2014, Professor McAlister also served as Director of Mount Wilson Observatory and CEO of the Mount Wilson Institute, a non-profit corporation that operates MWO under an agreement with its owner, the Carnegie Institution of Washington.

The author or co-author of some 250 scientific papers, Professor McAlister has also written a number of articles for the non-specialist in addition to editing several conference proceedings. During the 2009 station fire his blog attracted several hundred thousand hits while the fire threatened to destroy MWO for nearly a month. His reports during the fire crisis were compiled in the 2010 Amazon Kindle ebook *Diary of a Fire*. He started his first novel *Sunward Passage* in order to pass the time sitting on airplanes flying from his home base in Atlanta to Los Angeles. He is presently pursuing several other book projects to occupy his retirement. He and his wife Susan live in Decatur, Georgia, when they are not off exploring in their motorhome.

Brian D. Mason – United States Naval Observatory, Washington, DC, USA

Brian Mason was born in Florida in December 1961. He changed his major and the course of his life as a result of his first astronomy course, taught by Dick Miller, in his first term at Georgia State University (GSU). He graduated with a BS in physics and taught in secondary high school for five years, being reinvigorated by the enthusiasm of his very gifted students, then returning to GSU and entering graduate school under the direction of Hal McAlister. His work has always been in the area of double stars, observing primarily with speckle interferometry and working closely with Bill Hartkopf. He received his PhD in 1994.

In 1997 Brian Mason was hired by the US Naval Observatory (USNO) as the designated replacement for Charles Worley specialising in double star observing and research. He worked briefly with Charles before his untimely death. He is project manager of the Washington Double Star (WDS) programme, astronomer in charge of the historic

26-inch refractor, supervisor of the intern programme, and chair of the USNO History committee. While in charge of the WDS it has gone from 451,546 mean positions of 78,100 pairs to 1,583,945 measures of 141,765 pairs (as at June 2017). He is a member of the American Astronomical Society and the IAU. He is past (and last) president of Commission 26 (Double and Multiple Stars). He maintains web pages for various double star catalogues at the USNO (e.g., WDS, see `ad.usno.navy.mil/wds/wds.html`), the IAU web page for Commission G1 (Binary and Multiple Stars; see `ad.usno.navy.mil/wds/bsl/`) and serves as Advisory Editor of the *Journal of Double Star Observations* (`www.jdso.org`). He makes annual contributions to the double star chapter in the RASC *Observer's Handbook* and the USNO/HMNAO *Astronomical Almanac*.

While at GSU and the USNO he has used many large telescopes, three of which were, in turn, once the largest optical telescope in the world, always observing double and multiple stars: the 200-inch on Palomar (with Lewis Roberts), both the NOAO 4-metre telescopes on Cerro Tololo and Kitt Peak, the 3.6-metre CFHT on Mauna Kea, the 100-inch on Mount Wilson, the 82-inch Struve telescope at McDonald Observatory, the 61-inch Strand reflector at the USNO Flagstaff Station, the 1.5-metre Starfire AO telescope on Kirtland Air Force Base, and the 26-inch telescope of the USNO.

Marco Scardia – Osservatorio de Brera-Merate, Milan, Italy

Marco Scardia was born in Genoa, Italy, in October 1948. He developed his passion for astronomy after having watched the total solar eclipse on 15 February 1961, from the terrace of his home in Genoa. In 1962 he joined the amateur astronomers' association 'URANIA' in Genoa, established by Ing. Glauco de Mottoni y Palacios, great observer and scholar of the planet Mars, later studying with him. In 1972 he entered Brera Astronomical Observatory (OAB – Merate), where he remained until his retirement in 2012 and, in addition to carrying out his astronomical research, he holds important technical and administrative offices.

In 1976, a under a suggestion of Glauco de Mottoni, he started studying visual double stars and, in October 1980, he made his first observing run at l'Observatoire de la Côte d'Azur (OCA, Nice, France), where he met the great astronomers Paul Couteau, under whose guidance he learned to observe double stars with a filar micrometer, and Paul Muller. Back in Merate, he observed double stars with a filar micrometer using a 23-cm aperture refractor and

Figure 6.4 Marco Scardia (M. Scardia).

Figure 6.5 Brian Skiff (Lowell Observatory).

photographically using a 20-cm diameter Zen astrograph. In 1982 he invented and developed, with Renato Pannunzio, a new photographic method for the observation of visual double stars. Between 1986 and 1993 he made twice-annual visits to La Silla, Chile, to observe southern double stars using the 40-cm diameter GPO Astrograph and a filar micrometer.

In 1995, after a mission to the Pic du Midi Observatory to see how the speckle-camera PISCO installed on the telescope B. Lyot worked, he started a very valuable partnership, which continues until today, with Jean-Louis Prieur, who had designed and built it. After the decommissioning of PISCO from the telescope B. Lyot, he transported PISCO in January 2003 to Merate, and from 1 January 2004 to 4 June 2015, he systematically observed double stars with PISCO and the Zeiss telescope ($D = 102$ cm) of the OAB. Owing to the rapidly deteriorating observing conditions in Merate, in 2008 he proposed transferring PISCO to OCA, on the plateau de Calern behind the town of Grasse. Since 2014 Dr Scardia has been a collaborator of OCA (C2PU) and, after mounting the PISCO speckle camera on the telescope 'Epsilon' of Calern ($D = 104$ cm), fully modernized, in November 2015 he resumed observations of visual double stars. At the time of writing, he has made several thousand measurements, has discovered over 170 new double stars, and has calculated over 200 orbits.

Brian A. Skiff

Brian Skiff has spent over 40 years at the Lowell Observatory, Arizona, as an observer and research assistant. This has included taking the final photographic plates with the 33-cm 'Pluto Camera' astrograph, and measuring thousands of plates, films, and CCD images to improve the knowledge of the orbits of asteroids at a time when there was little activity in this area. He also spent 1200 nights over 15 years doing single-channel photoelectric photometry on Sun-like stars to explore long-term variations analogous to the 11-year sunspot cycle. This synoptic work continues to the present via the spectroscopic observation of chromospheric activity. He participated in the decade-long Lowell Observatory Near-Earth Object Search (LONEOS) survey for near-Earth asteroids, manning the telescope about 100 nights per year. More recently he has obtained rotational light curves via CCD for some 200 near-Earth and other asteroids, using several telescopes. He also monitors photometric activity in T Tauri stars in support of work involving accretion disks and planet formation around these stars. Having an abiding interest in bibliography and star-cataloguing, he maintains a comprehensive catalogue of stellar spectral classifications, which is among the most frequently used items in the VizieR catalogue-query service. A by-product of this work has been the identification of 2865 faint pairs of stars with common proper motion, which represents the compiling of 18571 individual positions from on-line plate archives. He has also discovered 12 comets and co-discovered two more.

Andrei Tokovinin – CTIO, Chile

Andrei Tokovinin graduated from the Moscow University in 1977, with a PhD following in 1980. His primary interest

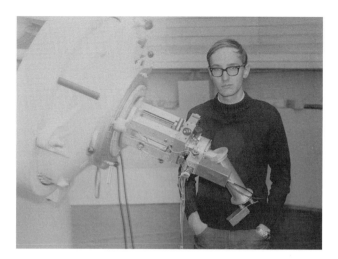

Figure 6.6 Andrei Tokovinin with a phase grating interferometer (A. Tokovinin).

Figure 6.7 Rainer Anton (R. Anton).

was, and still is, astronomical instrumentation. Binary stars were the objects of choice for interferometric observations with the small telescopes that he made in the 1980s, but he also developed a radial-velocity spectrometer to study close (spectroscopic) binaries, and eventually became interested in multiple (triple, etc.) stars that often combine a close (spectroscopic) and wide (resolvable) system. He compiled a catalogue of multiple stars in 1996.

Dr Tokovinin has worked at the Cerro Tololo Inter-American Observatory since 2001 and has been actively pursuing speckle interferometry at the 4.1-metre SOAR telescope since 2008. With the high spatial resolution and the high accuracy of these speckle measurements, it has been possible to resolve many new binaries or subsystems, more than 700 to date, and improve orbits. His goal is to understand the formation of binary and multiple systems through their statistics.

The Amateurs

(The word amateur as used here is based on its derivation from the Latin verb *to love*. Brian Mason has a better term – financially uncompensated professionals.)

Rainer Anton

Dr Rainer Anton has been interested in astronomy since school. He studied physics and astronomy at the University of Hamburg, where he became a professor of physics in 1984. During a two and a half years' stay as a post-doctoral fellow at NASA/Ames in California, he joined a local amateur astronomy club and built a 10-inch Newtonian, which is

still using at home. His interest in double stars was stimulated about 20 years ago by the appearance of affordable CCD cameras on the market, which made astrometry with a computer feasible. For recording double stars, he is using the technique of 'lucky imaging', by which seeing effects can largely be compensated, resulting in virtually diffraction-limited images. After retirement in 2007, Rainer Anton became a member of the Internationale Amateur Sternwarte (IAS), which operates an observatory in Namibia. Whilst there, he uses Cassegrain or Ritchey–Chrétien telescopes with apertures up to 50-cm. One of his goals is to assess the accuracy of measurements. Meanwhile, the sensitivity, resolution, and speed of CCD cameras has much improved, and error margins of separation measurements of the order of $0''.01$ and below can be reached. He has published his results in numerous articles in astronomy journals, including *Sterne und Weltraum*, *Sky & Telescope*, and the *Journal of Double Star Observations*.

David Arnold

Dave Arnold's involvement in performing telescopic work on the planets and double stars began in 1963, when he was in high school. He became motivated to become involved in this work by an article that Ronald Charles Tanguay had written for *Sky & Telescope* in February 1999 [32] and by Tanguay's website, in which he expressed the view that many entries in the WDS needed to be updated. The focus of this work was performing measurements for neglected pairs in the WDS catalogue that had not been observed for ten years or more. He was encouraged by Dr Brian Mason, who expressed his appreciation for this effort because not enough activity of this type was taking place. His methodology basically replicated what Tanguay presented in his *Sky & Telescope* article because he had the same equipment that Tanguay was using.

His Divinus Lux observatory is located in Flagstaff, Arizona. The telescope used was a Meade 8-inch Schmidt–Cassegrain equipped with a Celestron microguide eyepiece modified with a 360° protractor around the periphery and an attached exterior pointer, in order to enhance accuracy. He followed the methodology that Tanguay outlined in his *Sky & Telescope* article, which he preferred over the drift method or other techniques that were being used. Starting in April 2001 he was able to measure separations down to 3 arcseconds and magnitudes down to about 10.5 with this equipment. During the course of this work, 23 articles in Tanguay's *Double Star Observer* and 28 articles in the *Journal of Double Star Observations* were produced. This resulted in 5237 measurements and 121 new double stars, or additional components for known star systems, which have been incorporated into the WDS Catalogue.

David Arnold retired from doing formal double star work in the autumn of 2012.

Ernó Berkö

Born in 1955, Ernó Berkö has been a member of the Hungarian Astronomical Association since 1973. He has been making astronomical observations since the 1970s. He was an independent discoverer of the supernova SN 1999by. He has used several instruments, from smaller refractors to a 35.5-cm (14-inch) reflector. He was in charge of the Hungarian Astronomical Association's Deep Sky section between 1999 and 2005. He has published dozens of articles on different topics in the association's journal, *Meteor*. Since 1998 he has been working intensively on double stars. Initially, he carried out visual observations; then, for the sake of accuracy, he started making measurements on photographs taken with CCD and DSLR cameras. Up till now he has made and published more than 55,000 measures on more than 5500 pairs. His measurements appear in British and American publications including the *Webb Society Double Star Section Circular* and the *Journal of Double Star Observations* and are added to the *WDS Catalog* as well. At present, the *WDS Catalog* contains more than 1100 double stars with his discoverer designation BKO. He mostly measures USNO neglected pairs. Additionally, he is interested in observing and photographing meteors and atmospheric optical phenomena. His automatic meteor cameras are part of the Hungarian Video Meteor network and IMO (the International Meteor Organisation).

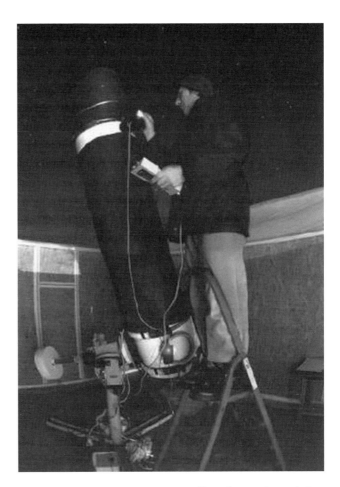

Figure 6.8 Ernó Berkö at the eyepiece of his telescope (E. Berkö).

José Luis Comellas

Born in El Ferrol (La Coruña) in 1928, J. L. Comellas is the grandson of the Galician writer Manuel Comellas Coimbra. During his university studies in Santiago de Compostela (1950) he made contact with Ramon Maria Aller and his team of collaborators, at that time pioneers of the observation of double stars in Spain.

He received his doctorate in history from the Complutense University of Madrid in 1953 with a thesis entitled 'The first pronouncements in Spain', which earned him the outstanding *cum laude* and for which he received the Menéndez Pelayo National Prize in 1954.

Thanks to his extraordinary abilities as observer and promoter, J. L. Comellas is the reference for several generations of amateur astronomers in Spain. In 1973 he published his first catalogue of double visual stars (1199 pairs), followed by his fundamental work, *Guia del Firmamento (Guide to the Night Sky)* in 1978, where he reviews data from his first catalogue. His second catalogue of double double stars was published in 1982 and contains more than 5000 pairs. He used Polarex Unitron refractors of 75-mm and 102-mm aperture.

The observatory in Mairena del Alcor had a hemispherical dome and several micrometers. He was founder of the Double and Multiple Stars Section of the Astronomical Association of Sabadell, in 1976, which he coordinated for more than ten years. He is still very close to many Spanish observers, who consider him a great teacher. They paid him a well-deserved tribute in 2008 to coincide with his 80th birthday. Professionally, he is a Doctor of History and a member of the Spanish Royal Academy of History. He has great international prestige and is the author of numerous research works on the history of Spain and astronomy.

His fondness for astronomy has been noted in several publications that have been written on this subject. He emphasises his catalogue on double stars. Among other works, he published the first edition in Spanish of the Messier catalogue. His most representative work is *Guia del Firmamento*, mentioned above, which has run to seven editions and which is considered the 'bible' for Spanish astronomy fans.

In 1963 he was appointed to the Chair of History of Modern and Contemporary Spain of the University of Seville and four years later he published his *History of Modern and Contemporary Spain*, a volume that has reached eight editions. The centre of this author's research is nineteenth-century Spanish work, about which his studies on the moderate decade and Cánovas stand out. In 2000 he became Professor Emeritus in the same university.

On 29 April 2011, Seville City Council agreed to nominate a street with his name, 'Calle Astrónomo José Luis Comellas', which is located between the Almendralejo and Mayor Luis Uruñuela roads of the said municipality.

Jean-François Courtot

Not long ago, light-pollution-free locations were common for Jean-Franços Courtot. Born in rural France in the fifties, the first lights in the night sky came from galactic distances. It is not surprising therefore how an interest in astronomy developed under those starry skies: even at magnification ×1, there was a lot to observe. He had little in the way of books or observing guides at first, but his grandfather's 8 × 26 World War I binoculars showed much more beautiful views than when they were used half a century before. An old 40-mm pair of field glasses discovered in the attic helped considerably, especially when it was found that a binocular eyepiece fitted to the antique tube boosted the magnification significantly. Jupiter and its moons and Saturn's ring (nearly) were visible, and Albireo was seen in its full glory. As *La Construction du Télescope d'Amateur* by Texereau was out of print, Ingalls' *Amateur Telescope Making* was ordered from the USA. Courtot did not have a real telescope for years to come, however, but, strangely, his English at school improved markedly. In 1993, a 205-mm Newtonian was eventually completed, but how good was it? One way to find out was the observation and measurement of double stars. Thousands of

Figure 6.9 José-Luis Comellas (Jesús & Pepa) (tiempodeestrellas.com).

Figure 6.10 Jean-Francois Courtot (J.-F. Courtot)

visual measurements later, he concludes that the poorest part of his viewing system is the atmosphere.

James Daley

James Daley has made substantial contributions to the *Journal for Double Star Observations*, using a 9-inch refractor at the Ludwig Schupmann Observatory in New Hampshire and a CCD camera used in conjunction with filters centred on the standard *B*, *V*, *R* and *I* bands. The telescope has a focal length of 100 inches, but a Barlow lens is used which increases the focal length to 278 inches giving a consequent pixel scale of $0''.3$. He has been particularly interested in the imaging and measurement of pairs of stars with very large differences in magnitude. This has been achieved by the use of an occulting bar close to the CCD window, which greatly reduces the level of light from the very bright primary star, allowing the faint close companions to be seen. An example of this is the field of Polaris, where he has detected and measured faint field stars which had not been measured for more than 100 years.

André Debackère

'As an amateur astronomer since the age of 18 living in the North of France, I was disappointed by observations with the small refractor that I had made. So I immersed myself in reading astronomy books for several years, and when in 1978 I moved to the Haute-Loire region, to the southeast of the Massif Central, I met an amateur astronomer who wanted to create an astronomy club. I hosted the club of the "MJC of Monistrol-sur-Loire", named "Les Gémeaux (The Twins)", for 25 years. When the sky started to be polluted by night lighting, I began serious telescopic observation. I joined the Astronomical Society of France (SAF) in 1980 and I signed up to a new venture: the Double Star Commission. It was the beginning of an adventure: I had contracted a tenacious virus, the measurement double stars. Subsequently I coordinated a small group of observers to validate methods of measuring visual double stars, such as timing pairs across eyepiece wires and the use of objective diffraction grids, but despite all the efforts of the observers, subjectivity remained. Then Jean Dommanget, professional astronomer at the Royal Observatory of Brussels (Uccle, Belgium) and scientific adviser to the Double Star Commission, asked me to participate in a work of identification; during these observations I discovered three new double stars, which were named DBR1, DBR2, and DBR3AB/AC in the *Washington Double Star (WDS) Catalog* listing. The advent of digital technology

Figure 6.11 André Debackère (A. Debackère).

in the world of amateur astronomy would change practices. The big advantage of digital image acquisition and the automatic processing of images is to remove subjectivity. It was a revival that came with the use of a webcam and my first steps with REDUC (double star measurement software), designed by Florent Losse. Thereafter, I had the opportunity to observe and measure double stars with the Celestron 14 telescope from the "Ecole Supérieure d'Optique", Saint-Etienne, France, with Thierry Lépine. By studying more closely the images obtained with the C14, I discovered nine other pairs, listed in the WDS as DBR6 to DBR14 (the observations corresponding to DBR4 and DBR5 have not been included in the WDS). My own profession of teaching physical science has led me to share with my students my passion for astronomy. I am involved in many activities including a sundial which I made with a group of pupils aged 15 that has been installed since 1980 on a wall of the college "Le Monteil" in Monistrol-sur-Loire, France. I created an astronomy workshop at the college in 2006 and was the initiator and coordinator from 2010 to 2013 of a European Comenius project having a scientific and astronomical character and entitled "In orbit with Europa". Since January 2010, through the good offices of Ferlet Roger and Anne-Laure Melchior of the European Hands-On Universe (EUHOU)

I have had access to the Las Cumbres Observatory Global Telescope (LCOGT) network of robotic telescopes in remote control via the internet, which allows me to do educational work on asteroids. From late 2014, thanks to Robert Mutel, Professor of Astronomy at the University of Iowa, USA, I have been able to access the robotic telescope of the university, located at the Winer Observatory in Arizona.'

Russell M. Genet

Russell Genet's early career involved working in rocket systems in California and he later took a position as a mathematical analyst for an aerospace guidance company in Ohio. He founded Fairborn Observatory in 1979 and moved the facility to Mount Hopkins, Arizona, in 1985, where he worked until 1993, the first four years as Director. Fairborn was the first fully automatic robotic observatory in the world and complemented Russell Genet's interest in photoelectric photometry. He established the magazine *IAPPP Communications*, the first international photometry periodical, and also served a term as President of the Astronomical Society of the Pacific.

For the last ten years or so his interest has broadened to include the observation of visual double stars and in particular the technique of speckle interferometry. He has organised observing times on several large telescopes, combining this with a wide interest in astronomical education, including organising several workshops on binary star astronomy. His recent publications include several books on the research potential of small telescopes.

René Gili[1]

A regular observer with the large refractors at Nice Observatory over about 20 years, Rene Gili very effectively applied the use of modern CCD cameras to the measurement of close visual binaries, and published more than 5500 mean measures. His work was published in *Astronomy & Astrophysics*. The WDS currently lists 109 discoveries, many of which are both faint and close. GII 109, for instance is a pair of stars of magnitude 15.3 separated by 0″.3.

Richard Harshaw

Richard holds a Bachelor of Science in Education (Mathematics/Physics) from Central Missouri State University, from where he graduated in 1973. Born and raised in Missouri, he now resides in Cave Creek (a suburb of Phoenix), Arizona, with his wife.

Figure 6.13 Richard Harshaw (R. Harshaw).

Figure 6.12 Russell Genet (G. Boyce).

[1] Died in August 2018.

His exploits in astronomy began 50 years ago when he purchased a second-hand 60-mm alt-azimuth refractor from a friend. Over the years, he worked his way up to a 4.5-inch reflector, then an 8-inch Schmidt–Cassegrain telescope (SCT), and his present scope, an 11-inch SCT. His main interest is the speckle interferometry of close double stars. But he is also an avid galaxy hunter and enjoys investigating star clusters and planetary nebulae.

He has had several articles on double star research and measurement published in various journals and has had a book published by Springer Publishing. He is also the author of the *Complete CD Guide To The Universe*, released late in 2006. He is presently working on the speckle interferometry of extremely close binaries, doing observing runs at Kitt Peak National Observatory and other observatories, as well as at his own observatory, Brilliant Sky Observatory. Asteroid 2000 EF116 was named for him (26586 Harshaw) by the Catalina Sky Survey team.

Wilfried Knapp

Wilfried Knapp was born in Lower Austria and, after finishing his studies in business administration, took different jobs in the IT industry, finally setting up his own small enterprise re-selling servers and storage systems to data centres.

He writes: 'Living in a large city at the price of heavy light pollution made visual astronomy not a first choice interest until about my mid-fifties.

The brilliant black sky with the impressive Milky Way over the Mediterranean island of Lesbos during a summer vacation 10 years ago sparked my interest in visual astronomy. The 400 year anniversary of the first use of a telescope for astronomy by Galileo Galilei in 2009 led to the purchase of my first small telescope with the intention to re-enact the discoveries he made using such a simple instrument. Despite rather frustrating first experiences (what I could see at the eyepiece had little to do with the fantastic images printed in astronomy journals) I got quickly hooked by the fascinating aspect of looking into the past with the help of a telescope.

Living in the suburbs I had to travel to dark spots for observation sessions but it soon became clear that I would prefer to avoid such efforts and would rather be able to use my telescope at short notice on the terrace in front of my flat whenever conditions were favourable. It was soon evident that double stars are, alongside open clusters, the targets most suited for visual observation under light-polluted skies and after my first double star sessions I got hooked again to locate an object by star hopping, and the resolution of a

seemingly single source of light into two objects by increasing the magnification re-enacts the discovery of such objects and is a quite rewarding experience.

Soon the question about the relationship between the parameters of a double star and the aperture required for resolution caught my attention. At the same time it was clear that even the most comprehensive and precise sources of double star data, especially the *Washington Double Star Catalog*, are often error-ridden when it comes to the magnitudes of faint secondaries like, for example, most of the Jonckheere pairs. This opened the field to activities beyond simply observing, by contributing to the reliability of such catalogues with the results of [one's] own observations. To be able to contribute as an amateur to a professionally maintained star catalogue is quite animating.

Odd results from visual observations combined with some catalogue research for visual magnitude data led to communication with the USNO staff in charge of the maintenance of the *WDS Catalog*, and to improved estimations. To have my own equipment to do photometry was, for several reasons, not an option, so I looked for access to remote telescopes able to deliver CCD images reliable enough to do differential photometry, starting with the use of AAVSO VPHOT software. Results were then submitted to the JDSO and as peer-reviewed publications found their way into the *WDS Catalog*, even if not counted as new observations. One ongoing long-term project in this regard is the measurement of all Jonckheere doubles, infamous for their unreliable magnitudes.

The next step was to report astrometry results as well, as software for doing this on your own is readily available with no or little costs. Via JDSO reports, these results then found their way into the *WDS Catalog* as new observations.

Discussions with peers about double star characteristics beyond their being close sources of light led to research about physical doubles using proper motion and parallax data from existing catalogues as my own astrometry results are, with the equipment available to me, not precise enough to be used for this purpose. The development of my own scheme for the evaluation of the probability for a double to be physical, using the given data, was a consequence of intense study of this topic.

The use of freely available tools like TAP VizieR and X-Match from CDS, as well as the Gaia archive, make possible the extensive use of large star catalogues by amateurs, so that they are able to feed the *WDS Catalog* with ever more recent precise measurements; looking for data allowing one to identify doubles as physical or optical is now a regular task.

Figure 6.15 Florent Losse (F. Losse).

Figure 6.14 Wilfried Knapp (W. Knapp).

Yet even with the most comprehensive and precise new star catalogues, like for example the coming Gaia DR2, there will remain a residue of several thousand unidentified WDS objects (most of them listed in the neglected subset), so there will be no shortage of new projects in the coming years.

Besides, there is also a new WDS catalogue category, called the WDS Supplemental Catalog for large and faint duplicity surveys, as a repository for close objects newly found in upcoming new catalogue releases of GAIA, etc. The never-ending learning curve when dealing with double stars is an ongoing fascinating aspect of this area of astronomy.'

Florent Losse

Florent Losse operates a roll-off-roof observatory in a small village in the southwest of France. Except for the optics everything is homemade. The observatory houses a cradle mount with an equatorial platform supporting a 16-inch/f5 Newtonian as the main instrument, an 8-inch/f6 as an auto-guider, and a 5-inch refractor used as a finder.

He became involved in double astrometry with Guy Soulié (1920–2015), who was a professional astronomer and a passionate and tireless observer, making observations until he was 92; he coached several skilled double star observers.

Starting with reticulated eyepieces, Florent Losse later adapted an old micrometer, historically used for reading the circles of a meridian instrument, for making measures.

By 2000 he had started imaging double stars on the 8-inch with a webcam and then added a CCD camera, making 1300 measures with these setups. When he built the 16-inch he discovered what the atmosphere really is! The lucky images were so exceptional that he decided to experiment with speckle interferometry. Using typical exposure times between 1 ms and 30 ms, the speckle pattern freezes the atmosphere and allows measurements up to the resolution limit. More than 1000 mean values for close systems were included in the *Fourth Catalogue of Interferometric Measurements of Binary Stars*.

Simultaneously with his first experiences with the webcams he wrote the first version of REDUC to reduce in an easy way the huge quantity of images taken every night. As the results were reliable he decided to share the software. REDUC is used by hundreds of observers and is continually updated with new features. The fifth version of REDUC was released for the Vilanova Pro-Am meeting in 2016.

Beside double star observations he is also very involved in the astrometry of near-Earth objects (NEOs). Up to date his observatory (IAU code 193) has provided 10,000 measures of minor planets and 1100 follow-ups or confirmations of newly discovered NEOs.

Xavier Miret Coll

Xavier Miret was born in Barcelona in 1963 and graduated in Medicine at the University of Barcelona, but his interest in astronomy started during his childhood in the context of activities related to various hiker centres in which workshops were held. After a long period dedicated to his formation and later professional activity, he joined the Observatori Astronómic del Garraf in 2000 as an active member from the point of view of both its management and also the development of the educational and teaching programmes that this observatory still carries out continuously. From a scientific perspective, he has actively participated in the project Neglected Visual Double Stars and has been a member of the coordination team of the Garraf Proper Motion Wide Pairs Survey mentioned below (see the section on T. Tobal). Also, he was part of the Organizing Committee of the II and IV International Pro-Am Meetings of Binary and Multiple Stars that took place in Spain, in Sabadell, 2010, and in Vilanova i la Geltrú, 2015.

Ignacio Novalbos Cantador

Ignacio Novalbos was born in Barcelona in 1966. He has been an amateur astronomer since 1980, when he discovered astronomy through the television programme *Cosmos* presented by Carl Sagan. After more than 25 years enjoying visual observation, in 2006 he discovered the world of double stars and decided to go a step further in astronomical observation, with the aim of providing useful data for the professional community. In 2008 he joined the Observatori Astronòmic del Garraf as an observer of the GWP project. Later, in 2009, he also became a member of the coordination team with T. Tobal and X. Miret. The author of about 80 articles and collaborations published in amateur magazines and research journals, he collaborates regularly with the research team of the Spanish Virtual Observatory (SVO) both in tests and in the implementation of new tools for the study and observation of binary stars.

He was a member of the organising committee of the IV International Pro-Am Meeting of Binary and Multiple Stars held in Vilanova i la Geltrú, 2015. He has recently collaborated with Josefina F. Ling (Universidad de Santiago de Compostela) in a review of the Spanish edition of the book *Esos Astrónomos Locos por el Cielo*, written in 1988 by Paul Couteau. He is currently coordinating the REDVO programme, dedicated to the identification, measurement, and study of forgotten pairs on UKIDSS images.

Francisco Rica Romero

One of the most active double star astronomers in Spain is Francisco Rica Romero. Between 1994 and 2000 he directed the Double Star Section of the magazine *Tribuna de Astronomía*. From 2001 he established the Double Star Section of LIADA (The Ibero-American Astronomy League). In 2008 he initiated a joint collaboration with the Instituto Astrofisica de Canarias in Tenerife and was able to use the 1.52-metre Carlos Sanchez Telescope for the electronic imaging of close binary stars. This has resulted in a number of papers which have been published in *Monthly Notices of the Royal Astronomical Society* and *Revista Mexicana de Astronomía y Astrofísica*.

Figure 6.17 Francisco Rica (Edgardo Masa).

Figure 6.16 (left to right) Ignacio Novalbos, Tofol Tobal, and Xavier Miret (X. Miret).

Juerg Schlimmer

Juerg Schlimmer lives in southern Germany, where he has observed visual double stars with his 12-inch telescope since 2012. He is a regular contributor to *JDSO*. He says 'I started my astronomical observations in 1996. At first, I was interested in the Messier objects and in particular in astrophotography. In addition, I directed my telescope on nights that were unfavourable for deep-sky observations towards double stars. Since 2004 they have been my favourite observation objects. The periastron passage of γ Virginis in 2005 was one of the first highlights for me. I followed the passage over several months with an interferometer attachment. In addition, I became increasingly interested in historical observation reports on double stars. From today's perspective it is hardly conceivable, but in the years 2005/2006 there were almost no historical observation reports on the internet, which was still at its beginning. So I started my own research in museum libraries, where I got insight into original documents. Astronomical institutes also copied desired documents that were of interest to me. They also offered the possibility of book sponsorships, and so became the godfather for the digitisation of a book and took over the costs. With my telescope I tried to understand these historical observations and carried out my own distance and position measurements. Even today my interest is in the observation and measurement of binary stars. The results of my measurements are published regularly in journals.'

Jean-Claude Thorel

Jean Claude Thorel became interested in astronomy for a time as a child. Some years later the interest returned and his early activities included lunar and planetary drawing and deep-sky observation. His interest in double stars began when he became involved in work to resolve some inconsistences in double star catalogues during the compilation of the Hipparcos Input Catalogue. He lives in Nice and for many years has been involved with the telescopes and personnel at the University of Nice. He has trained a number of observers to

Figure 6.18 Juerg Schlimmer (J. Schlimmer).

Figure 6.19 Jean-Claude Thorel (J.-C. Thorel).

use the micrometers and telescopes fitted to the 50-cm and 74-cm refractors. He has himself made more than 6000 measures with the large refractors.

For his work in this field he has received the following prizes: in 1997, the Prize of the Association for the International Development of the Nice Observatory, with his wife, and in 2002 the Isaac Roberts and Dorothy Klumpke-Roberts Prize for work on double stars from the Société Astronomique de France (SAF). He is a member of the Double Star Commission of the SAF, and has given several presentations and lectures. He has also written two works, the first on Robert Jonckheere, giving his genealogy and a small complement on his biography. Following the death of Paul Couteau, he was involved in sorting out his documents, archiving his works, and producing a summary of his work on double stars from discovery to publication.

Tòfol Tobal Conesa

Tòfol Tobal started his observations and measurements on double stars in 1976 as a member of the Double Star Section of the Agrupaciòn Astronómica de Sabadell (Barcelona), coordinated by José Luís Comellas, with whom he established a long period of collaboration. In 1985, along with other colleagues, he founded the Grup Astronòmic de Vilanova (GAV) and promoted the observation of double stars by its members. He established numerous contacts with Spanish and French observers, forming a working team with the objective of carrying out the revision and updating of the catalogues of J. L. Comellas (1973, 1978, 1982), so that their data could be included in the *Washington Double Star Catalog 2000.0*. Between 1994 and 2003 he was a member of the Commission des Etoiles Doubles de la Société Astronomique de France.

In 1997, along with J. Cairol, J. Planas, and A. Sánchez, and later X. Miret, he founded the Observatori Astronómic del Garraf (OAG), a centre that encourages professional–amateur collaboration in the study of visual double stars. Between 2003 and 2008 he obtained more than 4000 measurements on plates scanned within the programme Neglected Visual Double Stars (USNO), cataloguing also 400 wide anonymous pairs, designated with the code TOB in the *WDS*. From 2009, in collaboration with professional teams (J. A. Caballero, E. Solano, D. Valls-Gabaud) and with his colleagues at the OAG (X. Miret and I. Novalbos), he coordinated the long-term project (2009–2016) known as the Garraf Common Proper Motion Wide Pairs Survey on POSS-I/POSS-II digitised plates. The result of this programme is reflected in the GWP *Catalogue of 3382 Common Proper Motion Wide Pairs/Equatorial Zone (+20° to −20°)* published at the end of 2016. Early in 2017 a new survey of common proper motion systems in the North Circumpolar Zone was started and is still in progress.

Tòfol Tobal is a professional geographer who graduated at the University of Barcelona and is dedicated to teaching, dissemination, and research in geography, making trips and thematic expeditions around the world. His astronomical activities and geographic information can be followed at www.oagarraf.net and his blog can be found at www.portgeography.com. His fondness for astronomy started in 1969 when Apollo XI reached the Moon. He has published works on geography and astronomy in several national and international publications, including a Practical Astronomy Mountain Guide: *How to Observe the Milky Way* (2004). Together with his OAG colleagues he has promoted International Pro-Am Meetings in Barcelona (Castelldefels 2000, Sabadell 2010, and Vilanova i la Geltrú 2016), which have specialised in double stars.

7 Myths, Mysteries, and One-Offs

Given the nature of double star discovery it seems inevitable that there will be cases where a star is clearly seen to be double by one observer and yet not to appear so to another, given the variation of aperture, magnification, seeing, and keenness of eyesight. In this chapter there are a few case histories of stars where duplicity was seen by eye but where subsequent doubt has crept in about the reality of these observations. Not all the systems included here are doubtful – some are simply extremely difficult but have their own unique interest whilst others remain tantalisingly poised between existence and non-existence.

Mira Ceti = JOY 1

Alfred Joy [37], whilst looking at a spectrum of Mira taken with the 100-inch telescope at Mount Wilson, saw an unusual feature which led him to believe that Mira might have a tail, or a shell that was asymmetric on one side, or possibly a companion star at a distance of about $0''.25$ and PA $135°$, and so he wrote to R. G. Aitken and asked him to look at Mira (02 19 20.78 −02 59 39.5).

On 19 October 1923, Aitken [155] examined Mira with the Lick 36-inch. He immediately saw a star 0.5 to 0.75 magnitudes fainter than Mira at about $1''$ south following the variable. It was so easy to see that he thought that it would be visible in a 12-inch telescope. This observation coincided with the minimum of Mira, and the new companion appeared blue-white. This star itself turned out to be variable and was given the designation VZ Cet. A recent investigation by Sokoloski & Bilsen [157] concluded that it is a white dwarf, as the observed rapid variations in brightness on a timescale of minutes allied this star to known white dwarfs in cataclysmic variable binary systems.

Modern observations of Mira, especially with the ALMA array in Chile, describe a complex interaction between Mira A and B and between Mira A and the local interstellar medium. Mira B is thought to have a range of brightness of between 10 and 12.5. Mira A itself ranges between $V = 2$ and 10. The large baseline of ALMA (~ 15 km) means that the binary pair is well resolved at 95 and 229 GHz (Vlemmings et al. [156]), and the separation is measured to an accuracy of 1 mas. Vlemmings et al. noted that a significant decrease in separation over a two-week period was detected and further observations of this precision will soon constrain the visual orbit of A and B. Mira is moving relative to the local interstellar medium at 120 km s^{-1} and producing a bow shock ahead of it, whilst also creating a cometary wake as far as $2°$ (or about 12 light years) behind it.

Atlas = 27 Tau = STF 453

During his great survey at Dorpat, F. G. W. Struve observed 27 Tau = Atlas (03 49 09.73 +24 03 12.7), a member of the Pleiades cluster, and saw a faint star $0''.79$ distant. He noted 'mags. 5 and 8 but difficult with 600 power'. He observed the pair again in 1830, this time recording the distance as $0''.35$ and the observation as 'uncertain'. Despite subsequent attempts, no sign of the companion was found until 1876 when Hartwig (see Winnecke [40]), who was unaware of the star's duplicity, thought he had noted a half-second fading when the star was occulted by the Moon. No further evidence came to light until February 1904, when Thomas Lewis and William Bowyer were observing with the 28-inch at Greenwich. On two consecutive nights first Lewis, then Bowyer, recorded a companion star of magnitude 9 at around $0''.47$ at similar position angles.

In 1965 Abt *et al.* [42] announced that 27 Tau was a single-lined spectroscopic binary with a period of 1254 days, from nine velocities with a range of about 35 km s^{-1}. Four years later Nather and Evans [41] observed 27 Tau being occulted by the Moon and found a companion star 1.9 magnitudes fainter in the blue than Atlas at a vector separation of 0″.0061. In 1972, Bartholdi [44], using the 1-metre reflector Geneva telescope at Haute-Provence, also observed the occultation of the Pleiades by the Moon and recorded a faint extra feature in the light curve which led him to announce that there was a companion star 1.63 magnitudes fainter than 27 Tau and at a distance of 4 milliarcseconds and PA 124°. A year later, occultation photometry carried out at Tonantzintla Observatory in Mexico found a companion 1.5 magnitudes fainter than 27 Tau, this time at 4 milliarcseconds separation and a PA of 49°. The authors (McGraw *et al.* [43]) stated that 'It is unlikely that the alleged visual duplicity that led to the designation ADS 2786 can be real'.

In 1997, investigators belonging to the Hipparcos consortium had caused a bit of a stir by announcing that the satellite had measured the distance of the Pleiades to be 118 ± 3 pc, which did not agree with other determinations. Another way to determine distance is by accurately determining the orbit of a binary star. Short-period binaries are not very common in the Pleiades but, by using the resolving power of the NPOI a group [45] resolved 27 Tau into two components and were able to define an orbit quite quickly. The distance between the stars ranged from 4 to 15 mas and the period turned out to be 291 days. From this they were able to deduce a dynamical parallax which placed the star at a distance of 132 ± 4 pc. This would appear to rule out a visually resolvable component in the system.

104 Tau = A 3010

First resolved in 1912 by R. G. Aitken [46] with the Lick 36-inch refractor, this magnitude 6 G4 dwarf star in Taurus (05 07 27.00 +18 38 42.1) has been an enigmatic object ever since. Aitken's discovery note states that 'The measures of angle are all uncertain because of the small angular separation, but the image is certainly that of a very close double star'. It was resolved on several occasions between 1934 and 1971, but further attempts between 1977 and 1981 showed the star to be single. Dr Andrei Tokovinin then made a positive measurement from two separate observations with a 1-metre telescope in 1985 and it was also resolved in 1988 using a 4-metre aperture, but since then no further positive observations have been made.

Short- and long-term radial velocity monitoring, with precisions as good as 0.1 km s^{-1}, have showed no changes although this might be explained if the apparent orbit were exactly at 90° to the line of sight. In addition, one of the components would need to vary substantially in brightness and the combination of these two conditions does seem unlikely.

Dr Tokovinin [420] included 104 Tau in a recent paper in which he devoted a section to enigmatic or spurious binaries. His conclusion in the case of 104 Tau was 'If this star is single (as everything seems to suggest), we cannot dismiss its multiple resolutions with micrometer, eyepiece interferometer, and speckle as spurious; occasional image doubling (or at least elongation) must be real.'

From the UK, 104 Tauri was occulted by an almost full Moon on 21 December 2018, at 1925 UT. This provided a good opportunity to check this star's duplicity. There is a report from 13 October 1832, by Robert Snow [48], observing from London with a refractor of 45-inch focal length at ×180, who noted that the star 'vanished gradually, as if it had been a small planetary body'. His obituarist, writing in *Monthly Notices* [49], noted that 'his observations of the occultations of stars by the Moon may be safely relied on'.

Rigel B = BU 555

Whilst examining Rigel (05 14 32.27 −08 12 05.9) in 1871 with his 6-inch Clark refractor, S. W. Burnham [50] thought that he saw an elongation of the companion star, which is 9″.5 distant from Rigel itself. He estimated the position angle of the elongation as 200°. Five years later Herbert Sadler [51] also noted an elongation in PA, 170° to 180°.

Once Burnham had access to large telescopes such as the 18.5-inch at Dearborn and the 36-inch at Lick he revisited the system to try to get a positive confirmation and measure of BC. With the 18.5-inch in 1877 [245] he examined the star on many nights, '. . . only three of which were sufficiently perfect in definition for micrometer measures. The highest power only gave a small elongation but that appeared to be well defined and certain, and I have no doubt of the duplicity of this star'. It entered his catalogue as BU 555.

A few years later at Lick both R. G. Aitken and W. J. Hussey made positive measurements, but strangely Aitken did not list these in his ADS catalogue but, rather, included later less positive observations as though he had become very doubtful about the existence of C. When Lewis said in 1906 that there was no longer any doubt about the reality of the BC pair he was being somewhat premature, although he himself did measure BC at Greenwich in 1901.

In 1973 Frank Holden [53] measured BC using the 36-inch at Lick and found the stars 0″.12 apart and equally bright, and in 1971 G. van Biesbroeck [54] using the Kitt Peak 2.1-metre reflector also resolved BC, but on the whole several attempts to resolve the pair with 4-metre-class telescopes over the last 30 years have failed. The last positive observation was by Mason [55] and others in 2005, when the position angle was 29°.8 and the separation 0″.124 but the measure is marked as uncertain.

Star B is known to be a double-lined spectroscopic binary consisting of two stars of spectral class B9 orbiting in a period of 9.86 days (Sanford [211]) but, given the great distance of Rigel, then the angular separation subtended by this system is going to be substantially smaller than that encountered for the suspected visual pair.

Does BC have a very eccentric apparent orbit in which the stars are separated by 0″.2 for a short while but which soon close up and vanish beneath the resolving power of even 4-metre-class telescopes?

Capella = ANJ 1

In 1899 the newly developed science of stellar spectroscopy was under way in several astronomical centres. In that year both W. W. Campbell [163] at Lick Observatory and H. F. Newall [58], using the 25-inch refractor at Cambridge, were taking spectra of Capella (05 16 41.36 +45 59 52.8) and both happened on the fact that it is a spectroscopic binary. Two years later H. F. Reese [59] of Lick Observatory published a period of 104.022 days.

Newall did some calculations that showed that the two components might just be visible separately in a large enough refractor. He came up with a projected angular separation of between 0″.04 and 0″.1. This inspired a campaign of observation by observers at the Royal Observatory, Greenwich. They used the 28-inch refractor to observe Capella at magnifications of ×670 and ×1120. In 1976 Joseph Ashbrook [60] wrote a summary of this episode for *Sky and Telescope*. As many as ten members of the RO staff noted elongations in the image of the star. Ashbrook noted the arguments for and against these observations. Against was the fact that Aitken and Burnham using the Lick 36-inch in superb seeing with magnifications to ×2600 could see no sign of an elongation. Supporting evidence included the use of blue filters at Greenwich, which, it was suggested, would help cut down the glare of the star and it was also argued that the 28-inch had its objective adjusted for a part of the spectrum different from that of the 36-inch at Lick. The Greenwich observers also

noted that the position angles were decreasing with time, a fact that was not known until the interferometer measures were made at Mount Wilson.

In 1919 J. A. Anderson [61] and P. Merrill included Capella in a list of objects to investigate using the Michelson stellar interferometer which had been set upon the end of the 100-inch reflector tube at Mount Wilson. This consisted of a 20-ft-diameter iron frame with flat mirrors at the ends and again at the edges of the telescope aperture, oriented in the plane of the meridian. The light from both beams thus went down to the main mirror and into the eyepiece. Any source with structure greater than about 0″.03 would be detectable as a series of fringes which would come and go as the field of view slowly rotated with the Earth. From the time when the fringes disappeared and the separation of the fringes when they reappeared it was possible to determine the position angle and separation of the two components of Capella.

Since 1920 the components of Capella have made more than 330 revolutions and consequently our knowledge of this system is now as comprehensive as any binary star orbit can be. A recent paper by Torres *et al.* [62] defined the physical parameters of the Capella system to unprecedented accuracy. Star A is at 11.98 $R_\odot$ whilst the B star is at 8.83 $R_\odot$. The corresponding masses are 2.57 and 2.48 $M_\odot$ whilst the luminosities are 79 and 73 times that of the Sun. The orbital period of 104.0213 days is known to 15 seconds and the dynamical parallax gives a distance to the stars of 42.92 light years with an uncertainty of 0.05 light years. An additional interest lies in the faint star H (magnitude 10), which was found to share the proper motion of Capella by Furuhjelm [63] in 1895. This was first noted to be double by Carl L. Stearns [64] in 1935 when examining plates of the field of Capella taken with the 20-inch Clark refractor at Van Vleck Observatory, Middletown, Connecticut. The star H is 142° and 723″ from Capella; later observations confirmed the duplicity and subsequent measurements showed that the pair is undergoing orbital motion. A very provisional orbit gives a period of 388 years.

Sirius = AGC 1

One of the most famous double stars in the sky, Sirius (like Procyon) was known to have a companion long before it was found. Wilhelm Bessel in his investigations of the proper motion of bright stars noted that both stars showed significant irregularities in their tracks across the sky [257]. In 1862 Alvan G. Clark [255], whilst testing an 18.5-inch refractor for which his family firm were making the optics, pointed the telescope at Sirius (06 45 08.92 −16 42 58.0) and noticed a very faint star, quite close in. The position of this star corresponded closely to that predicted from the periodic oscillation in the proper motion of the primary star.

In 1920 Philip Fox [129] was using the same telescope, now at Dearborn Observatory in Illinois, for micrometer measures of double stars. In his observation of Sirius B he notes 'B appears persistently double in 231°: 0″.8'. In 1928/9, W. S. Finsen [66], using the 67-cm Innes refractor in Johannesburg, measured B as double on four nights but still noted 'The real existence of Sirius C is regarded as doubtful'. In all cases C was between 0″.8 and about 1″.8 from B. The WDS gives magnitudes of 8.5 and 12.6. If this star is real then it is unlikely to be a field star as the high proper motion of Sirius A and B would have left it behind long ago.

In 1995 Benest & Duvent [67] found a six-year perturbation in the motion of Sirius, which they ascribed to a low-mass M dwarf star, and they postulated that it is between 5 and 10 magnitudes fainter in the visual than Sirius B. Clearly this does not correspond to the visual candidate C and neither of these stars has been confirmed.

The most recent investigation of the Sirius system was by H. E. Bond *et al.* [256]. They were able to rule out the possibility of the Benest companion and found no sign of C, although the apparent visual magnitude of this object, if the WDS value is correct, may have been too faint for this particular survey.

Procyon = SHB 1

Here is a challenge for someone wanting something different to do. There is currently no one alive who has seen the companion of Procyon. The last person to do so was Charles Worley [292] (1935–1997). This pair (07 39 18.12 +05 13 30.0) needs both a large aperture and superlative seeing, probably substantial magnification, and preferably observation from a site where Procyon can be seen high in the sky. It might also help if the telescope aperture were fitted with a hexagonal mask. This has the effect of diffracting the light in such a way that the bright star is accompanied by six equi-spaced diffraction spikes, thus diverting light from the central disk and creating a darker area near the primary where the feeble image of the companion lies. The mask would need to be rotatable lest the faint star should coincide with one of the spikes. (Photographic images of Procyon B were obtained with the 24-inch refractor at Bosscha in Java using a lentil-shaped aperture.)

The orbit is well known thanks to some direct imaging using the HST [290]. At the time of writing (2018) the companion is at 4″.5 and the maximum separation of 5″.1 comes in 2029.

STF 1398/STF 1400

STF 1398 = ADS 7603 (10 01 32.06 +68 43 05.1) was discovered by F. G. W. Struve and measured in 1832. It is a rather difficult pair for the small aperture. The magnitudes are 8.1 and 11.4 and the position angle and separation then were 229°, 3″.7. A few months later Struve measured STF 1400 (ADS 7611) which is 1.4 minutes of RA following and 4′ N of STF 1398. The magnitudes given here are very similar to STF 1398 and the measurements by Struve showed the companion at 228° and 1″.8. Eventually, measurements of the latter pair were incorporated with measures of the former pair in the ADS by Aitken, who had accepted the (erroneous) identifications made by several observers. W. H. van den Bos [71] recorded a number of other instances where pairs of similar appearance and close to each other on the sky have been mistaken for each other. In 1925, using the Johannesburg refractor, he found two new pairs, B54 and B55, only 50 seconds of time apart in right ascension and at the same declination. He noted 'Strikingly similar in all respects'.

β Cen = VOU 31

A unique feature about the constellation Centaurus is that the three nominally brightest stars, α, β, and γ, are spectacular, bright, visual binaries. The components α and γ are dealt with in the text but β (14 03 49.20 −60 22 22.9) is probably a step too far, even for the well-equipped visual observer. In 1915 Wilson found that the radial velocity was variable. In 1935 J. G. E. G. Voûte [74] was using the 24-inch refractor at Bosscha Observatory at Lembang in Java and 'when I pointed the instrument on 16 June 1935, still by daylight, at β Centauri for determing [sic] an instrumental error and test the new mounting of the double wire grating (G2) I was astonished to notice a companion close to β Centauri, I could not remember having come across any statement about it before.' Voûte estimated the companion at $V = 5.0$ and made annual measures of the pair in the following decade or so. Angular motion was small and it was clear that this component could not be the star responsible for the periodic radial velocity shifts, because two spectra were involved and this implied that the stars in the spectroscopic system were equally bright and the period was in any case short. Whilst the variation of radial velocity in the A component had been known since 1915 [77], the first radial velocity orbit was done by Shobbrook and Robertson [72] and the stars were also first resolved by Hanbury Brown and others [75] using the stellar intensity interferometer at Narrabri, but they were unable to make any measurements of position angle or separation. Astrometry was first done in 1999 using the 3.9-metre Anglo-Australian Telescope, and six years later John Davis *et al.* [76], using the SUSI interferometer, were able to make a series of very accurate measurements of separation and position angle. In 2016 Pigulski *et al.* [73] used this data, along with more recent speckle data obtained by Andrei Tokovinin with 4-metre apertures, where this system is very close to the resolution limit, even at apastron. Combining this data with radial velocities they obtained a period of 357 days and derived masses of 12.01 ± 0.13 $M_\odot$ and 10.58 ± 0.18 $M_\odot$. The dynamical parallax places the stars at a distance of 361 light years with an accuracy of 1.6 light years. They considered the history of measures of the more distant Voûte companion and predicted that the apparent orbit around the central binary could be between 120 and 220 years long. Since discovery this star has been closing, and in 2014 was at $196°$ and $0''.4$. Tokovinin [177] estimates that the orbital period of this star is 166 years and the mass of B is about 6 $M_\odot$.

θ Sco = See 510

T. J. J. See is a controversial figure in astronomical history. He was a regular double star observer and several hundred of his discoveries, catalogued under SEE (or λ in earlier catalogues) can be found in the WDS. On 10 September 1896, using the 24-inch refractor at Flagstaff, he noted [79] a very faint companion to θ Scorpii (17 37 19.13 −42 59 52.2). He noted 'In spite of the low altitude, the rare atmosphere of Arizona and the splendid definition of the 24-inch Clark Refractor enabled me to see that the star is attended by a faint satellite of the 13th magnitude'. He measured it on each of the following five nights and gave a mean measure for six nights of $321°.5$, $6''.24$, 1896.702 later cataloguing it as λ 335. He called the primary 'reddish' and the companion 'greenish' and made the following comment: 'Magnificent system of surpassing interest'. T. J. J. See noted that this star had also been observed by Cogshall, another of the Lowell astronomers. In all See listed eight measures of θ Sco AB, all made at Flagstaff apart from one in 1897 when the 24-inch was in Mexico; the

remainder were done from Lowell Observatory in Arizona, where the star only rises to $12°$, as pointed out by Michael Kerr, David Frew, and Richard Jaworski in their analysis of 2008 [80]. Fortuitously, $1°$ N of θ lies HJ 4963, whose PA and separation are almost identical to See's mean measurement, but as the stars are magnitudes 8.5 and 10.5 it seemed very unlikely that this pair could be confused with See 510. R. T. A. Innes observed θ from Johannesburg in 1926 and found no sign of a companion, and there things remained until 1991.

The reduction of known double stars in the Hipparcos mission was based on a starting PA and separation to begin an iterative process. Unfortunately there were problems in this procedure when the primary was particularly bright. Arcturus, for instance, was announced to have a third magnitude companion at $0''.5$, later to be withdrawn. The same seems to have happened with θ Sco. Using See's measure of 1896 as a starting point, a published position was given for the companion of $314°.8$ and $6''.465$, whilst the derived magnitudes were 1.97 and 5.36, despite the fact that See's companion was estimated at $V = 13$. Whilst it is possible that the Hipparcos data on θ Sco may contain information on a close companion, such a star would need to be confirmed by another method before the data could be re-examined. However, a star at $V = 13$ is too faint to be detected by Hipparcos. One other fact might be relevant here. See often gave magnitude estimates of the companions in very unequal double stars which were significantly fainter than the estimates of other observers.

From Australia in 2005, Michael Kerr made a substantial effort to look for the See companion but was unable to do so even with a 25-inch in good seeing on two nights. RWA also used the Johannesburg 26-inch refractor in 2013 and was unable to see any companion. Could B have been a ghost image in the Lowell telescope, given that this was the only instrument to show it?

STF 2375 = FIN 332 AabBab

STF 2375 (18 45 28.36 +05 30 00.4) is a pair of sixth magnitude stars found about $2.5°$ N preceding the beautiful pair θ Serpentis. At discovery in 1825 F. G. W. Struve found $108°$ and $2''.2$. Orbital motion, for it appears to be a binary, has been rather slow. By 2010 the position angle had advanced to $120°$ and the separation to $2''.6$. Interest in the system was renewed in 1952 when Dr William Finsen [81] was observing the pair with his newly constructed eyepiece interferometer on the 26.5-inch refractor in Johannesburg. When an apparently single star is examined with the interferometer, if it is a close double then a set of fringes is formed which disappears when the instrument is rotated so that the slits are parallel to the line joining the stars, in other words the position angle. Finsen was somewhat surprised when he found that there were fringes on both stars and they disappeared at exactly the same angle of the interferometer. It transpired that both stars were equally close pairs with identical position angles. It led Finsen to call them Tweedledum and Tweedledee. Since then orbital motion has destroyed the symmetry of the pairs and Aa,Ab is currently at $0''.13$, whilst Ba,Bb is now only separated by $0''.08$. Aa,Ab has a period of 27 years whilst that of Ba,Bb is 38 years.

B 427 = HIP 94144

In September 1925, barely a month after arriving in South Africa, Willem van den Bos was observing with the 26.5-inch refractor at the Republic Observatory, Johannesburg. Using a power of $\times 420$ he was examining stars in Sagittarius and $1.5°$ N of π Sgr was the star CPD $-20°$ 7443 (19 09 48.14 -19 48 13.2). This is a K1 giant with $V = 6.1$ and van den Bos noted that the star was double, but only when he applied a very high power was he able to make a measure, noting that 'but only with 1680 on a very good night the measure is somewhat better than a guess'. The separation was given as $0''.11$ and subsequent measures over the next two years

showed position angles scattered by $\pm 30°$ and separations all below $0''.16$. He notified R. G. Aitken, who measured it from Lick with the 36-inch refractor and found the separation to be $0''.13$, although he noted that measures of the stars were 'very difficult'. Subsequent measures from the southern hemisphere over the next 40 or more years all showed a single image or an uncertain separation not much more than $0''.10$. In 1976 the first speckle measurement was made by H. A. McAlister, and with 3.8-metre aperture the star was single, i.e. closer than about $0''.035$ on four nights. In 1981, using a 1-metre, Andrei Tokovinin made an apparently secure measurement at $236°.1$ and $0''.105$, but since 2007 the same observer has made more than 20 attempts with a 4-metre aperture to measure the stars without success. Voronov [82] produced an orbit in 1934 with a period of 2.38 years predicting separations of up to $0''.14$, which would have been measureable given the large apertures used, but it appears to be spurious. (Voronov published many orbits in the 1930s, most of which were simply revisions of existing calculations; these had come to the attention of van den Bos, who delivered a stinging critique [83]). Unless one of the stars is a large-amplitude variable, the evidence seems to be against the duplicity of this star. It is not discussed in the list of 'spurious and enigmatic pairs' enumerated by Tokovinin (see A 3010). A mean radial velocity of $+29.7 \pm 0.5$ km s^{-1} was recorded from six nights' observations at the Erwin Fick Observatory in 1976/7 [84] whilst Gaia DR2 gives $+28.58 \pm 0.13$ km s^{-1}.

8 Catalogue Lists and Charts

Table Headings

WDS, Washington Double Star Catalog number
Disc., discoverer
Comps., components
HIC, Hipparchus number

Epoch, year of latest measure
θ, position angle in degrees
ρ, separation of components in arcseconds
V_A, V_B, magnitudes of components A and B

I   The Catalogue in Right Ascension Order

| No. | WDS | Disc. | Comps. | HIC | Latest measure | | | V_A | V_B |
					Epoch	θ	ρ		
1	00057+4549	STT 547	AB	473	2016	189	6.0	8.98	9.15
2	00063+5826	STF 3062		518	2017	1	1.7	6.42	7.32
3	00094−2759	BU 391	AB	761	2003	258	1.5	6.13	6.24
4	00184+4401	GRB 34	AB	1475	2015	64	34.3	8.31	11.36
5	00315−6257	LCL 119	AC	2484	2016	168	27.3	4.28	4.51
6	00373−2446	BU 395		2941	2015	113	0.8	6.60	6.20
7	00491+5749	STF 60	AB	3821	2018	327	13.5	3.52	7.36
8	00550+2338	STF 73	AB	4288	2018	335	1.3	6.12	6.54
9	01061−4643	SLR 1	AB	5165	2016	84	0.7	4.10	4.19
10	01084−5515	RMK 2	AB,C	5348	2016	239	6.8	4.00	8.23
11	01137+0735	STF 100	AB	5737	2016	63	22.9	5.22	6.26
12	01158−6853	HJ 3423	AB	5896	2016	316	4.8	5.00	7.74
	01158−6853	I 27	CD	5896	2016	345	1.2	7.84	8.44
13	01361−2954	HJ 3447		7463	2015	194	0.8	5.97	7.35
14	01398−5612	DUN 5		7751	2016	186	11.5	5.78	5.90

contd

No.	WDS	Disc.	Comps.	HIC	Latest measure				
					Epoch	θ	ρ	V_A	V_B
15	01535+1918	STF 180	AB	8832	2018	1	7.5	4.52	4.58
16	02020+0246	STF 202	AB	9487	2017	264	1.9	4.10	5.17
17	02037+2556	STF 208	AB	9621	2018	348	1.3	5.82	7.87
18	02039+4220	STF 205	A,BC	9640	2016	63	9.4	2.31	5.02
19	02193−0259	JOY 1	Aa,Ab	10826	2014	98	0.5	6.8	10.4
20	02291+6724	STF 262	AB	11569	2016	227	2.7	4.63	6.92
21	02318+8916	STF 93	AB	11767	2013	233	18.1	2.1	9.1
22	02433+0314	STF 299	AB	12706	2015	299	2.0	3.54	6.18
23	02583−4018	PZ 2		13847	2013	91	8.6	3.20	4.12
24	02592+2120	STF 333	AB	13914	2016	211	1.4	5.17	5.57
25	03121−2859	HJ 3555		14879	2013	301	5.4	3.98	7.19
26	03124−4425	JC 8	AB	14913	2015	150	0.6	6.42	7.36
27	03184−0056	AC 2	AB	15383	2015	261	1.2	5.60	7.97
28	03401+3407	STF 425	AB	17129	2015	59	1.9	7.52	7.60
29	03543−0257	STF 470	AB	18255	2016	349	6.9	4.80	5.89
30	04153−0739	STF 518	A,BC	19849	2016	102	83.7	4.51	9.7
31	04301+1538	STF 554		20995	2014	14	1.5	5.70	8.12
32	04400+5328	STF 566	AB,C	21730	2016	170	0.8	5.56	7.49
33	05079+0830	STT 98		23879	2017	291	1.0	5.76	6.67
34	05145−0812	STF 668	A,BC	24436	2017	204	9.4	0.3	6.8
35	05226+7914	STF 634	AB	25110	2016	142	31.1	5.14	9.14
36	05245−0224	DA 5	AB	25281	2015	77	1.8	3.56	4.87
37	05248−5219	DUN 20	AB,C	25303	2008	288	38.3	6.24	6.74
38	05320−0018	STFA 14	Aa-C	25930	2015	2	53.3	2.41	6.83
39	05407−0157	STF 774	AB	26727	2013	167	2.4	1.88	3.70
40	05597+3713	STT 545	AB	28380	2017	305	4.1	2.60	7.2
41	06149+2230	BU 1008		29655	2017	258	1.8	3.52	6.15
42	06221+5922	STF 881	AB	30272	2014	149	0.6	6.13	7.71
43	06238+0436	STF 900	AB	30422	2017	29	12.0	4.42	6.64
44	06288−0702	STF 919	AB	30867	2017	133	7.0	4.62	5.00
45	06298−5014	DUN 30	AB-CD	30953	2014	311	12.0	5.97	7.98
	06298−5014	R 65	AB		2014	257	0.5	5.97	6.15
	06298−5014	HDO 195	CD		2014	184	0.4	7.98	8.73
46	06410+0954	STF 950	AB	31978	2015	214	3.0	4.66	7.9
47	06451−1643	AGC 1	AB	32349	2016	75	10.7	−1.46	8.5

contd

| No. | WDS | Disc. | Comps. | HIC | Latest measure | | | V_A | V_B |
					Epoch	θ	ρ		
48	06462+5927	STF 948	AB	32438	2017	68	1.9	5.44	6.00
49	06546+1311	STF 982	AB	33202	2017	143	7.3	4.75	7.80
50	07033−5911	DUN 39		34000	1997	86	1.4	5.83	6.78
51	07087−7030	DUN 42		34481/73	2002	296	14.4	3.86	5.43
52	07166−2319	HJ 3945	AB	35210/3	2008	52	26.4	5.00	5.84
53	07201+2159	STF 1066		35550	2016	230	5.5	3.55	8.18
54	07247−3149	DUN 47	A,CD	35957	2009	342	97.6	5.40	7.58
55	07346+3153	STF 1110	AB	36850	2018	53	5.3	1.93	2.97
56	07401+0514	STF 1126	AB	—	2016	177	1.0	6.55	6.96
57	08095−4720	DUN 65	AB	39953	2009	221	40.3	1.79	4.14
58	08122+1739	STF 1196	AB	—	2018	10	1.2	5.30	6.25
	08122+1739	STF 1196	AB-C	—	2018	65	6.0	5.30	6.25
59	08198−7131	BSO 17	AB	40817/34	2011	60	63.8	5.31	5.59
60	08447−5443	I 10	AB	42913	2013	263	0.4	1.99	5.57
61	08467+2846	STF 1268		43103	2016	308	31.3	4.13	5.99
62	08468+0625	STF 1273	AB,C	43109	2017	309	2.9	3.49	6.66
63	09144+5241	STF 1321	AB	120005	2016	97	17.2	7.79	7.88
64	09188+3648	STF 1334	AB	45688	2017	224	2.6	3.92	6.09
65	09210+3811	STF 1338	AB	45858	2018	318	1.0	6.72	7.08
66	09285+0903	STF 1356		46454	2016	112	0.9	5.69	7.28
67	09307−4028	COP 1		46651	2016	122	1.0	3.91	5.12
68	09471−6504	RMK 11		48002	2010	126	5.0	3.02	6.00
69	09525−0806	AC 5	AB	48437	2016	42	0.5	5.43	6.41
70	10062−4722	I 173		49485	2016	9	1.0	5.32	7.10
71	10200+1950	STF 1424	AB	50583	2018	127	4.7	2.37	3.64
72	10209−5603	RMK 13	AB	59676	2000	102	7.1	4.49	7.19
73	10393−5536	DUN 95	AB	52154	2000	105	51.7	4.38	6.06
74	10468−4925	R 155		52727	2013	56	2.3	2.82	5.65
75	10451−5941	DUN 98	AB		2000	17	60.6	6.58	8.14
76	10535−5851	DUN 102	AB		2000	204 1	159.4	3.88	6.23
77	10556+2445	STF 1487		53417	2017	112	6.4	4.48	6.30
78	11182+3132	STF 1523	AB	55203	2018	161	2.0	4.33	4.80
79	11190+1416	STF 1527		55254	2013	224	0.3	7.01	7.99
80	11239+1032	STF 1536	AB	55642	2018	96	2.2	4.06	6.71
81	11323+6105	STT 235	AB	56290	2017	46	0.9	5.69	7.55

contd

| No. | WDS | Disc. | Comps. | HIC | Latest measure | | | | |
					Epoch	θ	ρ	V_A	V_B
82	11363+2747	STF 1555	AB	56601	2016	148	0.8	6.41	6.78
83	11551+4629	STF 1579	AB	58112	2012	42	3.7	6.68	8.72
84	12140−4543	RMK 14		59654	2008	243	2.7	5.78	6.98
85	12244+2535	STF 1639	AB	60525	2017	326	1.7	6.74	7.83
86	12266−6306	DUN 252	AB	60718	2016	112	4.2	1.25	1.55
87	12312−5707	DUN 124	AB	61084	2010	26	128.9	1.83	6.45
88	12351+1823	STF 1657		61418/5	2016	272	20.4	5.11	6.33
89	12415−4858	HJ 4539	AB	61932	2014	199	0.2	2.82	2.88
90	12417−0127	STF 1670	AB	61941	2018	359	2.9	3.48	3.53
91	12463−6806	R 207	AB	62322	2016	53	1.1	3.52	3.98
92	12533+2115	STF 1687	AB	62886	2015	198	1.2	5.15	7.08
93	12546−5711	DUN 126	AB	63003/5	2016	17	34.9	3.94	4.95
94	12560+3819	STF 1692		63125/1	2016	229	20.0	2.85	5.52
95	13081−6518	RMK 16	AB	64094	2016	189	5.5	5.65	7.55
96	13100+1732	STF 1728	AB	64241	2015	193	0.0	4.85	5.53
97	13226−6059	DUN 133	AB,C	65271	2016	345	60.4	4.49	6.15
98	13239+5456	STF 1744	AB	65378	2016	152	14.4	2.23	3.88
99	13375+3618	STF 1768	AB	66458	2017	88	1.7	4.98	6.95
100	13491+2659	STF 1785		67422	2016	189	3.2	7.36	8.15
101	13518−3300	H 3 101		67669	2013	104	7.9	4.50	5.97
102	14135+5147	STF 1821	AB	69483/1	2017	235	13.7	4.53	6.62
103	14396−6050	RHD 1	AB	71683/1	2016	313	4.1	−0.01	1.33
104	14411+1344	STF 1865	AB	71795	2015	287	0.5	4.46	4.55
105	14450+2704	STF 1877	AB	72105	2016	344	2.8	2.58	4.81
106	14514+1906	STF 1888	AB	72659	2018	300	5.3	4.76	6.95
107	15038+4739	STF 1909		73695	2016	72	0.9	5.20	6.10
108	15051−4703	HJ 4728		73807	2016	65	1.8	4.56	4.60
109	15185−4753	HJ 4753	AB	74911	2016	299	0.8	4.93	4.99
110	15227−4441	DUN 182	AB,C	74376/38	2010	143	26.5	3.83	5.52
111	15232+3017	STF 1937	AB	75312	2017	231	0.6	5.64	5.95
112	15234−5919	HJ 4757		75323	2016	0	1.0	4.94	5.73
113	15245+3723	STF 1938	Ba,Bb	75415/1	2017	4	2.2	7.09	7.63
114	15351−4110	HJ 4786	AB	76297	2016	277	1.0	2.95	4.45
115	15360+3948	STT 298	AB	76382/75	2017	187	1.3	7.16	8.44
116	15394+3638	STF 1965		76669	2016	307	6.2	4.96	5.91

contd

No.	WDS	Disc.	Comps.	HIC	Latest measure			V_A	V_B
					Epoch	θ	ρ		
117	15549−6045	DUN 194	AB	77927	2016	47	44.1	6.35	9.97
	15549−6045	SLR 11	AB		2016	97	1.1	6.35	8.09
118	16044−1122	STF 1998	AB		2016	10	1.1	5.16	4.87
	16044−1122	STF 1998	AC		2016	45	7.6	5.16	4.87
119	16120−1928	H 5 6	AC	79374	2016	338	41.6	4.35	6.60
119	16120−1928	BU 120	AB	79374	2016	2	1.6	4.35	5.31
	16120−1928	MTL 2	CD		2016	56	2.5	6.60	7.23
120	16147+3352	STF 2032	AB	79607	2016	241	7.4	5.62	6.49
121	16294−2626	GNT 1		80763	2016	276	3.2	0.96	5.4
122	16309+0159	STF 2055	AB	80883	2017	44	1.5	4.15	5.15
123	16413+3136	STF 2084		81693	2017	121	1.6	2.95	5.40
124	17053+5428	STF 2130	AB	83608	2017	3	2.5	5.66	5.69
125	17104−1544	BU 1118	AB	84012	2016	231	0.6	3.05	3.27
126	17146+1423	STF 2140	AB	84345	2016	104	4.8	3.48	5.40
127	17150+2450	STF 3127	AB	84379	2017	291	13.1	3.12	8.3
128	17153−2636	SHJ 243	AB	84405	2016	143	5.4	5.12	5.12
129	17190−3459	MLO 4	AB	84709	2016	124	0.9	6.37	7.38
130	17191−4638	BSO 13	AB	84720	2016	258	10.8	5.61	8.88
131	17237+3709	STF 2161	AB	85112	2016	320	4.1	4.50	5.40
132	17269−4551	DUN 216	AC	85389	2016	312	102.5	5.63	7.12
133	17322+5511	STFA 35		85829/19	2016	310	62.5	4.87	4.90
134	18015+2136	STF 2264		88267	2016	259	6.4	4.85	5.20
135	18031−0811	STF 2262	AB	88404	2017	291	1.7	5.27	5.86
136	18055+0230	STF 2272	AB	88601	2018	125	6.5	4.22	6.17
137	18068−4325	HJ 5014		88726	2016	3	1.8	5.65	5.68
138	18096+0400	STF 2281	AB	88964	2016	285	0.7	5.97	7.52
139	18369+3846	H 5 39	AB	91262	2015	184	82.3	0.09	9.5
140	18428+5938	STF 2398	AB	91768	2015	180	11.6	9.11	9.96
141	18443+3940	STF 2382	AB	91919	2017	346	2.2	5.15	6.10
	18443+3940	STF 2383	CD	91926	2017	77	2.4	5.25	5.38
142	18501+3322	STFA 39	AB	92420	2016	148	45.6	3.63	6.69
143	18562+0412	STF 2417	AB	92951	2016	104	22.6	4.59	4.93
144	19026−2953	HDO 150	AB	93506	2016	252	0.6	3.27	3.48
145	19064−3704	HJ 5084		93825	2016	343	1.6	4.53	6.42
146	19172−6640	GLE 3		94789	2016	351	0.6	6.12	6.42

contd

No.	WDS	Disc.	Comps.	HIC	Latest measure			V_A	V_B
---	---	---	---	---	Epoch	θ	ρ		
147	19226–4428	DUN 226		95241	2010	76	28.4	3.98	7.21
148	19307+2758	STFA 43	AB	95947/51	2016	54	34.7	3.19	4.68
149	19418+5032	STFA 46	AB	96895/901	2016	133	40.0	6.00	6.23
150	19450+4508	STF 2579	AB	97165	2017	217	2.6	2.89	6.27
151	20099+2055	STF 2637	AB	99352	2016	331	11.6	6.56	8.85
152	20136+4644	STFA 50	AC	99675/6	2016	173	108.6	3.93	6.97
153	20375+1436	BU 151	AB	101769	2012	74	0.2	4.11	5.02
154	20467+1607	STF 2727	AB	102532	2016	266	9.0	4.36	5.03
155	20474+3629	STT 413	AB	102589	2017	5	1.0	4.73	6.26
156	20516–6226	RMK 26		102962	2006	82	2.7	6.23	6.58
157	20591+0418	STF 2737	AB	103569	2012	284	0.4	5.96	6.31
158	21069+3845	STF 2758	AB	104214/7	2017	153	31.5	5.20	6.05
159	21137+6424	H 1 48		104758	2012	246	0.7	7.21	7.33
160	21148+3803	AGC 13	AB	104887	2017	190	1.0	3.83	6.57
161	21199–5327	HJ 5258		105319	2010	268	7.3	4.50	6.93
162	21287+7034	STF 2806	AB	106032	2016	251	13.5	3.17	8.63
163	21441+2845	STF 2822	AB	107310	2017	322	1.7	4.75	6.18
164	22038+6438	STF 2863	AB	108917	2017	274	8.0	4.45	6.40
165	22266–1645	SHJ 345	AB	110778	2017	81	1.3	6.29	6.39
166	22288–0001	STF 2909	AB	110960	2017	161	2.3	4.34	4.49
167	22280+5742	KR 60	AB	110893	2013	326	1.5	9.93	11.41
168	22292+5825	STFA 58	AC	110991	2016	192	40.9	4.21	6.11
169	22359+3938	STF 2922	AB	111546	2017	185	22.3	5.66	6.29
170	22478–0414	STF 2944	AB	112559	2016	307	1.8	7.30	7.68
171	23069–4331	JC 20	AB	114131	2003	115	1.5	4.45	6.60
172	23079+7523	STT 489	AB	114222	2016	354	1.1	4.61	6.80
173	23340+3120	BU 720		116310	2016	108	0.6	5.67	6.11
174	23460–1841	H 2 24		117281	2016	135	7.1	5.65	6.46
175	23595+3343	STF 3050	AB	118281	2018	343	2.5	6.46	6.72

II   The Catalogue in Constellation Order

No.	WDS	Disc.	Comps.	HIC	Latest measure Epoch	θ	ρ	V_A	V_B
Andromeda									
1	00057+4549	STT 547	AB	473	2016	189	6.0	8.98	9.15
4	00184+4401	GRB 34	AB	1475	2015	64	34.3	8.31	11.36
8	00550+2338	STF 73	AB	4288	2018	335	1.3	6.12	6.54
18	02039+4220	STF 205	A,BC	9640	2016	63	9.4	2.31	5.02
175	23595+3343	STF 3050	AB	118281	2018	343	2.5	6.46	6.72
Aquarius									
165	22266−1645	SHJ 345	AB	110778	2017	81	1.3	6.29	6.39
167	22280+5742	KR 60	AB	110893	2013	326	1.5	9.93	11.41
170	22478−0414	STF 2944	AB	112559	2016	307	1.8	7.30	7.68
174	23460−1841	H 2 24		117281	2016	135	7.1	5.65	6.46
Ara									
130	17191−4638	BSO 13	AB	84720	2016	258	10.8	5.61	8.88
132	17269−4551	DUN 216	AC	85389	2016	312	102.5	5.63	7.12
Aries									
15	01535+1918	STF 180	AB	8832	2018	1	7.5	4.52	4.58
17	02037+2556	STF 208	AB	9621	2018	348	1.3	5.82	7.87
24	02592+2120	STF 333	AB	13914	2016	211	1.4	5.17	5.57
Auriga									
40	05597+3713	STT 545	AB	28380	2017	305	4.1	2.60	7.2
Boötes									
100	13491+2659	STF 1785		67422	2016	189	3.2	7.36	8.15
102	14135+5147	STF 1821	AB	69483/1	2017	235	13.7	4.53	6.62
104	14411+1344	STF 1865	AB	71795	2015	287	0.5	4.46	4.55
105	14450+2704	STF 1877	AB	72105	2016	344	2.8	2.58	4.81
106	14514+1906	STF 1888	AB	72659	2016	302	5.5	4.76	6.95
107	15038+4739	STF 1909		73695	2016	72	0.9	5.20	6.10
113	15245+3723	STF 1938	Ba,Bb	75415/1	2017	4	2.2	7.09	7.63
115	15360+3948	STT 298	AB	76382/75	2017	187	1.3	7.16	8.44
Camelopardalis									
32	04400+5328	STF 566	AB,C	21730	2016	170	0.8	5.56	7.49
35	05226+7914	STF 634	AB	25110	2016	142	31.1	5.14	9.14
Cancer									
58	08122+1739	STF 1196	AB	—	2017	16	1.2	5.30	6.25
	08122+1739	STF 1196	AB-C	—	2018	65	6.0	5.30	6.25
61	08467+2846	STF 1268		43103	2016	308	31.3	4.13	5.99

contd

No.	WDS	Disc.	Comps.	HIC	Latest measure			V_A	V_B
					Epoch	θ	ρ		
Canes Venatici									
94	12560+3819	STF 1692		63125/1	2016	229	20.0	2.85	5.52
99	13375+3618	STF 1768	AB	66458	2017	88	1.7	4.98	6.95
Canis Major									
47	06451−1643	AGC 1	AB	32349	2016	75	10.7	−1.46	8.5
52	07166−2319	HJ 3945	AB	35210/3	2008	52	26.4	5.00	5.84
54	07247−3149	DUN 47	A,CD	35957	2009	342	97.6	5.40	7.58
Canis Minor									
56	07401+0514	STF 1126	AB	—	2016	177	1.0	6.55	6.96
Carina									
50	07033−5911	DUN 39		34000	1997	86	1.4	5.83	6.78
68	09471−6504	RMK 11		48002	2010	126	5.0	3.02	6.00
74	10468−4925	R 155		52727	2013	56	2.3	2.82	5.65
76	10535−5851	DUN 102	AB		2000	204	159.4	3.88	6.23
Cassiopeia									
2	00063+5826	STF 3062		518	2017	1	1.7	6.42	7.32
7	00491+5749	STF 60	AB	3821	2018	327	13.5	3.52	7.36
20	02291+6724	STF 262	AB	11569	2016	227	2.7	4.63	6.92
Centaurus									
84	12140−4543	RMK 14		59654	2008	243	2.7	5.78	6.98
89	12415−4858	HJ 4539	AB	61932	2014	199	0.2	2.82	2.88
97	13226−6059	DUN 133	AB,C	65271	2016	345	60.4	4.49	6.15
101	13518−3300	H 3 101		67669	2013	104	7.9	4.50	5.97
103	14396−6050	RHD 1	AB	71683/1	2016	313	4.1	−0.01	1.33
Cepheus									
159	21137+6424	H 1 48		104758	2012	246	0.7	7.21	7.33
162	21287+7034	STF 2806	AB	106032	2016	251	13.5	3.17	8.63
164	22038+6438	STF 2863	AB	108917	2017	274	8.0	4.45	6.40
166	22288−0001	STF 2909	AB	110960	2017	161	2.3	4.34	4.49
168	22292+5825	STFA 58	AC	110991	2016	192	40.9	4.21	6.11
172	23079+7523	STT 489	AB	114222	2016	354	1.1	4.61	6.80
Cetus									
6	00373−2446	BU 395		2941	2015	113	0.8	6.60	6.20
19	02193−0259	JOY 1	Aa,Ab	10826	2014	98	0.5	6.8	10.4
22	02433+0314	STF 299	AB	12706	2015	299	2.0	3.54	6.18
27	03184−0056	AC 2	AB	15383	2015	261	1.2	5.60	7.97

contd

No.	WDS	Disc.	Comps.	HIC	Latest measure			V_A	V_B
					Epoch	θ	ρ		
Circinus									
112	15234–5919	HJ 4757		75323	2016	0	1.0	4.94	5.73
Coma Berenices									
85	12244+2535	STF 1639	AB	60525	2017	326	1.7	6.74	7.83
88	12351+1823	STF 1657		61418/5	2016	272	20.4	5.11	6.33
92	12533+2115	STF 1687	AB	62886	2015	198	1.2	5.15	7.08
96	13100+1732	STF 1728	AB	64241	2015	193	0.0	4.85	5.53
Corona Australis									
137	18068–4325	HJ 5014		88726	2016	3	1.8	5.65	5.68
145	19064–3704	HJ 5084		93825	2016	343	1.6	4.53	6.42
Corona Borealis									
111	15232+3017	STF 1937	AB	75312	2017	231	0.6	5.64	5.95
116	15394+3638	STF 1965		76669	2016	307	6.2	4.96	5.91
120	16147+3352	STF 2032	AB	79607	2016	241	7.4	5.62	6.49
Crux									
86	12266–6306	DUN 252	AB	60718	2016	112	4.2	1.25	1.55
87	12312–5707	DUN 124	AB	61084	2010	26	128.9	1.83	6.45
93	12546–5711	DUN 126	AB	63003/5	2016	17	34.9	3.94	4.95
Cygnus									
148	19307+2758	STFA 43	AB	95947/51	2016	54	34.7	3.19	4.68
149	19418+5032	STFA 46	AB	96895/901	2016	133	40.0	6.00	6.23
150	19450+4508	STF 2579	AB	97165	2017	217	2.6	2.89	6.27
152	20136+4644	STFA 50	AC	99675/6	2016	173	108.6	3.93	6.97
155	20474+3629	STT 413	AB	102589	2017	5	1.0	4.73	6.26
158	21069+3845	STF 2758	AB	104214/7	2017	153	31.5	5.20	6.05
160	21148+3803	AGC 13	AB	104887	2017	190	1.0	3.83	6.57
163	21441+2845	STF 2822	AB	107310	2017	322	1.7	4.75	6.18
Delphinus									
153	20375+1436	BU 151	AB	101769	2012	74	0.2	4.11	5.02
154	20467+1607	STF 2727	AB	102532	2016	266	9.0	4.36	5.03
Draco									
124	17053+5428	STF 2130	AB	83608	2017	3	2.5	5.66	5.69
133	17322+5511	STFA 35		85829/19	2016	310	62.5	4.87	4.90
140	18428+5938	STF 2398	AB	91768	2015	180	11.6	9.11	9.96

contd

| No. | WDS | Disc. | Comps. | HIC | Latest measure | | | V_A | V_B |
					Epoch	θ	ρ		
Equuleus									
157	20591+0418	STF 2737	AB	103569	2012	284	0.4	5.96	6.31
Eridanus									
14	01398−5612	DUN 5		7751	2016	186	11.5	5.78	5.90
23	02583−4018	PZ 2		13847	2013	91	8.6	3.20	4.12
26	03124−4425	JC 8	AB	14913	2015	150	0.6	6.42	7.36
29	03543−0257	STF 470	AB	18255	2016	349	6.9	4.80	5.89
30	04153−0739	STF 518	A,BC	19849	2016	102	83.7	4.51	9.7
Fornax									
25	03121−2859	HJ 3555		14879	2013	301	5.4	3.98	7.19
Gemini									
41	06149+2230	BU 1008		29655	2017	258	1.8	3.52	6.15
49	06546+1311	STF 982	AB	33202	2017	143	7.3	4.75	7.80
53	07201+2159	STF 1066		35550	2016	230	5.5	3.55	8.18
55	07346+3153	STF 1110	AB	36850	2017	54	5.2	1.93	2.97
Grus									
171	23069−4331	JC 20	AB	114131	2003	115	1.5	4.45	6.60
Hercules									
123	16413+3136	STF 2084		81693	2017	121	1.6	2.95	5.40
126	17146+1423	STF 2140	AB	84345	2016	104	4.8	3.48	5.40
127	17150+2450	STF 3127	AB	84379	2017	291	13.1	3.12	8.3
131	17237+3709	STF 2161	AB	85112	2016	320	4.1	4.50	5.40
134	18015+2136	STF 2264		88267	2016	259	6.4	4.85	5.20
Hydra									
62	08468+0625	STF 1273	AB,C	43109	2017	309	2.9	3.49	6.66
Indus									
161	21199−5327	HJ 5258		105319	2010	268	7.3	4.50	6.93
Lacerta									
169	22359+3938	STF 2922	AB	111546	2017	185	22.3	5.66	6.29
Leo									
66	09285+0903	STF 1356		46454	2016	112	0.9	5.69	7.28
71	10200+1950	STF 1424	AB	50583	2017	127	4.7	2.37	3.64
77	10556+2445	STF 1487		53417	2017	112	6.4	4.48	6.30
79	11190+1416	STF 1527		55254	2013	224	0.3	7.01	7.99
80	11239+1032	STF 1536	AB	55642	2017	97	2.1	4.06	6.71
82	11363+2747	STF 1555	AB	56601	2016	148	0.8	6.41	6.78

contd

| No. | WDS | Disc. | Comps. | HIC | Latest measure | | | V_A | V_B |
					Epoch	θ	ρ		
Lupus									
108	15051–4703	HJ 4728		73807	2016	65	1.8	4.56	4.60
109	15185–4753	HJ 4753	AB	74911	2016	299	0.8	4.93	4.99
110	15227–4441	DUN 182	AB,C	74376/38	2010	143	26.5	3.83	5.52
114	15351–4110	HJ 4786	AB	76297	2016	277	1.0	2.95	4.45
Lynx									
42	06221+5922	STF 881	AB	30272	2014	149	0.6	6.13	7.71
48	06462+5927	STF 948	AB	32438	2017	68	1.9	5.44	6.00
64	09188+3648	STF 1334	AB	45688	2017	224	2.6	3.92	6.09
65	09210+3811	STF 1338	AB	45858	2017	312	1.2	6.72	7.08
Lyra									
139	18369+3846	H 5 39	AB	91262	2015	184	82.3	0.09	9.5
141	18443+3940	STF 2382	AB	91919	2017	346	2.2	5.15	6.10
	18443+3940	STF 2383	CD	91926	2017	77	2.4	5.25	5.38
142	18501+3322	STFA 39	AB	92420	2016	148	45.6	3.63	6.69
Monoceros									
43	06238+0436	STF 900	AB	30422	2017	29	12.0	4.42	6.64
44	06288–0702	STF 919	AB	30867	2017	133	7.0	4.62	5.00
46	06410+0954	STF 950	AB	31978	2015	214	3.0	4.66	7.9
Musca									
91	12463–6806	R 207	AB	62322	2016	53	1.1	3.52	3.98
95	13081–6518	RMK 16	AB	64094	2016	189	5.5	5.65	7.55
Ophiuchus									
122	16309+0159	STF 2055	AB	80883	2017	44	1.5	4.15	5.15
125	17104–1544	BU 1118	AB	84012	2016	231	0.6	3.05	3.27
128	17153–2636	SHJ 243	AB	84405	2016	143	5.4	5.12	5.12
135	18031–0811	STF 2262	AB	88404	2017	291	1.7	5.27	5.86
136	18055+0230	STF 2272	AB	88601	2017	124	6.4	4.22	6.17
138	18096+0400	STF 2281	AB	88964	2016	285	0.7	5.97	7.52
Orion									
33	05079+0830	STT 98		23879	2017	291	1.0	5.76	6.67
34	05145–0812	STF 668	A,BC	24436	2017	204	9.4	0.3	6.8
36	05245–0224	DA 5	AB	25281	2015	77	1.8	3.56	4.87
38	05320–0018	STFA 14	Aa-C	25930	2015	2	53.3	2.41	6.83
39	05407–0157	STF 774	AB	26727	2013	167	2.4	1.88	3.70

contd

No.	WDS	Disc.	Comps.	HIC	Latest measure			V_A	V_B
					Epoch	θ	ρ		
Pavo									
146	19172–6640	GLE 3		94789	2016	351	0.6	6.12	6.42
156	20516–6226	RMK 26		102962	2006	82	2.7	6.23	6.58
Pegasus									
173	23340+3120	BU 720		116310	2016	108	0.6	5.67	6.11
Perseus									
28	03401+3407	STF 425	AB	17129	2015	59	1.9	7.52	7.60
Phoenix									
9	01061–4643	SLR 1	AB	5165	2016	84	0.7	4.10	4.19
10	01084–5515	RMK 2	AB,C	5348	2016	239	6.8	4.00	8.23
Pictor									
37	05248–5219	DUN 20	AB,C	25303	2008	288	38.3	6.24	6.74
Pisces									
11	01137+0735	STF 100	AB	5737	2016	63	22.9	5.22	6.26
16	02020+0246	STF 202	AB	9487	2017	264	1.9	4.10	5.17
Puppis									
45	06298–5014	DUN 30	AB-CD	30953	2014	311	12.0	5.97	7.98
	06298–5014	R 65	AB		2014	257	0.5	5.97	6.15
	06298–5014	HDO 195	CD		2014	184	0.4	7.98	8.73
Sagitta									
151	20099+2055	STF 2637	AB	99352	2016	331	11.6	6.56	8.85
Sagittarius									
144	19026–2953	HDO 150	AB	93506	2016	252	0.6	3.27	3.48
147	19226–4428	DUN 226		95241	2010	76	28.4	3.98	7.21
Serpens									
143	18562+0412	STF 2417	AB	92951	2016	104	22.6	4.59	4.93
Sculptor									
3	00094–2759	BU 391	AB	761	2003	258	1.5	6.13	6.24
13	01361–2954	HJ 3447		7463	2015	194	0.8	5.97	7.35
Sextans									
69	09525–0806	AC 5	AB	48437	2016	42	0.5	5.43	6.41

contd

No.	WDS	Disc.	Comps.	HIC	Latest measure			V_A	V_B
					Epoch	θ	ρ		
Scorpio									
118	16044–1122	STF 1998	AB		2016	10	1.1	5.16	4.87
	16044–1122	STF 1998	AC		2016	45	7.6	5.16	4.87
119	16120–1928	H 5 6	AC	79374	2016	338	41.6	4.35	6.60
	16120–1928	BU 120	AB	79374	2016	2	1.6	4.35	5.31
	16120–1928	MTL 2	CD		2016	56	2.5	6.60	7.23
121	16294–2626	GNT 1		80763	2016	276	3.2	0.96	5.4
129	17190–3459	MLO 4	AB	84709	2016	124	0.9	6.37	7.38
Taurus									
31	04301+1538	STF 554		20995	2014	14	1.5	5.70	8.12
Triangulum Australe									
117	15549–6045	DUN 194	AB	77927	2016	47	44.1	6.35	9.97
	15549–6045	SLR 11	AB		2016	97	1.1	6.35	8.09
Tucana									
5	00315–6257	LCL 119	AC	2484	2016	168	27.3	4.28	4.51
12	01158–6853	HJ 3423	AB	5896	2016	316	4.8	5.00	7.74
	01158–6853	I 27	CD	5896	2016	345	1.2	7.84	8.44
Ursa Major									
63	09144+5241	STF 1321	AB	120005	2016	97	17.2	7.79	7.88
78	11182+3132	STF 1523	AB	55203	2017	164	1.9	4.33	4.80
81	11323+6105	STT 235	AB	56290	2017	46	0.9	5.69	7.55
83	11551+4629	STF 1579	AB	58112	2012	42	3.7	6.68	8.72
98	13239+5456	STF 1744	AB	65378	2016	152	14.4	2.23	3.88
Ursa Minor									
21	02318+8916	STF 93	AB	11767	2013	233	18.1	2.1	9.1
Vela									
57	08095–4720	DUN 65	AB	39953	2009	221	40.3	1.79	4.14
60	08447–5443	I 10	AB	42913	2013	263	0.4	1.99	5.57
67	09307–4028	COP 1		46651	2016	122	1.0	3.91	5.12

contd

No.	WDS	Disc.	Comps.	HIC	Epoch	θ	ρ	V_A	V_B
					Latest measure				
70	10062–4722	I 173		49485	2016	9	1.0	5.32	7.10
72	10209–5603	RMK 13	AB	59676	2000	102	7.1	4.49	7.19
73	10393–5536	DUN 95	AB	52154	2000	105	51.7	4.38	6.06
75	10451–5941	DUN 98	AB		2000	17	60.6	6.58	8.14

Virgo

| 90 | 12417–0127 | STF 1670 | AB | 61941 | 2017 | 3 | 2.5 | 3.48 | 3.53 |

Volans

| 51 | 07087–7030 | DUN 42 | | 34481/73 | 2002 | 296 | 14.4 | 3.86 | 5.43 |
| 59 | 08198–7131 | BSO 17 | AB | 40817/34 | 2011 | 60 | 63.8 | 5.31 | 5.59 |

III   All-Sky Finder Charts

In these all-sky finder charts, numbered 1–6, the shape enclosing a peripheral number indicates the shape of the corresponding chart.

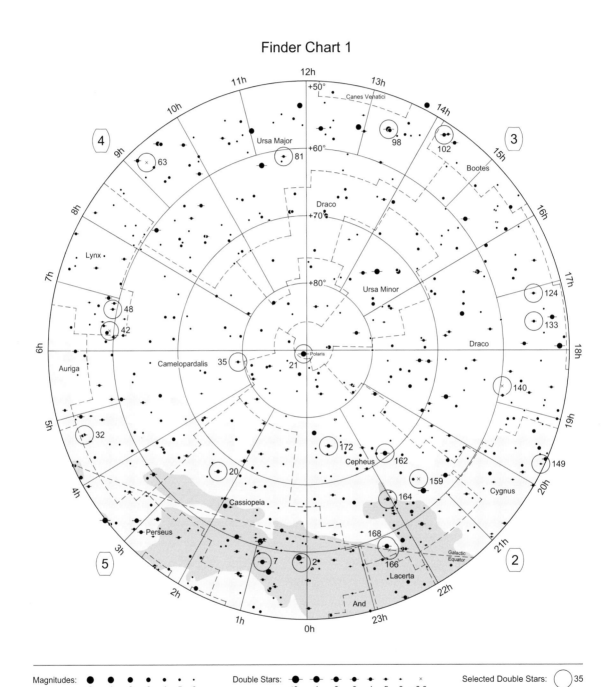

Finder Chart 1

Finder Chart 2

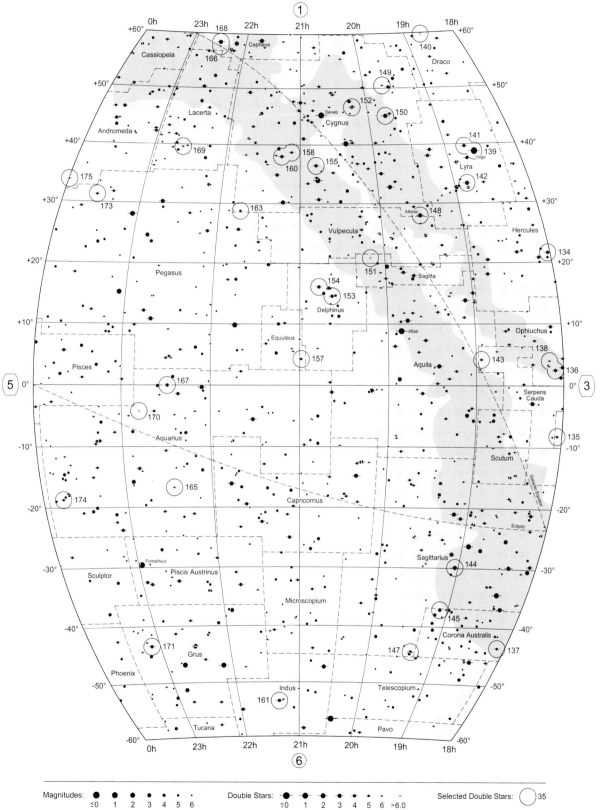

Finder Chart 3

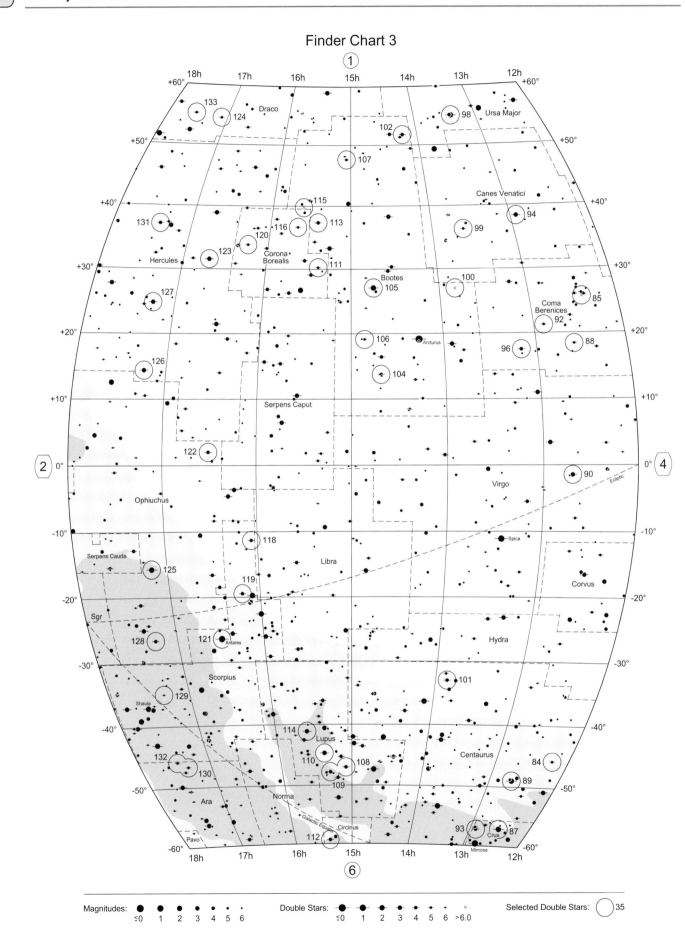

Magnitudes: ● ● ● ● ● ● ·
≤0 1 2 3 4 5 6

Double Stars: ● ● ● ● ● · ×
≤0 1 2 3 4 5 6 >6.0

Selected Double Stars: ◯ 35

Finder Chart 4

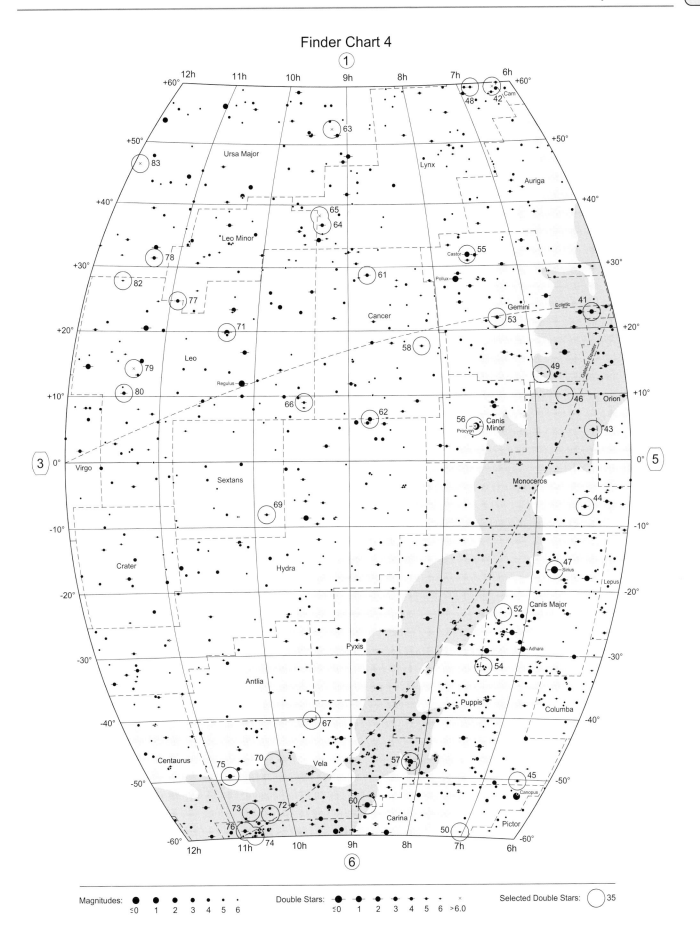

Finder Chart 5

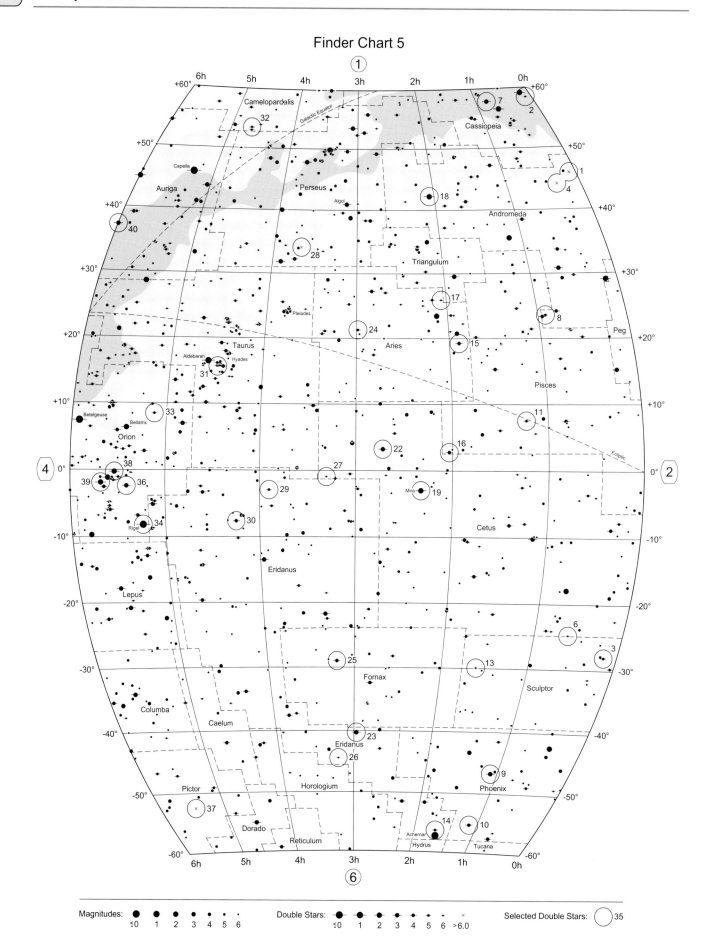

Magnitudes: ● ● ● ● ● ● ⋅
　　　　　　≤0　1　2　3　4　5　6

Double Stars: ●— ●— ●— ●— +— +— ×
　　　　　　　≤0　1　2　3　4　5　6　>6.0

Selected Double Stars: ◯ 35

Finder Chart 6

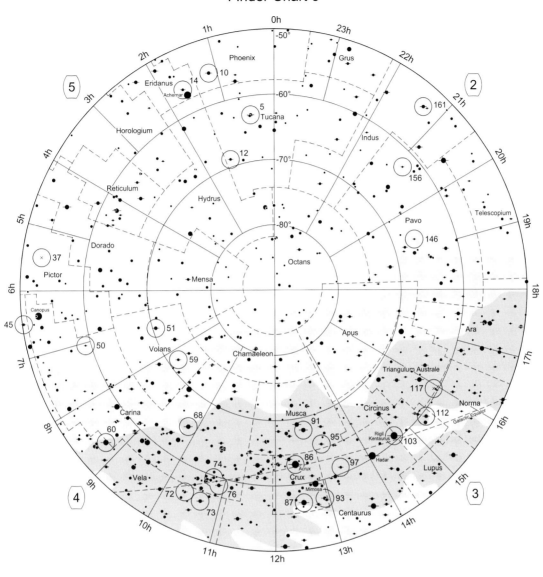

Magnitudes: ● ● ● ● • ·
≤0 1 2 3 4 5 6

Double Stars: ● ● ● ● • · ×
≤0 1 2 3 4 5 6 >6.0

Selected Double Stars: ◯ 35

9 The Catalogue

Introduction to the Catalogue

As a young boy, and before I acquired an interest in astronomy, I collected the cards which were given free with packets of tea. They ranged over many subjects and even included astronomy and space [4]. Each card gave a potted description of its subject and, of course, the idea was to collect all of them before the tea company lost interest and changed the subject.

This book is by way of being an album of sorts with visual double stars as its subject. Galaxies, star clusters, and planetary nebulae all have their adherent enthusiasts but over the last 10 or 20 years the number of people observing double stars has also increased significantly. There is a decided increase in the number of observers who are measuring visual double stars, be it by using a telescope and camera or by accessing the raw data from the large-scale surveys available on the internet. Reducing these objects to streams of numbers means that we lose sight of the objects themselves, in all their variety, and it is partially to remedy this that the book has been compiled.

This is not a how-to book. There is practical advice to be had in other publications and on the internet forums. Here we describe 175 double and multiple stars in all parts of the sky. Each object is interesting in its own right, in the view of the authors, and for a variety of different reasons. It might be a challenge – the pair is particularly close, or the companion is faint and elusive, the stars may be of significantly different colours, there may be significant orbital motion, or the star may just be a glorious sight in the telescope. These are mostly subjective qualities, so the list of targets must also be subjective.

Each story starts with the discovery of an object and moves on to see how the accumulating knowledge in the following years helps to build up a picture of the system. The age of the visual discoverer finished around 40 years ago, and since then the increase in aperture and particularly in techniques such as speckle interferometry, adaptive optics, ground-based interferometry, and astrometric satellites has led, and continues to lead, a revolution in resolution and light-gathering power. The atmosphere, which used to dominate proceedings, can now be countered.

Nor are the observers themselves to be denied a place here. Some of the stories of the more famous exponents are reasonably well known but some have been unjustly neglected. The recent activity by non-professional astronomers in the field also justifies some biographical notes of the more active observers.

The main part of the book is the catalogue. This has been prepared in such a way as to supply the observer with all that he or she might need to find the double star in question and to be aware of any difficulties that might be encountered. The book is aimed at the visual observer. The stars in the catalogue range in difficulty from naked-eye to the limits of a 20-inch (50-cm) telescope, but most can be seen with moderate apertures, say 8–12 inches (20–80 cm). Each catalogue entry comes with a find chart and, where relevant, a plot of the apparent orbit and ephemeris for future years.

There are two different sets of finder charts. The first consists of six charts covering the whole sky. Four of these each cover a six-hour interval of RA between declinations +60° and −60°, whilst the remaining two cover the North and South Celestial Poles. The stars in the catalogue are numbered. Each individual entry in the catalogue has its own finder chart, which is centred on that star; concentric circles with radii 1°, 3°, and 5° are plotted in each case. The field stars go down to $V = 8$ and other labelled double stars are discussed in the text.

As is the case for the finder charts, the diagrams showing the apparent orbit of each binary have all been compiled by Mike Swan, whose description of them follows later in this section. The angular scale of the field of view is indicated in each case. The resolving limit for three separate apertures, 10 cm ($1''.16$), 20 cm ($0''.56$), and 40 cm ($0''.30$), in each case based on the Dawes limit, is represented on the diagram by three shaded grey circles (smallest, 40-cm limit; middle, 20-cm limit), and the largest outer circle represents the limit of resolution of a 10-cm telescope. The ephemerides are calculated from the orbital elements given in the *Sixth Catalog of Elements of Visual Binary Orbits* [6], take from the USNO website at the time of going to press. In general the validity of an orbit calculation is more short-lived in the case of pairs of shorter period. The catalogue [6] is constantly being updated and added to, so, in critical cases where quickly moving binaries are being observed, it is advisable to check with the current version of the online catalogue at the USNO website.

The history of each system is traced back to the original observation, where possible, in order to give due credit to the discoverer. In many cases, stars with a Struve (STF) designation were actually found by Sir William Herschel, who, in turn, re-observed many of the pairs found by Christian Mayer. References to the original sources, which have been consulted, are given wherever possible.

'The Modern Era' section for each star system attempts to summarise what has been found out about the pair in the last few decades, in terms of membership in particular. Current high-resolution observations with large telescopes are constantly adding new components, often, in the case of infrared work, at the faint end of the main sequence.

Exoplanets seem to be no respecters of stellar singularity. Multi-planetary systems are known to be present in binary star systems and examples are even known where each component of a binary star is accompanied by exoplanets. Where exoplanets are known to exist, or have been looked for around the systems in this catalogue, details are given.

For the visual observer the most relevant section is 'Observing and Neighbourhood'. Here there are more details on how to star hop in order to reach the system being described from bright stars, and, having made the effort to find a particular pair, it may be of interest to know what other pairs are in the immediate neighbourhood. Details are given of any such pairs to be found within a few degrees of the current target.

Some of the systems in this volume are bright stars and visible to the naked eye; nevertheless, finder charts are supplied for each system. The field sizes are given and the orientation is always with N to the top and E to the left. Nearby double stars of interest are also marked on the finder charts.

The data tables start with the RA and Dec for J2000 taken from the *WDS* catalogue. Although more accurate positions are available, it was felt that RA to two decimal places (d.p.) and Dec to one d.p. was more than adequate for finding any of the stars in the catalogue using GOTO (computer controlled) telescopes. For those without this facility, all-sky finder charts are also given, at the end of Chapter 8. Next to the position is the order in the *WDS* observations catalogue, in terms of the total number of observations made followed in brackets by the number of observations.

The V magnitudes and $(B - V)$ colours are taken where possible from the Tycho catalogue produced by the Hipparcos satellite.

Proper motions and parallaxes as determined by Hipparcos are presented except when Gaia (DR2) data is available. The definitive final catalogue of this mission is not due out until 2022 and, unfortunately from our point of view, many of the double and multiple systems in this catalogue will not be dealt with until DR3 is issued in 2020. Occasionally, dynamical parallaxes, which are derived from the orbits of the binary pairs, are available which in some cases (Sirius, for example) rival the astrometric values. In each case the proper motion in RA is followed by the formal error, then the proper motion in Declination and its formal error, usually in milliarcseconds. The precision of the astrometry is reproduced as given in the original source catalogues. How accurate these values are can be judged from the size of the accompanying error. Proper motions (μ, in milliarcseconds per year) and parallaxes (π, in milliarcseconds) from DR2 are marked thus. Unsourced astrometry can be assumed to come from Hipparcos and represents the centre of the light measures of the two components.

Spectra are taken from SIMBAD or *WDS*.

Occasionally, individual masses for components in binary systems are available. This is also true for radii, although directly observed radii are rather uncommon.

Luminosity is derived in each case from the absolute magnitude, which in turn is calculated from the distance as found by the trigonometrical parallax measured by Hipparcos or Gaia. No correction is made for reddening. The Gaia DR2 parallax is known to have a systematic error of 0.1 mas. This has not been applied, as it has a negligible effect even on the smallest measured parallaxes.

The 'official' name for a double star is taken to be its classification in the *WDS* catalogue (i.e. WDS J12345+1234). Although rather long, it does have the advantage of being both uniform and useful, containing as it does the J2000 coordinates within its structure. The double star catalogue numbers are given in chronological order, starting with Mayer, then William Herschel, F. G. W. Struve (or O. Struve), and the Burnham and Aitken double star catalogues. Many books adopt the official *WDS* nomenclature and shorten the catalogue name to three letters, representing the discoverer, followed by a running number, e.g. STF 1523. (A short list of the common three-letter codes used is given in the Appendix.) These are followed by the relevant star catalogue numbers, again listed chronologically – Flamsteed/Bayer, HR, HD, SAO, and Hipparcos numbers. Some wider and brighter pairs have catalogue numbers for both components.

Radial velocities, derived largely from SIMBAD, are quoted for the combined system in most part or, in the case of wider pairs, where each star can be observed separately, the velocities are those of the individual stars. Radial velocities derived by the Gaia DR2 project are marked '(DR2)'. Radial velocities that are negative indicate a motion away from the Solar System, and positive velocities represents motion towards the Sun.

Description of the Charts

All-Sky Finder Charts

These six charts (see the end of the previous chapter) cover the entire night sky. The polar charts use a zenithal equidistant projection, whilst the equatorial charts use a pseudoazimuthal Aitoff projection. These seemed the most suitable projections considering the large areas involved. The scale at the centre of each chart is 18 mm to 10° of arc but will vary away from the centre point owing to the projections used. There are no overlapping areas between the four equatorial charts but there is a 10° overlap between the polar and equatorial charts.

The stars are plotted in AutoCad^{TM} from the Hipparcos catalogue. The limiting visual magnitude of the charts is 6.00. Stars are allocated in one-magnitude bins, e.g. the magnitude-3 bin contains stars in the 2.50 to 3.49 range but the magnitude-6 bin just contains the stars from 5.50 to 6.00.

Constellation names have been added and constellation boundaries plotted with dashed lines. No other information, apart from the galactic equator, the ecliptic, and the names of some of the brightest or most popular stars, has been included.

To make it easier to find the positions of the double stars listed in this book, each one is centred in a circle and numbered 1 to 175, according to their listing in the catalogue.

Double stars are plotted using information from the *WDS* catalogue. All doubles are shown by the conventional symbol. Those doubles that feature in this book, and are fainter than magnitude 6.00, are shown by a cross.

All variable stars are plotted at their maximum magnitude with an ordinary star symbol, but are not annotated as a variable.

Individual Finder Charts

Every double star is accompanied by its own finder chart. Each chart covers 8° square, has a scale of 8.35 mm to 1° of arc and has north to the top; as it covers such a small area, and could be at any latitude (declination, Dec), a gnomic projection was deemed the most suitable.

The RA and Dec for the double star centred on each chart is given, and these are the same as the *WDS* code. The 1°, 3°, and 5° 'field of view' circles may assist in locating the double star in the night sky.

The stars are plotted in AutoCad^{TM} using data from the Hipparcos and Tycho catalogues. Stars are allocated in one-magnitude bins; e.g. magnitude-3 contains stars in the 2.50 to 3.49 range and magnitude-8 contains stars from 7.50 to 8.49.

All stars with Greek letters and Flamsteed numbers are annotated. Constellation names have been added and constellation boundaries plotted with dashed lines.

Double stars are plotted using information from the WDS catalogue. Most doubles are shown by the conventional symbol, but where the separations of the individual components are large enough, and they can be plotted separately, then a unique symbol is used. Double stars are only labelled if they are mentioned in the 'Observing and Neighbourhood' paragraph for each double in the main section of this book.

Variable stars are only shown and labelled if their magnitude range is greater than 2. These consist mostly of Mira-type variables. They are annotated with a circle symbol indicating the maximum magnitude and sometimes (if the minimum is 8 or brighter) with a dot at the centre that shows the minimum magnitude.

The most prominent deep-sky objects are plotted and annotated, but, in the case of open star clusters, they are only shown when a number of the stars in the cluster are brighter than magnitude 8.5.

Orbital Diagrams

Orbital diagrams for those double stars that are also binary stars have been plotted in AutoCad^TM using data extracted from the USNO *Sixth Catalog of Orbits of Visual Binary Stars*. The full references can all be found in the *WDS Reference Notes Catalog*.

The orbit has been plotted through one complete revolution. Each diagram has North (0°) to the bottom with the position angle (PA) increasing in an anticlockwise direction. The scale (in arcseconds) is shown beneath each diagram. Close binaries are shown with three levels of grey shading, which indicate the Dawes limit when one is splitting pairs of equal magnitude, using respectively 10-cm, 20-cm, and 40-cm aperture telescopes.

The orbital ephemeris, computed using the orbital elements from the *Sixth Catalog*, is included with each diagram, showing the PAs and separations (Sep) of the binary in future years. Also listed is the abbreviation for the name of the author who computed the orbital elements, the period of the binary in years, and the grade (1 to 5) of the elements, '1' being the most accurate and usually a binary that has been observed through at least one complete revolution of its orbit.

Orbital Measures

The diagrams of seven binaries that are regularly measured by author RWA have been included in the individual sections on those stars. They are STF 1196, 1523, 1670, 1937, 2084, and 2272 and SHJ 345. They are copies of the orbital diagrams but with the addition of his own observations made with the 8-inch refractor at Cambridge.

Legend for Individual Finder Charts

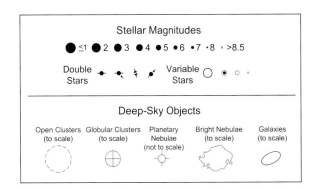

1. STT 547 AND = WDS 00057+4549AB

Table 9.1 Physical parameters for STT 547 And

STT 547	RA: 00 05 41.03	Dec: +45 48 43.3	WDS: 77(399)	
V magnitudes	A: 8.98	B: 9.15	F: 11.8	
$(B - V)$ magnitudes	A: +1.61	B: +1.61		
μ(A)	888.62 mas yr^{-1}	± 0.07	−162.47 mas yr^{-1}	± 0.05 (DR2)
μ(B)	845.89 mas yr^{-1}	± 0.17	−148.54 mas yr^{-1}	± 0.11 (DR2)
π(A)	86.87 mas	± 0.05	37.55 light yr	± 0.02 (DR2)
π(B)	86.94 mas	± 0.06	37.52 light yr	± 0.03 (DR2)
Spectra	A: K6V	B: M0V		
Luminosities (L$_\odot$)	A: 0.03	B: 0.02		
Catalogues	HD 38	HIP 473		
DS catalogues	STT 547	BDS 12740	ADS 48	
Radial velocity (A/B)	1.15 km s^{-1}	± 0.6	−2.26 km s^{-1}	± 0.22 (DR2)
Galactic coordinates	114°.651	−16°.323		

History

The observations for the Pulkovo catalogue compiled by Otto Struve were carried out at Pulkova Observatory between August 1841 and December 1842. They were started by F. G. W. Struve using the 15-inch refractor and after a month his son Otto took over and completed the work. The result was the 'Catalogue of 514 double and multiple stars discovered on the northern celestial hemisphere by the great telescope of the central observatory of Pulkovo and a catalogue of 256 double stars the distances between the principal stars of which lie between 32 arc seconds and 2 arc minutes [and] which have been found in the northern hemisphere'. The principal achievement was the 514 new pairs which had been missed during the Dorpat survey with the smaller telescope and included some particularly close and interesting objects. Number 38 is the companion to γ And; number 535 is δ Equ, with a period of 5.7 years. It is still an exceptionally short-period visual system. The secondary catalogue is of 256 wide pairs

Finder Chart

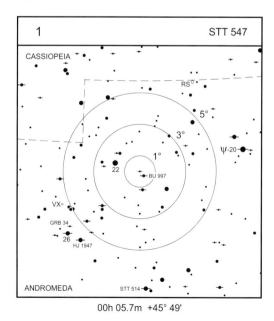

00h 05.7m +45° 49'

Orbit

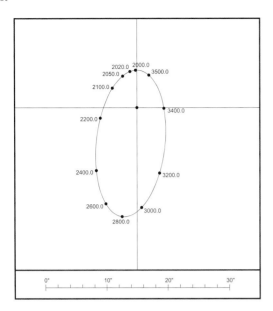

Ephemeris for STT 547 AB (2010 to 2100)

Orbit by Pop (1996b) Period: 1550.637 years, Grade: 4

Year	PA(°)	Sep(")	Year	PA(°)	Sep(")
2010.0	186.4	5.94	2060.0	209.9	5.43
2020.0	190.8	5.88	2070.0	215.2	5.31
2030.0	195.3	5.79	2080.0	220.7	5.21
2040.0	199.9	5.68	2090.0	226.4	5.13
2050.0	204.8	5.56	2100.0	232.3	5.07

of little interest to modern astronomers but which do afford some attractive sights in small telescopes and binoculars. They are denoted by $O\Sigma\Sigma$ in older observing handbooks but the modern appellation is STTA. Until 1878 Otto continued to make small additions to the catalogue, and the last set of eight pairs brought the total to 547. The pair STT 547 is a particularly interesting system as it contains two red dwarfs in orbit and which are relatively close to the Solar System. The large proper motion means that the field attracted the attention of people such as Burnham, who added further faint components to assist the astrometry of the pair. Component C is magnitude 11.6 at 116″ (distance increasing) and D is 12.65 at 110″ (incr.); E is 11.8 at 58″ (decr.) whilst a distant companion F (at 254° and 328″) has the same proper motion as AB, and according to Shaya & Olling [270] has a near 100% probability of being physically connected to AB.

The Modern Era

There are two extant orbits for AB. Popovic [107] calculated the period to be 1550.6 years whilst Kiyaeva *et al.* in 2001 [108] found 509.7 years. The current USNO *Sixth Orbit Catalog* prefers Popovic's calculation. Both A and B have been found to have variable radial velocity (up to 45 km s^{-1} for A and 25 km s^{-1} for B). The 10.2 magnitude star F at 254° and 327″ has common proper motion and Tokovinin [177] in the MSC estimated a period of 169.2 kiloyears for the rotation around AB.

Observing and Neighbourhood

STT 547 is not plotted in the *Cambridge Double Star Atlas (2nd edition)* (*CDSA2*) because it is too faint; the finding chart is given here. I found both stars to be orange and equally bright in 1971 with a 10-inch mirror ×80 and ×300, and two distant field stars were also noted. The WDS gives stars of 13.7 at 263°, 116″, 12.5 at 230°, 110″, and 11.8 at 344°, 58″ (all at 2015), but these values are changing quickly owing to the large proper motion of AB. Close by (8′ S and 35″ of RA preceding) is the similar pair BU 997, which has, on occasion, been mistaken for STT 547. The magnitudes are 7.6 and 9.4 and the current position is 337° and 3″.8, but the colours are yellow and blue. By chance, the last pair in the original Pulkovo catalogue, STT 514, is about 4° S. The stars are 6.2 and 9.7 at 170°, 5″.1, 2015. Just 30′ WSW of 26 And is HJ 1947 (6.2, 9.8, 75°, 9″.1, 2015), which is a long-period binary. DR2 finds almost identical parallaxes and proper motions and that the system is 320 light years distant.

Measures

Early measure (D)	111°.4	4″.24	1876.29
(Orbit	110°.7	4″.30)	
Recent measure (SER)	189°.7	6″.01	2016.92
(Orbit	189°.4	5″.90)	

2. STF 3062 CAS = WDS J00063+5826

Table 9.2 Physical parameters for STF 3062 Cas

STF 3062	RA: 00 06 15.81	Dec: +58 26 12.5	WDS: 34(604)	
V magnitudes	A: 6.42	B: 7.32		
$(B - V)$ magnitudes	A: +0.70	B: +0.96		
μ(A)	237.90 mas yr^{-1}	± 0.06	36.96 mas yr^{-1}	± 0.05 (DR2)
μ(B)	286.74 mas yr^{-1}	± 0.50	37.92 mas yr^{-1}	± 0.21 (DR2)
π(A)	47.80 mas	± 0.04	68.23 light yr	± 0.06 (DR2)
π(B)	46.21 mas	± 0.19	70.58 light yr	± 0.29 (DR2)
Spectra	A: G3V	Ba: G9V	Bb: M2V?	
Masses (M$_\odot$)	A: 0.98	Ba: 0.95	Bb: 0.22	
Luminosities (L$_\odot$)	A: 1	B: 0.5		
Catalogues	HD 123	HR 5	SAO 21085	HIP 518
DS catalogues	H 1 39	STF 3062	BDS 12755	ADS 61
Radial velocity	−13.79 km s^{-1}	± 0.14		
Galactic coordinates	137°.075	−3°.917		

History

Noted by William Herschel on 25 May 1782, this pair was sadly neglected by double star observers until measured by F. G. W. Struve at Dorpat in 1831, when the separation had reached 0″.5 and the position angle was changing at 10° per year. Harper [110], using the 72-inch reflector at Victoria, suggested that one of the stars in the system was itself a spectroscopic binary.

The Modern Era

In the 1990s the separation of the two stars had reached 1″.5, enabling Roger Griffin to measure the radial velocity of each component with the 36-inch reflector at Cambridge. He was able to say that the radial velocity of A was sensibly constant whilst that of B varied with a period of 47 days. In his paper Griffin [111] also discussed the announcement by

Finder Chart

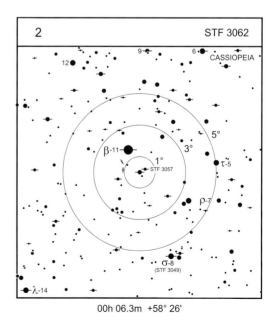

00h 06.3m +58° 26'

Orbit

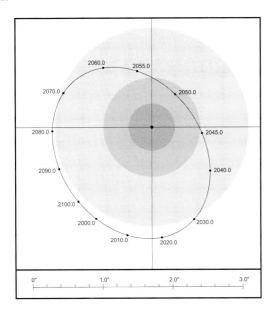

Ephemeris for STF 3062 (2016 to 2034)

Orbit by Sod (1999) Period: 106.7 years, Grade: 2

Year	PA(°)	Sep(")	Year	PA(°)	Sep(")
2016.0	357.9	1.55	2026.0	16.7	1.49
2018.0	1.5	1.55	2028.0	20.8	1.46
2020.0	5.2	1.54	2030.0	25.0	1.42
2022.0	9.0	1.53	2032.0	29.6	1.37
2024.0	12.8	1.51	2034.0	34.6	1.30

Brettmann *et al.* [112] that the star was a periodic variable star, and showed by reference to the data taken by Hipparcos that the brightness of the system was constant to around 0.008 magnitudes and that the GCVS name V640 Cas was incorrectly applied.

Observing and Neighbourhood

The pair STF 3062 will be separable in 10-cm apertures for at least another 20 years. The two yellow stars will then close up and reach the minimum distance of 0″.56 in 2049. STF 3062 shares a field with STF 3057 about 1° SW of β Cas; STF 3062 is the brighter of the two stars. STF 3057 is not particularly easy. The stars have magnitudes of 6.7 and 9.3 and are only 3″.8 apart. Further south is σ Cas (STF 3049), which is similarly challenging. In this case stars of magnitudes 5.0 and 7.2 are 3″.1 apart. At 2°.5 due W of τ (not on the chart) is SHJ 355, as marked on the *Cambridge Double Star Atlas (2nd edition) – CDSA2*. This is a multiple group of 10 stars which contains components that range from the telescopically supremely difficult to those visible in binoculars. The group is also known as STT 496. The South–Herschel pair have magnitudes of 4.9 and 7.2 and the separation is 76″.

Measures

Early measure (STF)	82°.5	0″.82	1831.71
(Orbit	84°.3	0″.72)	
Recent measure (ARY)	2°.5	1″.66	2018.01
(Orbit	1°.6	1″.55)	

3. κ^1 SCL = BU 391 = WDS J00094–2759AB

Table 9.3 Physical parameters for κ^1 Scl

BU 391	RA: 00 09 21.02	Dec: −27 59 16.5	WDS: 1027(75)	
V magnitudes	A: 6.20	B: 6.26		
$(B - V)$ magnitudes	A: +0.47	B: +0.42		
μ(A)	69.58 mas yr^{-1}	± 0.19	−8.22 mas yr^{-1}	± 0.46 (DR2)
μ(B)	72.77 mas yr^{-1}	± 1.03	−12.25 mas yr^{-1}	± 0.98 (DR2)
π(A)	14.51 mas	± 0.15	224.8 light yr	± 2.3 (DR2)
π(B)	13.91 mas	± 0.60	234.5 light yr	± 10.1 (DR2)
Spectra	A: F4	B:		
Luminosities (L$_\odot$)	A: 13	B: 14		
Catalogues	HD 493	HR 24	SAO 166083	HIP 761
DS catalogues	BU 391	BDS 30	ADS 111	LDS 2095 (AC)
Radial velocity	+7.70 km s^{-1}	± 1.6		
Galactic coordinates	25°.242	−80°.636		

History

This object was found by Burnham [113] with his 6-inch Clark on 5 November 1875, one of three new pairs that he discovered that night. Without a micrometer Burnham estimated the PA at 110° and the separation at 0″.75. Modern observations suggest that Burnham's quadrant should be reversed.

The Modern Era

The star κ^1 is a pale yellow mid-F dwarf whilst κ^2 is a K5 giant displaying a clear orange colour. This is a slow moving binary whose period is rather uncertain; κ^1 has been separating since discovery, reaching a maximum distance of 1″.5 in the early 1970s, but the stars are now beginning to close slowly. During his extensive survey of stars with proper motion, W. J. Luyten [115] noted a very faint star near κ^1 some 73″ distant and of magnitude 18.6. The system is known as LDS 2095.

Finder Chart

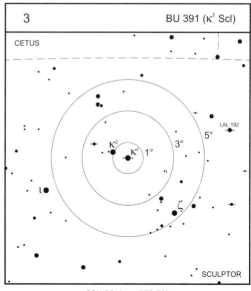

00h 09.4m -27° 59′

Orbit

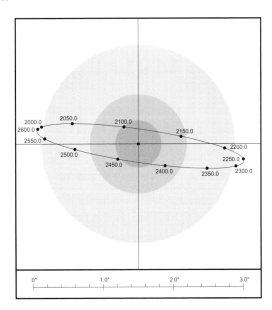

Ephemeris for BU 391 AB (2015 to 2060)

Orbit by Zir (2010) Period: 616.04 years, Grade: 5

Year	PA(°)	Sep(")	Year	PA(°)	Sep(")
2015.0	258.4	1.32	2040.0	255.0	1.10
2020.0	257.9	1.28	2045.0	254.1	1.05
2025.0	257.2	1.24	2050.0	253.1	0.99
2030.0	256.5	1.20	2055.0	252.0	0.93
2035.0	255.8	1.15	2060.0	250.7	0.87

Van Belle *et al.* [545] noted that A is a fast rotator with a $v \sin i$ velocity noted to be 170 km s^{-1}. Coincidentally, Luyten also found an even fainter star close to κ^2. This companion, now known as LDS 2099B, is magnitude 21.

Observing and Neighbourhood

The pair κ^1 Scl is easily seen by the naked eye, with κ^2 some 32' to the east. In binoculars the significant difference in spectral types will be clear. BU 391 is a good resolution test for 10-cm apertures. It is nominally below the Rayleigh limit but above the Dawes limit. There are two interesting star systems in the neighbourhood. A star which the WDS calls ϕ Scl was found to be double by James Dunlop and called DUN 253. When the pair was included in the Lick IDS double star catalogue it became LAL 192 and is thus attributed to Jérôme de Lalande. This bright and easy pair consists of stars with magnitudes 6.8 and 7.4 separated by 6''.6 in position angle 272° (2013). Not on the finding chart but 4°.5 W of κ^1 is δ Scl, which is an interesting group of four stars, two of which can be seen in 10-cm, another glimpsed in 20-cm, and a fourth detectable only in a large telescope with infrared optics. The components A and B are of magnitudes 4.6 and 9.4, separated by 74'' at 297° (2010), whilst a third star of $V = 11.6$ (BU 1013) sits only 3''.4 from A and was discovered by Burnham [116] with the 12-inch refractor at Lick Observatory in October 1881. K-band imaging of this group shows that B is a close (0''.2) equally bright pair called DRS 24.

Measures

Early measure (HWE)	277°.2	0''.78	1876.80
(Orbit	276°.6	0''.84)	
Recent measure (ARY)	258°.3	1''.48	2013.70
(Orbit	258°.6	1''.33)	

4. GRB 34 AND = WDS J00184+4401AB

Table 9.4 Physical parameters for GRB 34 And

GRB 34	RA: 00 18 22.88	Dec: +44 01 22.6	WDS: 508(123)	
V magnitudes	A: 8.31	B: 11.36		
$(B - V)$ magnitudes	A: +1.78	B: +1.80		
$\mu(A)$	2891.53 mas yr^{-1}	± 0.06	411.90 mas yr^{-1}	± 0.03 (DR2)
$\mu(B)$	2863.28 mas yr^{-1}	± 0.07	336.53 mas yr^{-1}	± 0.04 (DR2)
$\pi(A)$	280.69 mas	± 0.04	11.620 light yr	± 0.001 (DR2)
$\pi(B)$	280.79 mas	± 0.05	11.616 light yr	± 0.002 (DR2)
Spectra	A: M2V	B: M3.5V		
Masses (M$_\odot$)	A: 0.38	B: 0.19		
Radii (R$_\odot$)	A: 0.386	± 0.002	B: 0.163	
Luminosities (L$_\odot$)	A: 0.005	B: 0.0003		
Catalogues	GRB 34	HD 1326	HIP 1475	
DS catalogues	ADS 246			
Radial velocity	11.51 km s^{-1}	± 0.14		
Galactic coordinates	116°.677	−18°.447		

History

The name of this pair is derived from the catalogue of stars by Stephen Groombridge, formed from his numerous observations made with a transit circle in Blackheath between 1806 and 1817 [121]. The first observation of a companion was that by Otto Struve in 1860. S. W. Burnham did not include it in his General Catalogue because it was too wide and faint, but R. G. Aitken in the ADS recognised the system as a genuine binary.

The Modern Era

The observed motion to date has been just 12° in position angle and 5″ in separation over 150 years. S. L. Lippincott [117] produced a preliminary orbit with a period of 2600 years, which has recently been superseded in the orbit

Finder Chart

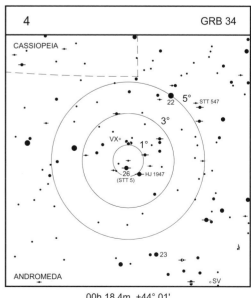

00h 18.4m +44° 01'

Orbit

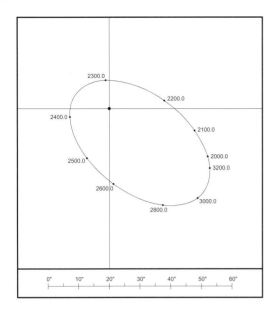

Ephemeris for GRB 34 AB (2010 to 2100)

Orbit by PkO (2014c) Period: 1253 years, Grade: 5

Year	PA(°)	Sep(")	Year	PA(°)	Sep(")
2010.0	65.4	34.65	2060.0	70.7	31.49
2020.0	66.4	34.08	2070.0	71.9	30.76
2030.0	67.4	33.48	2080.0	73.1	30.00
2040.0	68.5	32.85	2090.0	74.4	29.20
2050.0	69.6	32.19	2100.0	75.8	28.36

catalogue by a period which is about half as long. This is one of the closest known visual binaries to the Sun. The Gaia DR2 parallax puts it 11.62 light years away. The proper motion is correspondingly large. The current linear separation of the stars is 146 AU. Both components of GRB 34 are variable stars: A is GX And and B is GQ And, a BY Dra-type variable with star spots. Star A has radial velocity variation at the metres per second level due to an exoplanet in orbit discovered by Howard *et al.* [118]. Star B is known to be a spectroscopic binary. Tanner *et al.* [119] reported deep exposures in the K band around both stars, and they revealed that there were three stars at around $K = 15$, two within $12''$ of B and one at $6''$ to A, none of which appeared to be connected to the components of the binary. The 11th magnitude component C is also a background object and is rapidly being left behind by GRB 34. In 1925 it was at $232°$, $36''$ from A whilst by 2011 it was at $259°$, $286''$.

Exoplanet Host?

A planetary body was found in 2014 [118]. Known as GJ 15 Ab, it had a period of 11.4433 days and a minimum mass of 5.35 ± 0.75 Earth masses. However, Trifonov *et al.* [120] do not find the 11.4-day periodicity; rather they see a 52 Earth-mass planet with a period of about 7000 days.

Observing and Neighbourhood

The difficulty with this pair is both the difference in magnitudes and the faintness of the B component. The finder chart shows that the star can be found in the same low-power field as 26 And (STT 5) and HJ 1947 (see Star 1). The majestic slowness with which the companion moves can be seen in the ephemeris.

Measures

Early measure (STT)	$52°.9$	$40''.05$	1860.18
(Orbit	$53°.2$	$39''.87$)	
Recent measure (LOC)	$64°.4$	$34''.27$	2015.92
(Orbit	$66°.0$	$34''.31$)	

5. β TUC = LCL 119 = WDS J00315–6257AC

Table 9.5 Physical parameters for β Tuc

LCL 119	RA: 00 31 32.69	Dec: −62 57 29.6	WDS: 1861(41)		
I 260			WDS: 1044(74)		
V magnitudes	A(β^1): 4.29	C(β^2): 4.51	E(β^3): 5.09		
$(B-V)$ magnitudes	A: −0.04	C: +0.14	E: +0.05		
μ(A)	82.74 mas yr^{-1}	± 0.41	−54.58 mas yr^{-1}	± 0.40 (DR2)	
μ(B)	105.61 mas yr^{-1}	± 1.47	−48.06 mas yr^{-1}	± 1.77 (DR2)	
μ(C)	86.42 mas yr^{-1}	± 0.21	−50.36 mas yr^{-1}	± 0.21 (DR2)	
π(A)	23.26 mas	± 0.25	140.2 light yr	± 1.5 (DR2)	
π(B)	19.59 mas	± 0.97	166.5 light yr	± 8.2 (DR2)	
π(C)	21.79 mas	± 0.13	149.7 light yr	± 0.7 (DR2)	
Spectra	A: B8/A0	C: A2V	E: A0V		
Masses	A: 2.2	B: 0.5	C: 2.0	D: 1.5	Ea: 1.6
	Eb: 1.6				
Luminosities (L$_\odot$)	A: 32	B: 35	E: 18		
Catalogues (A/C/E)	HD 2484/5/3003	HR 126/7/36	SAO 248201/2/8	HIP 2484/7/2578	
DS catalogues	LCL 119 (AC)	DUN 1 (AC)	I 260 (CD)	B7 (AB)	B8 (Ea,Eb)
Radial velocity	+7.70 km s^{-1}	± 1.6			
Galactic coordinates	25°.242	−80.636			

History

Nicolas-Louis de Lacaille spent a year between 1751 and 1752 surveying the southern skies from an observing site near Cape Town in South Africa. His observations led to the first systematic southern star catalogue, although he was using a telescope of only 0.5-inch (13.5-mm) aperture and 866-mm focal length, with a magnification of ×8 and a field of almost 3°. Even so he was able to note a number of bright double stars, one of which was β Tucanae. It was next observed in 1826 by James Dunlop during his observations from Parramatta in New South Wales, and it is the first entry in his catalogue of double stars although he does acknowledge

that de Lacaille had already noted it. R. T. A. Innes, who had started a programme of double star discovery whilst an amateur observer in Sydney, continued to search for new double stars when he moved to South Africa and joined the staff of the Royal Observatory at the Cape. With the 7-inch Merz refractor there he found a number of new pairs but as there was no micrometer he was only able to make estimates of position angle and separation. One of the new objects was β^1 Tuc [122], which revealed itself as a rather unequal pair separated by about 0″.7. It was also discovered independently, and announced two years later, by Solon Bailey [248]. In 1925, using the 26.5-inch refractor at Johannesburg, W. H.

Finder Chart

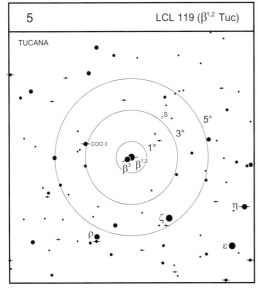

00h 31.5m -62° 57'

Orbit

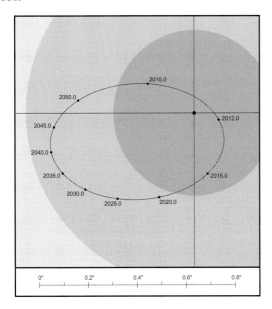

Ephemeris for I 260 CD (2018 to 2036)

Orbit by Tok (2015c) Period: 44.73 years, Grade: 2

Year	PA(°)	Sep(")	Year	PA(°)	Sep(")
2018.0	348.1	0.32	2028.0	309.1	0.51
2020.0	336.8	0.36	2030.0	304.3	0.54
2022.0	327.9	0.40	2032.0	300.0	0.56
2024.0	320.6	0.44	2034.0	295.9	0.58
2026.0	314.4	0.48	2036.0	292.1	0.60

van den Bos noted that both β^1 and β^3 were very difficult double stars: β^1 (B7) had a 14th magnitude companion about 2″ away whilst β^3 (B8) appeared to be an almost equally bright pair of stars separated by about 0″.15. Tokovinin assigns a

period of 5.05 years to this pair, but, despite many efforts by van den Bos and Finsen, there are times when the pair is only marginally resolved and the orbit may be highly inclined. This system needs speckle interferometry with large apertures.

The Modern Era

This appears to be a physical septuple as the A component is also a spectroscopic binary. The three bright stars all appear to be at the same distance, allowing for the large formal error on the DR2 parallax of β^2, and the whole group would appear to be part of the moving group of young stars that forms the Tucana–Horologium association.

Observing and Neighbourhood

There is something for every telescope in this beautiful multiple system. The separation between β^1 and β^3 (549″) makes it a naked-eye object, that between β^1 and β^2 is ideal for the small telescope or large binoculars, and the close binary which forms β^2 (I 260CD) is a challenge for 30-cm and, although the separation is now widening, the stars will only reach 0″.6 in 2040 before starting to close again. Then there are the surpassingly difficult double stars that form B7 and B8, neither of which has been seen for the last 50 years or so. At 1°.5 following β is COO 3, a fine binary pair of magnitudes 6.3 and 8.0 at 73° and 2″.4 (2016) and closing slowly. Andrew James calls it 'an uncommonly beautiful pair'. About 8° south of COO 3 are the bright pairs κ and $\lambda^{1,2}$ Tuc. The star κ will be dealt with more thoroughly elsewhere (see Star 12) whilst λ is the second entry in James Dunlop's catalogue. It appears to be a binary pair with magnitudes of 6.7 and 7.4 at 82° and 20″ (2013). Andrew James notes the that primary is deep yellow and the pair looks magnificent in binoculars or with a small telescope.

Measures

LCL 119

Early measure (DUN)	174°.1	25″.0	1826.50
Recent measure (ANT)	168°.4	27″.05	2009.71

I 260CD

Early measure (I)	297°.9	0″.76	1900.36
(Orbit	294°.7	0″.59)	
Recent measure (ANT)	15°.7	0″.25	2013.70
(Orbit	15°.5	0″.24)	

6. BU 395 CET = WDS J00373−2446

Table 9.6 Physical parameters for BU 395

BU 395	RA: 00 37 20.68	Dec: −24 46 02.1	WDS: 270(187)	
V magnitudes	A: 6.25	B: 6.62		
(*B* − *V*) magnitudes	A: +0.78	B: +0.85		
μ	1450.34 mas yr^{-1}	± 3.77	−19.38 mas yr^{-1}	± 1.73
π	64.93 mas	± 1.85	50.2 light yr	± 1.4
Spectra	A: K1V	B: G		
Masses (M$_\odot$)	A: 0.90	± 0.01	B: 0.87	± 0.01
Radii (M$_\odot$)	A: 0.87		B: 0.88	
Luminosities (L$_\odot$)	A: 0.6	B: 0.4		
Catalogues	HR 159	HD 3443	SAO 166418	HIP 2941
DS catalogues	BU 395	BDS 335	ADS 520	
Radial velocity	+18.63 km s^{-1}	± 0.11		
Galactic coordinates	68°.846	−86°.049		

History

This pair was found by Burnham with his 6-inch Clark on 5 November 1875, along with BU 391 and BU 393. His note in AN [113] reads 'The third close pair tonight'.

The Modern Era

The current orbit by Hartkopf is given Grade 1. The two stars are in their sixth revolution since discovery. The separation varies from about 0″.8 to just over 0″.1 (2022). This pair is not included in Gaia DR2.

Finder Chart

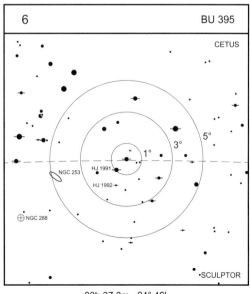

00h 37.3m -24° 46'

Orbit

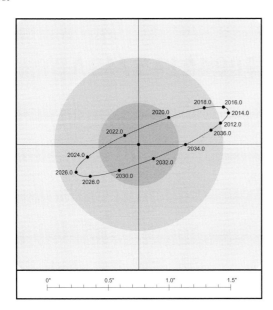

Ephemeris for BU 395 (2017 to 2026)

Orbit by Hrt (2010a) Period: 25.02 years, Grade: 1

Year	PA(°)	Sep(")	Year	PA(°)	Sep(")
2017.0	115.4	0.69	2022.0	237.3	0.13
2018.0	118.4	0.60	2023.0	273.0	0.28
2019.0	122.8	0.48	2024.0	283.4	0.43
2020.0	130.8	0.33	2025.0	289.0	0.52
2021.0	154.6	0.17	2026.0	293.3	0.56

Observing and Neighbourhood

The pair BU 395 is located right on the border between Cetus and Sculptor. It is one of the shortest-period visual binaries which can be seen with a medium aperture. It is now closing rapidly but will reappear for apertures of 20-cm after 2025. Nearby are two wide pairs from the John Herschel catalogue. The pair HJ 1991 (6.6, 9.7, 95°, 47″, 2010) is 30′ away to the SE whilst a further 30′ S from here brings you to HJ 1992 (7.8, 8.9, 247°, 46″, 2013). Six degrees WSW is the pair κ^1 and κ^2 Sculptoris, the first of which is BU 391 (see Star 3). The magnificent spiral galaxy NGC 253 is 2°.5 to the E and slightly S.

Measures

Early measure (LV)	104°.7	0″.65	1886.85
(Orbit	103°.9	0″.67)	
Recent measure (TOK)	112°.8	0″.76	2015.91
(Orbit	112°.9	0″.76)	

7. η CAS = STF 60 = WDS J00491+5749AB

Table 9.7 Physical parameters for η Cas

STF 60	RA: 00 49 05.14	Dec: +57 48 59.4	WDS: 11(1067)	
V magnitudes	A: 3.52	B: 7.36	E: 10.2	
$(B - V)$ magnitudes	A: +0.62	B: +1.39		
μ(A)	1073.52 mas yr^{-1}	$\pm$ 1.34	$-$558.83 mas yr^{-1}	$\pm$ 1.39 (DR2)
μ(B)	1144.57 mas yr^{-1}	$\pm$ 0.05	$-$469.38 mas yr^{-1}	$\pm$ 0.05 (DR2)
π(A)	171.29 mas	$\pm$ 0.58	19.04 light yr	$\pm$ 0.06 (DR2)
π(B)	168.75 mas	$\pm$ 0.04	19.328 light yr	$\pm$ 0.005 (DR2)
Spectra	F9V	M0V		
Masses (M$_\odot$)	A: 1.08	B: 0.63		
Radii (R$_\odot$)	A: 1.06	$\pm$ 0.002		
Luminosities (L$_\odot$)	A: 1.1	B: 0.03		
Catalogues	24 Cas	HR 219	HD 4614	HIP 3821
DS catalogues	H 3 3	STF 60	BDS 426	ADS 671
Radical velocity (A)	+8.44 km s^{-1}	$\pm$ 0.09		
Radical velocity (B)	+10.44 km s^{-1}	$\pm$ 0.14 (DR2)		
Galactic coordinates	122°.620	-5°.055		

History

This pair was found by Herschel on 17 August 1779 and he noted colours of fine white and fine garnet, 'both beautiful colours'. Both stars were also seen by Giuseppe Piazzi, who included them in his *Palermo Catalogue* of 1814.

The Modern Era

In 1955 van de Kamp & Flathers [124] mentioned that a perturbation had been suspected in η Cas A. Their paper found no supportive evidence for this claim. The WDS also notes that A has been reported to be a spectroscopic binary,

and SIMBAD flags it as an SB, but this has not been confirmed. DR2 quotes a significant error in its determination of the bright star's parallax, which may be supportive evidence for higher multiplicity.

Exoplanet Host?

In 2014 Fischer *et al.* [125] announced the results of a 25-year programme to measure the accurate radial velocities of stars which may harbour a planet or planetary system. For η Cas they found that over a period of 17 years the variation amounted to 10 ($\pm$ 38) metres per second, thus discouraging the notion of an exoplanet or binary companion.

Finder Chart

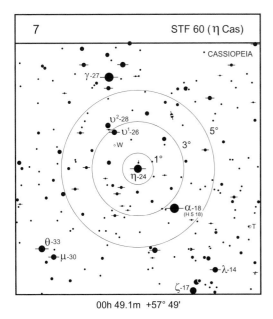

00h 49.1m +57° 49'

Orbit

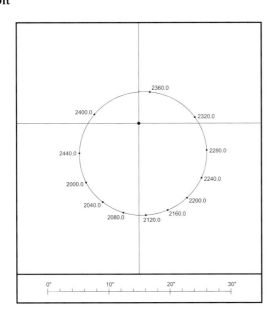

Ephemeris for STF 60 AB (2010 to 2100)

Orbit by Pru (2017) Period: 479.27 years, Grade: 3

Year	PA(°)	Sep(")	Year	PA(°)	Sep(")
2010.0	321.9	13.12	2060.0	342.4	14.20
2020.0	326.3	13.41	2070.0	346.1	14.33
2030.0	330.5	13.65	2080.0	349.8	14.43
2040.0	334.6	13.87	2090.0	353.5	14.52
2050.0	338.5	14.05	2100.0	357.1	14.58

Observing and Neighbourhood

The pair is an easy object in 7.5-cm although B is considerable fainter than A and the colours will be seen better in a larger aperture. The stars will be furthest apart in their orbit at 14″.7 in 2130 or so, whilst the closest approach is 4″.9, around 2370. This is a splendid object for the small aperture and, with the stars differing by two whole spectral classes, the colours offer a contrast to the observer. Smyth found pale white and violet, whilst T. W. Webb noted yellow and garnet. In 1968, using a 21-cm reflector, I made the colours yellow and purple. There are eight additional stars listed in the WDS, all of which appear to be optical. The considerable proper motion of AB is reflected in the change of separation of star E (magnitude 10.15) between 1856 and 2015 (see below). The Star Splitters website [126] gives details of John Nanson's 2011 search with a 6-inch refractor for all ten of the components listed in the WDS. He thinks that a 5-inch under a clear dark sky would suffice to see them all. About 2° SW is α Cas (H 5 18), which boasts a bright orange primary and a distant comes (2.4, 9.0, 283°.2, 70″.36, 2018).

Measures

AB			
Early measure (STF)	81°.1	10″.68	1820.16
(Orbit	79°.9	10″.80)	
Recent measure (ARY)	324°.2	13″.59	2014.64
(Orbit	324°.0	13″.26)	
AE			
Early measure (JC)	118°.0	275″.7	1856.12
Recent measure (ARY)	126°.0	75″.57	2015.09

8. 36 AND = STF 73 = WDS J0550+2338AB

Table 9.8 Physical parameters for 36 And

STF 73	RA: 00 54 58.11	Dec: +23 37 42.9	WDS: 22(717)		
V magnitudes	A: 6.12	B: 6.54	C: 11.0	D: 10.9	
$(B - V)$ magnitudes	A: +1.22	B: +1.88			
μ(A)	129.03 mas yr^{-1}	± 0.26	−39.17 mas yr^{-1}	± 0.24 (DR2)	
μ(B)	139.67 mas yr^{-1}	± 0.12	−19.37 mas yr^{-1}	± 0.16 (DR2)	
π(A)	23.31 mas	± 0.24	140.0 light yr	± 1.4 (DR2)	
π(B)	24.04 mas	± 0.07	135.7 light yr	± 0.4 (DR2)	
Spectra	G6IV	K6IV			
Masses (M$_\odot$)	1.86	± 0.15 (dyn.)			
Luminosities (L$_\odot$)	A: 5	B: 3			
Catalogues	36 And	HR 258	HD 5286	SAO 74359	HIP 4288
DS catalogues	STF 73 (AB)	BDS 482	FOX 115(AB,C)	ADS 755	TOK 447 (AD)
Radial velocity	−0.84 km s^{-1}	± 0.12			
Galactic coordinates	123°.976	−39°.236			

History

Discovered by F. G. W. Struve during his great survey between February 1825 and February 1827 with the 9-inch Fraunhofer refractor at Dorpat, W. H. Smyth found it difficult enough to measure this pair that he often used a power ×600 on his 5.9-inch refractor. At the time, almost a complete orbital cycle ago, the separation was close to 1″. Smyth noted that he employed a paper disk in the centre of his objective in order to see better the disks of the stars. This early attempt at apodising an aperture was originated by Sir John Herschel, who recommended the method to W. R. Dawes. In an early paper, Dawes [127] says 'The separating power of the telescope is increased, but the concentric rings accompanying bright stars are multiplied, and rendered more luminous and are thrown further from the disc. Hence small stars may often be obscured or distorted by the ring passing through them.' Philip Fox [129] measured a

Finder Chart

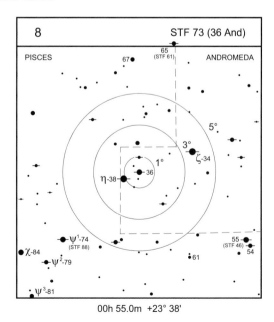

00h 55.0m +23° 38'

Orbit

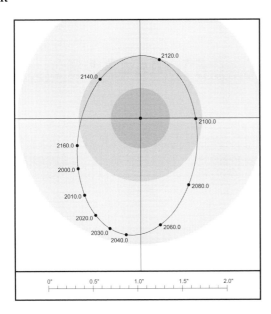

Ephemeris for STF 73 AB (2015 to 2060)

Orbit by Mut (2016b) Period: 167.5135 years, Grade: 2

Year	PA(°)	Sep(")	Year	PA(°)	Sep(")
2015.0	329.2	1.12	2040.0	352.7	1.29
2020.0	334.5	1.18	2045.0	357.0	1.28
2025.0	339.4	1.22	2050.0	1.4	1.26
2030.0	344.0	1.26	2055.0	5.9	1.23
2035.0	348.4	1.28	2060.0	10.7	1.19

magnitude 11 companion at 162″ (distance increasing) but this had already been seen by Burnham [130].

The Modern Era

Tokovinin & Lépine [128] found a faint star 1407 arcseconds away with a common proper motion (TOK 447 AD). The Hipparcos catalogue puts the bright pair at a distance of 124 light years but, with a relatively high proper motion, this is a high-velocity star. The star AB has been suspected of variability and is catalogued as NSV 343.

Observing and Neighbourhood

About $1°$ N preceding the $V = 4.4$ magnitude star η And, there is a beautiful pair of yellow stars, 36 And, which is clearly resolved in 15-cm and forms a resolution test for 10-cm. I have measured this pair for over 25 years. In 1991 it was found at $289°.5$, $0''.68$ but it is now becoming much easier, and is just resolvable with 10-cm at present. The separation will continue to ease for several decades to come. Within four degrees are the pairs ψ^1 Psc (STF 88), 65 Psc (STF 61), and 55 Psc (STF 46). The pair ψ^1 Psc is a bright wide pair of hot, blue-white dwarf stars (magnitudes 5.3, 5.5) separated by $29''.7$ in position angle $159°$ (2016). The primary is a $0''.14$ pair currently just past widest separation (2017) in a 14.4-year orbit. Despite the large separation, AB,C appears to be a physical pair. 65 Psc (magnitudes 6.3 and 6.4 and $1°.5$ NW of 67 Psc) is a neat pair which is perfect for small apertures and provides colour contrast (Webb, very yellow, very blue; Hartung, orange yellow and ashy). The position in 2015 was $116°$, $4''.2$. The pair 55 Psc is more of a test, the magnitude 8.5 companion to the magnitude 5.6 primary being only $6''.6$ away in PA $192°$ (2014). The spectral types are K3 and F3 and they appear yellowish to several observers including Smyth and Haas.

Measures

Early measure (STF)	307°.8	0″.85	1832.14
(Orbit	307°.6	0″.89)	
Recent measure (ARY)	330°.2	1″.08	2015.50
(Orbit	329°.8	1″.08)	

9. β PHE = SLR 1 = WDS J01061–4643AB

Table 9.9 Physical parameters for β Phe

SLR 1	RA: 01 06 05.10	Dec: −46 43 06.4	WDS: 883(85)	
V magnitudes	A: 4.10	B: 4.19		
$(B − V)$ magnitudes	A: +1.03	B: ?		
μ	−80.81 mas yr^{-1}	± 13.24	34.97 mas yr^{-1}	± 9.51
π	17.93 mas	± 0.74	182 light yr	± 8 (dyn.)
Spectra	A: G8III	B: G8III(?)		
Masses (M$_\odot$)	A: 2.5	B: 2.5		
Luminosities (L$_\odot$)	A: 60	B: 55		
Catalogues	HD 6595	HR 322	SAO 215365	HIP 5165
DS catalogues	SLR 1 (AB)	HJ 3417 (AB-C)		
Radial velocity	−1.10 km s^{-1}	± 0.74		
Galactic coordinates	295°.506	−70°.198		

History

During one of his sweeps at Feldhausen, John Herschel noted a very faint companion to β Phoenicis, which he characterized as 'a mere blot'. From occasional measures of the distance and separation, both of which have increased since, we know that this is a background star. In 1891 R. P. Sellors [131] at Sydney Observatory found that β itself was a close, equal, double with separation about 0″.9. The stars slowly widened over the next few decades and attracted occasional measurements, but the brightness has clearly created problems, even for experienced micrometrists with substantial apertures, and thus measures of separation could differ significantly.

The Modern Era

In 2002 RWA and colleagues Andreas Alzner and Elliott Horch [310] wrote a paper which highlighted five southern binaries of interest, of which SLR 1 was one. An orbit by Andreas Alzner using all the available data found a period of

Finder Chart

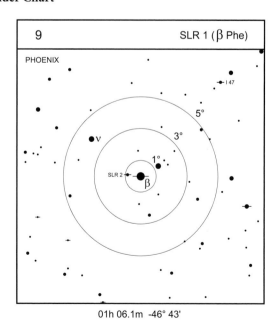

01h 06.1m -46° 43'

Orbit

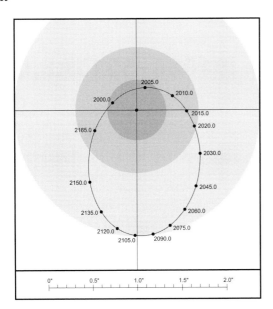

Ephemeris for SLR 1 AB (2015 to 2060)

Orbit by Ary (2015b) Period: 170.7 years, Grade: 3

Year	PA(°)	Sep(")	Year	PA(°)	Sep(")
2015.0	88.9	0.56	2040.0	43.4	1.00
2020.0	74.6	0.67	2045.0	38.4	1.06
2025.0	64.1	0.76	2050.0	33.9	1.12
2030.0	55.9	0.85	2055.0	29.8	1.17
2035.0	49.2	0.93	2060.0	26.1	1.22

195 years, but at that time the stars were close to periastron and less than 0″.3 apart. The next measures were not made

until 2008, and in 2015 Alzner revised the orbit to 170.7 years. The total mass of the system was 5 M⊙ and, assuming the stars are similar in all aspects, then they are G8 giants. Interpolating the orbit back by one revolution we see that in 1833 the stars were also at their closest, fully explaining why John Herschel failed to see them. The system β Phe is not included in DR2.

Observing and Neighbourhood

The pair β Phe is a bright naked-eye star about 12° NW of Achernar. It should be resolvable in 25–30-cm aperture and will continue to widen for about another 80 years. The widest separation, 1″.38, comes in 2100. Another binary pair (SLR 2) can be found in the same field some 25′ E. It was found by Sellors on the same night in 1891 as β (7.1, 8.7, 180°, 1″.3, 2016) having been at 0″.6 at discovery. Nine degrees due S of β is ζ Phe – a fine but difficult triple star. It contains the bright pair RMK 2 (see Star 10). Four degrees NW of β is I 47 (7.5, 8.0, 32°, 0″.8, 2010). This is an orbital pair with period 517 years and is currently at 0″.6 and closing slowly.

Measures

Early measure (SLR)	24°.4	0″.92	1892.93
(Orbit	23°.3	1″.25)	
Recent measure (ARY)	83°.8	0″.70	2016.69
(Orbit	83°.5	0″.60)	

10. ζ PHE = RMK 2 = WDS J01084−5515AB,C

Table 9.10 Physical parameters for ζ Phe

RMK 2	RA: 01 08 23.08	Dec: −55 14 44.7	WDS: 2332(41)	
V magnitudes	A: 4.13	B: 6.29	C: 8.23	
$(B − V)$ magnitudes	A: -0.14	B: +0.19		
μ	20.87 mas yr^{-1}	± 0.36	30.64 mas yr^{-1}	± 0.38
π	10.92 mas	± 0.39 mas	300 light yr	± 11
μ(A)	19.35 mas yr^{-1}	± 0.99	29.45 mas yr^{-1}	± 1.04 (DR2)
μ(B)	18.17 mas yr^{-1}	± 0.25	26.40 mas yr^{-1}	± 0.30 (DR2)
π(A)	14.68 mas	± 0.73	222 light yr	± 11 (DR2)
π(B)	12.21 mas	± 0.22	267.1 light yr	± 4.8 (DR2)
Spectra	A: B6V	B: A7V	C: F3V	
Luminosities (L$_\odot$)	A: 85	B:15	C: 2.5	
Catalogues	HR 338	HD 6882	SAO 232306	HIP 5348
DS catalogues	RST 1205 (AB)	RMK 2 (AB,C)		
Radial velocity	15.40 km s^{-1}	± 0.90		
Galactic coordinates	297°.833	−69°.714		

History

Brian Warner [340] gives a small list of double stars which were noted by Fearon Fallows using the mural and transit circles at the Cape of Good Hope in 1829/30; about a dozen of them predate other discoverers. In the case of ζ Phe, Fallows may just have been later than Rümker in recording this beautiful pair, but he would have had little knowledge of the activity of the Australian astronomers and so it counts as an independent discovery. John Herschel noted it as 'Extremely delicate but bears a fair illumination' and estimated the magnitudes as 5 and 9. In 1913 R. E. Wilson [134] took a spectrum of ζ Phe using the two-prism spectrograph from the D. O. Mills expedition to Chile and noted that it showed two spectra. Periodic measurements of the radial velocities of both components led to an orbit by Colacevich [135] which gave a period of 1.66958 days. The eclipsing nature of this system was found by Hogg [136], who was aware that

Finder Chart

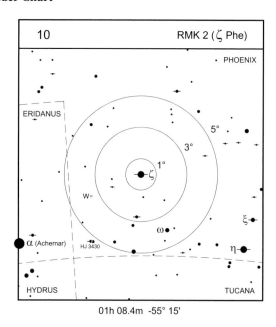

01h 08.4m -55° 15'

Orbit

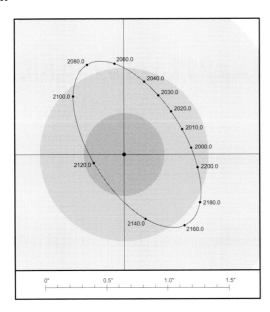

Ephemeris for RST 1205 AB (2015 to 2060)

Orbit by Lin (2004a) Period: 210.37 years, Grade: 5

Year	PA(°)	Sep(")	Year	PA(°)	Sep(")
2015.0	122.2	0.51	2040.0	164.0	0.60
2020.0	131.5	0.51	2045.0	170.3	0.64
2025.0	140.6	0.53	2050.0	176.1	0.67
2030.0	149.1	0.55	2055.0	181.3	0.70
2035.0	156.9	0.57	2060.0	186.1	0.73

Rossiter [137] had added another visual component to the system in 1931 when using the Lamont–Hussey 27-inch refractor at Bloemfontein. He recorded a star some 2.5 magnitudes fainter than ζ, about 0″.6 away. Hogg's variable was ζ itself and it has turned out to be an Algol-type system. The range of brightness is 3.9 to 4.4 and the mass of the eclipsing pair is 6.5 M⊙.

The Modern Era

The close visual pair RST 1205 has moved through about one-third of its orbit since discovery, so the orbital elements are still somewhat provisional; in fact the USNO orbital catalogue classifies them as Grade 5, i.e. indeterminate, because the stars have not yet been observed through periastron and so the apastron part of the orbit is not yet defined. Anderson [139] added as a postscript to his paper on eclipsing binaries in multiple systems that spectra of the Rümker star indicated that it could also be a binary.

Observing and Neighbourhood

Easily found 5° NW of Achernar, the Rümker pair is a fine sight in small telescopes, both stars being white, according to Malin & Frew [138]. Seeing the Rossiter companion (RST 1205), however, will probably take 30-cm and a fine night. The stars are only slowly widening, and that, combined with their significant magnitude difference, will make the pair a real test. In 2020 they should be found at 132°, 0″.51. The period is 211 years.

Measures

Early measure (I)	21°.6	0″.58	1931.91
(Orbit	22°.1	0″.58)	
Recent measure (TOK)	119°.7	0″.56	2015.91
(Orbit	123°.9	0″.51)	

11. ζ PSC = STF 100 = WDS J01137+0735AB

Table 9.11 Physical parameters for ζ Psc

STF 100	RA: 01 13 43.80	Dec: +07 34 31.8	WDS: 99(326)		
V magnitudes	A: 5.19	B: 6.32	C: 9		
(*B − V*) magnitudes	A: +0.27	B: +0.48			
μ(A)	143.25 mas yr^{-1}	± 0.33	−53.28 mas yr^{-1}	± 0.23 (DR2)	
μ(B)	139.36 mas yr^{-1}	± 0.09	−53.37 mas yr^{-1}	± 0.07 (DR2)	
π(A)	24.71 mas	± 0.20	132.0 light yr	± 1.1 (DR2)	
π(B)	24.25 mas	± 0.05	134.5 light yr	± 0.3 (DR2)	
Spectra	A7IV	F7V			
Luminosities (L$_\odot$)	A: 11	B: 4	C: 0.3		
Catalogues (A/B)	86 Psc	HR 361/2	HD 7344/5	HIP 5737/43	SAO 109739/40
DS catalogues	Mayer 3	H 4 8 (AB)	BU 1029 (BC)	BDS 648	ADS 996
Radial velocity (A/B)	15 km s^{-1}	± 4.9	10.9 km s^{-1}	± 0.9	
Galactic coordinates	132°.565	−54°.879			

History

William Herschel noted this pair on 19 October 1779 and found them 'Double. Pretty unequal L(arge). w(hite).; S(mall).w(hite) inclining to blue'. They were swept up in the catalogue of F. G. W. Struve at Dorpat, having shown little motion during the intervening 40 years, and remained of little interest until 1888 when S. W. Burnham, using the 36-inch refractor at Lick, noted a faint companion close to B. He measured a position angle of 249° and a separation of 0″.93 and also observed that C must belong to the system as a whole, as the proper motion of A and B would have left it behind fairly quickly had it been a field star.

The Modern Era

In 1963 Charles Worley [141] measured BC (BU 1029) with the 36-inch refractor at Washington. He found that C was not particular difficult and that the values of Δ*m* given by

Finder Chart

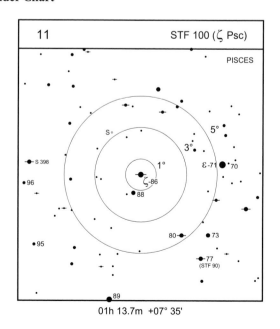

Burnham (11 and 13.5) were considerably overestimated. Worley made it 2.6. If this is still true and the separation is currently 1″.8 then the pair must be within range of a 30-cm aperture. The WDS notes that the most recent measure of BC is 70° and 1″.8, which implies it has moved 180° since discovery, but no orbit exists. Halbwachs [140] confirmed that the two bright stars have a common proper motion although the values for proper motion given by Hipparcos differ considerably more than the values listed in the WDS; possibly this is again complicated by the multiplicity of the system. The pair ζ Psc is in the Zodiacal catalogue and hence both components have been subject to lunar occultations. Both stars are occultation and spectroscopic binaries. Star B is an SB2 with a period of 9.075 days.

Observing and Neighbourhood

This bright, wide, pair is a popular target for small apertures. In 1968 with a 21-cm mirror, ×96, RWA made them yellow and reddish-white. Move 3°.5 SW and you'll alight on the bright pair 77 Psc (STF 90, 6.4, 7.3, 84°, 33″, 2016) which Webb noted as white and bluish. Franks in 1914 called them yellow and ruddy. RWA made them yellowish-white and blue in 1968 with the 21-cm ×96. Four degrees E of ζ is S 398 (6.3, 8.0, 100°, 69″, 2010), which has an orange primary and bluish companion. Another 2° in the same direction (but off the chart) brings you to STF 138 (7.5, 7.6, 58°, 1″.8, 2015), a twin pair of yellowish-white stars which forms a stiff test for 7.5-cm.

Measures

Early measure (STF)	63°.7	23″.46	1832.83
Recent measure (WSI)	62°.8	22″.69	2015.85

12. κ TUC = HJ 3423/I 27
= WDS J01158−6853AB/CD

Table 9.12 Physical parameters for κ Tuc

HJ 3423 I 27	RA: 01 15 46.16 01 15 00.78	Dec: −68 52 33.3 −68 49 08.1	WDS (AB): 1028(75) WDS (CD): 1010(76)	
V magnitudes (ABCD)	A: 5.00	B: 7.74	C: 7.84	D: 8.44
(B − V) magnitudes	A: +0.47	B:	C: +1.10	D: +1.22
μ(A)	409.24 mas yr⁻¹	± 1.22	107.00 mas yr⁻¹	± 1.19 (DR2)
μ(B)	386.27 mas yr⁻¹	± 0.11	82.38 mas yr⁻¹	± 0.07 (DR2)
μ(C)	360.48 mas yr⁻¹	± 0.19	95.36 mas yr⁻¹	± 0.15 (DR2)
μ(D)	426.22 mas yr⁻¹	± 0.23	121.33 mas yr⁻¹	± 0.19 (DR2)
π(A)	46.67 mas	± 0.71	68.9 light yr	± 1.1 (DR2)
π(B)	47.53 mas	± 0.05	68.62 light yr	± 0.07 (DR2)
π(C)	47.66 mas	± 0.11	68.4 light yr	± 0.2 (DR2)
π(D)	47.80 mas	± 0.12	68.2 light yr	± 0.2 (DR2)
Spectra	A: F6IV	B:	C: K2V	D: K3V
Masses	A: 1.3	B: 0.9	C: 0.8	D: 0.7
Luminosities (L⊙)	A: 4	B: 0.3	C: 0.25	D: 0.15
Catalogues (AB)	HR 377	HD 7788	SAO 248345	HIP 5896
Catalogues (CD)		HD 7693	SAO 248342	HIP 5842
DS catalogues	HJ 3423 (AB)	I 27 (CD)		
Radial velocity	7.70 km s⁻¹	± 1.7		
Radial velocity (A/B)	7.48 km s⁻¹	± 1.1	7.91 km s⁻¹	± 0.14 (DR 2)
Galactic coordinates (AB)	299°.654	−48°.097		
Galactic coordinates (CD)	299°.743	−48°.163		

History

It is hard to believe that this beautiful object was not picked up by either Dunlop or Rümker at Parramatta but perhaps the components were then considerably closer than they are today. It was left to John Herschel [142] using his 18-inch reflector to sweep it up in 1834. His first distance was given as 2″ and in a subsequent sweep he made it 2″.5. 'Very beautiful' was his description and although the CD pair would have been well within range of his telescope he makes no mention of it, presumably because it was outside the field of view of his ×320 power eyepiece. In 1836 using the 7-foot (5-inch aperture) refractor he measured 4″.75 from two nights' observations. Since then the stars widened to about 5″.2 in the middle of the twentieth century and now appear to be slowly closing again. The CD pair was found from Sydney by Robert Innes in 1895 [187] using a borrowed reflecting telescope.

Finder Chart

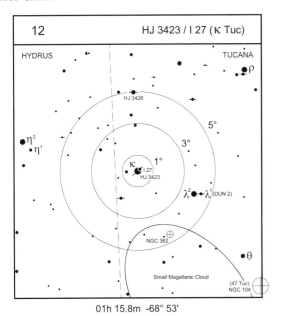

01h 15.8m -68° 53'

The Modern Era

The two pairs, which are 319″ apart, are physically linked – they have almost identical parallaxes and similar proper motions. Two spectra, taken earlier this century 1529 days apart, showed the radial velocity of AB varying by more than 30 km s^{-1}, but no confirming measurements appear to have been made since. Baize & Petit [144] noted that the A component is NSV 454. Tokovinin notes that Hipparcos found an astrometric perturbation in A, making the system quintuple. The relatively large formal error on the DR2 parallax of A supports this notion of the structure.

Observing and Neighbourhood

This is a spectacular quadruple for the small telescope. A 15-cm at about ×200 should show all four stars clearly. The close pair has the plane of its orbit tilted towards us, so the distance varies only between 0″.9 and 1″.1. The orbit of AB is much less certain but it is likely to be in range of 7.5-cm apertures for the next century or more. In 2016 with the Johannesburg 67-cm refractor, the colours of AB were noted as yellow and bluish whilst those of CD appeared yellow and lilac. κ Tucanae is about 4° NNE of the Small Magellanic Cloud and about 5° NE of the fine globular cluster 47 Tuc. Also worth searching out nearby are the bright binocular pair λ^{1,2} Tuc, the fainter component of which is DUN 2, magnitudes 6.7, 7.4 and separated by 20″ at 82°, 2013, and the globular cluster NGC 362. Due N by 2°.5 is HJ 3426, a fine pair (6.4, 8.3, 328°, 2″.5, 2013).

Orbit for HJ 3423AB

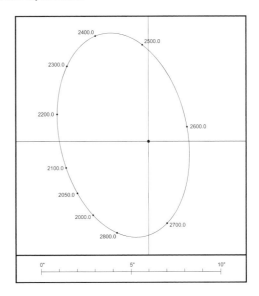

Ephemeris for HJ 3423 AB (2010 to 2100)

Orbit by Sca (2005b) Period: 857.0 years, Grade: 5

Year	PA(°)	Sep(″)	Year	PA(°)	Sep(″)
2010.0	318.7	4.99	2060.0	301.4	4.84
2020.0	315.3	4.95	2070.0	297.9	4.83
2030.0	311.9	4.92	2080.0	294.3	4.82
2040.0	308.5	4.89	2090.0	290.7	4.82
2050.0	305.0	4.86	2100.0	287.2	4.83

Orbit for I 27 CD

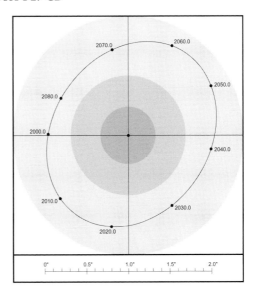

Ephemeris for I 27 CD (2016 to 2034)

Orbit by Sod (1999) Period: 85.2 years, Grade: 3

Year	PA(°)	Sep(″)	Year	PA(°)	Sep(″)
2016.0	333.9	1.10	2026.0	14.3	1.00
2018.0	341.4	1.09	2028.0	23.5	0.98
2020.0	349.2	1.07	2030.0	33.0	0.96
2022.0	357.2	1.05	2032.0	42.8	0.96
2024.0	5.5	1.02	2034.0	52.6	0.96

contd

Measures

HJ 3423AB			
Early measure (HJ)	16°.4	4″.75	1836.75
(Orbit	15°.0	4″.54)	
Recent measure (ARY)	315°.6	4″.77	2016.68
(Orbit	316°.5	4″.98)	
I 27CD			
Early measure (SLR)	185°.4	1″.55	1893.10
(Orbit	179°.7	1″.05)	
Recent measure (ARY)	324°.3	1″.05	2013.70
(Orbit	325°.5	1″.11)	

13. τ SCL = HJ 3447 = WDS J01361–2954AB

Table 9.13 Physical parameters for τ Scl

HJ 3447	RA: 01 36 08.50	Dec: −29 54 26.5	WDS: 551(115)	
V magnitudes	A: 6.03	B: 7.34		
$(B - V)$ magnitudes	A: +0.34	B: +0.49		
μ	117.37 mas yr^{-1}	± 0.92	46.72 mas yr^{-1}	± 0.48
π	14.42 mas	± 0.81	226 light yr	± 13
μ(A)	127.16 mas yr^{-1}	± 0.47	34.77 mas yr^{-1}	± 0.23 (DR2)
π(A)	18.53 mas	± 0.31	176 light yr	± 3 (DR2)
Spectra	A: F2V	B		
Masses (M$_\odot$)	A: 1.70	± 0.05	B: 1.45	± 0.03
Luminosities (L$_\odot$)	A: 9	B: 3		
Catalogues	HR 462	HD 9906	SAO 193201	HIP 7463
DS catalogues	HJ 3447			
Radial velocity	3.00 km s^{-1}	± 4.5		
Galactic coordinates	231°.360	−79°.800		

History

This object was discovered by John Herschel [142] on 19 November 1835, during his survey of the southern sky from Feldhausen. 'A fine double star' he recorded, assigning it magnitudes of 6 and 8.

The Modern Era

The system τ Scl is a pair of yellow dwarfs that form a long-period binary. If the current catalogue orbit is correct then the closest separation (0″.80) was passed in 2003 and the stars will continue to widen for nine centuries, until the maximum separation of 4″.5 is reached, although the orbit is essentially indeterminate and will remain so for many years.

Finder Chart

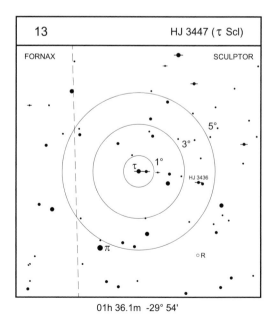

01h 36.1m -29° 54'

Orbit

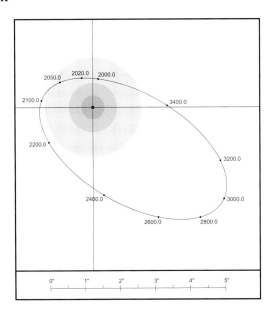

Ephemeris for HJ 3447 (2010 to 2100)

Orbit by Cve (2006e) Period: 1503.58 years, Grade: 5

Year	PA(°)	Sep(")	Year	PA(°)	Sep(")
2010.0	185.9	0.81	2060.0	240.5	1.26
2020.0	201.2	0.87	2070.0	246.9	1.34
2030.0	214.1	0.96	2080.0	252.6	1.40
2040.0	224.6	1.06	2090.0	257.9	1.45
2050.0	233.2	1.17	2100.0	262.9	1.49

Observing and Neighbourhood

At least 15-cm is needed to see this pair clearly resolved, as the difference in brightness adds to the difficulty of observing it. Hartung notes that both stars are yellow dwarfs, although the distance of 40 pc he quotes has been significantly increased by the Hipparcos satellite findings. HJ 3436 (6.9, 9.6, 128°, 9″.6, 2013) is about 2° to the W, a little S. DR2 shows that the stars have similar proper motions but their parallaxes are respectively 5.59 mas ± 0.06 and 4.98 mas ± 0.08.

Measures

Early measure (HJ)	75°.5	3″.12	1837.11
(Orbit	80°.0	2″.81)	
Recent measure (TOK)	193°.6	0″.82	2015.91
(Orbit	195°.2	0″.84)	

14. p ERI = Δ 5 = WDS J01398−5612

Table 9.14 Physical parameters for p Eri

DUN 5	RA: 01 39 47.83	Dec: −56 11 36.0	WDS: 326(166)	
V magnitudes	A: 5.78	B: 5.90		
$(B − V)$ magnitudes	A: +0.99	B : +1.02		
μ(A)	309.10 mas yr^{-1}	± 0.08	10.69 mas yr^{-1}	± 0.07 (DR2)
μ(B)	262.38 mas yr^{-1}	± 0.08	15.33 mas yr^{-1}	± 0.07 (DR2)
π(A)	122.13 mas	± 0.05	26.71 light yr	± 0.01 (DR2)
π(B)	122.06 mas	± 0.05	26.72 light yr	± 0.01 (DR2)
Spectra	A: K2V	B: K2V		
Luminosities (L$_\odot$)	A: 0.3	B: 0.2		
Catalogues (A/B)	HR 486/7	HD 10360/1	SAO 232490	HIP 7751
DS catalogues	DUN 5 (AB)			
Radial velocity (A/B)	+22.5 km s^{-1}	± 0.9	+19.5 km s^{-1}	± 0.9
Radial velocity (A/B)	+20.14 km s^{-1}	± 0.16	+21.59 km s^{-1}	± 0.16 (DR2)
Galactic coordinates	289°.595	−59°.662		

History

This near-equally-bright rich yellow southern pair was discovered by James Dunlop [12] in 1824 when the components were almost at their closest distance, estimating separation by the two Airy disk sizes as 2″.5. Later, during December 1825, he measured the position angle as 343°.1. Early micrometric measures found the stars slowly widening but could not determine any true attachment, and p Eridani was deemed important because few binary stars were then known in the southern skies. During 1850, Captain William S. Jacob [145] (1813–1862) in Madras was first to publish a premature orbit. Yet by June 1880, Henry Chamberlain Russell (1836–1907) boldly published several papers claiming that the system was optical. Several inconclusive orbital solutions then followed, until in 1956 G. B. van Albada [146] found a moderately long period of 454 years.

Finder Chart

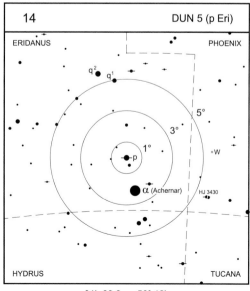

01h 39.8m -56° 12'

Orbit

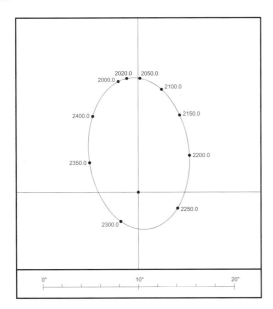

Ephemeris for DUN 5 (2010 to 2100)

Orbit by Sca (2015c) Period: 475.2 years, Grade: 4

Year	PA(°)	Sep(")	Year	PA(°)	Sep(")
2010.0	188.1	11.56	2060.0	176.7	11.50
2020.0	185.8	11.63	2070.0	174.4	11.36
2030.0	183.6	11.66	2080.0	172.0	11.19
2040.0	181.3	11.65	2090.0	169.5	10.98
2050.0	179.0	11.59	2100.0	166.8	10.73

The Modern Era

A recent orbit by Scardia and colleagues [147], making use of 60 years of additional observations, lengthened the period to 475 years. Currently the stars are nearing their widest apparent separation, 11″.7, now predicted to occur in 2033.

They will then close until about 2300 when they will be 3″.5 apart, one orbital period on from their discovery by Dunlop.

Exoplanet Host?

There has been some recent interest in this solar-like system owing to both its high proper motion and its relatively close proximity to us, and it is a candidate for searching for exoplanets orbiting either star.

Observing

The pair p Eri is quickly found as it lies just 1°.1 N of the first magnitude star α Eridani. Components are presently easily visible in 5-cm, and it is a ideal starter for southern measures. Both stars were adjudged yellow by the ASNSW observers. Moving 2°.2 west from Achernar, HJ 3430 (7.2, 9.5, 224°, 3″.2, 2016) is a long-period binary which is gradually opening. DR2 finds identical parallaxes for the two bright stars, placing them 162 light years away. Andrei Tokovinin adds a distant $V = 17$ star (TOK 454 AC), which has very similar proper motions to the binary pair.

Measures

Early measure (HJ)	302°.3	3″.65	1835.00
(Orbit	302°.1	3″.54)	
Recent measure (ARY)	186°.2	11″.45	2016.70
(Orbit	186°.7	11″.62)	

15. $\gamma^{1,2}$ ARI = STF 180 = WDS J01535+1918AB

Table 9.15 Physical parameters for γ Ari

STF 180	RA: 01 53 31.76	Dec: +19 17 38.6	WDS ranking: 82(383)		
V magnitudes	A (γ^2): 4.52	B (γ^1): 4.58	C: 8.63	D: 13.6	
$(B-V)$ magnitudes	A: −0.03	B: −0.03			
μ(A)	78.11 mas yr^{-1}	± 0.72	−97.50 mas yr^{-1}	± 0.62 (DR2)	
μ(B)	77.85 mas yr^{-1}	± 0.91	−106.97 mas yr^{-1}	± 0.73 (DR2)	
π(A)	19.98 mas	± 0.35	163.2 light yr	± 2.9 (DR2)	
π(B)	18.88 mas	± 0.43	172.8 light yr	± 3.9 (DR2)	
Spectra	A: A1pSi	B: B9V			
Luminosities (L$_\odot$)	A: 32	B: 34			
Catalogues (A/B)	5 Ari	HR 545/6	HD 11502/3	SAO 92681/2	HIP 8832
DS catalogues	Mayer 5	H 3 9	STF 180	BU 512(CD)	BDS 993
	ADS 1507				
Radial velocity (A/B)	−0.6 km s^{-1}	± 2	+3.7 km s^{-1}	± 2	
Galactic coordinates	142°.548	−41°.201			

History

This binary was discovered by Robert Hooke in 1664 as he was observing the comet Hevelius. He noted 'I took notice that it consisted of two small stars very close together; a like instance to which I have not else met with in all the heavens.' William Herschel re-observed it on 27 September 1779 and noted that the stars were almost equal: 'Equal, or if any difference the following is the largest'. Both stars were recorded as white although he thought the larger star 'inclined a little to red'. His observed position gave 356°.1, 10″.17.

The Modern Era

The distant field star C, which is currently 217″ away from A at position angle 81°, is unconnected with the system and the change in distance is due to the proper motion of AB,

Finder Chart

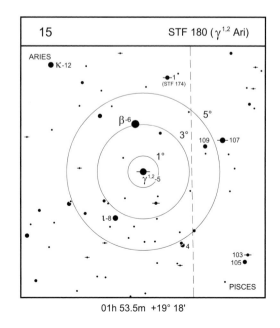

01h 53.5m +19° 18'

which is around $0''.1$ per year. In 1878, Burnham [219] found a very faint and close companion (the WDS gives $V = 13.6$) at a distance of $1''.45$ and the most recent measure in 1975 shows little change in either coordinate. This pair is catalogued as BU 512 CD. The two bright stars certainly form a binary system but of indeterminately long period. In more than 200 years of observation the position angle has changed by a few degrees and the stars are now $2''.8$ closer than they were in Herschel's day.

Observing and Neighbourhood

Easily found about $1°.5$ due S of β Arietis ($V = 2.65$), this is one of the finest pairs in the autumn sky; γ Arietis is a beautiful sight in apertures of 6-cm or more. Both stars appear white to many observers, including Haas, Webb, Hartung, and RWA. Smyth thought the B star was 'pale grey'. Three degrees NW of β is 1 Ari (STF 174, 6.3, 7.2, 165°, $2''.9$, 2016). The primary has a composite spectrum of K1III and A6V and is thus most likely a physical triple star. F. G. W. Struve reports gold and blue whilst RWA found yellow and blue in 21-cm at $\times 96$ and $\times 216$.

Measures

Early measure (STF)	360°.0	8″.63	1830.84
Recent measure (ARY)	1°.4	7″.50	2016.02

16. α PSC = STF 202 = WDS J02020+0246

Table 9.16 Physical parameters for α Psc

STF 202	RA: 02 02 02.80	Dec: +02 45 49.4	WDS: 29(616)	
V magnitudes	A: 4.10	B: 5.17	C: 8.25	D: 8.59
$(B - V)$ magnitudes	A: −0.05	B: +0.31		
μ	32.45 mas yr^{-1}	± 1.01	0.04 mas yr^{-1}	± 0.72
π	21.66 mas	± 1.06	151 light yr	± 7
μ(A)	32.69 mas yr^{-1}	± 0.94	−2.90 mas yr^{-1}	± 0.82 (DR2)
π(A)	19.80 mas	± 0.67	164.7 light yr	± 5.6 (DR2)
Luminosities (L$_\odot$)	A: 50	B: 20		
Catalogues (A/B)	HR 596/5	HD 12447/6	SAO 110291	HIP 9487
DS catalogues	H 2 12 (AB)	STF 202 (AB)	BDS 1061	ADS 1615
Spectra	A: A0p	B: A2p		
Radial velocity	+7.50 km s^{-1}	± 1.8		
Galactic coordinates	155°.351	−55°.600		

History

This object was noted by William Herschel on 19 October 1779. He recorded that both stars were white and 'with ×222 the separation was not quite 2 diameters of L' (L was the large or primary star). The two faint and very distant companions (C is at 63°, 405″ and D is at 335°, 435″) listed in the WDS were first measured by Eyre B. Powell [149] (1819–1904). An amateur astronomer working from Madras in the 1850s, he used a 4-inch refractor and a Simms micrometer to make the observations.

The Modern Era

The observed arc of the apparent orbit now amounts to about 75°, and the two stars are almost as close as they will ever get. The use of modern observational techniques has meant that relative positions are now measured very accurately.

Finder Chart

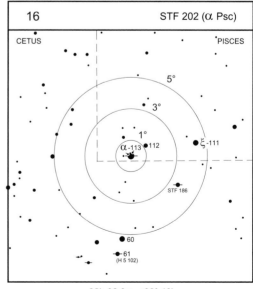

02h 02.0m +02° 46′

Orbit

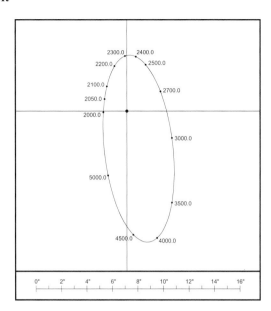

Ephemeris for STF 202 AB (2010 to 2100)

Orbit by Pru (2017) Period: 3267.4 years, Grade: 4

Year	PA(°)	Sep(")	Year	PA(°)	Sep(")
2010.0	266.5	1.83	2060.0	237.0	2.04
2020.0	260.1	1.84	2070.0	232.1	2.13
2030.0	253.9	1.86	2080.0	227.7	2.23
2040.0	248.0	1.91	2090.0	223.6	2.33
2050.0	242.3	1.97	2100.0	219.8	2.44

The previous orbit of 933 years, published in 1983, has now been superseded by a much longer, but still highly uncertain, value. There have been reports that one or both components is an SB but there is no entry in the *Ninth Catalogue of Spectroscopic Binary Orbits* [150] for either star. The primary is a peculiar A star and varies on a period of 0.738 days. Star B is also suspected of being an ACV. Shaya & Olling [270] include the AC pair in their catalogue (as SHY 132) and find a

very high probability that they are physically connected. The linear distance between A and C is 0.15 light years. Gaia DR2 does not include the B component.

Observing and Neighbourhood

The pair α Psc is close to the celestial equator and is found at the easternmost end of a stream of six fourth and fifth magnitude stars beginning with δ Psc. Note also the binary STF186, found 1.5° S preceding (6.8, 6.8, 72°, 0″.7, 2017.5). It is a highly inclined binary, with a period of 165.7 years, which is closing quite quickly and already needs 20-cm to get a good view. The separation of α Psc is opening slowly, and if the current long period is correct then the pair will be within range of a 10-cm aperture for many centuries yet. Smyth, in the *Bedford Catalogue*, makes the stars pale green and blue, whilst the ASNSW observers made the colours yellowish and white, and RWA found both stars to be blue-white. About 3° S and slightly E is 61 Cet, a William Herschel discovery (H 5 102), 6.0, 10.8. 194°, 44″, 2010. DR2 shows that these stars have common parallax and proper motion. They are 405 light years away. The C component (12.8 at 324°, 91″) is four times more distant according to Gaia. Tokovinin finds the primary to be a very close pair.

Measures

Early measure (STF)	335°.5	3″.65	1832.50
(Orbit	331°.6	3″.72)	
Recent measure (ARY)	265°.1	1″.82	2016.03
(Orbit	262°.7	1″.83)	

17. 10 ARI = STF 208 = WDS J02037+2556AB

Table 9.17 Physical parameters for 10 Ari

STF 208	RA: 02 03 39.26	Dec: +25 56 07.6	WDS: 191(229)	
V magnitudes	A: 5.82	B: 7.87		
(*B − V*) magnitudes	A: +0.54	B: +0.59		
μ(A)	126.53 mas yr^{-1}	± 0.14	13.21 mas yr^{-1}	± 0.15 (DR2)
μ(B)	132.99 mas yr^{-1}	± 0.63	28.90 mas yr^{-1}	± 0.52 (DR2)
π(A)	19.38 mas	± 0.75	168.3 light yr	± 6.5 (DR2)
π(B)	22.81 mas	± 0.57	143.0 light yr	± 3.6 (DR2)
Spectra	F8IV/V			
Luminosities (L$_\odot$)	A: 10	B: 2		
Catalogues	10 Ari	HR 605	HD 12558	SAO 75114	HIP 9621
DS catalogues	STF 208	BDS 1074	ADS 1631	
Radial velocity	+13.0 km s^{-1}	± 0.2		
Radial velocity (A)	+13.22 km s^{-1}	± 0.23 (DR2)		
Galactic coordinates	142°.615	−34°.161		

History

The star 10 Arietis revealed itself to be double during the survey at Dorpat by F. G. W. Struve. John Herschel called it a miniature of ϵ Bootis (Pulcherrima). Near 2″ when first seen by Struve, the stars closed steadily throughout the nineteenth century, reaching a minimum separation of 0″.4 around 1910 before slowly widening.

The Modern Era

The USNO *Sixth Orbital Catalog* prefers to include the Grade 2 orbit by Heintz [309], but this predicts separations about 0″.3 wider than those actually observed at the time of going to press.

Finder Chart

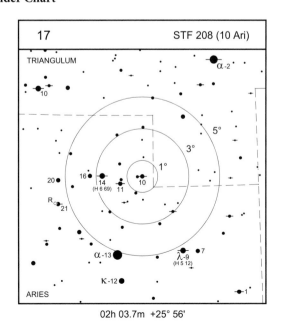

02h 03.7m +25° 56'

Orbit

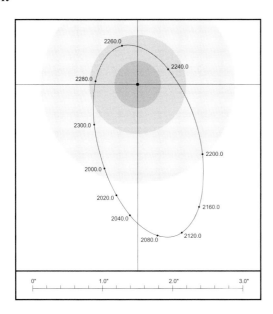

Ephemeris for STF 208 AB (2010 to 2100)

Orbit by Hei (1996a) Period: 325 years, Grade: 3

Year	PA(°)	Sep(")	Year	PA(°)	Sep(")
2010.0	344.0	1.42	2060.0	2.5	2.00
2020.0	348.9	1.57	2070.0	5.1	2.07
2030.0	353.0	1.70	2080.0	7.6	2.12
2040.0	356.5	1.81	2090.0	10.0	2.16
2050.0	359.6	1.91	2100.0	12.3	2.18

Observing and Neighbourhood

The pair 10 Arietis is the westernmost star in a line of four naked-eye stars which can be found 3° N of α Arietis (*V* magnitude 2.00). This unequally bright pair of stars should be comfortably visible with a 15-cm aperture for some decades to come, providing the night has good transparency and steady seeing. Colours: Struve, yellow and blue; Smyth, yellow and pale grey; Burnham, yellow and ash. An observation in February 2018 showed the companion well separated at ×450 on a 20-cm refractor. Nearby are 14 (H 6 69) and λ Arietis (H 5 12). The former was found by William Herschel in 1781 and is a coarse triple (5.0, 8.0, 8.0) with the two companions at 34°, 93″ and 279°, 106″ (2011). The three stars have significant and similar proper motions and may constitute a physical group. They are 1°.5 E of 10 Ari. The star λ is a wide pair with two bright components (4.6, 6.7, 48°, 37″, 2017). Two distant tenth magnitude stars are unrelated.

Measures

Early measure (STF)	35°.2	1″.98	1833.05
(Orbit	26°.5	1″.94)	
Recent measure (ARY)	344°.8	1″.24	2014.80
(Orbit	346°.5	1″.49)	

18. $\gamma^{1,2}$ AND = STF 205 = WDS J02039+4220A,BC

Table 9.18 Physical parameters for γ And

STF 205	RA: 02 03 53.92	Dec: +42 19 47.5	WDS: 84(374) (A-BC)		
			WDS: 88(360) (BC)		
V magnitudes	A: 2.31	B: 5.3	C: 6.5		
μ(A)	+43.08 mas yr^{-1}	$\pm$ 0.71	-50.05 mas yr^{-1}	$\pm$ 0.52	
μ(B)	+43.1 mas yr^{-1}	$\pm$ 0.7	-50.8 mas yr^{-1}	$\pm$ 0.6	
π	9.19 mas	$\pm$ 0.73	355 light yr	$\pm$ 28	
μ(B)	30.02 mas yr^{-1}	$\pm$ 1.43	-51.64 mas yr^{-1}	$\pm$ 0.97 (DR2)	
π(B)	12.57 mas	$\pm$ 0.59	259.5 light yr	$\pm$ 12.1 (DR2)	
Spectra	A: K3IIb	B: B8V+B9V	C: A0V		
Masses (M$_\odot$)	A: 14.5	Ba: 3.1	Bb: 2.5	C: 2.9	
Luminosities (L$_\odot$)	A: 620	B: 40	C: 15	D	
Catalogues (A/B)	57 And	HR 603	HD 12533	SAO 37734	HIP 9640
DS catalogues	Mayer 7	H 3 5 (A-BC)	STF 205 (A-BC)	STT 38 (BC)	
	BDS 1070	ADS 1630			
Radial velocity (A/B)	-11.7 km s^{-1}	$\pm$ 0.9	14 km s^{-1}	$\pm$ 5	
Galactic coordinates	136°.965	$-18°.559$			

History

Thomas Lewis [194] noted that Charles Messier was comparing the light of γ Andromedae with the Andromeda nebula on a fine night in August 1764 but noticed neither that it was coloured nor that it was double. The wide pair was discovered by Christian Mayer on 29 January 1777 [152], although he too made no special remarks about the colours of the stars; the first measure was made by the elder Herschel on 17 August 1779. He noted the colours as reddish-white and a fine, light, sky-blue tending to green. 'A most beautiful object.' Subsequently, Otto Struve [153] at Pulkovo using the new 15-inch (38-cm) refractor in 1842 found that B was itself double (STT 38). Hermann Struve considered the bright star itself to be a close double and measured an elongation on three nights in 1887 and 1890, but there have been no confirming observations of this. In the last century, star B was discovered to be a spectroscopic binary, making the system a quadruple; the period is 2.670 days.

The Modern Era

The BC pair has been below 0''.4 since 2004 and will not reach that value again until sometime in 2026 (see the ephemeris). The bright component is not in DR2 but the wide visual companion has been measured and shows a significantly greater parallax than that of Hipparcos for the bright star.

Finder Chart

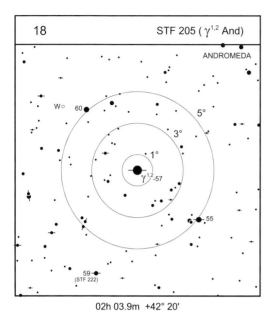

18   STF 205 ($\gamma^{1,2}$ And)

ANDROMEDA

W○ 60

5°

3°

1°

$\gamma^{1,2}$-57

55

59
(STF 222)

02h 03.9m +42° 20'

Orbit

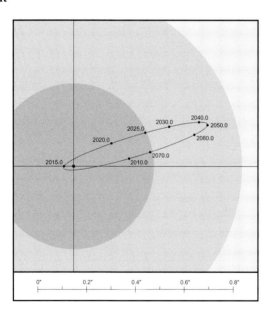

0"   0.2"   0.4"   0.6"   0.8"

Ephemeris for STT 38 BC (2018 to 2036)

Orbit by Doc (2017a) Period: 62.63 years, Grade: 2

Year	PA(°)	Sep(")	Year	PA(°)	Sep(")
2018.0	128.0	0.11	2028.0	112.8	0.39
2020.0	120.6	0.18	2030.0	111.9	0.42
2022.0	117.3	0.24	2032.0	111.2	0.46
2024.0	115.3	0.30	2034.0	110.5	0.48
2026.0	113.9	0.34	2036.0	110.0	0.51

Observing and Neighbourhood

A glorious object, whose full splendour is currently denied to the small- or medium-aperture telescope user; nevertheless it deserves its reputation for being one of the most beautiful pairs in the sky. Colours (A,BC): John Herschel and South, orange and emerald green; Struve, orange and emerald; Smyth, yellow and pale grey; Dembowski, golden and blue; Burnham, golden and blue. The star 59 And (STF 222) is 3°.5 SSE of γ. There has been little motion since the pair was discovered by the elder Herschel in 1783. The stars have magnitudes 6.1 and 6.7 and they are 17″ apart in PA 37° (2016): STF 162 is 6° N and 3° preceding γ. This is a fine close pair (magnitudes 6.4, 7.2 at 199° and 1″.9, 2016) with two tenth magnitude stars 20″ N and 139″ to the W.

Measures

A-BC			
Early measure (STF)	62°.4	10″.33	1830.02
Recent measure (ARY)	64°.7	9″.60	2013.06

BC			
Early measure (DA)	118°.2	0″.45	1843.04
(Orbit	110°.6	0″.51)	
Recent measure (PRI)	95°.7	0″.219	2010.09
(Orbit	99°.4	0″.24)	

19. o CET = JOY 1 = WDS J02193–0259AaAb

Table 9.19 Physical parameters for o Cet

JOY 1	RA: 02 19 20.79	Dec: −2 58 39.5	WDS: 551(115)		
V magnitudes	Aa: 2–10	Ab: 10–12.5	B: 14.1	C: 9.7	
$(B − V)$ magnitudes	A: +1.10				
μ	9.33 mas yr^{-1}	± 1.99	−237.36 mas yr^{-1}	± 1.58	
π	10.91 mas	± 1.22	300 light yr	± 33 (Hipparchos)	
π	9.09 mas	± 0.41	360 light yr	± 16 (PL relation)	
Spectra	A: M5–9 III	B: DA			
Masses (M$_\odot$)	A: 2.0		B: 0.6		
Radii (R$_\odot$)	A: 176–201	B:			
Luminosities (L$_\odot$)	A: 1–1600	B: 0.1–1			
Catalogues	68 Cet	HR 681	HD 14386	SAO 129825	HIP 10826
DS catalogues	H 6 1 (AC)	JOY 1 (AaAb)	BDS 1209	ADS 1778	
Radial velocity	63.50 km s^{-1}	± 0.6			
Galactic coordinates	167°.755	−57°.983			

History

Omicron Ceti, better known as Mira Ceti, was first noticed on 3 August 1596 by David Fabricius, who thought it was a nova but then re-observed it in 1618. The period of variability was first determined by Johannes Holwarda in 1638 and the current value of 332 days is very close to that found by Holwarda. The discovery of the companion is discussed elsewhere. William Herschel added a distant star of magnitude 9 on 20 October 1777 and made two separate sets of measures about five years apart, giving mean distances of 104″ and 113″. He commented that 'I can hardly doubt the motion of this star', but subsequent observations show little actual motion and that there must be an error in the first distance. In 1877 Burnham [656], using the 18.5-inch at Dearborn, found a 14th magnitude object, which is now catalogued as B and appears to be a field star.

Finder Chart

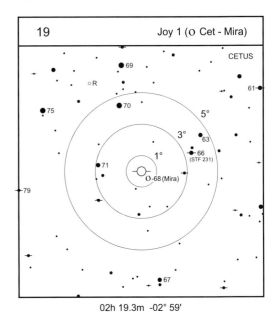

02h 19.3m -02° 59'

Orbit

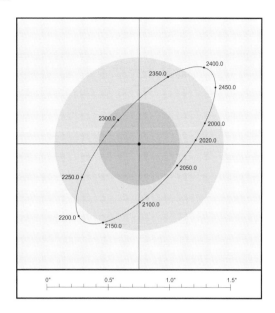

Ephemeris for JOY 1 Aa,Ab (2015 to 2060)

Orbit by Sca (2001f) Period: 497.88 years, Grade: 5

Year	PA(°)	Sep(")	Year	PA(°)	Sep(")
2015.0	97.8	0.49	2040.0	74.0	0.38
2020.0	93.9	0.46	2045.0	67.8	0.36
2025.0	89.7	0.44	2050.0	61.2	0.35
2030.0	85.0	0.42	2055.0	54.2	0.35
2035.0	79.7	0.40	2060.0	47.1	0.35

The Modern Era

Few Mira stars are in binary systems, and those that are are of long period, so astrometric measurements of the two components need to be very precise to allow even an approximate estimate of the mass of the variable. Since the first observations by Aitken [155] in 1923 the position angle has decreased by only 32° whilst the separation has almost halved. Observations in 2014 using the ALMA array [156] produced extremely accurate relative positions for Mira A and B and continuing these over 10 or 20 years may be enough to constrain the orbit significantly. In 2010 Sokoloski confirmed that B was a white dwarf, because it was seen to vary on timescales of minutes, pointing to a compact object, and the spectrum resembled those of known white dwarves in cataclysmic variable systems. See Chapter 7 for more details. The Hipparcos parallax is not likely to be very reliable since the stellar diameter is three times the measured parallax. Recently, studies of the period–luminosity and period–colour–luminosity relationships for Mira have more or less agreed on a value of 110 parsecs with an uncertainty of about 5 pc. Gaia DR2 contains an entry for Mira but gives only a G magnitude.

Observing and Neighbourhood

Visually resolving Mira is going to be difficult. Attempts need to be made when Mira is faint and should be carried out as soon as possible, as the current orbit shows the stars closing over the next 50 years; however, for the next few years it may be possible to succeed using 40-cm. The two components are significantly different colours – orange-red and white. An added complication is that Mira B is also a variable (VZ Ceti with a range of 10–12.5) and would need to be near maximum at the same time that Mira is near minimum to minimise the apparent magnitude difference. Much easier is 66 Cet (STF 231), which is 1°.7 to the WNW (5.7, 7.7, 235°, 16″.9, 2016). Star C is 11.5 at 53°, 147″; the distance is decreasing owing to the proper motion of AB, which amounts to more than 0″.3 per year. DR2 finds the distance to A and B to be 124 and 125 light years respectively. Star A is a close pair (TOK 39).

Measures

Early measure (A)	130°.3	0″.91	1923.84
(Orbit	133°.2	0″.82)	
Recent measure (ALM)	98°.7	0″.47	2014.81
(Orbit	98°.0	0″.49)	

20. ι CAS = STF 262 = WDS J02291+6724AB

Table 9.20 Physical parameters for ι Cas

STF 262	RA: 02 29 03.95	Dec: +67 24 08.9	WDS: 147(263) (AB)		
			WDS: 189(230) (AC)		
V magnitudes	Aa: 4.65	Ab: 8.63	B: 6.89	Ca: 9.1	Cb: 11.8
(B − V) magnitudes	A: +0.08	Aa: +0.72	B: +0.43	C: +1.0	
μ(A)	−12.59 mas yr^{-1}	± 0.37	6.54 mas yr^{-1}	± 0.42 (DR2)	
μ(B)	−20.64 mas yr^{-1}	± 0.19	−4.25 mas yr^{-1}	± 0.23 (DR2)	
μ(C)	−43.91 mas yr^{-1}	± 0.08	10.30 mas yr^{-1}	± 0.11 (DR2)	
π(A)	21.96 mas	± 0.33	148.5 light yr	± 2.2 (DR2)	
π(B)	21.70 mas	± 0.14	150.3 light yr	± 1.0 (DR2)	
π(C)	22.22 mas	± 0.08	146.8 light yr	± 0.5 (DR2)	
Spectra	A: A5p + G6V?	B: F5	C: G7V+M2V?		
Masses (M$_\odot$)	Aa: 1.98	Ab: 0.90	B: 1.2	Ca: 0.8	Cb: 0.6
Luminosities (L$_\odot$)	Aa: 24	Ab: 0.6	B: 3	Ca: 0.4	Cb: 0.03
Catalogues	HR 707	HD 15089	HIP 11569	SAO 12298	
DS catalogues	H 1 34 (AB)	H 3 4 (AC)	STF 262 (ABC)	BDS 1262	ADS 1860
	CHR 6 (AaAb)				
Radial velocity	+1.2 km s^{-1}	± 2			
Galactic coordinates	132°.163	+6°.290			

History

On 11 June 1782 William Herschel observed ι Cas and noted that it was a triple star, with the closer companion very unequal and its more distant neighbour 'extremely unequal'. He placed them in different categories, with the closer of the stars at 2″.5 apart in class I whilst the more distant of the companions was allocated class III status. Otto Struve at Pulkovo had shown that there was significant motion in the AB pair and it was followed more assiduously. In 1906 Thomas Lewis [194] considered the astrometric history of the pair. The motion of AB caused some concern; Lewis at first attributed an unusual variation, amounting to a loop in the motion, as a function of observational personality. After looking more closely and 'correcting' the measures of Otto Struve, the loop still remained and he concluded that this was a genuine feature and due to the presence of an invisible body and that the period was about 40 years. It was clear that the perturbation was around the A component as there was a corresponding loop in the motion of AC. Neither Burnham nor Aitken in their general catalogues noted anything unusual.

Finder Chart

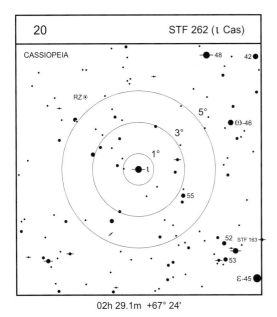

02h 29.1m +67° 24'

The Modern Era

The analysis of Lewis has proved to be correct. McAlister and colleagues [158], using the Kitt Peak 4-metre reflector, found that A was double at a distance of 0″.50 and PA 174° in 1982.75, and a later observation at 1985.85 showed the star had moved to 160° and 0″.41, commensurate with an orbital period of several decades. W. D. Heintz [309] used photographic plates to map the loop and to derive an orbit with a period of 52 years. Heintz thought that the perturbing star should be somewhat less than two magnitudes fainter than A and therefore visible with sufficient aperture, but no direct visual observations have been recorded. He was somewhat doubtful about connecting the McAlister component with the perturbing star. The close AaAb pair was first directly imaged by Drummond *et al.* [159] using the adaptive optics system on the 3.6-metre telescope at Haleakala on the Hawaiian island of Maui using the *I*-band (0.9 μm) and *H*-band (1.6 μm). An orbit yielded a period of 47.1 years. They also noted that B appears to be moving linearly and that C is almost half as close to the Sun as Aa.

Orbit

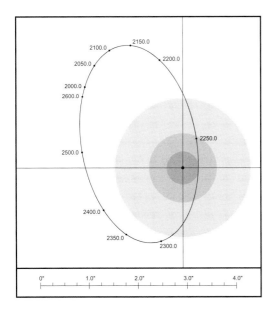

Ephemeris for STF 262 AB (2010 to 2100)

Orbit by Hei (1996b) Period: 620 years, Grade: 5

Year	PA(°)	Sep(")	Year	PA(°)	Sep(")
2010.0	229.1	2.60	2060.0	219.7	2.74
2020.0	227.1	2.64	2070.0	217.9	2.76
2030.0	225.2	2.67	2080.0	216.1	2.77
2040.0	223.3	2.69	2090.0	214.4	2.77
2050.0	221.5	2.72	2100.0	212.6	2.77

Observing

The pair AB is continuing to widen from its discovery position but it is still quite close and the significant difference in

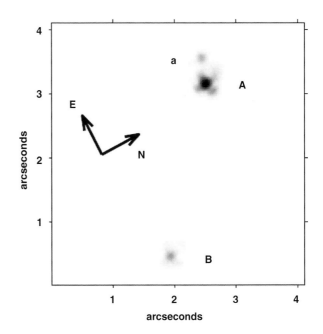

Figure 9.1 Three components of ι Cas were clearly separated with AO on the 3.6-metre AEOS telescope on Haleakala on 25 July 2001. This was the first time that the close astrometric/spectroscopic companion Aa was imaged (courtesy Dr J. Drummond).

magnitude makes it a delicate object for the small aperture. Whilst C is three times as distant from A, its relative faintness can also render it quite hard to see unless the night is good. A 10-cm aperture will show the three stars but a 20-cm would be better. A severe challenge for the larger aperture is Aa. Drummond indicates that the difference in brightness in the visual band is about 2.8 magnitudes. The separation remains greater than 0″.5 until 2030 or so. About 1.5° ENE of ϵ Cas is the wide pair STF 163 (magnitudes 6.8, 9.1, 38°, 34″.5, 2016). This rather unequal pair of stars boasts a nice colour contrast. The components are spectral types K5Ia and B5V and show colours of orange and pale blue. Gaia DR2 places the primary star at a distance of 820 light years.

Measures

AB			
Early measure (STF)	276°.7	1″.86	1829.66
(Orbit	277°.9	1″.92)	
Recent measure (ARY)	232°.0	2″.93	2015.13
(Orbit	229°.1	2″.62)	

AC			
Early measure (STF)	107°.3	7″.63	1829.65
Recent measure (ANT)	115°.4	7″.11	2010.80

21. α UMI = POLARIS = STF 93 = WDS J02318+8926AB

Table 9.21 Physical parameters for α UMi

STF 93	RA: 02 31 47.09	Dec: +89 15 15.50	WDS: 1351(63)		
V magnitudes	Aa: 2.30	Ab: 4.30	B: 9.1		
(B − V) magnitudes	A: +0.67	B: +0.49			
μ(A)	44.48 mas yr^{-1}	± 0.11	−11.85 mas yr^{-1}	± 0.13	
μ(B)	42.09 mas yr^{-1}	± 0.05	−13.64 mas yr^{-1}	± 0.06 (DR2)	
π(A)	7.54 mas	± 0.12	433 light yr	± 6	
π(B)	6.26 mas	± 0.24	521 light yr	± 20 (HST)	
π(B)	7.29 mas	± 0.03	447.4 light yr	± 1.8 (DR2)	
Spectra	Aa: F7Ib-IIv	Ab: F3V			
Masses (M$_\odot$)	A: 4.5	(+2.2 −1.4)	B: 1.26	(+0.14 −0.07)	
Radii (R$_\odot$)	A: 46	± 0.3			
Luminosities (L$_\odot$)	A: 2500	B: 4			
Catalogues	1 UMi	HR 424	HD 8890	SAO 308	HIP 11767
DS catalogues	H 4 1	STF 93	BDS 713	ADS 1477	WRH39 (AaAb)
Radial velocity	−16.42 km s^{-1}	± 0.03			
Galactic coordinates	123°.281	+26°.461			

History

Polaris is the bright star which is currently close to the North Celestial Pole. Closest approach occurs around 2100 when the separation will be less than half a degree. The star was observed by William Herschel on 17 August 1779; he noted the colours as 'Pale r(ed) or nearly r(ed). Garnet or deeper red than the other'. In 1852 Seidel [161], and later Schmidt [162] in 1856, reported a small photometric variation in Polaris. In 1899 Campbell [163] announced that Polaris was an SB with a period of 3.96809 days but, according to Aitken, this was later dismissed as a measurement of the pulsation period of the star. However, Campbell [164] did also note a small change in the systemic velocity of the star, evidence of a longer-period orbital motion. An astrometric companion with a period of 30 years was detected by Gerasimovič [169]. Raymond

Finder Chart

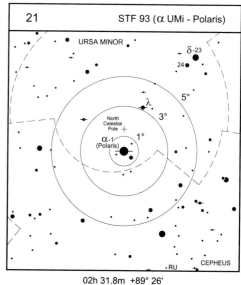

02h 31.8m +89° 26'

H. Wilson [165] at Flower Observatory in the USA then used a double-slit interferometer of his own design on the 18.5-inch refractor to report that the Pole Star was a very close visual pair, and he made three measurements of the position angle and separation.

The Modern Era

Hamilton Jeffers used an interferometer on the Lick 36-inch refractor but could not confirm Wilson's observations. The amplitude of pulsation has decreased markedly over the observational history of the star [166]. In 1899 it was about 6 km s^{-1} whilst by 1995 it was about 1.5 km s^{-1}, since when it has started to increase in size again. The spectra taken at the DDO between 1980 and 1994 were combined with earlier ones from Lick Observatory to give spectroscopic orbits with periods of 30.46 years (Lick) and 29.59 years (DDO). The radius of the primary star was determined by Nordgren et $al.$ [167]. This pair was not resolved until imaged by HST (Evans et $al.$ [168] and two observations of the position angle and separation were made. The ensuing orbit, which appears on the USNO website, appears highly unconvincing, and more observations are needed. The dynamical mass for the Cepheid itself derived from this work is 4.5 M$_\odot$. There are three more distant and much fainter stars in the field. The William Herschel companion (B) is 18$''$ away but appears to have a common proper motion with the bright primary. Jim Daley [170] revisited the Polaris system using a masking technique to image the sky close to Polaris whilst cutting out the considerable light from the primary star. In this way he measured the faint companions C and D for the first time for over a century (Burnham was the last to see star C, for instance). Bond et $al.$ [171] find a trigonometrical parallax for the B star of 6.26 $\pm$ 0.24 mas, which seems to contradict previous evidence that it is physically connected to A. They wonder if the Hipparcos parallax has been overestimated and expect that Gaia will help to resolve this problem. In fact DR2 agrees much more with Hipparcos on the parallax of B than with Bond et $al.$, assuming the two stars are physical, which seems likely given the similarity of the proper motions. An accurate parallax for A awaits further processing.

Observing

Easily found with the naked eye by following a line between the Pointers (α and β UMa) and continuing for a further 28$°$. F. G. W. Struve found the colours to be yellow and white, and Smyth, a decade or so later, came to a similar conclusion – 'topaz yellow and pale white'.

Measures

| Early measure (STF) | 210$°$.1 | 18$''$.27 | 1834.14 |
| Recent measure (FYM) | 233$°$ | 18$''$.12 | 2013.28 |

22. γ CET = STF 299 = WDS J02433+0314AB

Table 9.22 Physical parameters for γ Cet

STF 299	RA: 02 43 18.04	Dec: +03 14 08.9	WDS: 181(237)		
V magnitudes	A: 3.54	B: 6.18			
$(B - V)$ magnitudes	A: +0.09	B: +0.52			
μ	-146.10 mas yr^{-1}	$\pm\,0.71$	-146.12 mas yr^{-1}	$\pm\,0.55$	
π	40.97 mas	$\pm\,0.63$	79.6 light yr	$\pm\,1.2$	
μ(A)	-151.26 mas yr^{-1}	$\pm\,1.39$	-147.57 mas yr^{-1}	$\pm\,1.38$ (DR2)	
π(A)	43.60 mas	$\pm\,0.82$	74.82 light yr	$\pm\,1.4$ (DR2)	
Spectra	A2Vn	F4V			
Masses (M$_\odot$)	A: 2.0	B: 1.0			
Luminosities (L$_\odot$)	A: 20	B: 1.5			
Catalogues	86 Cet	HR 804	HD 16970	HIP 12706	SAO 110707
DS catalogues	STF 299	BDS 1401	ADS 2080	ALD 124 (AC)	
Radial velocity	-4.90 km s^{-1}	$\pm\,0.9$			
Galactic coordinates	$168°.919$	$-49°.382$			

History

Alden & van de Kamp [172], working with the 26-inch refractor at McCormick Observatory, used photography to derive a parallax for γ Ceti. In the course of this work they noted that a star of magnitude 10.15 at a distance of 843″ turned out to have the same proper motion as the close pair. This angular distance converts to a linear distance of 21,000 AU. In the same paper Alden quoted the parallax of γ from five determinations as being 40 mas. This is in remarkably good agreement with the Hipparcos value given above.

The Modern Era

Hartkopf [173], using the 4-metre KPNO telescope for speckle photography in 1984, found that the bright component was single at the 30-mas level. Fuhrmann *et al.* [641] gave the

Finder Chart

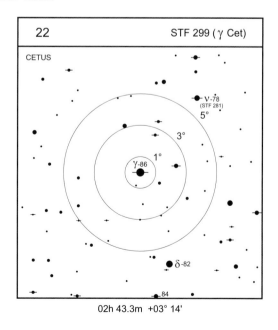

02h 43.3m +03° 14'

masses in the table above. The companion, although more than 2″ distant, is not in DR2, so this may indicate possible multiplicity.

Exoplanet Host?

The star has been the subject of a number of surveys looking at the possibility of exoplanetary or brown dwarf companions. The VAST (Volume-limited A-STar) A-star survey found no evidence of an exoplanetary system, nor did a radio survey conducted with VLBI by Katarzynski *et al.* [174]. They had postulated that the strong stellar wind from young A stars combined with the magnetic field around exoplanets might produce radiation at a frequency which could be detectable by radio telescopes. They too did not find any evidence of an exoplanetary system.

Observing and Neighbourhood

Found by moving 5° W from α Ceti (V = 2.5), Ceti γ is a beautiful, but not easy, pair with components giving a fine colour contrast. F. G. W. Struve thought the stars were yellow and blue, whilst Smyth noted pale yellow and lucid blue. From Australia, Ernst Hartung found A to be brilliant white and B to be deep yellow, which is more in agreement with the most recent estimate by John Nanson. He found the primary was very bright white with just a tinge of yellow whilst the colour of B was 'uncertain'. Three degrees NW of γ is ν Cet (STF 281, 5.0, 9.1, 80°, 8″.4, 2011).

Measures

Early measure (STF)	287°.4	2″.59	1832.48
Recent measure (DRS)	298°.5	2″.31	2008.80

23. $\theta^{1,2}$ ERI = PZ 2 = WDS J02583−4018

Table 9.23 Physical parameters for θ Eri

PZ 2	RA: 02 58 15.68	Dec: −40 18 16.8	WDS: 1030(75)	
V magnitudes	A: 3.20 (θ^1)	B: 4.12 (θ^2)		
($B - V$) magnitudes	A: +0.17	B: +0.11		
μ(A)	−54.46 mas yr^{-1}	± 0.55	23.50 mas yr^{-1}	± 0.59 (DR2)
μ(B)	−51.66 mas yr^{-1}	± 0.36	16.18 mas yr^{-1}	± 0.38 (DR2)
π(A)	19.56 mas	± 0.35	166.8 light yr	± 3.0 (DR2)
π(B)	20.34 mas	± 0.22	160.4 light yr	± 1.7 (DR2)
Spectra	A: A4III + ?	B: A1V		
Masses (M$_\odot$)	Aa: 2.4	Ab: 1.3	C: 3.0	
Radii (R$_\odot$)	A: 6.0	B: 4.9		
Luminosities (L$_\odot$)	A: 110	B: 45		
Catalogues (A/B)	HR 897/8	HD 18622/3	SAO 216113	HIP 13847
DS catalogues	PZ 2			
Radial velocity	11.9 km s^{-1}	± 2.6		
Galactic coordinates	247°.856	−60°.736		

History

Giuseppe Piazzi (1746–1826) was born in Valtellina in northern Italy and after becoming Professor of Astronomy in Malta he was made Professor in Palermo in 1781. He visited Lalande in Paris to conduct joint observations and then went to England to buy instruments for the Palermo Observatory. Although perhaps more well known for his subsequent work on minor planets, he also produced a star catalogue in 1814 containing 7500 entries, which earned him the Lalande Prize. This catalogue contains a number of bright double stars, some of which appear in this volume and six of which still appear in the WDS catalogue. Piazzi observed both components of θ Eri using his transit instrument. Timing both stars, which crossed the meridian less than one second of time apart, must have been tricky, added to which the stars were low down in the sky and reached an altitude of only 21° from Palermo. In 1897

Finder Chart

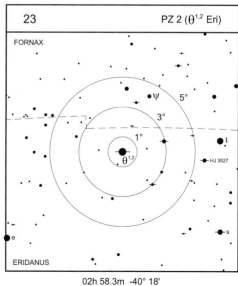

02h 58.3m −40° 18'

Annie Cannon [176] classified the spectrum of θ Eri A as A2, although noting that some contamination from the B star had been seen.

The Modern Era

The brighter component is widely referred to as a spectroscopic binary but there are no obvious references to radial velocity measurements and there seems to have been no attempt to publish an orbit. Hurly & Warner [178] used an area scanner for the photometry of southern double stars using the telescopes at Sutherland, South Africa, but they were not able to confirm a report that the primary star was variable. Tokovinin notes in his MSC [177] that the period of θ^1 and θ^2 is expected to be about 1.1 million years.

Observing and Neighbourhood

'One of the gems of the southern sky' says Hartung in *Astronomical Objects for Southern Telescopes*. This is a brilliant pair of white stars which is clearly a long-period binary system. It sits in a visually sparse area of sky. Five degrees SSE is the triple star JC 8/HJ 3556 (see Star 26). There is an attractive John Herschel pair to be found 2°.5 W, and slightly S: HJ 3527, 7.0, 7.2, 40°, 2″.3, 2013. RWA found a distant faint star (ARY 95), magnitude 11.6 at 80°, 133″, whilst observing at Johannesburg in 2013.

Measures

Early measure (JC)	82°.4	8″.11	1851.76
Recent measure (ARY)	91°.1	8″.63	2013.67

24. ε ARI = STF 333 = WDS J02592+2120AB

Table 9.24 Physical parameters for ε Ari

STF 333	RA: 02 59 12.73	Dec: +21 20 25.6	WDS: 57(452)		
V magnitudes	A: 5.17	B: 5.51			
(*B* − *V*) magnitudes	A: +0.06	B: +0.07			
μ	−13.5 mas yr^{-1}	± 0.93	−5.0 mas yr^{-1}	± 0.64	
π	9.54 mas	± 0.72	342 light yr	± 26	
μ(B)	−14.09 mas yr^{-1}	± 0.56	−7.35 mas yr^{-1}	± 0.46 (DR2)	
π(B)	8.51 mas	± 0.34	383 light yr	± 15 (DR2)	
Spectra	A2Vs	A2Vs			
Luminosities (L$_\odot$)	A: 100	B: 70			
Catalogues (A/B)	48 Ari	HR 888/7	HD 18520/19	SAO 75673	HIP 13914
DS catalogues	STF 333	BDS 1512	ADS 2257		
Radial velocity	−7.9 km s^{-1}	± 0.9			
Galactic coordinates	158°.693	−32°.514			

History

This double star was discovered by F. G. W. Struve at Dorpat. According to Smyth, Struve regarded ε Ari as perhaps the closest of his double stars and actually recorded them as in contact in his 1827 catalogue. Smyth also says that W. R. Dawes first observed this pair with Smyth's 5.9-inch refractor at Bedford (now on view at the Science Museum in London). William Henry Smyth (1788–1865) was a direct descendant of Captain John Smith of Virginia. He went to sea early, and later saw action with large British fighting ships against the French and Spanish during the Napoleonic period. He learned hydrography and carried out a survey of Sicily, and later other areas in the Mediterranean, and during this time made the acquaintance of Piazzi, for whom he proofread some of the pages of the Palermo catalogue. In 1830 he acquired a 5.9-inch Tulley refractor with which he carried out the work for *A Cycle of Celestial Objects* [179], often known as the Bedford catalogue of 1844, which contained extensive notes on

Finder Chart

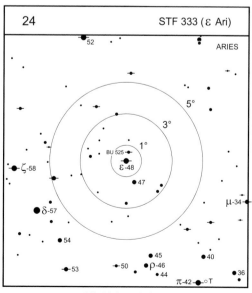

02h 59.2m +21° 20'

Orbit

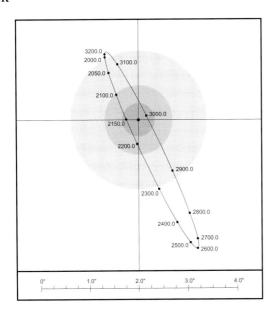

Ephemeris for STF 333 AB (2010 to 2100)

Orbit by FMR (2012g) Period: 1215.913 years, Grade: 4

Year	PA(°)	Sep(")	Year	PA(°)	Sep(")
2010.0	209.3	1.37	2060.0	214.0	1.02
2020.0	210.1	1.32	2070.0	215.4	0.94
2030.0	210.9	1.25	2080.0	217.1	0.85
2040.0	211.8	1.18	2090.0	219.2	0.76
2050.0	212.8	1.10	2100.0	221.9	0.66

680 double and multiple stars. He later wrote *Sidereal Chromatics* [180], a treatise on double star colours.

The Modern Era

An orbit for this pair with a period of 1215 years was calculated by Rica [181]. This is rather premature since the motion in position angle amounts to only 20° over 200 years. The B component is in DR2 but with a *G* magnitude only.

Observing

To find ε Ari, which forms a faint naked-eye pair with 47 Ari, start with β Arietis (*V* = 2.65) and move eastwards by 16°. This is an excellent test for a 10-cm aperture. Hartung finds both components to be pale yellow, although RWA has always seen them as white. The pair should be in the range of a 10-cm aperture for some decades to come. In the same field 17′ N is BU 525, a stiff test for 25-cm on a good night. The stars have an orbital period of 242 years and are both magnitude 7.5. The positions are 0″.47 and 279° (2020) and the stars will continue to edge closer for the next 50 years or so.

Measures

Early measure (STF)	189°.7	0″.58	1831.16
(Orbit	191°.5	0″.62)	
Recent measure: (ARY)	209°.9	1″.41	2012.66
(Orbit	209°.5	1″.36)	

25. α FOR = HJ 3555 = WDS J03121–2859

Table 9.25 Physical parameters for α For

HJ 3555	RA: 03 12 04.53	Dec: −28 59 18.4	WDS: 710(97)		
V magnitudes	A: 3.98	B: 7.19			
$(B − V)$ magnitudes	A: +0.53	B: +0.00			
μ(A)	359.97 mas yr^{-1}	± 0.33	618.52 mas yr^{-1}	± 0.42 (DR2)	
μ(B)	343.23 mas yr^{-1}	± 0.10	636.65 mas yr^{-1}	± 0.14 (DR2)	
π(A)	71.68 mas	± 0.31	45.50 light yr	± 0.20 (DR2)	
π(B)	71.13 mas	± 0.09	45.85 light yr	± 0.06 (DR2)	
Spectra	A: F6V	Ba: K0V	Bb: WD		
Masses	A: 1.2	Ba: 0.8	Bb: 0.5		
Luminosities (L$_\odot$)	A: 4	B: 0.2			
Catalogues	12 Eri	HR 963	HD 20010	SAO 168373	HIP 14879
DS catalogues	HJ 3555	BDS 1612	ADS 2402		
Radial velocity	−17.14 km s^{-1}	± 0.2			
Radial velocity (A)	−17.52 km s^{-1}	± 0.73 (DR2)			
Galactic coordinates	224°.730	−59°.030			

History

'12 Eridani. Very fine D(ouble) star' recorded John Herschel after Sweep 643 with the 18-inch telescope at Feldhausen. The stars began to close during the remainder of the nineteenth century and between 1902 and 1924 were not fully measured anywhere, being recovered by Robert Innes using the 9-inch refractor at Johannesburg in 1925 when the separation was close to 0″.8. W. H. van den Bos [366] observed the B component to fade in brightness between 1925 and 1926 and was of the opinion that it varied between magnitudes 7.0 and 9.0. The star is in the NSV catalogue as number 1074. After 1930, the companion then moved rapidly around the primary and made another close approach around 1950 when the separation was close to 0″.20; it has been widening ever since.

Finder Chart

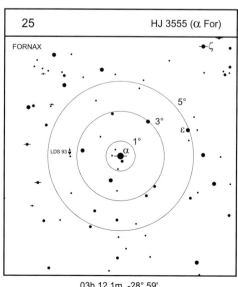

03h 12.1m -28° 59'

Orbit

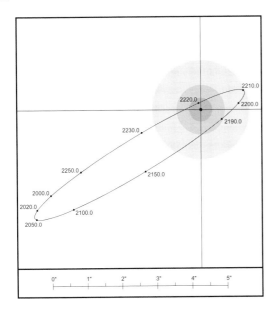

Ephemeris for HJ 3555 (2010 to 2100)

Orbit by Sod (1999) Period: 269 years, Grade: 4

Year	PA(°)	Sep(")	Year	PA(°)	Sep(")
2010.0	299.9	5.23	2060.0	303.6	5.51
2020.0	300.6	5.45	2070.0	304.3	5.36
2030.0	301.4	5.58	2080.0	305.2	5.15
2040.0	302.1	5.63	2090.0	306.1	4.89
2050.0	302.8	5.60	2100.0	307.1	4.58

Exoplanet Host?

The star α For A was observed with the HARPS spectrograph on the 3.6-metre spectrograph in Chile as part of a search for exoplanets by Zechmeister *et al.* [528]. The conclusion was that the trend found in the change of the radial velocity of A could be accounted for by the effect of the companion.

Observing and Neighbourhood

This comprises a fine but very unequal pair that would be seen well in 20-cm. The stars will be at their widest in the next few decades, giving a good opportunity to see them clearly. Two degrees E of α is LDS 93, a wide pair, readily accessible to binoculars (7.4, 8.4, 358°, 253", 2015), in which Luyten found a common proper motion in the stars amounting to over 0".3 per year and which, as DR2 confirms, are both at a distance of 117 light years.

Measures

Early measure (JC)	310°.0	3".32	1856.16
(Orbit	309°.9	3".58)	
Recent measure (ARY)	300°.5	5".40	2013.68
(Orbit	300°.2	5".32)	

The Modern Era

Fuhrmann *et al.* [182] observed the stars of this system and found that the B star has radial velocity variations on a period of 3.75 days. They concluded that the pair consists of a barium star (Ba) and the companion Bb is likely to be a white dwarf.

26. JC 8 ERI = WDS J03124−4425AB

Table 9.26 Physical parameters for JC 8 Eri

JC 8	RA: 03 12 25.68	Dec: −44 25 10.8	WDS: 543(117)	
V magnitudes	A: 6.42	B: 7.36	C: 8.76	
$(B - V)$ magnitudes	A: +0.40	B: +0.57		
μ	81.63 mas yr^{-1}	± 0.55	−4.57 mas yr^{-1}	± 0.98
π	23.53 mas	± 0.62	139 light yr	± 4
μ(C)	99.95 mas yr^{-1}	± 0.06	−14.29 mas yr^{-1}	± 0.07 (DR2)
π(C)	22.75 mas	± 0.04	143,37 light yr	± 0.25 (DR2)
Spectra	A: F7III	B: A0V	C: ?	
Masses (M$_\odot$)	A: 1.41	± 0.05	B: 1.31	± 0.05
Luminosities (L$_\odot$)	A: 4	B: 2	C: 0.5	
Catalogues	HR 968	HD 20121	SAO 216209	HIP 14913
DS catalogues	JC 8(AB)	HJ 3556 (AB-C)		
Radial velocity	17.00 km s^{-1}	± 7.4		
Galactic coordinates	253°.823	−56°.958		

History

In 1834 John Herschel observed the star Brisbane 501. He noted that 'With the utmost difficulty I get a glimpse of the star as wedge-shaped, but the definition is growing very bad' and he estimated the PA as 230° and the separation as 1″.5. The star was entered in Herschel's published list as HJ 3556. William Stephen Jacob was born in 1813 and joined the Indian Army, becoming a Lieutenant in the Bombay Engineers from 1833 to 1848. In 1842 he established a private observatory at Pune. From 1848 to 1859 he was Director of the Madras Observatory and measured double stars with a 6.33-inch refractor by Lerebours as well as observing the Sun and planets. The climate in India did not agree with him and he died just after returning there in August 1862, aged only 49. His studies of the binary 70 Ophiuchi, which led him to believe that perceived irregularities in its orbital motion were due to the perturbing influence of a planet, mean that he was the first scientist to

Finder Chart

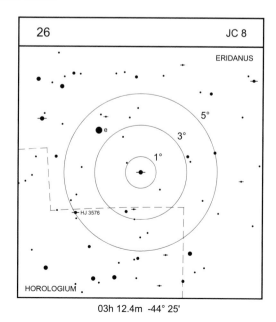

03h 12.4m -44° 25'

Orbit

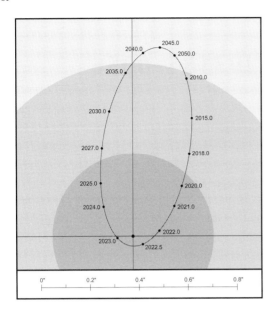

Ephemeris for JC 8 AB (2017 to 2026)

Orbit by Sod (1999) Period: 45.2 years, Grade: 3

Year	PA(°)	Sep(")	Year	PA(°)	Sep(")
2017.0	147.9	0.45	2022.0	101.2	0.11
2018.0	144.8	0.40	2023.0	276.9	0.06
2019.0	140.7	0.35	2024.0	225.5	0.17
2020.0	134.9	0.28	2025.0	211.9	0.25
2021.0	125.3	0.21	2026.0	204.7	0.31

suggest the detection of a planet outside the Solar System. The star HJ 3556 was observed by Jacob in 1856 and he noted 'a strong suspicion that A is a close double in a direction nearly S., but the definition is not good enough to be certain'. Over the next 40 years or so the star was unobserved until it was 're-discovered' by the Harvard observers [248] in Chile in 1895, by Innes [187] in 1896, who acknowledged that Jacob had suspected the duplicity of A in 1856 but nevertheless assigned his discovery number 55 to it, and by Sellors [186] in 1897 who was aware of Innes' observation the year before but called it SLR 25.

The Modern Era

Olin Eggen [189], in a study of the radial velocities of double and multiple systems, noted that there was a significant variation in the radial velocities of this system and that the system was quadruple. This conclusion was questioned by Heintz in 1979; he argued that the observed mass ratio in the close pair does not support a fourth component. Heintz also estimated [185] that the period of the Herschel component was close to 1000 years. Recent radial velocity measures of the close AB and C stars by Tokovinin using the 100-inch Du Pont telescope at Las Campanas suggest that the stars are fast rotators and that they have 'similar' velocities. The current orbit by Söderhjelm gives the period as only 45.2 years, yet, surprisingly, the orbit is graded only as 3 despite having been observed through almost three complete revolutions. However, the apparent ellipse takes the companion from a distance of around 0″.8 to 0″.04 when angular motion is rapid, and there is no coverage of this part of the orbit.

Observing and Neighbourhood

The JC 8 stars are easily found about 5° SSE of the magnitude 3.0 θ Eri, a very bright pair, details of which can be found in the section on Star 23. The pair JC 8 will require 40-cm for the years to 2020 (135°, 0″.28), after which it closes rapidly and will not exceed 0″.30 again until 2027 or so. The bright Herschel companion is a much easier object and is accessible to 7.5-cm. Another John Herschel pair can be found 2°.5 SE. This is HJ 3576, 7.3, 8.8, 342°, 2″.8, 2015, closing since discovery.

Measures

JC 8			
Early measure (I)	182°.4	0″.87	1900.72
(Orbit	180°.1	0″.68)	
Recent measure (TOK)	149°.6	0″.48	2015.91
(Orbit	150°.8	0″.49)	
HJ 3556			
Early measure (HJ) :	233°.3	1″.5	1835.80
Recent measure (ANT) :	188°.6	3″.75	2013.72

27. 95 CET = AC 2 = WDS J03184−0056

Table 9.27 Physical parameters for 95 Cet

AC 2	RA: 03 18 22.43	Dec: −00 55 49.0	WDS: 681(100)		
V magnitudes	A: 5.60	B: 7.97	C: 16.2		
$(B - V)$ magnitudes	A: +1.26	B: +0.64			
μ	−253.18 mas yr^{-1}	±0.95	−60.32 mas yr^{-1}	±0.78	
π	14.89 mas	± 0.84 mas	219 light yr	± 12	
$\mu(A)$	249.67 mas yr^{-1}	± 0.24	−60.71 mas yr^{-1}	± 0.24 (DR2)	
$\mu(C)$	250.77 mas yr^{-1}	± 0.22	−57.77 mas yr^{-1}	± 0.21 (DR2)	
$\pi(A)$	15.63 mas	± 0.18	208.7 light yr	± 2.4 (DR2)	
$\pi(C)$	15.71 mas	± 0.16	207.6 light yr	± 2.1 (DR2)	
Spectra	A: G9IV	B: ?	C: DA		
Luminosities (L$_\odot$)	A: 20	B: 2	C: 0.001		
Catalogues	95 Cet	HR 992	HD 20559	SAO 130408	HIP 15383
DS catalogues	AC 2 (AB)	BDS 1650	ADS 2459	LDS 3472 (AC)	
Radial velocity	31.2 km s^{-1}	± 0.2			
Radial velocity (A)	31.97 km s^{-1}	± 0.12 (DR2)			
Galactic coordinates	123°.976	−39°.236			

History

The names of Alvan Clark (AC) and Alvan G. Clark (AGC) are associated with some rather difficult and interesting binary systems. The most famous example is Sirius (AGC1) – another is τ Cygni (AGC 13). Neither man was a double star observer, but during the course of testing the firm's famous telescope objectives they often resorted to using stars to make the final assessment. In 1853 Alvan was checking the performance of a 7.5-inch objective on the sky when he looked at the star 95 Cet and noted that it was a close and unequal double (*Burnham Double Star Catalogue*, 1906). Clark happened to come to Europe later that year and met the Reverend W. R. Dawes. When Dawes heard about this difficult pair he became very keen to obtain such an objective for his own observatory. In 1854 the lens was duly delivered and Dawes was delighted

with its performance. At the end of the nineteenth century the pair was difficult to observe (it was then separated by about 0″.4) and Aitken [191] noted that Burnham called 95 Ceti 'the most mysterious and strange double star in the heavens.' Burnham added 'I have tried it, first and last, perhaps hundreds of times with apertures all the way from 6 to 36 inches without being able to see any trace of the little star.' Aitken was reluctant to ascribe this to variability, pointing out it could be due to poor seeing or a poor telescope.

The Modern Era

Baize & Petit [144] included 95 Cet in their catalogue of 1989 and marked the A component as an RS CVn variable, whilst Willem Luyten [192], in a search for common proper motion

Finder Chart

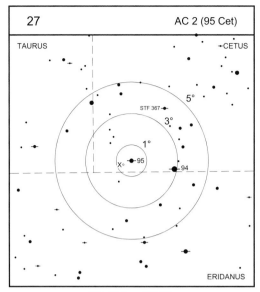

27	AC 2 (95 Cet)

TAURUS • •← CETUS

5°
STF 367 •←
3°
1°
X° • 95
• 94

ERIDANUS

03h 18.4m -00° 56'

Orbit

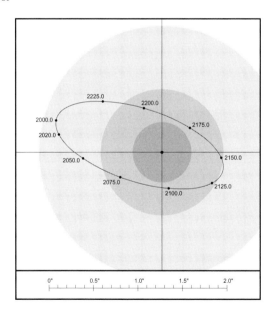

Ephemeris for AC 2 AB (2015 to 2060)

Orbit by Pop (1997f) Period: 282.42 years, Grade: 4

Year	PA(°)	Sep(")	Year	PA(°)	Sep(")
2015.0	258.7	1.20	2040.0	268.8	1.01
2020.0	260.5	1.17	2045.0	271.4	0.95
2025.0	262.4	1.14	2050.0	274.4	0.89
2030.0	264.3	1.10	2055.0	277.8	0.82
2035.0	266.5	1.06	2060.0	281.8	0.76

systems, notes a magnitude 16.2 star at 30° and 49″ sharing the same space motion as AB. This is a DA white dwarf (LP 592-50) and physically connected to AB. Olin Eggen [190] also noted a nearby star, BD, at −01° 474, magnitude 10.1, PA 101°, some 11′ distant, which also has a common proper motion.

Observing and Neighbourhood

The present time gives a good opportunity to resolve this pair. The stars have just passed widest separation and in 2020 will be at 261°, 1″.2. If the orbit by Popovic [193] is correct, and it is graded as preliminary, they will close slowly until the end of the current century, when the distance reaches 0″.4. Six degrees due S of 95 Cet is a wide pair of sixth magnitude stars. The southwestern component is β 84 (6.4, 7.9, 9°, 1″.0, 2013) which is slowly widening. Five degrees ENE is 10 Tau (V = 4.30) and lying just 10′ N of that star is the fine pair STF 422 (6.0, 8.9, 274°, 7″.1, 2013), which has a premature orbit of 2101 years calculated for it. The primary is an RS CVn star (V711 Tau) and an SB. A good test of resolution and light gathering is STF 367, 2° NW. The stars of magnitudes 8.1, 8.2 are orbital with a period of 420 years. The 2020 position is 130°, 1″.27.

Measures

Early measure (DA)	73°.2	0″.73	1854.81
(Orbit	71°.2	0″.71)	
Recent measure (TOK)	260°.9	1″.12	2015.03
(Orbit	258°.8	1″.20)	

28. STF 425 PER = WDS J03401+3407AB

Table 9.28 Physical parameters for STF 425 Per

STF 425	RA: 03 40 07.24	Dec: +34 06 59.3	WDS: 184(236)		
V magnitudes	Aa: 7.60	Ab: 14.5	C: 7.60		
$(B - V)$ magnitudes	A: +0.56	B: +0.58			
μ(A)	-66.99 mas yr^{-1}	± 0.10	8.08 mas yr^{-1}	± 0.08 (DR2)	
μ(B)	-78.78 mas yr^{-1}	± 0.11	17.60 mas yr^{-1}	± 0.08 (DR2)	
π(A)	21.84 mas	± 0.07	149.3 light yr	± 0.5 (DR2)	
π(B)	21.77 mas	± 0.06	149.8 light yr	± 0.4 (DR2)	
Spectra	Aa: F5	Ab: ?	B: ?		
Masses (M$_\odot$)	Aa: 1.5	Ab: 0.02	B: 1.5		
Luminosities (L$_\odot$)	Aa: 2	Ab:	C: 2	(DR2)	
Catalogues	HD 22692	SAO 56613	HIP 17129		
DS catalogues	H 3 36 (AB)	STF 425 (AB)	BDS 1799	ADS 2668	RBR 26Aa,Ab
Radial velocity	-5.7 km s^{-1}	± 0.5			
Galactic coordinates	158°.440	-16°.879			

History

This pair was discovered by Sir William Herschel on 7 September 1782. It was noticed by Lewis [194], who commented that 'the motion shown by the micrometer measures is very curious'. Over the nineteenth century the companion gradually approached A, closing from about 3″ to 2″.5, but after about 1870 the distance then remained constant. Neither Burnham nor Aitken, in their respective catalogues, noticed any unusual behaviour.

The Modern Era

In 2014 Zirm and Rica Romero [196] laid out similar orbital parameters for the putative third body but they were unable to say around which of the two visual components this body rotated. In 2014 Russell Genet *et al.* [195] published an analysis of the motion and sub-motions within the

Finder Chart

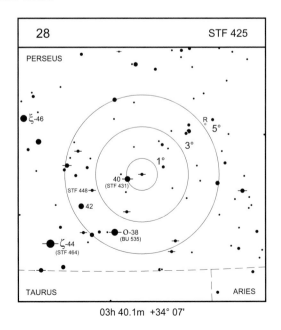

03h 40.1m +34° 07'

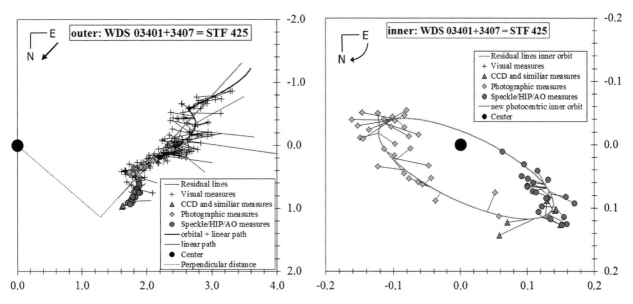

Figure 9.2 The astrometric history of STF 425 is shown in the left-hand figure; the perturbation of the star A by its much fainter and closer companion can be clearly seen. The right-hand figure shows the apparent orbit of star A and the perturbing star a.

STF 425 system, and the accompanying Figure 9.2 is reproduced from that paper. In 2013 Roberts *et al.* [197], using the PALM 3000 adaptive optics system on the Hale 5-metre telescope when they observed STF 425 and using the *K* band detected a star at a distance of 0″.50 to A and a position angle of 44°. This agreed very well with the orbital prediction of Zirm and Rica Romero, who derived a period of 107 years for the third body.

Observing

The pair STF 425 can be found 0°.5 preceding 40 Per and slightly N, but curiously it is not plotted in *CDSA2*. The pairs STF 431 = 40 Per (5.0, 10.0, 244°, 19″.8, 2010), STF 448 (6.7, 9.4, 13°, 3″.5, 2016), o (BU 535) (3.9, 6.7, 23°, 1″.0, 2009), and ζ Per (STF 464) (2.9, 9.2, 209°, 13″.9, 2012), with three further stars of magnitudes 10 and 11 within a radius of 120″, are all nearby.

Measures

Early measure (STF)	104°.7	2″.87	1830.16
Recent measure (ARY)	60°.4	2″.00	2016.09

29. 32 ERI = STF 470 = WDS J03543−0257

Table 9.29 Physical parameters for 32 Eri

STF 470	RA: 03 54 17.49	Dec: −02 57 17.0	WDS: 434(138)	
V magnitudes	A: 4.45	B: 5.86		
$(B − V)$ magnitudes	A: +0.68	B: +0.19		
μ(A)	25.89 mas yr^{-1}	± 0.37	−0.23 mas yr^{-1}	± 0.36 (DR2)
μ(B)	26.93 mas yr^{-1}	± 0.10	0.85 mas yr^{-1}	± 0.09 (DR2)
π(A)	10.11 mas	± 0.24	322.6 light yr	± 7.7 (DR2)
π(B)	9.65 mas	± 0.06	338.0 light yr	± 2.1 (DR2)
Spectra	A: G8III	B: A2V		
Luminosities (L$_\odot$)	A: 130	B: 40		
Catalogues (A/B)	HR 1212/1	HD 24555/4	SAO 130806/5	HIP 18255
DS catalogues	H 2 36	STF 470	BDS 1939	ADS 2850
Radial velocity (A/B)	26.9 km s^{-1}	± 0.2	17.6 km s^{-1}	± 2
Galactic coordinates	192°.094	−40°.099		

History

This double star was noted by William Herschel on 22 October 1781. 'Double. Considerably unequal. L(arge). reddish w(hite).; S(mall). blue. Distance 4″ 19‴. Position 73° 23′ n(orth). preceding.' In 1886, J. Baillaud [198] noted a distant star of magnitude 10.5 at 5°, 165″ which shows little or no change.

The Modern Era

SIMBAD gives the spectral type of the primary star as G8III and A1IV.

Finder Chart

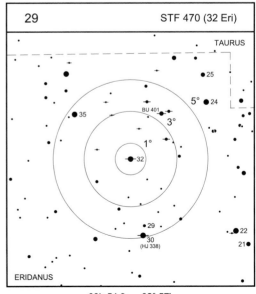

03h 54.3m -02° 57'

Observing and Neighbourhood

The star 32 Eri is a bright, unequal pair and a tempting target for the small telescope user. F. Struve noted yellow and purple, whilst Admiral Smyth recorded the colours as topaz yellow and sea-green using his 5.9-inch refractor. In Italy Secchi called them 'magnifici, superbi'. Olcott and Putnam in *Field Book of the Heavens* record 'Topaz and bluish, fine contrast' whilst RWA found yellow and blue with a 25-cm reflector at ×80 in 1972. E. J. Hartung, observing from Australia, noted the colours were deep yellow and white. Two and a half degrees S is 30 Cet (HJ 338, 5.5, 10.4, 134°, 8″.3, 2000), whilst 1°.7 NW is BU 401 (6.5, 10.5, 253°, 4″.6, 2015, fixed).

Measures

Early measure (STF)	347°.3	6″.70	1833.15
Recent measure (WSI)	347°.9	6″.84	2015.07

30. o² ERI = STF 518 = WDS J04153−0739BC

Table 9.30 Physical parameters for o² Eri

STF 518	RA: 04 15 07.57	Dec: −07 38 41.5	WDS: 280(181)		
V magnitudes	A: 4.51	B: 10.02	C: 11.47		
$(B − V)$ magnitudes	A: +0.93	B: +0.11	C: +1.68		
μ(A)	−2240.52 mas yr^{-1}	± 0.45	−3421.43 mas yr^{-1}	± 0.42 (DR2)	
μ(B)	−2250.12 mas yr^{-1}	± 1.59	−3408.28 mas yr^{-1}	± 0.55 (DR2)	
π(A)	198.57 mas	± 0.51	16.43 light yr	± 0.04 (DR2)	
π(B)	199.46 mas	± 0.32	16.35 light yr	± 0.03 (DR2)	
Spectra	A: K1V	B: DA4	C: M4.5Ve		
Masses (M$_\odot$)	A: 0.9	B: 0.57	± 0.02	C: 0.204	± 0.006
Radii (R$_\odot$)	A: 0.81	± 0.01			
Luminosities (L$_\odot$)	A: 0.3	B: 0.002	C: 0.0005		
Catalogues (A/B)	40 Eri	HD 26965	HR 1235	SAO 131063/5	HIP 19849
DS catalogues	H 2 80(BC)	STF 518 (BC)	STFB1 (AB)	BDS 2109	ADS 3093
Radial velocity (A/B)	−42.32 km s^{-1}	± 0.08	−21 km s^{-1}	± 10	
Radial velocity (A)	−42.62 km s^{-1}	± 0.42 (DR2)			
Galactic coordinates	200°.753	−38°.048			

History

Another binary star, discovered by William Herschel, which has been subsumed into the Struve Dorpat catalogue, o² Eri was first seen by Herschel on 31 January 1783 as a difficult object. 'Very unequal. Both d(usky)r(ed). With 227, hardly visible; with 460, very obscure'. This was the only pair in its class in the *Mensurae Micrometricae* that Struve was unable to measure – he estimated the position angle only. It was left to Otto Struve in 1850 to commence a series of accurate measures, which quickly showed that the stars formed a binary system. Herschel also noted the bright nearby star 40 Eri and measured AC as 107°.9 and 89″. The earliest measures of F. G. W. Struve were quite scattered but for the last 175 years or so the distance has remained almost fixed at around 82″, reflecting the almost identical and large

proper motions of A and BC, which amount to more than 4″ per annum. This is the amount by which the distant faint stars D (magnitude 12.6) and E (magnitude 13.0) are being left further behind each year. This is therefore a physical triple star, which is located a little over 16 light years from the Sun.

The Modern Era

The system 40 Eri itself is a bright orange star whilst the brighter of the two stars in the binary system BC is the fifth nearest white dwarf known and is the easiest of the five to observe, whilst its partner is a red dwarf and a flare star (DY Eri). Tokovinin [177] estimates the period of A-BC to be about 6100 years, whilst the recent orbit by Mason *et al.* [201]

Finder Chart

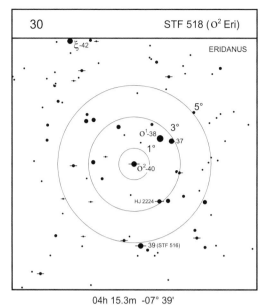

30	STF 518 (o² Eri)

ERIDANUS

ξ-42

5°

3°

o¹-38
37

1°

o²-40

HJ 2224

39 (STF 516)

04h 15.3m -07° 39'

Orbit

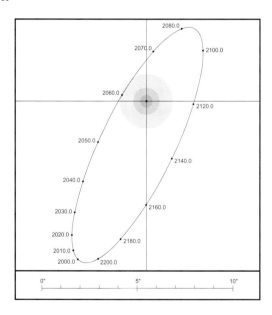

Ephemeris for STF 518 BC (2015 to 2060)

Orbit by Msn (2017a) Period: 224.28 years, Grade: 3

Year	PA(°)	Sep(")	Year	PA(°)	Sep(")
2015.0	331.7	8.23	2040.0	320.7	5.29
2020.0	330.1	7.86	2045.0	316.3	4.34
2025.0	328.4	7.39	2050.0	309.1	3.28
2030.0	326.3	6.81	2055.0	294.7	2.17
2035.0	323.9	6.11	2060.0	256.2	1.31

gives a periodic time of 230.30 years for BC. The stars were widest in 1995 at 8″.9 and will close up to 1″.3 in 2060. The stars in the BC system have been examined for any sign of exoplanets [199]. The angular diameter of the star was determined by Boyarjian [200] and colleagues and found to be 1.504 ± 0.006 mas, equivalent to 0.8 R⊙. The Gaia DR2 data gives a parallax for stars A and C (but not B) which is in close agreement with the results from Hipparcos.

Observing and Neighbourhood

Starting at Rigel, move the telescope 15° due W and you will alight on a naked-eye pair of stars called o² and, 70′ further W and N, o¹ Eri. The star o¹ is an F giant of visual magnitude 4.0 whilst its neighbour is a an orange star at $V = 4.4$. The BC pair is located 83″ away in PA 107°. Hartung notes that both B and C can be seen in 7.5-cm and that the separation is favourable for the smaller aperture for at least the next 20 years or so. Observers in northern Europe are at a disadvantage because of the low declination. Any attempt to use field illumination in the 20-cm at Cambridge renders star C invisible. Moving 3° due S of this star brings you to 39 Eri (STF 516), magnitudes 5.0 and 8.5, at 144° and 6″.3, 2011, with the primary an early K giant. One point five degrees N and slightly W of 39 is HJ 2224, whose components are of magnitudes 6.6 and 9.8 at 306° and 57″, 2011.

Measures

BC			
Early measure (STT)	158°.0	4″.11	1855.06
(Orbit	153°.6	4″.06)	
Recent measure (WSI)	332°.8	8″.68	2010.72
(Orbit	333°.8	8″.48)	
A-BC			
Early measure (STF)	107°.3	83″.48	1836.04
Recent measure (FYM)	103°.7	83″.49	2011.89

31. 80 TAU = STF 554 = WDS J04301+1538

Table 9.31 Physical parameters for 80 Tau

STF 554	RA: 04 30 08.60	Dec: +15 38 16.2	WDS 301(175)	
V magnitudes	A: 5.70	B: 8.12	C(81): 5.45	
$(B-V)$ magnitudes	A: +0.33	B: +0.65	C(81): +0.28	
μ(A)	108.92 mas yr^{-1}	± 0.25	−22.14 mas yr^{-1}	± 0.16 (DR2)
π(A)	21.12 mas	± 0.12	154.43 light yr	± 0.88 (DR2)
μ(B)	102.57 mas yr^{-1}	± 0.17	−31.43 mas yr^{-1}	± 0.09 (DR2)
π(B)	20.93 mas	± 0.07	155.84 light yr	± 0.52 (DR2)
μ(81)	103.43 mas yr^{-1}	± 0.25	−24.49 mas yr^{-1}	± 0.17 (DR2)
π(81)	21.36 mas	± 0.14	153 light yr	± 1 (DR2)
Spectra	A: F0V	B: G0V	C: Am	
Masses (M$_\odot$)	A: 0.86	B: 0.70		
Luminosities (L$_\odot$)	A: 10	B: 1		
Catalogues (A/C)	HR 1422/8	HD 28485/546	SAO 93970/978	HIP 20995/21039
DS catalogues	STF554	BDS 2230	ADS 3264	
Radial velocity	29.3 km s^{-1}	± 2.3		
Galactic coordinates	180°.784	−21°.878		

History

One of F. G. W. Struve's discoveries at Dorpat, 80 Tau, is a member of the Hyades and is also called van Bueren 80. Admiral Smyth observed it at Bedford with his 5.9-inch refractor and found it 'of no very easy measurement'; between 1837 and 1843 his observations showed an increase in position angle, but the orbital motion is, in fact, retrograde. The pair then closed steadily, Dembowski finding it at 1″.1 in 1862. It was visually unresolved in the large refractors between 1888 and 1900 and then seen at 0″.6 by See in 1900, having traversed 300° of position angle in 20 years.

Finder Chart

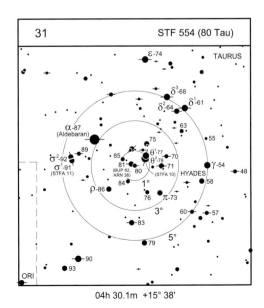

04h 30.1m +15° 38'

Orbit

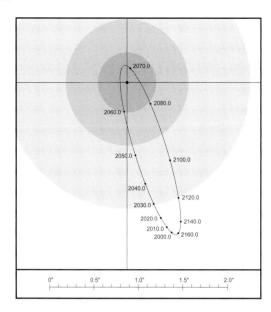

Ephemeris for STF 554 (2015 to 2060)

Orbit by Baz (1980a) Period: 180.0 years, Grade: 4

Year	PA(°)	Sep(")	Year	PA(°)	Sep(")
2015.0	15.1	1.59	2040.0	10.5	1.13
2020.0	14.4	1.53	2045.0	8.9	0.98
2025.0	13.6	1.45	2050.0	6.7	0.80
2030.0	12.7	1.36	2055.0	3.1	0.59
2035.0	11.7	1.25	2060.0	354.2	0.32

The Modern Era

The primary star is a spectroscopic binary with a period of 30.5 days according to the WDS but it does not appear in SB9 [150]. It has so far resisted attempts to resolve it by Hartkopf & McAlister [202] and McAlister *et al.* [482]. The pair 80 Tau shares a common proper motion with 81 Tau, which is 485″ distant and which is a spectroscopic binary.

Observing and Neighbourhood

This is not an easy object to resolve. It needs a night of good seeing and an aperture of 15-cm. It is 1.5° from Aldebaran but such is the profusion of bright stars that RWA occasionally has difficulty in identifying it. The finder chart makes it clear. If the rather preliminary orbit by Baize is correct, the pair will remain near 1″.5 for the next decade but then will close rapidly, reaching 0″.09 in 2065. The star 81 Tau is BUP 62 (5,5, 8.9, 339°, 164″, 2014) whilst 80 and 81 Tau together form ARN 36 AC (5.5, 5.7, 246°, 485″, 2014). Aldebaran (magnitude 0.9) is 1.5° to the NE. This is a challenge for the smaller-aperture user. There are several faint, distant stars in the field, the brightest of which is C (magnitude 11.3). This star is now 135″ away from Aldebaran and the distance is getting wider thanks to the proper motion of the bright primary. Burnham (BU 1031) doubled C (the companion is of magnitude 13.7) and by 1962 it was only 1″.3 apart but has not been measured since then. There are several good binocular doubles in the wider field: θ Tau (STFA 10) (3.4, 3.9, 339°, 348″, 2016) and σ Tau (STFA 11) (4.7, 5.1, 194°, 444″, 2014), for example.

Measures

Early measure (STF)	12°.9	1″.74	1831.18
(Orbit	15°.3	1″.63)	
Recent measure (ARY)	12°.6	1″.61	2013.59
(Orbit	15°.9	1″.61)	

32. 2 CAM = STF 566 = WDS J04400+5328AB,C

Table 9.32 Physical parameters for 2 Cam

STF 566	RA: 04 39 58.03	Dec: +53 28 23.7	WDS: 338(163)	
V magnitudes	A: 5.56	B: 7.35	C: 7.49	
$(B - V)$ magnitudes	AB: +0.35	C: +0.54		
μ	+34.51 mas yr^{-1}	± 4.19	−84.70 mas yr^{-1}	± 3.66
π	22.49 mas	± 4.69	145 light yr	± 30
Spectra	A: A8V	B: ?	C: F9V	
Masses	A: 2.3	B: 1.5	C: 1.6	
Luminosities (L$_\odot$)	A: 9	B: 2	C: 1	
Catalogues (A/B)	HR 1466	HD 29316	SAO 24744	HIP 21730
DS catalogues	STF 566 (AB,C)	BU 1295 (AB)	BDS 2279	ADS 3358
Radial velocity	20.10 km s^{-1}	± 3.2 km s^{-1}		
Galactic coordinates	153°.031	+4°.528		

History

AC was discovered by Struve during his great survey at Dorpat. The pair showed slow direct motion through the nineteenth century and then started to close during the twentieth century. Over the last few years it has reached 0″.8, and the orbit by the USNO astronomers shows that it is gradually widening and will reach a maximum separation of just over 2″.3 around 2200. The D component was discovered by Burnham in 1888 with the 36-inch refractor. At that time it was 23″.8 distant from AB. A measure in 2013 places it 18″.8 away (in PA 243°) owing to the proper motion of the multiple system ABC. Using the 40-inch Yerkes refractor in 1901, Burnham found that AB was a very close and unequal pair (BU 1295); he does not appear to have made any measures himself, but subsequent measures listed in the ADS seemed to indicate that direct, if rather slow, orbital motion was taking place. In fact between 1902 and 1921 few measures were made and it is now clear that the pair

Finder Chart

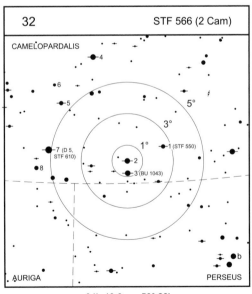

04h 40.0m +53° 28'

Orbit

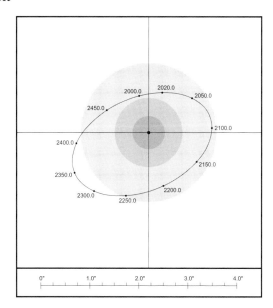

Ephemeris for STF 566 AB,C (2015 to 2060)

Orbit by Cve (2008a) Period: 480.75 years, Grade: 4

Year	PA(°)	Sep(")	Year	PA(°)	Sep(")
2015.0	168.4	0.80	2040.0	136.7	1.03
2020.0	160.8	0.84	2045.0	132.0	1.08
2025.0	153.8	0.89	2050.0	127.7	1.12
2030.0	147.5	0.93	2055.0	123.7	1.16
2035.0	141.8	0.98	2060.0	119.9	1.19

underwent almost a whole revolution in that time. It is a little strange that Burnham did not see the close pair in 1888, whilst he was measuring STF 566, as the predicted separation was 0″.24, somewhat wider than Hussey's pioneering measure of 0″.17 in 1901. The latest ephemeris by Heintz shows that the maximum separation is less than 0″.3; the next time maximum separation happens will be around 2030, but it is the difference in magnitude between the two stars that probably takes this system out of the realm of amateur apertures.

The Modern Era

The distance to the system is still rather uncertain. The Hipparcos (HIP) parallax has a large error, no doubt caused by the complex nature of the star and the motions within the multiple system. According to Tokovinin [177] in his MSC catalogue the dynamical parallax is considerably smaller than the HIP2 parallax. Gaia DR2 offers no further insights – the close AB pair needs further processing and so, presumably, will not appear until DR3. Heintz [205] found a period of 26.7 years for this system. The separation ranges from 0″.04 (2016) to 0″.28 (2028).

Observing and Neighbourhood

The components AB,C form a good test for 20-cm and were measured by RWA in 2013, with the results shown below. The system now appears to be slowly widening and will get very gradually easier over the next century or two. Many observers, including Struve, Secchi and Smyth, have made the colours of the stars yellow and blue, whilst Dembowski thought them blue and ashy. Whilst in the area, it is worth looking at 1 Cam (STF 550). Found just 1°.5 WNW of 2 Cam, this pair of stars, of magnitudes 5.78 and 6.82, is a beautiful pair for any aperture. Admiral Smyth called them white and sapphire blue whereas Lewis, reflecting F. G. W. Struve's colours, noted white and bluish-white. Another neat pair nearby was discovered by Dembowski in 1867 [203]. RWA measured D 4 on 2013.30 at 262° and 6″.86. The brightest components are magnitudes 9.0 and 10.3. It is only 230″ following 2 Cam, but it is not physically connected to that system according to the WDS. The object D 5 is the primary star of 7 Cam, a difficult pair (4.5, 7.9, 202°, 0″.6, 2011) – closing, possibly optical. Another star of magnitude 11.3 at 242°, 26″ forms STF 610. The pair 3 Cam is also a difficult double (BU 1043, magnitudes 5.2, 12.3, 295°, 3″.7, 1916). Star A is a W UMa eclipsing binary, according to the WDS. SIMBAD calls it a Cepheid. John Percy [204] suspected a periodic variation of 0.05 magnitudes in a report published in 1975. The AAVSO website notes that it is an SB with an orbital period of 121 days.

Measures

AB-C			
Early measure (STF)	310°.9	1″.65	1831.31
(Orbit	311°.0	1″.61)	
Recent measure (ARY)	169°.7	0″.89	2013.30
(Orbit	172°.0	0″.83)	

33. 14 ORI = STT 98 = WDS J05079+0830

Table 9.33 Physical parameters for 14 Ori

STT 98	RA: 05 07 52.89	Dec: +08 29 54.3	WDS: 52(474)		
V magnitudes	A: 5.76	B: 6.67	P: 9.44	Q: 9.51	
$(B - V)$ magnitude	A: +0.26	B: +0.53	P:	Q:	
π:	16.84 mas	± 1.32	194 light yr	± 15 (Davidson)	
μ(A)	30.35 mas yr^{-1}	± 0.45	−57.04 mas yr^{-1}	± 0.35 (DR2)	
π(A)	16.02 mas	± 0.31	203.6 light yr	± 3.9 (DR2)	
μ(P)	23.42 mas yr^{-1}	± 0.08	−61.66 mas yr^{-1}	± 0.06 (DR2)	
π(P)	18.71 mas	± 0.05	174.3 light yr	± 0.5 (DR2)	
μ(Q)	18.17 mas yr^{-1}	± 0.11	−60.39 mas yr^{-1}	± 0.09 (DR2)	
π(Q)	18.59 mas	± 0.05	175.5 light yr	± 0.5 (DR2)	
Spectra	A: F0IV	B: F5V			
Masses (M$_\odot$)	Aab: 1.82	B: 1.46	P: 0.9	Q: 0.9	
Luminosities (L$_\odot$)	A: 15	B: 7	P: 0.4	Q: 0.4	
Catalogues	14,i Ori	HR 1664	HD 33054	SAO 112440	HIP 23879
DS catalogues	STT 98	BDS 2535	ADS 3711		
Radial velocity (A/B)	10.70 km s^{-1}	± 0.3 km s^{-1}			
Galactic coordinates	192°.672	−18°.478			

History

This object was found by Otto Struve in 1843 with the Pulkovo 15-inch refractor.

The Modern Era

Eggen [207] notes that the nearby pair STF 643 (PQ in the above table) has a similar proper motion to STT 98. This pair is 6′ S and the WDS magnitudes are 9.6 and 9.6. The current PA and separation of STF 643 are 124° and 2″.4, so this probably needs at least 20-cm to be clearly resolved. Tokovinin regards STF 643 as a physical member of the group, which makes the star quintuple, since the WDS notes that one component, possibly A, is a spectroscopic binary. DR2 shows that there is a statistically significant difference in parallax between 14 Ori and the components of STF 643. Recent speckle observations of 14 Ori AB by Davidson et al. [206] concluded that the mass of the system derived from photometry is 3.28 ± 0.28 M$_\odot$ but that when the dynamical parallax is considered the mass sum is 5.45 ± 1.28 M$_\odot$, and this does not agree with their conclusion that the spectra are expected to be F0 and F5 with the primary possibly slightly evolved. The Davidson et al. [206] paper has given an improved value of the distance to the system.

Finder Chart

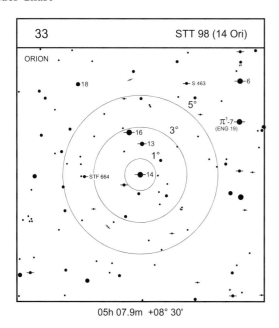

| 33 | STT 98 (14 Ori) |

ORION

05h 07.9m +08° 30'

Observing and Neighbourhood

The pair 14 Ori has been followed for a large portion of the orbit, which is relatively open. The separation varies between $0''.7$ and $1''.1$ and so will be within reach of 15-cm for much of the time and certainly for the next century or so. 14 Ori lies between γ Orionis (Bellatrix) and a vertical stream of stars, all of which are called π Orionis and represent the Hunter's arm. Two of these, π^1 and π^3, are coarse doubles. The pair π^1 Ori was found by Engelmann (ENG 19); the primary is magnitude 5.7 and it has a magnitude 9.9 companion at $254°$ and $172''$, 2011. Three degrees NW is S 463 (7.2, 10.1, $29°$, $32''.7$, 2013), whilst $2°$ directly E is STF 664 (7.8, 8.4, $177°$, $4''.7$, 2011).

Measures

Early measure (STT)	$248°.9$	$1''.12$	1844.53
(Orbit	$254°.3$	$1''.09$)	
Recent measure (ARY)	$290°.6$	$1''.02$	2017.16
(Orbit	$287°.4$	$0''.96$)	

Orbit

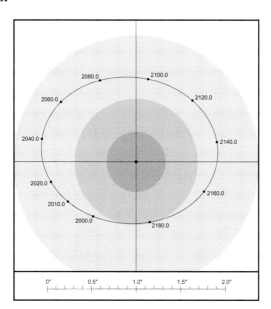

Ephemeris for STT 98 (2015 to 2060)

Orbit by Sca (2008d) Period: 197.45 years, Grade: 2

Year	PA(°)	Sep(")	Year	PA(°)	Sep(")
2015.0	290.8	0.94	2040.0	256.7	1.09
2020.0	283.0	0.98	2045.0	250.6	1.09
2025.0	275.8	1.02	2050.0	244.6	1.09
2030.0	269.1	1.06	2055.0	238.4	1.08
2035.0	262.8	1.08	2060.0	232.1	1.06

34. β ORI = RIGEL = STF 668 = WDS J05145–0812AB

Table 9.34 Physical parameters for β Ori

STF 668	RA: 05 14 32.27	Dec: −08 12 05.9	WDS: 526(120)		
V magnitudes	A: 0.13	B: 6.8			
$(B - V)$ magnitudes	A: −0.03	B: 0.0?			
μ(A)	1.31 mas yr^{-1}	± 0.40	0.50 mas yr^{-1}	± 0.30	
π(A)	3.78 mas	± 0.34	860 light yr	± 80	
μ(B)	−0.40 mas yr^{-1}	± 0.14	−0.15 mas yr^{-1}	± 0.12 (DR2)	
π(B)	2.92 mas	± 0.08	1120 light yr	± 31 (DR2)	
Spectra	A: B8Iae	B: B9V + B9V?			
Masses (M$_\odot$)	A: 23 M$_\odot$				
Radii (R$_\odot$)	A: 79	± 0.3			
Luminosities (L$_\odot$)	A: 43000	B: 185			
Catalogues (A)	19 Ori	HD 34085	HR 1713	SAO 131907	HIP 24436
DS catalogues	H 2 33 (AB)	STF 668 (AB)	BU 555 (BC)	BDS 2605	ADS 3823
Radial velocity (A)	17.80 km s^{-1}	± 0.4			
Galactic coordinates (A)	209°.241	−25°.245			

History

This binary was found by William Herschel on 1 October 1781. He noted 'Extremely unequal. L(arge). w(hite). S(mall) inclining to (r)ed. With 227 2$\frac{1}{4}$ or 2$\frac{1}{2}$ diameters of Rigel. With 460, more than 3 diameters of Rigel'. He never attempted to measure the distance to the magnitude 6.8 companion star but it seems that since then the distance between A and B has changed very little in any case. The B star was also independently found by Fearon Fallows at the Cape of Good Hope around 1830. In 1846, using the Cincinnati 11-inch refractor, O. M. Mitchel noted a very faint star 44″ distance from the primary, which was recovered independently by Burnham, who made it magnitude 13.5. The WDS gives V = 15.4. In 1875 Burnham noted an elongation in B using his 6-inch Clark refractor, thus initiating a mystery that has survived to this day. Details can be found in Chapter 7.

Finder Chart

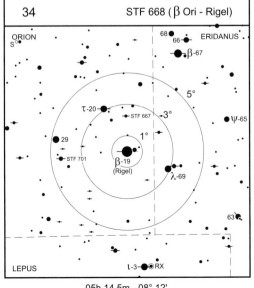

05h 14.5m -08° 12'

The spectroscopic binary nature of the companion was reported by Plaskett [208] in 1909. In 1942, R. Sanford [211] wrote a paper detailing the nature of the orbit. He found that it was a double-lined spectroscopic binary with a period of 9.86 days.

The Modern Era

Rigel is a blue supergiant with a mass conservatively estimated to be 23 $M_\odot$ and consequently has very high luminosity. Assuming the Hipparcos parallax is approximately correct, the distance to the star is 860 light years and the bolometric absolute magnitude is −11.8 (assuming no interstellar absorption), equivalent to 120,000 $L_\odot$. In 1977 Blazit *et al.* [210] reported that Rigel was a 'possible binary' at six different wavelengths, using the 200-inch reflector at Palomar in 1975. The approximate separations were around 0″.04 but no confirmation of this has ever been given. The effect of such a relatively distant companion would probably not make itself apparent in changes in Rigel's radial velocity. An intense study, resulting in more than 2300 spectra, taken with the 2-metre telescope at the University of Tennessee shows variations in radial velocity between 11 and 25 km s^{-1} and that these are typical of a multi-periodic pulsating star. Gaia is expected to deliver results even on first magnitude stars, but the complexity of the reduction procedure may relegate these stars to the final iteration of the catalogue (2022). The companion does seem to have been reliably pinned down to a distance of about 1120 light years by Gaia DR2; this, if the companion is physically connected to Rigel, will increase the luminosity of the B8 supergiant, which corresponds to the Hipparcos parallax.

Observing and Neighbourhood

The companion should be visible in 7.5-cm on any reasonably dark and clear night. The Reverend T. W. Webb [209], in his book, describes successful attempts to see it with apertures as small as 1.25 inches. The WDS catalogue gives the magnitude of B as 6.8, whilst the MSC lists 7.0. The *V* magnitude given in SIMBAD (10.2) is clearly far too faint. It certainly seems to be between 7 and 8 to RWA. Colours of bluish and blue-white were recorded by members of the Astronomical Society of New South Wales (ASNSW) Double Star Section. There are two other Struve pairs nearby. Directly N about 1°.5 is STF 667 (7.2, 8.8, 315°, 4″.2, 2004), and 2° E is STF 701 (6.1, 8.1, 139°, 6″.3, 2016).

Measures

Early measure (STT)	188°.0	9″.14	1831.53
Recent measure (ARY)	203°.5	9″.43	2017.16

35. STF 634 CAM = WDS J05226+7914AB

Table 9.35 Physical parameters for STF 634 Cam

STF 634	RA: 05 22 33.53	Dec: +79 13 52.1	WDS: 797(90)		
V magnitudes	A: 5.14	B: 9.14			
$\mu(A)$	-78.39 mas yr^{-1}	± 0.23	162.12 mas yr^{-1}	± 0.30 (DR2)	
$\mu(B)$	51.91 mas yr^{-1}	± 0.19	-156.20 mas yr^{-1}	± 0.29 (DR2)	
$\pi(A)$	47.70 mas	± 0.17	68.4 light yr	± 0.2 (DR2)	
$\pi(B)$	5.17 mas	± 0.19	631 light yr	± 23 (DR2)	
Spectra	F6Ve				
Masses (M$_\odot$)	A: 1.25				
Luminosities (L$_\odot$)	A: 3	B: 7			
Catalogues	19H Cam	HR 1686	HD 33564	SAO 5486	HIP 25110
DS catalogues	STF 634	BDS 2548	ADS 3864		
Radial velocity	-10.8 km s^{-1}	± 0.3			
Radial velocity (A/B)	-11.03 km s^{-1}	± 0.23	-46.11 km s^{-1}	± 0.35 (DR2)	
Galactic coordinates	133°.735	+22°.648			

Introduction

With its location only 10° from the Celestial Pole, STF 634 is an awkward star to find for the equatorially-mounted-refractor user but it can be found by locating the pretty, wide pair β Cam and then moving due N by two-thirds of the distance to Polaris.

History

This star appears in the catalogue of Hevelius as 19 Cam and the nomenclature was carried forward by Admiral Smyth, but when transferred to Webb's *Celestial Objects for Common Telescopes* it became P IV 269. In the sixth version of *Celestial Objects* it was simply called 19 but should not be confused with Flamsteed 19 Cam, which is a different star; as luck would have, that star is also double but much more difficult to resolve. Smyth considered that its observed motion up to 1840

Finder Chart

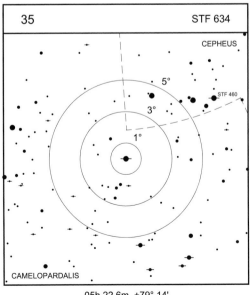

05h 22.6m +79° 14'

represented orbital motion with a period of 'not less than 1000 or 1200 years'. In fact the stars are completely unassociated and the relative motion is due to the large proper motions of A and B, each amounting to $0''.18$ per year and in virtually opposite directions on the sky.

The Modern Era

The stars are now separating quite quickly. The closest approach occurred in 1929 when they were $9''.55$ apart in position angle $66°$. Gaia DR2 indicates that the companion is twice as far away, as measured by Hipparcos, and the radial velocities confirm that the two stars are totally unconnected.

Exoplanet Host?

Observations in 2005 using the Élodie echelle spectrograph on the 1.9-metre reflector at Haute-Provence [212] revealed that there was a variation in the radial velocity of the primary star with a period of 388 days, due to the presence of an exoplanet whose minimum mass is estimated at 9.1 Jupiter masses assuming that the mass of the host star is 1.25 $M_\odot$.

Observing and Neighbourhood

This is a fine, if very unequal, pair for the small telescope. Smyth noted colours of light yellow and pale blue though the companion is sufficiently faint for its colour not to be obvious. There is a good resolution test for a 15-cm aperture $3°.5$ to the NW. This is the 372-year orbital pair STF 460 with components of magnitudes 5.6, 6.3, to be found at $159°$, $0''.67$, 2020.

Measures

Early measure (STF)	$348°.2$	$34''.53$	1832.10
(Linear prediction	$347°.8$	$34''.59$)	
Recent measure (ARY)	$140°.8$	$30''.32$	2013.30
(Linear prediction	$138°.7$	$30''.07$)	

36. η ORI = DA 5 = WDS J05245–0224AB

Table 9.36 Physical parameters for η Ori

DA 5	RA: 05 24 28.62	Dec: −02 23 49.7	WDS: 174(240)		
V magnitudes	A: 3.37	B: 4.71	C: 11.0		
$(B − V)$ magnitudes	A: +0.20	B: +0.21			
μ	A: −0.71 mas yr^{-1}	± 1.13	−3.46 mas yr^{-1}	± 0.85	
π	3.34 mas	± 1.03	980 light yr	± 55	
μ(A)	3.26 mas yr^{-1}	± 1.00	−2.36 mas yr^{-1}	± 0.78 (DR2)	
π(A)	2.23 mas	± 0.50	1460 light yr	± 330 (DR2)	
Spectra	Aa: B1V	Ab: B3V	Ac: B3V	B: B2V	
Masses (M$_\odot$)	Aa: 9.8	Ab: 5.9	Ac: 6.8	B: 8.0	
Luminosities (L$_\odot$)	A: 7400	B: 2200			
Catalogues	28 Ori	HR 1788	HD 35411	SAO 132071	HIP 25281
DS catalogues	H 6 67 (AC)	DA 5 (AB)	MCA 18 (AaAb)	BDS 2712	ADS 4002
Radial velocity	19.8 km s^{-1}	± 0.9			
Galactic coordinates	204°.866	−20°.392			

History

On 27 December 1781, William Herschel observed η Orionis and noted 'Double. Excessively unequal' and gave the separation as 110″57‴. In 1848 W. R. Dawes [213] was also looking at η and he noted 'On 15 January 1848, η Orionis was discovered to be a close double, 1″; the magnitudes of the components being 4 and 5, and distance 1″. It is a beautiful object of its class and can scarcely fail to prove a *binary* star.' He also pointed out that neither Herschel nor Struve had noticed it. In 1901, spectra taken at Yerkes Observatory showed that the radial velocity of η Orionis was varying. W. S. Adams [215] analysed the plates and derived a period of 7.9 days.

Finder Chart

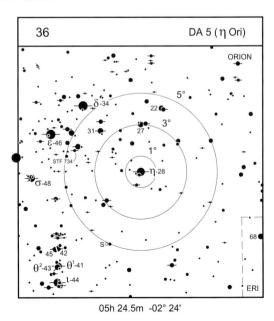

05h 24.5m -02° 24'

The Modern Era

Ironically, considering Dawes' comment, it now appears that the two bright visual components may not be connected. Hartkopf *et al.* [214] calculated a linear ephemeris for the stars which shows a close approach of 0″.39 in late 1675 and, at the time that William Herschel was observing, the separation of AB would have been around 0″.67, sufficiently close for Herschel to have overlooked the pair. However, it is not impossible that the stars are connected and that the apparent orbit is both long and tilted near the line of sight. The amount of apparent motion to date is simply not sufficient to be certain, except to say that the linear ephemeris seems to be a good fit to the data. Even ignoring the status of component B, η Orionis is a massive quadruple star. The brighter visual component A is a tight pair (Aa and Ab) rotating around a third star, Ac, in 9.4 years. The orbital period of Aa, Ab is 7.99 days. The Ac component itself shows radial velocity variations which may be due to β Cepheid-like pulsations on a timescale of 0.864 days or to another component. Rather surprisingly there is no data for B in Gaia DR2 even though the stars are almost 2″ apart.

Observing and Neighbourhood

The system η Ori is 3° from the westernmost star of Orion's Belt and about 5° from M42. There is an attractive triple star 2°.5 ENE in the shape of STF 734 (H 5 119). Herschel found the 8.4 magnitude companion to the 6.7 magnitude primary, which is now 29″ distant in PA 244°, 2014. It was left to F. G. W. Struve to add a close companion of magnitude 8.2, which is now 1″.7 distant in PA 357°, 2008. Later still S. W. Burnham doubled C (BU 1049) but since discovery the pair has closed to below 0″.5 and has not been measured since 2002.

Measures

Early measure (DA)	86°.8	0″.94
(Linear	87°.4	0″.95
Recent measure (TOK)	76°.5	1″.81
(Linear	77°.3	1″.80)

37. θ PIC = Δ 20 = WDS J05248–5219AB,C

Table 9.37 Physical parameters for θ Pic

DUN 20	RA: 05 24 46.29	Dec: −52 18 58.2	WDS: 2728(37) (AB)	
V magnitudes	A: 6.8	B: 7.4	C: 6.84	
$(B − V)$ magnitudes	A: +0.08	B: +0.08		
μ(AB)	−4.00 mas yr^{-1}	± 0.27	−27.79 mas yr^{-1}	± 0.39
π(AB)	6.36 mas	± 0.32	513 light yr	± 26
μ(C)	−7.19 mas yr^{-1}	± 0.07	−28.14 mas yr^{-1}	± 0.06 (DR2)
π(C)	6.30 mas	± 0.03	517.7 light yr	± 2.5 (DR2)
Spectra	Aa: A1III	Ab: Am	B: A2V?	C: A0V
Masses (M$_\odot$)	Aa: 3.1	Ab: 2.4	B: 2.3	C: 2.8
Luminosities (L$_\odot$)	A: 40	B: 20	C: 40	
Catalogues (AB/C)	HR 1818	HD 35860/59	SAO 233965/4	HIP 25303/298
DS catalogues	DUN 20 (AB,C)	I 345 (AB)		
Radial velocity (AB/C)	−3.1 km s^{-1}	± 0.8	−8.20 km s^{-1}	± 3.7
Galactic coordinates	259°.563	−34°.176		

History

The wide pair in θ Pictoris was found by Dunlop [12] at Parramatta, and his measures indicate that there has been virtually no relative motion between the stars since then. In 1901 Robert Innes [527] found that A was a close pair (I 345).

The Modern Era

The visual binary was observed occasionally during the twentieth century and the stars continued to close, reaching 0″.1 in 1993. There were no further observations until 2008, when the position angle had decreased a further 170°. Hipparcos verified that the proper motions and parallaxes agree closely. Verramendi *et al.* [462] observed the main components and found that the brighter component of I 345 was a double-lined spectroscopic binary. They found the spectral types to be A1III

Finder Chart

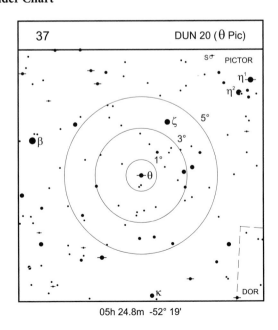

Orbit

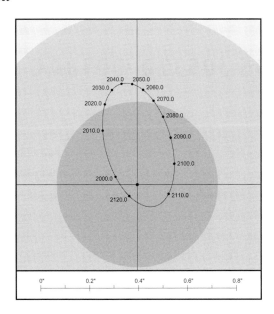

Ephemeris for I 345 AB (2015 to 2060)

Orbit by Alz (2010) Period: 123.2 years, Grade: 3

Year	PA(°)	Sep(")	Year	PA(°)	Sep(")
2015.0	207.1	0.31	2040.0	188.9	0.41
2020.0	202.4	0.34	2045.0	185.9	0.41
2025.0	198.6	0.37	2050.0	182.9	0.40
2030.0	195.1	0.39	2055.0	179.7	0.39
2035.0	192.0	0.40	2060.0	176.3	0.38

and Am but were surprised not to see any sign of the B star. This may be due to rotational broadening or it could be due to another close pair of stars. The radial velocity of C is the same as that of the AB system but it was found to vary with a range of 11 km s^{-1}. The C star is a fast rotator ($v \sin i = 200$ km s^{-1}), and it is not clear, from the very broadened spectral lines, whether this variation is due to orbital motion or just scatter. Thus θ Pic is certainly a quadruple system and may be sextuple.

Observing and Neighbourhood

Andrew James says that the bright stars appear more yellow than the spectral types would appear to suggest. This corroborates Hartung, who calls the stars 'pale yellow'. The stars in the close visual binary that is A are now slowly widening to a maximum of 0″.4 around 2040. To divide them at present will require about 35-cm but 30-cm should show an elongated image. Hipparcos found the stars to differ by 0.7 magnitude in V.

Measures

AB			
Early measure (I)	197°.2	0″.45	1901.91
(Orbit	197°.6	0″.33)	
Recent measure (TOK)	207°.6	0″.30	2014.04
(Orbit	208°.2	0″.30)	
AB-C			
Early measure (HJ)	285°.9	38″.20	1836.40
Recent measure (ANT)	287°.9	38″.3	2008.75

38. δ ORI = STFA 14 = WDS J05320−0018AC

Table 9.38 Physical parameters for δ Ori

STFA 14	RA: 05 32 00.40	Dec: −00 17 56.7	WDS: 1036(75)		
V magnitudes	A: 2.23	B: 3.8	C: 6.83		
$(B-V)$ magnitudes	A: −0.39	B: ?	C: −0.16		
μ	+0.64 mas yr^{-1}	± 0.56	−0.69 mas yr^{-1}	± 0.27	
π	4.71 mas	± 0.58	693 light yr	± 85	Hipparcos
π	2.63 mas	—	1239 light yr		(Shenar *et al.* [223])
μ(C)	1.64 mas yr^{-1}	± 0.13	−1.76 mas yr^{-1}	± 0.12 (DR2)	
π(C)	2.57 mas	± 0.08	1270 light yr	± 40 (DR2)	
Spectra	Aa1: O9II	Aa2: ?	Ab: ?	Ca: B2V	Cb: ?
Masses (M$_\odot$)	Aa1: 19.6	Aa2: 17.7	Ab: 19.6	Ca: 6.0	Cb: 3.0
Luminosities (L$_\odot$)	A: 15300	B: 3600	C: 250		
Catalogues (A/C)	34 Ori	HR 1852/1	HD 36486/5	SAO 132220/1	HIP 25930
DS catalogues	Mayer 15 (AC)	H 5 10 (AC)	STFA 14 (AC)	HEI 42 (AaBb)	BDS 2796
	ADS 4134				
Radial velocity (A/B)	18.50 km s^{-1}	± 0.5	21 km s^{-1}	± 5	
Galactic coordinates	207°.855	−17°.740			

History

William Herschel recorded the distant sixth magnitude companion in his first published catalogue, observing the pair on 6 October 1779 and noting that the large star was white and the smaller bluish, distance 52″.968. F. G. W. Struve included this pair in his first appendix catalogue (as ΣI 14 – now STFA 14 in the WDS), whilst T. W. Webb noted in 1859 that John Herschel thought the bright star varied between magnitudes 2.2 and 2.7. In 1892, S. W. Burnham [219], observing the pair with the 18.5-inch refractor at Dearborn Observatory, noted a very faint and distant companion to A some 38″ away and of magnitude 14. In his *General Catalogue* of 1906, Burnham notes that the bright star was found to be an SB with period 1.92 days by Delandres [221]. The star was then observed by Hartmann [225] at Potsdam, and in 1904 he found that the true period was 5.73 days. At the same time he discovered a single line in the spectrum of δ which did not move in concert with the stellar lines as the two stars moved in their orbit. He correctly postulated that this was due to an intervening cloud of material lying between us and the star. In 1978 Wulff Heintz [549], observing visually with the 24-inch Sproul refractor, found that δ was resolved into two unequal stars at a distance of 0″.15, now known as HEI 42. However, this is not the spectroscopic component: the star has moved only 20° in the interval between discovery and this publication.

Finder Chart

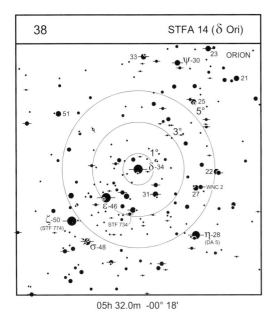

38   STFA 14 (δ Ori)

05h 32.0m -00° 18'

The Modern Era

The system δ Orionis is a group of massive, luminous, hot, and rapidly rotating O and B stars. The distance to this system is still rather uncertain. The Hipparcos parallax, which equates to a distance of 212 pc, has a nominal error of about 15% and the complex nature of star A no doubt contributes to this uncertainty. However, Shenar *et al.* [223] argue that the masses which they derive in their investigation make sense only if the true distance to δ is much greater, nearer the value 380 pc for the massive σ Orionis cluster. The HST and the STIS spectrograph have been used to obtain radial velocity curves of the close P = 5.7 day binary for the first time (Richardson *et al.* [224]. Such is the breadth of the lines due to fast rotation, though, that only the largest radial velocity shifts can be clearly measured, at the extrema of the orbit.

The orbit of this pair is also edge-on to the line of sight, leading to Algol-like eclipses. The primary eclipse is about 0.1 magnitude whilst the secondary eclipse is 0.04 magnitude. The rotation period of HEI 42 has been estimated at 326 years. Interpolating back to 1892, the stars would have been just 0″.21 apart, which possibly explains why Burnham did not see them at that time. Star C is also an SB2 and Tokovinin [177] gives a period of 29.96 days for C. The brightness and complexity of δ Ori A has kept it from inclusion in DR2 but the C component is included and appears to be at a very similar distance to the A star as that found by Shenar.

Observing and Neighbourhood

Easily found as the westernmost star in the Belt of Orion, δ is almost on the celestial equator and hence has often provided a means of determining telescopic eyepiece fields by allowing it to drift from one side of the field of view to the other, noting the time taken in seconds, and dividing by 4 to get the eyepiece field in arcminutes. The relatively small motion between the stars A and C also mean that this star can be used as an astrometric calibrator. The current separation is 52″.4 and the position angle is 0°. Another fine binary can be found in ζ = STF 774 (see Star 39), the easternmost star in the belt. Also worth a look is WNC 2 – a binary pair – magnitudes 6.9, 7.0 at 158° and 3″.1 (2020). One point five degrees SSE is STF 734, a fine triple in a beautiful field (6.7, 8.2, 357°, 1″.7, 2008) with magnitude 8.4 at 244°, 29″, 2014.

Measures

| Early measure (STF) | 359°.2 | 52″.58 | 1835.75 |
| Recent measure (SMR) | 0°.3 | 52″.7 | 2010.17 |

39. ζ ORI = STF 774 = WDS J05408−0156AB

Table 9.39 Physical parameters for ζ Ori

STF 774	RA: 05 40 45.52	Dec: −01 56 33.3	WDS: 154(257)		
V magnitudes	Aa: 2.0	Ab: 4.0	B: 3.7	C: 9.6	
$(B - V)$ magnitudes	Aab: −0.11	B: −0.20			
μ	3.19 mas yr^{-1}	± 0.59	2.03 mas yr^{-1}	± 0.26	
π	4.43 mas	± 0.64	736 light yr	± 106	Hipparcos
π	2.58 mas	± 0.36	1264 light yr	± 176	Photometric
μ(C)	2.36 mas yr^{-1}	± 0.11	−2.02 mas yr^{-1}	± 0.10 (DR2)	
π(C)	2.62 mas	± 0.07	1245 light yr	± 33 (DR2)	
Spectra	A: O9.5Ibe	Ab: B0.5IV	B: B0III		
Masses (M$_\odot$)	Aa: 19.6	Ab: 9.9	B: 10.9		
Luminosities (L$_\odot$)	Aa: 19600	Ab: 3100	B: 4100	C: 18	
Catalogues (A/B)	50 Ori	HR 1948/9	HD 37742/3	SAO 132444	HIP 26727
DS catalogues	H 4 21 (AC)	STF 774 (AB)	BDS 2902	ADS 4263	NOI1(AaAb)
Radial velocity	18.50 km s^{-1}	± 1.3			
Galactic coordinates	206°.452	−16°.585			

History

William Herschel noted a distant 9.6 magnitude star associated with ζ Ori on 10 October 1780, noting that the two stars were extremely unequal and that the distance between them was about 25″. By late 1781 he had found a PA of 7° and a distance of about 60″. Eleven years later he re-examined ζ and pronounced the primary star 'distinctly round'. It seems strange that he did not see the close-in companion found by F. G. W. Struve in 1822 during the course of his first survey for visual double stars. Over the last 200 years the pair has remained close to 2″.5. The angular motion in that time amounts to only 15° but this has not prevented an orbit being calculated for it.

The Modern Era

The binary ζ Ori sits in the Orion OB1 association and in common with the stars in this group it is very young, hot, and luminous. Its spectral classification is O9.5Ibe, making it a rare example of an O-type supergiant. In 2000, observations with the Navy Prototype Optical Interferometer (NPOI), revealed [226] that the bright star was a close pair with a companion, some 2.2 magnitudes fainter in the V band, resolved at a distance of 0″.040. Even better, it turned out that the period was only seven years, and an astrometric orbit was obtained for it [227], giving a handle for the first time on the mass of an O supergiant. The orbit yields a dynamical parallax of 3.4 mas, corresponding to 294 pc, but this does not agree

Finder Chart

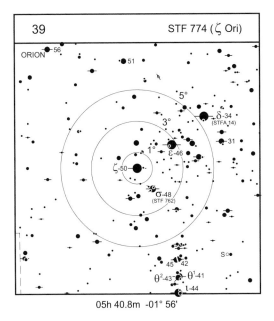

39   STF 774 (ζ Ori)

ORION 56

51

5°

3°

δ-34
(STFA 14)

31

1° ε-46

ζ-50

σ-48
(STF 762)

S

45 42

θ²-43 θ¹-41

L-44

05h 40.8m -01° 56'

Orbit

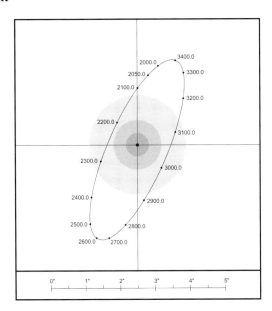

3400.0
2000.0 3300.0
2050.0 3200.0
2100.0
3100.0
2200.0
3000.0
2300.0
2900.0
2400.0
2800.0
2500.0
2600.0 2700.0

0"   1"   2"   3"   4"   5"

Ephemeris for STF 774 AB (2010 to 2100)

Orbit by Hop (1967) Period: 1508.6 years, Grade: 5

Year	PA(°)	Sep(")	Year	PA(°)	Sep(")
2010.0	165.9	2.23	2060.0	172.3	1.89
2020.0	167.0	2.17	2070.0	173.9	1.82
2030.0	168.2	2.10	2080.0	175.6	1.74
2040.0	169.5	2.04	2090.0	177.5	1.66
2050.0	170.9	1.97	2100.0	179.6	1.58

with the observed spectral type of the STF companion. The photometric parallax of 2.6 mas both eliminates the spectral type mismatch and also places the system at the distance of the Orion OB1 association. If this is true then the close binary is about one million times brighter than the Sun.

Observing and Neighbourhood

The binary ζ Ori is the easternmost star in the Belt of Orion. The primary is white but the colour of star B has been variously described as 'olive tending to brown', by F. G. W. Struve, and 'light purple', by W. H. Smyth. The nebulosity it which it is enmeshed also contains the Horsehead Nebula. ζ Ori can be resolved in 7.5-cm, and possibly in smaller apertures. About 1° to the SW is the multiple star σ Orionis (STF 762), in which Jose Caballero finds 16 components, the brightest being of magnitudes 3.8 and 8.8 at 12″, with a magnitude 6.6 star at 13″ and a magnitude 6.3 at 42″. The primary star of σ is a close visual binary (BU 1032), which has stars of V magnitudes 4.0 and 5.2 separated by 0″.25. Four degrees SSW is the famous multiple star θ Ori (The Trapezium), which is enmeshed in the Orion Nebula. The four brightest components can be seen in small apertures whilst two of the fainter stars are available to 15-cm.

Measures

Early (STF) measure	151°.3	2″.35	1831.22
(Orbit	149°.7	2″.53)	
Recent measure (ARY)	167°.1	2″.36	2013.61
(Orbit	166°.9	2″.20)	

40. θ AUR = STT 545 = WDS 05597+3713AB

Table 9.40 Physical parameters for θ Aur

STT 545	RA: 05 59 43.24	Dec: +37 29 45.9	WDS: 518(122)	
V magnitudes	A: 2.7	B: 7.2	C: 11.1	D: 10.10
(B − V) magnitudes	A:	B:		
μ	47.63 mas yr^{-1}	± 0.21	−73.79 mas yr^{-1}	± 0.08
π	19.70 mas	± 0.16	165.6 light yr	± 1.4
Spectra	A0pSi			
Luminosities (L$_\odot$)	A: 180	B: 3		
Catalogues	37 Aur	HIP 28380	HD 40312	SAO 58636
DS catalogues	H 5 89 (AC)	H VI 34 (AD)	STT 545 (AB)	BDS 3074	ADS 4566
Radial velocity	29.30 km s^{-1}	± 1.7		
Galactic coordinates	174°.338	+6°.729		

Introduction

Although θ is the eighth letter of the Greek alphabet, the star itself is the fourth brightest in the constellation of Auriga, although strangely missing from the list of 200 brightest stars in the book by Bakich [228]. It is easily found, being the next star around the pentangle of Auriga, anticlockwise from Capella.

History

William Herschel added a star at a 'distance about 2.5 min' from θ Aur on 26 September 1780 (now C) and later, when he was engaged on his second survey, noted the star now called D, on 5 September 1782: 'distance with 460 35″ 18‴, narrow measure'. It was not until 1871 that Otto Struve found the close companion at 2″.1 and included it in his *Pulkovo Catalogue*. This seems to indicate

Finder Chart

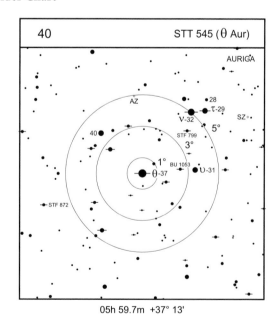

05h 59.7m +37° 13'

that the separation of B was somewhat less when William Herschel and F. G. W. Struve were carrying out their surveys, and hence they missed it, as did James South, John Herschel, and W. H. Smyth. Indeed, since discovery it has continued to widen and is now an easier object for the small aperture.

The Modern Era

The object θ Aur is an example of the chemically peculiar (CP) stars, and is classified as an α^2 CVn variable (see Star 94). In this case, there are stellar regions consisting of unusually large abundances of heavier elements such as Si. The effect of these is to act as star spots, and recent observations [229] have shown that the observed light curve, which is sinusoidal and has a period of 3.6187 days and an amplitude of about 0.04 magnitude in V, can be explained by the existence of two star spots on the star's surface. The magnetic field also varies on the same period, which is that of the star's axial rotation. The photometric variations were found in 1974 by Winzer [230]. The component AB is clearly binary – the relatively large proper motion of A, which amounts to about $0''.1$ per year, would have separated the two stars by about $15''$ since discovery were B to be just an optical companion. The period is still indeterminate and the angular motion to date amounts to $60°$. It is likely that the period is over 1000 years. The object C is certainly a background star.

Observing and Neighbourhood

Smyth noted that the bright star was 'brilliant lilac' whilst C was pale yellow. The star is now relatively easy to measure with a filar micrometer and red-field illumination in the 20-cm refractor at Cambridge. The primary appears white and no colour has been seen in the companion. There are numerous examples of sightings in smaller apertures. John Nanson saw it in 2012 with a 5-inch f/15 refractor at ×229 whilst Bill Green saw it with his 4-inch Apo at ×77, but the star appeared clearer at ×154. There are three other pairs of interest in the finder chart field. STF 799 is the most difficult and challenges measurement with the 8-inch at Cambridge. It consists of stars of magnitudes 7.3 and 8.3 separated by $0''.8$ in $158°$ (2013). A late Burnham find, which has widened from $0''.4$ at discovery and is now within the range of a 15-cm aperture is BU 1053 (6.9, 8.8, $0°$, $1''.91$, 2020) – the orbital period is 511 years. The easiest of the three is STF 872 (6.9, 7.4, $215°$, $11''.3$, 2017).

Measures

Early measure (STT)	$7°.4$	$2''.12$	1871.42
Recent measure (ARY)	$311°.4$	$3''.72$	1992.26
	$303°.6$	$4''.03$	2014.97

41. η GEM = BU 1008 = WDS J06149+2230

Table 9.41 Physical parameters for η Gem

BU 1008	RA: 06 14 52.69	Dec: +22 30 24.6	WDS: 52(474)	
V magnitudes	A: 3.15 - 3.90	B: 6.15		
$(B - V)$ magnitudes	A: +1.94	+1.21		
μ	-62.46 mas yr^{-1}	± 1.06 mas yr^{-1}	-12.12 mas yr^{-1}	± 0.70 mas yr^{-1}
π	8.48 mas	± 1.23 mas	385 light yr	± 56 light yr
$\mu(A)$	-59.52 mas yr^{-1}	± 1.45	-7.77 mas yr^{-1}	± 1.39 (DR2)
$\pi(A)$	4.73 mas	± 1.02	690 light yr	± 150 (DR2)
Spectra	M2III			
Radius (R$_\odot$)	A: 317	± 1		
Luminosities (L$_\odot$)	A: $1000 - 2000$	B: 125		
Catalogues	7 Gem	HIP 29655	HD 42995	SAO 78135
DS catalogues	BU 1008 (AB)	BDS 2796	ADS 4841	
Radial velocity (A/B)	22.39 km s^{-1}	± 0.36 km s^{-1}		
Galactic coordinates	$188°.853$	$+2°.519$		

History

Gilliss [231] in 1852 observed an occultation of η Geminorum and saw a double event, which he ascribed to the star disappearing behind a projecting lunar mountain. Gilliss also observed the occultation of η on another occasion, according to Tatlock [232]. The companion of η Gem was first seen by S. W. Burnham using the 12-inch refractor at Mount Hamilton in 1881. His measure showed the stars to be just below 1 arcsecond apart, and Burnham gave magnitudes of 3 and 8.8. About nine years later he assigned magnitudes of 10.5 and 10.7 to B. When he found very unequal and close pairs, it seems that Burnham would tend to underestimate the brightness of the companion. A similar case is α UMa (BU 1077), another subarcsecond pair, for which Burnham gave the discovery magnitudes as 2.0 and 11.1. The current WDS has 2.02 and 4.95. Another is example is BU 648 – the discovery magnitudes assigned by Burnham were 6.0 and 9.5. The current WDS gives 3.52 and 6.15.

Finder Chart

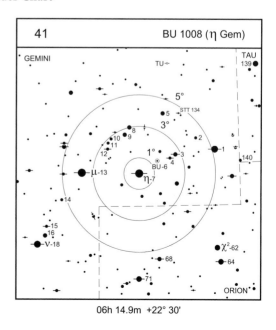

06h 14.9m +22° 30'

Orbit

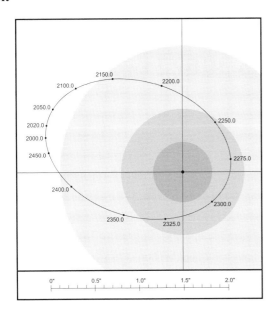

Ephemeris for BU 1008 (2010 to 2100)

Orbit by Baz (1980a) Period: 473.7 years, Grade: 5

Year	PA(°)	Sep(")	Year	PA(°)	Sep(")
2010.0	253.8	1.60	2060.0	242.3	1.60
2020.0	251.5	1.61	2070.0	240.0	1.58
2030.0	249.2	1.62	2080.0	237.6	1.56
2040.0	246.9	1.62	2090.0	235.1	1.54
2050.0	244.6	1.61	2100.0	232.5	1.50

The Modern Era

The primary star is a well-known semi-regular variable and the AAVSO website [233] gives the period as 234 days and an observed amplitude of 3.15 to 3.90. In 1944 McLaughlin & van Dyke [235] found that the spectroscopic binary period of the primary star was 8.17 years. They concluded that an astrometric perturation at the level of 0″.12 might be measurable. The primary has not yet been resolved by ground-based interferometers. Richichi & Calamai [234] measured the diameter of η Gem A from a lunar occultation and obtained a value which converts to a stellar diameter of 156 R$_\odot$. The orbit of Baize predicts that the stars will reach a maximum separation of 1″.62 in around 2030 and they will then close to 0″.45, which will be reached in 2305. A speckle measure from 2008 indicates that the stars are falling behind the ephemeris and that the orbit is likely to be longer than currently thought. The 2017 measure by the RWA confirms this discrepancy. R. Wasson [236] and M. Scardia *et al.* [237] suggest that the motion of B is linear. Gaia DR2 almost doubles the distance to η Gem, with a consequent increase for the observed stellar diameter.

Observing and Neighbourhood

The star η Gem is easily found, about 2° W of the bright star μ Gem ($V = 1.9$). The AAVSO predicts the following epochs of maximum brightness for η: 21 July 2019; 11 March 2020; and 29 October 2020. It is well seen in 20-cm at most attempts to observe it, although a night of good seeing would be an advantage. The primary is strong orange. To the NW about 2°.5 is the superb open cluster M35, in which is embedded the easy pair STT 134 (7.5, 9.1, 189°, 31″, 2009). An aperture of 25-cm might reveal the duplicity of the B star, first found by Paul Couteau. The component BC is COU 85 (9.1, 10.0, 334°, 1″.9, 1991).

Measures

Early measure (BU)	301°.7	0″.93	1882.09
(Orbit	300°.2	0″.87)	
Recent measure (ARY)	257°.5	1″.78	2017.22
(Orbit	252°.8	1″.61)	

42. 4 LYN = STF 881 = WDS J06221+5922AB

Table 9.42 Physical parameters for 4 Lyn

STF 881	RA: 06 22 03.57	Dec: +59 22 19.5	WDS: 387(148)	
V magnitudes	A: 6.43	B: 7.52		
$(B - V)$ magnitudes	A: +0.16	B: +0.18		
μ	-5.85 mas yr^{-1}	± 0.66	0.03 mas yr^{-1}	± 0.56
π	6.57 mas	± 0.67	496 light yr	± 51
$\mu(A)$	15.33 mas yr^{-1}	± 0.50	-15.56 mas yr^{-1}	± 0.60 (DR2)
$\pi(A)$	8.22 mas	± 0.50	397 light yr	± 24 (DR2)
Spectra	A3V			
Luminosities (L$_\odot$)	A: 33	B: 12		
Catalogues	HD 43812	SAO 25678	HIP 30272	
DS catalogues	STF 881	BDS 3277	ADS 4950	CHR 128 (AaAb)
Radial velocity	-20.70 km s^{-1}	± 1 km s^{-1}		
Galactic coordinates	155°.354	+19°.537		

History

Burnham [238], in 1908, whilst carrying out his proper motion project; measured a magnitude 12.9 star at 96° and 26″. Previously Robert Ball [239] had observed a more distant magnitude 11.9 at 100″ separation and almost due N. Wallenquist added a third faint star of magnitude 12.2 at 32° and 75″.

The Modern Era

Observations of 4 Lyn with a speckle interferometer on a 3.8-metre telescope by McAlister *et al.* [240] showed that A was a very close and rather unequal pair (CHR 128), the separation being 0″.187 in PA 109°.5. No subsequent measures have been made of this pair. The orbit by Zirm [243] of the STF pair is based on 60° of direct motion in the apparent orbit and the stars closing from around 0″.9, so it is only a preliminary attempt until the stars pass periastron.

Finder Chart

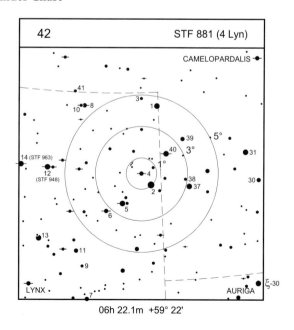

06h 22.1m +59° 22'

Orbit

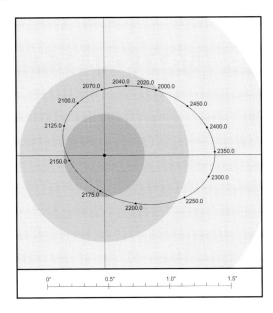

Ephemeris for STF 881 AB (2015 to 2060)

Orbit by Zir (2013a) Period: 503 years, Grade: 4

Year	PA(°)	Sep(")	Year	PA(°)	Sep(")
2015.0	147.7	0.64	2040.0	161.7	0.58
2020.0	150.3	0.62	2045.0	164.8	0.57
2025.0	153.0	0.61	2050.0	168.0	0.56
2030.0	155.8	0.60	2055.0	171.4	0.55
2035.0	158.7	0.59	2060.0	174.8	0.54

Observing and Neighbourhood

This is a binary pair which will test telescopes of 20-cm aperture. It is a useful resolution object because the separation is well known, and it will stay more or less the same for at least the next 15 years. The stars are not equal, which adds to the difficulty of splitting them, but they were seen well enough in the Cambridge 20-cm in 2013, with the following result below from three nights of measures. Quite how unequal the stars are is open to debate. The WDS values are given at the top of the page and indicate that Δm is about 1.6. Other authorities, such as Fabricius & Makarov [241], using the Tycho data from the Hipparcos satellite, get a value of 1.09, in fair agreement with that of Pluzhnik [242], 1.11. The star 4 Lyn is located in a stream of fifth and sixth magnitude stars that can be found about 15° N of β Aurigae. In this stream you can also find the beautiful triple 12 Lyncis (STF 948; see Star 48), and there are two, more difficult, binaries in 14 Lyn (STF 963) (6.0, 6.5, 358°, 0″.32, 2020 – slowly widening) and 15 Lyn (STT 159) (4.5, 5.5, 236°, 0″.71, 2020). This latter pair is not on the chart but is just 1°.2 SSE of 14 Lyn.

Measures

Early measure (STF)	89°.0	0″.82	1830.28
(Orbit	85°.4	0″.90)	
Recent measure (ARY)	149°.1	0″.63	2013.23
(Orbit	147°.0	0″.64)	

43. ϵ MON = STF 900 = WDS J06238+0436AB

Table 9.43 Physical parameters for ϵ Mon

STF 900	RA: 06 23 46.10	Dec: +04 35 34.2	WDS: 860(85)	
V magnitudes	A: 4.40	B: 6.60		
$(B - V)$ magnitudes	A: +0.18	B: +0.39		
μ(A)	-20.77 mas yr^{-1}	± 0.68	13.16 mas yr^{-1}	± 0.59 (DR2)
μ(B)	-19.01 mas yr^{-1}	± 0.07	0.38 mas yr^{-1}	± 0.06 (DR2)
π(A)	24.28 mas	± 0.43	134.3 light yr	± 2.4 (DR2)
π(B)	25.01 mas	± 0.05	130.4 light yr	± 0.3 (DR2)
Spectra	A: A8V	B: F5V		
Luminosities (L$_\odot$)	A: 25	B: 3		
Catalogues	HR 2298	HD 44769	SAO 113810	HIP 30419
DS catalogues	H 3 29	STF 900	BDS 3349	ADS 5012
Radial velocity (A/B)	13.10 km s^{-1}	± 0.8	12.40 km s^{-1}	± 0.9
Galactic coordinates	205°.675	-4°.033		

History

'Double. Distance about 12″', reported William Herschel after observing this pair on 15 February 1781. Burnham [244] using the 40-inch refractor at Yerkes added a third faint star, C, magnitude 12.2 at 254°, 92″.4, 2013. The distance is decreasing as the AB pair moves towards it.

The Modern Era

Star A is a spectroscopic binary.

Observing and Neighbourhood

According to Lewis, Struve recorded the colours of ϵ Mon as golden yellow and lilac. Hartung noted that the colours were pale and deep yellow, lying in a rich field with other wide pairs and a red star 5′ distant, in PA 200°. John Nanson, using

Finder Chart

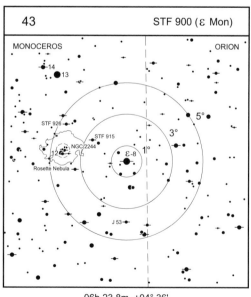

06h 23.8m +04° 36′

a 6-inch refractor at ×84 in February 2011, found yellow-white primary and pale-white secondary that 'seemed to have a trace of orange in it at times'. The later versions of Webb record 'Glorious low power field'. The cluster NGC 2264 is nearby and centred in the Rosette Nebula (NGC 2237). Five arcminutes NW from ε Mon is a 9.7 magnitude star variously described as red-orange and yellow-orange. STF 915 (7.6, 8.5, 41°, 6″, 2012) is 1°.5 NE with a third star of magnitude 11.8 at 138°, 39″.4 from A. The primary is V648 Mon. Just 1° NE of STF 915 is STF 926 (7.2, 8.6, 288°, 10″.7, 2013). For aficionados of the double stars found by Robert Jonckheere (1882–1974), J 53 (2° S) has stars of magnitudes 7.0 and 10.5 at 105° and 35″, which will be visible in 10-cm. The magnitude 10.3 companion to A, just 2″.1 away in PA 128°, will probably require 20-cm. Robert Jonckheere was born in Belgium and set up his first observatory, fitted with a 22-cm reflector, in 1905, later moving to the village of Hem in 1908 with a 35-cm refractor. Although self-taught he was allowed to use the Greenwich 28-inch refractor during World War I and later joined the staff of Marseilles Observatory, where he used the 80-cm reflector for double star work. His catalogue was published in 1962 and contains 3350 pairs.

Measures

Early measure (STF)	25°.9	13″.87	1831.74
Recent measure (ARY)	29°.5	12″.12	2012.19

44. β MON = STF 919 = WDS J06288–0702A,BC

Table 9.44 Physical parameters for β Mon

STF 919	RA: 06 28 49.01	Dec: −07 01 59.0	WDS (AB): 435(138)	
			WDS (AC): 460(132)	
V magnitudes	A: 4.62	B: 5.00	C: 5.39	
$(B-V)$ magnitudes	A: −0.10	B: −0.19	C: −0.11	
μ(A)	−5.02 mas yr^{-1}	± 0.25	−2.86 mas yr^{-1}	± 0.47 (DR2)
μ(B)	−7.36 mas yr^{-1}	± 0.46	−2.27 mas yr^{-1}	± 0.53 (DR2)
μ(C)	−7.25 mas yr^{-1}	± 0.22	−3.77 mas yr^{-1}	± 0.26 (DR2)
π(A)	5.22 mas	± 0.25	625 light yr	± 30 (DR2)
π(B)	4.89 mas	± 0.20	557 light yr	± 23 (DR2)
π(C)	5.05 mas	± 0.13	621 light yr	± 16 (DR2)
Spectra	A: B3Ve	B: B3ne	C: B3e	
Masses (M$_\odot$)	A: 11.0	B: 11.0	Ca: 11.0	Cb: 5.4
Luminosities (L$_\odot$)	A: 430	B: 350	C: 220	
Catalogues (A/B/C)	11 Mon	HR 2356/7/8	HD 45725/6/7	SAO 133316/7
	HIP 30867			
DS catalogues	Mayer 18	H 1 10 (BC)	H 2 17 (AB)	STF 919 (A,BC)
	BDS 3402	ADS 5107	CHR 167 CaCb	
Radial velocity	A: 17.2 km s^{-1}	± 2.9	B: 18 km s^{-1}	± 5
	C: 23 km s^{-1}	± 5		
Galactic coordinates	216°.661	−8°.214		

History

This is one of the earliest telescopic double star discoveries. It was found by Benedetto Castelli on 30 January 1617, using a telescope borrowed from Galileo. It next appeared in the catalogue of Christian Mayer as number 18, and two years later in 1779 William Herschel examined AB and noted the close pair BC. 'A curious treble star; may appear double at first sight; but with some attention we see that one of them is again double.' He noted that all three stars were white and went on to say 'As perfect as I have seen this treble star with 460, it is one of the most beautiful sights in the heavens; but requires a very fine evening.' Burnham [245], using the 18.5-inch Dearborn refractor, added a distant 12th magnitude star (BU 570).

Finder Chart

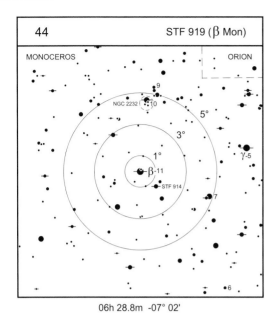

44   STF 919 (β Mon)

MONOCEROS   ORION

NGC 2232

5°

3°

1°

γ-5

β-11

STF 914

06h 28.8m -07° 02'

The Modern Era

In 1988 H. A. McAlister *et al.* [482] found that C was double in PA 141° at a separation of 0″.26 (CHR 167). The motion between the bright stars is very small and any orbital period is bound to be long. Tokovinin, in his MSC, estimates the period of BC, for instance, at 2800 years whilst CaCb could be 100 years; it may well be shorter, but the speckle companion has gone to ground – it has not been resolved since 1988.

Observing and Neighbourhood

Smyth noted 'Golden yellow and lilac', whilst Webb drew attention to 'the glorious low-power field'. More recently Hartung found pale yellow and deep yellow in a rich field with other wide pairs and a red star 5′ away in PA 200°. RWA has always seen the three stars as white, but Greg Stone [126], using a 60-mm f/6 refractor on a night good enough to use a magnification of ×180, resolved the close pair comfortably. He saw a hint of yellow in the primary and light blue in the other two stars. Also on the Star Splitters website, John Nanson, using 72-mm at ×108 sees three very white stars. North of β Mon by 2.5° is the open cluster NGC 2232, whilst 1° SW is the pair STF 914 (6.3, 9.3, 299°, 21″), which was first seen by William Herschel as (H 3 43). The Burnham companion is as at 55°, 26″.6 (2012).

Measures

AB			
Early measure (STF)	130°.0	7″.25	1831.23
Recent measure (ARY)	132°.2	7″.33	2014.10

AC			
Early measure (STF)	101°.7	2″.46	1831.23
Recent measure (ARY)	108°.9	2″.81	2014.10

45. △ 30 PUP = WDS J06298–5014

Table 9.45 Physical parameters for △ 30 Pup

DUN 30	RA: 06 29 49.03	Dec: −50 14 20.7	WDS: 3200(33)	
R 65			WDS: 1093(72)	
HDO 195			WDS: 1497(58)	
V magnitudes	A: 5.97	B: 6.15	C: 7.98	D: 8.73
μ	A: −67.94 mas yr^{-1}	±1.03	B: −49.01 mas yr^{-1}	± 0.72
π	19.44 mas	± 0.66	168 light yr	± 6
Spectra	A: F2	B: ?	C: K3?	D: ?
Masses (M$_\odot$)	A: 1.54	B: 1.50	C: 0.63	D: 0.63
Luminosities (L$_\odot$)	A: 9	B: 8	C: 1.4	D: 0.7
Catalogues	HR 2384	HD 46273	SAO 234539	HIP 30953
DS catalogues	DUN 30 (AB,CD)	R 65 (AB)	HDO 195 (CD)	
Radial velocity	−3.40 km s^{-1}	± 0.3		
Galactic coordinates	258°.815	−23°.837		

History

Dunlop recorded this pair in 1826. John Herschel observed and measured it in 1834 and 1835 with his 7-foot telescope but did not use the 20-foot. Had he done so he might have seen the duplicity of both components. The primary was divided by Russell [525] in 1879 with the 11.5-inch refractor at Sydney when the separation was 0″7. Again the duplicity of CD was missed – the stars were actually also separated by 0″.58 at that time according to a recent orbit. The pair CD was finally resolved by Solon Bailey [248] at Arequipa in 1894. Using the 13-inch refractor which had been brought from Harvard for photography, Bailey realized the quality of the high-altitude site for the discovery of close and faint double stars and initiated a programme of observation. The first observations were estimates and the group later sent for a micrometer, which arrived a few months later. On 15 April 1894, DUN 30 appeared as a quadruple star to Bailey. The Russell pair were estimated as magnitudes 6.5, 6.5 at 0″.5 whilst C appeared as a new pair, with two 9.5 magnitude components 0″.3 apart.

Finder Chart

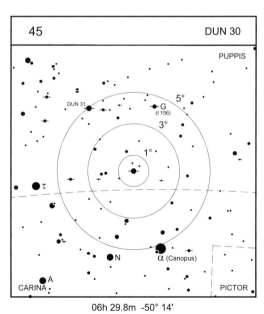

06h 29.8m -50° 14'

Orbit for R 65 AB

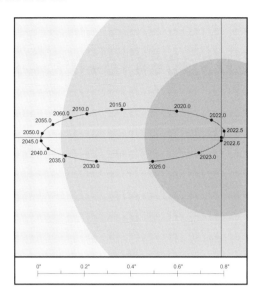

Ephemeris for R 65 AB (2016 to 2034)

Orbit by Hrt (2010a) Period: 53.13 years, Grade: 3

Year	PA(°)	Sep(")	Year	PA(°)	Sep(")
2016.0	253.0	0.41	2026.0	286.3	0.37
2018.0	248.5	0.32	2028.0	282.9	0.47
2020.0	240.2	0.22	2030.0	280.6	0.55
2022.0	210.2	0.08	2032.0	278.8	0.60
2024.0	293.0	0.23	2034.0	277.3	0.65

Orbit for HDO 195 AB

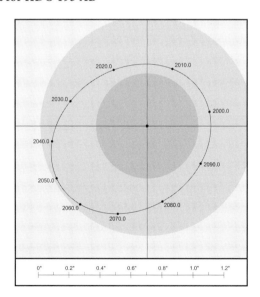

Ephemeris for HDO 195 AB (2017 to 2044)

Orbit by Doc (2014d) Period: 99.22 years, Grade: 3

Year	PA(°)	Sep(")	Year	PA(°)	Sep(")
2017.0	195.5	0.40	2032.0	258.7	0.54
2020.0	211.5	0.41	2035.0	267.0	0.58
2023.0	225.9	0.44	2038.0	274.4	0.61
2026.0	238.4	0.47	2041.0	281.2	0.63
2029.0	249.3	0.51	2044.0	287.5	0.65

The Modern Era

A recent study of binaries by Trilling *et al.* [249] using the Spitzer Space Telescope at 24 and 70 μm revealed a dust disk at a radius of 16 AU, presumably circling the brightest star in the system. Again, in this paper the authors noted that A was a spectroscopic binary but gave no further details.

Observing and Neighbourhood

For the small telescope the wide pair, located 2°.5 NNE of the brilliant Canopus, offers an attractive sight but, for those who wish to see the two sub-systems, this is a rather difficult 'double–double'. Neither pair exceeds 1″ even at apastron and they are often substantially less. R 65 will be a challenge for the next few years up to 2025 or so and even then will need 40-cm to be split. The orbit is extremely eccentric (*e* = 0.977). Andrew James reported that the colours of DUN 30 are yellow and orange-red. G Pup is in the finder chart as I 156, now at half its discovery separation but still within range of 15- to 20-cm (5.9, 8.1, 125°, 1″, 1999). No-one has observed it for 20 years; where is the companion now? Due E, 2°.5 distant, is DUN 31, 5.1, 7.3, 326°, 13″.5, 2011.

Measures

DUN 30			
Early measure (R)	318°.0	12″.50	1893.12
Recent measure (ANT)	311°.1	12″.03	2014.25
R 65			
Early measure (SLR)	85°.2	0″.68	1893.12
(Orbit	266°.3	0″.75)	
Recent measure (ANT)	256°.9	0″.46	2014.24
(Orbit	255°.9	0″.47)	
HDO 195			
Early measure (I)	117°.8	0″.40	1903.19
(Orbit	114°.1	0″.41)	
Recent measure (TOK)	160°.7	0″.39	2010.89
(Orbit	160°.9	0″.39)	

46. 15 MON = STF 950 = WDS J06410+0954AB

Table 9.46 Physical parameters for 15 Mon

STF 950	RA 06 40 58.66	Dec: +09 53 44.7	WDS: 901(83)		
V magnitudes	A: 4.37	B: 7.52			
$(B - V)$ magnitudes	A: +0.27	B: +0.27			
μ	-2.61 mas yr^{-1}	± 0.50	-1.61 mas yr^{-1}	± 0.39	
π	3.55 mas	± 0.50	920 light yr	± 130	
μ(A)	-4.39 mas yr^{-1}	± 0.99	-6.58 mas yr^{-1}	± 0.85 (DR2)	
μ(B)	-1.80 mas yr^{-1}	± 0.47	-6.55 mas yr^{-1}	± 0.40 (DR2)	
π(A)	-0.78 mas	± 0.62			
π(B)	-6.36 mas	± 0.45			
Spectra	Aa: O7V	Ab: ?	B: B7V	C: B8V	
Masses (M$_\odot$)	Aa: 31.0	Ab: 10.7	B: 7.0	C: 3.0	
Luminosities (L$_\odot$)	A: 10,000	B: 550			
Catalogues	S Mon	HR 2456	HD 47839	SAO 114258	HIP 31978
DS catalogues	H 6 65	STF 950	BDS 3542	ADS 5322	CHR 168 (AaAb)
Radial velocity	+22.00 km s^{-1}	± 0.3			
Galactic coordinates	202°.936	+2°.199			

History

William Herschel observed S Mon on 22 October 1781. He noted 'Multiple. It is one star with at least 12 around it, all within the field of my telescope.'

The Modern Era

Centred within the open cluster NGC 2264, the Christmas-Tree cluster, 15 Mon (also known as S Mon) is an extremely bright and hot O7 star, which dominates the northern part of the stellar group. It is thought to be the source of a bubble of hot gas which influences the kinematic properties of the surrounding gas and stars. Next to the Orion Nebula,

NGC 2264 is the nearest significant region of star formation (Tobin *et al.* [250]). The WDS lists 17 components within 190 arcseconds of 15 Mon, whose magnitudes range from 7.8 to 12.8. There is a substantial discrepancy between the accepted distance to the cluster (2500–3000 light years; Sung *et al.* [252], Baxter *et al.* [253]) and the distance to S Mon derived by Hipparcos (920 light years). The duplicity of the A component was found by McAlister *et al.* [482] (CHR 168). The dynamical parallax derived from the orbit of AaAb is 1.2 mas, which is 2700 light years, so this favours the larger value for the distance to S Mon. The results from Gaia DR2 do not settle the argument but indicate that the stars are indeed extremely distant – considerably more so than that found by Hipparcos.

Finder Chart

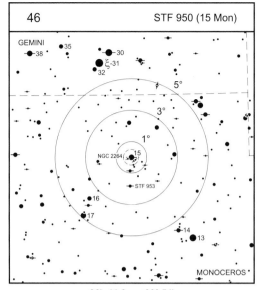

06h 41.0m +09° 54'

Observing and Neighbourhood

STF 953 is a neat pair 1° S (7.1, 7.7, 329°, 7″.1, 2017). The luminosities in the table are based on the 2700-light-year distance. For the small telescope user, the principal pair is STF 950 itself. The stars differ by more than three magnitudes in the V band and the separation has widened a small amount in the last 200 years or so. Between 2008 and 2011, RWA measured a number of the fainter components in the group NGC 2264, as in the following table.

Aa,B	213°.7	2″.87	2008.72	3 nights
Aa,E	139°.9	73″.56	2008.57	2 nights
Aa,H	168°.0	88″.46	2008.57	3 nights
Aa,G	264°.4	39″.35	2010.24	2 nights
Aa,F	223°.5	155″.00	2008.74	2 nights
STF 952 MN	116°.6	14″.02	2011.21	4 nights

Measures of Principal Pair

Aa,B			
Early measure (R)	318°.0	12″.50	1893.12
Recent measure (ANT)	311°.1	12″.03	2014.25

47. SIRIUS = α CMA = AGC 1 = WDS J06451–1643

Table 9.47 Physical parameters for α CMa

AGC 1	RA: 06 45 08.92	Dec: −16 42 58.0	WDS: 26(640)		
V magnitudes	A: −1.44	B: 8.44			
$(B - V)$ magnitudes	A: 0.00	B: −0.04			
μ	A: −546.01 mas yr^{-1}	± 1.33	−1223.07 mas yr^{-1}	± 1.24	
π	379.21 mas	± 1.58 mas	8.60 light yr	± 0.04	
μ(B)	−459.68 mas yr^{-1}	± 0.54	−915.02 mas yr^{-1}	± 0.53 (DR2)	
π(B)	376.68 mas	± 0.45	8.66 light yr	± 0.01 (DR2)	
Spectra	A: A1V	B: DA2			
Masses (M$_\odot$)	A: 2.06	± 0.02	B: 1.02	± 0.11	
Radii (R$_\odot$)	A: 1.711	B: 0.0084			
Luminosities (L$_\odot$)	A: 22	B: 0.002			
Catalogues	9 CMa	HR 2491	HD 48915	SAO 151881	HIP 32349
DS catalogues	AGC 1	BDS 3596	ADS 5423		
Radial velocity	−5.50 km s^{-1}	± 0.4			
Galactic coordinates	227°.230	−8°.890			

History

In 1834, F. W. Bessel, working at Konigsberg Observatory, noticed that the proper motion of Sirius was variable. Further observations led him, in 1844, to write to Alexander von Humboldt – 'I adhere to the conviction that Procyon and Sirius are genuine binary systems, each consisting of a visible and an invisible star. We have no reason to suppose that luminosity is a necessary property of cosmical bodies. The visibility of countless stars is no argument against the invisibility of countless others' [257]. Sirius B was first measured by G. P. Bond with the Harvard College 15-inch refractor on February 19th, 1862. His report to *Astronomische Nachrichten* [255] notes that 'An interesting discovery of a companion to Sirius was made on the evening of Jan 31, 1862 by Mr. Clark with his new object-glass of eighteen and one half inches aperture'.

Finder Chart

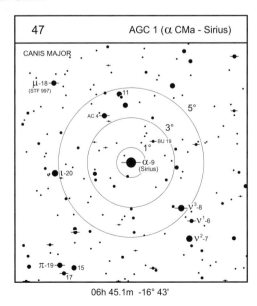

06h 45.1m -16° 43'

Orbit

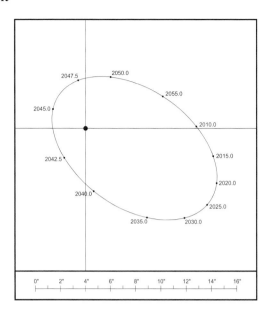

Ephemeris for AGC 1 AB (2016 to 2034)

Orbit by BdH (2017) Period: 50.1284 years, Grade: 2

Year	PA(°)	Sep(")	Year	PA(°)	Sep(")
2016.0	76.0	10.57	2026.0	57.0	11.16
2018.0	71.9	10.94	2028.0	53.1	10.86
2020.0	68.1	11.19	2030.0	48.9	10.39
2022.0	64.4	11.32	2032.0	44.3	9.74
2024.0	60.7	11.31	2034.0	38.9	8.88

The Modern Era

Bond *et al.* [256] published a paper using HST observations which fixes the orbit of AB to unprecedented accuracy. The error on the period is 1.5 days whilst the masses are now known to around 1%. The various extra bodies that have been postulated (and reportedly observed) are thought not to exist if the mass is greater than 15 to 25 times that of Jupiter. The third star seen occasionally by visual observers cannot be explained away as a background object because the large proper motion of Sirius (5".5 per year) would leave it behind rapidly whereas the reported sightings over a period of 80 years always have the star close to Sirius B. Gaia DR2 gives a result for the white dwarf B, which at the time of observation is as well placed for observation as it is likely to be in the 50-year orbital period.

Exoplanet Host?

The paper by Bond [256] found no evidence for additional bodies whose masses exceed 15–25 Jupiter masses.

Observing and Neighbourhood

Now is the best time to observe Sirius B. It reaches maximum separation in 2023 (11".3), after which it closes up, reaching a minimum value of 2".6 in 2043 – a point in the orbit when even Burnham could not see the star with the 36-inch at Lick. Reports of seeing the Pup with apertures as small as 10-cm have been recorded and RWA readily admits to never having seen Sirius B despite occasional efforts in the last 25 years. The altitude of Sirius in the sky is a key element for success – the higher you can observe it the better. A large aperture is not essential but sufficient magnification to give some separation to the two stars is needed. Some recommend the use of a hexagonal diaphragm mask over the objective or the end of the telescope tube. This spreads the light of Sirius into six diffraction spikes and shrinks the apparent size of the bright star's disk. The mask should be rotatable in case the Pup should happen to be on one of the spikes. For more help try http://www.rasc.ca/sirius-b-observing-challenge. Nearby is BU 19 – 7.1, 9.0, 169°, 3".9 a good test for 7.5-cm, whereas AC 4 (2° NE) (5.6, 7.2, 354°, 0".31, 2018 would test 35-cm. A fine sight, worth seeking out 4° NE of Sirius, is μ (STF 997) (5.3, 7.1, 346°, 3".1, 2015).

Measures

Early measure (BND) :	85°.3	10".33	1862.11
(Orbit :	83°.1	9".63)	
Recent measure (ANT) :	83°.2	9".64	2013.72
(Orbit :	81°.1	10".00)	

48. 12 LYN = STF 948 = WDS J06462+5927AB,C

Table 9.48 Physical parameters for 12 Lyn

STF 948	RA: 06 46 14.15	Dec: +59 26 30.1	WDS: 81(384) (AB)		
			WDS: 162(251) (AC)		
V magnitudes	A: 5.44	B: 6.00	C: 7.05		
$(B - V)$ magnitudes	A: +0.04	B: +0.12	C: +0.25		
μ	-19.63 mas yr^{-1}	± 0.80	-7.23 mas yr^{-1}	± 0.49	
π	15.19 mas	± 0.78	215 light yr	± 11	
$\mu(B)$	-19.55 mas yr^{-1}	± 0.13	3.90 mas yr^{-1}	± 0.11 (DR2)	
$\mu(C)$	-23.94 mas yr^{-1}	± 0.07	-1.59 mas yr^{-1}	± 0.06 (DR2)	
$\pi(B)$	13.32 mas	± 0.11	244.9 light yr	± 2.0 (DR2)	
$\pi(C)$	13.64 mas	± 0.06	239.1 light yr	± 1.1 (DR2)	
Spectra	A: A3V	B: ?	C: A9?		
Masses (M$_\odot$)	A: 2.2	B: 1.9	C: 1.4		
Radii (R$_\odot$)	A: 1.5				
Luminosities (L$_\odot$)	A: 24	B: 18	C: 7		
Catalogues	HR 2470	HD 48250	SAO 25939	HIP 32438	
DS catalogues	H 1 6 (AB)	H 3 22 (AC)	STF 948	BDS 3559	ADS 5400
Radial velocity	-3.0 km s^{-1}	± 4.2			
Galactic coordinates	156°.305	+22°.481			

History

William Herschel included both the close and wide companion in his catalogues when he observed 12 Lyncis on 6 October 1780 and pronounced it 'a curious treble star'. He allocated colours of white and 'white inclining to rose' to the close pair whilst star C was noted as 'pale red'. A distant magnitude 10.5 (259°, 182″) was found by Ball [347].

The Modern Era

In a recent study, an orbital period of 907.6 years was derived for the closer pair, AB. The brighter component is an A3 dwarf with a projected rotational velocity of 150 km s^{-1}. Tokovinin, in his *Multiple Star Catalogue*, gives a suggested period of 8.5 kiloyears for the rotation of C around AB.

Finder Chart

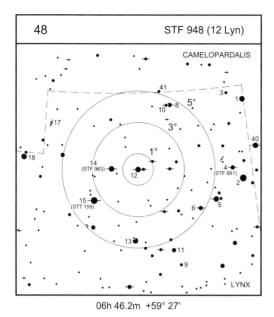

48   STF 948 (12 Lyn)

06h 46.2m +59° 27'

Orbit

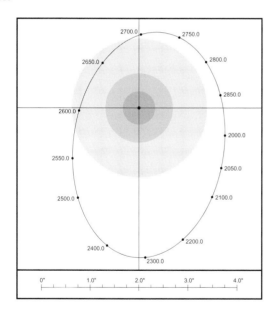

Ephemeris for STF 948 AB (2010 to 2100)

Orbit by WSI (2006b) Period: 907.6 years, Grade: 4

Year	PA(°)	Sep(")	Year	PA(°)	Sep(")
2010.0	68.7	1.88	2060.0	51.3	2.11
2020.0	64.9	1.92	2070.0	48.3	2.16
2030.0	61.2	1.96	2080.0	45.4	2.21
2040.0	57.8	2.01	2090.0	42.7	2.26
2050.0	54.4	2.06	2100.0	40.1	2.32

Observing and Neighbourhood

This beautiful triple is visible in 75-mm and above. The close pair has traced out about one-quarter of the apparent orbit and will continue to widen to 3″ in 2300 before closing to 1″.1 in 2630 or so. The triple 12 Lyncis is part of a group of faint naked-eye stars which can be found about 6° NE of δ Aurigae. There is a wealth of testing binaries in the neighbourhood. Nearby are the considerably more difficult pairs 4 (see Star 42), 14, and 15 Lyncis (see below). The pair 14 Lyn (STF 963) is given in Webb with the intriguing colours of gold and purple. It has so far escaped the attentions of RWA. It is a 316-year binary which is now slowly widening. In 2020 the relative positions will be 358° and 0″.32 and the stars have visual magnitudes 6.0 and 6.5, making them a severe test for 30–40-cm. Meanwhile, 15 Lyn was a discovery by Otto Struve at Pulkovo (STT 159) and needs 20-cm to be seen well. The stars are relatively unequal (4.5 and 5.5) and in 2020 the position will be 236° and 0″.71.

Measures

AB			
Early measure (STF)	152°.5	1″.60	1832.60
(Orbit	154°.8	1″.60)	
Recent measure (ARY)	67°.9	1″.95	2016.28
(Orbit	66°.3	1″.90)	
AC			
Early measure (STF)	304°.2	8″.67	1831.10
Recent measure (ARY)	310°.3	8″.70	2016.28

49. 38 GEM = STF 982 = WDS J06546+1311AB

Table 9.49 Physical parameters for 38 Gem

STF 982	RA: 06 54 38.63	Dec: +13 10 40.2	WDS: 115(301)		
V magnitudes	A: 4.75	B: 7.80	C: 11.32	D: 15.50	
$(B - V)$ magnitudes	A: +0.28	B: +1.25			
μ(A)	68.75 mas yr^{-1}	± 1.32	−86.83 mas yr^{-1}	± 1.15 (DR2)	
μ(B)	89.23 mas yr^{-1}	± 0.21	−77.17 mas yr^{-1}	± 0.15 (DR2)	
π(A)	33.92 mas	± 0.82	96.2 light yr	± 2.3 (DR2)	
π(B)	33.33 mas	± 0.08	97.9 light yr	± 0.2 (DR2)	
Spectra	A: F0Vp	B: G6V	D: M4?		
Masses (M$_\odot$)	A: 1.5	B: 0.9	D: 0.2		
Radii (R$_\odot$)	A: 1.2	B: 1.3			
Luminosities (L$_\odot$)	A: 9	B: 0.6		D: 0.0005	
Catalogues	38 Gem	HR 2654	HD 50635	SAO 92965	HIP 33202
DS catalogues	H 3 47	STF 982	BDS 3692	ADS 5559	TOK 261
Radial velocity (A/B)	15.90 km s^{-1}	± 2.3			
Galactic coordinates	201°.511	+06°.666			

History

The star 38 (or e) Gem was noted to be double by William Herschel on 27 December 1781. He noted 'Extremely unequal', and his colours were reddish-white for A and red for the companion. He also mentioned two other stars in the field, forming a rectangle, one of which was 'perhaps 40″'. Apparently neither of these two stars is in the WDS. The star appears as 38 E Geminorum in Piazzi's 1814 Palermo catalogue. Smyth, writing in *Cycle of Celestial Objects*, noted light yellow and purple and went on to say 'This is a very fine object, and the colours so marked, that they cannot be entirely imputed to the illusory effect of contrast'. In more recent times Ernst Hartung found yellowish and pale orange with his 12-inch reflector from Victoria, and Tim Leese, observing from Spain, found yellow-white and pale yellow-orange. Thomas Lewis [194] lists 14 estimates of AB colours by 10 observers between 1829 and 1887 and the recorded hues range

Finder Chart

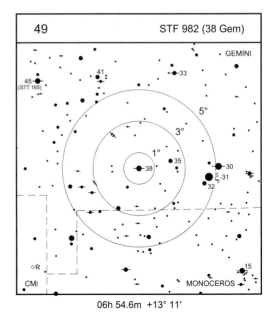

49   STF 982 (38 Gem)

06h 54.6m +13° 11′

Orbit

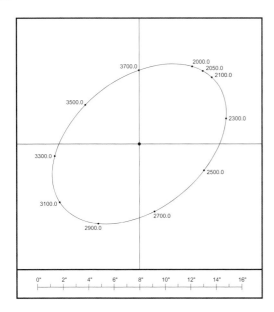

Ephemeris for STF 982 AB (2010 to 2100)

Orbit by Msn (2014b) Period: 1898.12 years, Grade: 4

Year	PA(°)	Sep(")	Year	PA(°)	Sep(")
2010.0	143.6	7.28	2060.0	136.9	7.52
2020.0	142.2	7.33	2070.0	135.6	7.55
2030.0	140.9	7.38	2080.0	134.4	7.58
2040.0	139.5	7.43	2090.0	133.1	7.61
2050.0	138.2	7.48	2100.0	131.8	7.63

from white, yellow, or orange for A and white, blue, greenish, red, or purple for B. Star C was first measured by Burnham in 1913 as part of his project to measure proper motions.

The Modern Era

A speckle interferometry study of the primary star by McAlister [482] using the 3.6-metre CFHT reflector on Hawaii concluded that the star was single at the 0″.038 level. The Hipparcos satellite data supplies a distance of 83 light years and shows that the annual proper motion of the two brighter stars is almost exactly 0″.1 per year and that the fainter component C is being rapidly left behind. In 2012, Andrei Tokovinin [261] noted a very faint star (D) some 150″ distant which appears to be moving through space with the same proper motion as AB. This has been catalogued in the WDS as TOK 261AD. Brian Skiff derived position angles and separations for AB and D from both the POSS-R plate from 1950 and the 2010 WISE catalogue and showed there had been no net motion over 60 years.

Observing and Neighbourhood

The pair 38 Gem can be easily found almost 2.3° E of the V = 3.4 star ξ Gem. F. G. W. Struve found colours of yellow and blue, whilst Smyth noted light yellow and purple. Five degrees NE is 45 Gem (STT 165), a very unequal pair which is becoming easier as the stars are separating due to the proper motion of the bright primary (5.5, 10.9, 6°, 16″, 2015). The stars were only 5″ when discovered by Otto Struve in 1847. A magnitude 13.5 star lies at 333°, 60″.

Measures

Early measure (STF)	174°.9	5″.74	1829.24
(Orbit	174°.3	5″.84)	
Recent measure (ARY)	143°.8	7″.33	2014.86
(Orbit	142°.9	7″.30)	

50. Δ 39 CAR = WDS J07033–5911

Table 9.50 Physical parameters for Δ 39 Car

DUN 39	RA: 07 03 15.12	Dec: −59 10 41.1	WDS: 2539(39)	
V magnitudes	A: 5.64	B: 6.79		
(*B* − *V*) magnitudes	A: −0.14	B: −0.08		
μ(A)	−9.90 mas yr^{-1}	± 0.26	12.43 mas yr^{-1}	± 0.46 (DR2)
μ(B)	−17.01 mas yr^{-1}	± 0.15	10.13 mas yr^{-1}	± 0.30 (DR2)
π(A)	6.49 mas	± 0.11	503 light yr	± 9 (DR2)
π(B)	4.82 mas	± 0.09	677 light yr	± 13 (DR2)
Spectra	A: B9III + B8V	B: B8V		
Luminosities (L$_\odot$)	A: 110	B: 70		
Catalogues	HR 2674	HD 53921	SAO 234890	HIP 34000
DS catalogues	DUN 39			
Radial velocity	7.00 km s^{-1}	± 4.3		
Galactic coordinates	269°.542	−21.598		

History

This binary was found by Dunlop at Parramatta but his observation was criticized by John Herschel, who first observed the star on sweep 664 at the Cape of Good Hope: 'Fine ∗'. On 1838.110 he notes 'practice wire set to 72° 40′, but it cannot be borne out. The star has certainly changed (Mr Dunlop's position is 11° 12′, which is probably a mistake of reading).'

The Modern Era

Gaia DR2 would seem to indicate that the stars are at significantly different distances and that they are merely an optical pair. Star A is the variable star V450 Car, a member of the 53 Per class of B stars which pulsate but in a non-radial fashion. The period is 1.6518 days and the range is between *V* = 5.45 and 5.47 [265]. de Cat & Aerts [263] found that one component is a single-lined spectroscopic binary with an eccentric orbit, but they did not specify which star.

Finder Chart

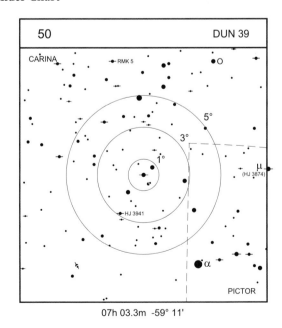

07h 03.3m -59° 11'

Observing and Neighbourhood

Ross Gould [264] says 'A test for 8-cm refractors, and not difficult with 10-cm scopes'. He found both stars to be pale yellow. DUN 39 has been underobserved and should be kept under occasional surveillance. Four degrees preceding is μ Pic – HJ 3874 (5.6, 9.3, 221°, 2″.5, 2005), which, Hartung says, '75-mm shows with close attention', and 1°.5 SE is HJ 3941 (7.3, 8.3, 273°, 0″.5, 1991). This system has appeared to be closing but has not been observed for nearly 20 years. An observation with 30- to 40-cm might be useful to estimate the nature of the motion. At the edge of the finder chart, 3°.5 NNE from DUN 39, is RMK 5 (7.7, 7.8, 227°, 6″.9, 2010).

Measures

Early measure (HJ)	74°.9	2″.40	1836.19
Recent measure (HOR)	85°.9	1″.41	1997.12

51. $\gamma^{1,2}$ VOL = Δ 42 = WDS J07087-7030

Table 9.51 Physical parameters for γ Vol

DUN 42	RA: 07 08 44.87	Dec: −70 29 56.2	WDS: 2540(39)	
V magnitudes	A: 3.86 (γ^2)	B: 5.43 (γ^1)		
$(B - V)$ magnitudes	A: +1.10	B: +0.59		
μ(A)	24.26 mas yr^{-1}	± 0.63	106.88 mas yr^{-1}	± 0.60 (DR2)
μ(B)	5.94 mas yr^{-1}	± 0.59	112.93 mas yr^{-1}	± 0.54 (DR2)
π(A)	24.76 mas	± 0.30	131.7 light yr	± 1.6 (DR2)
π(B)	23.20 mas	± 0.32	140.6 light yr	± 1.9 (DR2)
Spectra	A: K0III	B: F2V		
Radii (R$_\odot$)	A: 2.45 mas	± 0.06		
Luminosities (L$_\odot$)	A: 40	B: 10		
Catalogues (A/B)	HR 2736/5	HD 55865/4	SAO 256374/3	HIP 34481/73
DS catalogues	DUN 42			
Radial velocity (A/B)	2.80 km s^{-1}	± 0.7	4.50 km s^{-1}	± 0.8
Radial velocity (B)	4.97 km s^{-1}	± 1.17 (DR2)		
Galactic coordinates	281°.556	−24°.042		

History

This binary was noted by Dunlop from Parramatta; he recorded magnitudes of 5 and 8 with the bright star yellow and the companion greenish. John Herschel observed this pair four times with the 20-foot telescope and called it a 'superb double star'. His estimated distance ranged from 18″ to 25″ but when he later measured it with the 7-foot telescope he obtained close to 12″.8 and a position angle consistently near 300°.

The Modern Era

The B component is known to be a spectroscopic binary and a variable. This is almost certainly a physical pair, the uncertainty in the distance of B as found by Hipparcos probably a result of its binary nature. The common proper motion and

Finder Chart

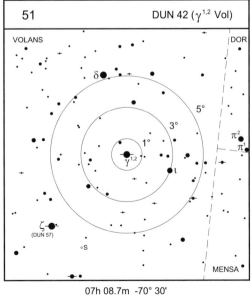

07h 08.7m -70° 30'

similar parallax confirmed by Gaia DR2 marks this pair out as a binary of long period, confirmed by the very small change in position angle since observations began.

Observing and neighbourhood

Richard Jaworski [378] counts it amongst the finest southern double stars. He notes a 'stunning bright gold and pale-yellow pair in a field of scattered faint stars'. Ross Gould [267] with a 350-mm calls it 'Beautiful, bright and easy'. He sees colours of deep yellow and dull yellow, and the ASNSW observers make the components orangey-yellow and white. Four degrees to the SE is ζ Vol (DUN 57), a fine unequal pair with an orange primary and white comes (4.1, 9.3, 123°, 15″.6).

Measures

Early measure (TEB)	299°.7	13″.28	1886.30
Recent measure (ANT)	296°.1	14″.4	2002.70

52. 145 CMA = HJ 3945 = WDS J07166–2319AB

Table 9.52 Physical parameters for 145 CMa

HJ 3945	RA: 07 16 36.84	Dec: −23 18 56.1	WDS: 2023(46)		
V magnitudes	A: 5.00	B: 5.84	C: 9.15		
(*B* − *V*) magnitudes	A: +1.92	B: +0.44	C: +0.21		
μ(A)	−3.94 mas yr^{-1}	± 0.25	B: 2.25 mas yr^{-1}	± 0.26 (DR2)	
μ(B)	−33.79 mas yr^{-1}	± 0.08	B: 40.71 mas yr^{-1}	± 0.08 (DR2)	
μ(C)	−29.903 mas yr^{-1}	± 0.021	B: +39.081 mas yr^{-1}	± 0.027 (Hipparchus)	
π(A)	1.43 mas	± 0.16	2280 light yr	± 255 (DR2)	
π(B)	9.46 mas	± 0.06	344.8 light yr	± 2.2 (DR2)	
π(C)	10.83 mas	± 0.46	301 light yr	± 13 (Hipparchus)	
Spectra	A: K3Ib	B: A5m	C: A7m		
Luminosities (L$_\odot$)	A: 4000	B: 40	C: 1.5		
Catalogues (A/B)	145 CMa	HR 2764	HD 56577/8	SAO 173349/53	HIP 35210/3
Catalogues (C)	HD 57527	SAO 173505	HIP 35578		
DS catalogues	HJ 3945 (AB)	SHY 508 (BC)	BDS 3954	ADS 5951	
Radial velocity (A)	29.00 km s^{-1}	± 0.5			
Radial velocity (B)	36.90 km s^{-1}	± 1.78			
Radial velocity (C)	40.60 km s^{-1}	± 0.7			
Galactic coordinates	236°.499	−5°.214			

History

This beautiful pair was swept up by John Herschel on 23 January 1835. He noted 'Large star orange, small pale blue'. Two years later he re-observed it and called the colours very high yellow and contrasted blue. Burnham, in his *General Catalogue*, notes that Birmingham thought the A component was variable and Aitken also notes that, whilst its radial velocity is small, it may be variable.

The Modern Era

The two bright components of 145 CMa are completely unrelated, as shown by the parallax and proper motion recently obtained by the Gaia satellite for the A and B components. Hartkopf [269] published a linear ephemeris which predicts that A and B are closest together in 2023 and then begin to widen. Component B is actually co-moving with a distant star C (HIP 35578), which is 3°.5 away to the SE. Shaya & Olling

Finder Chart

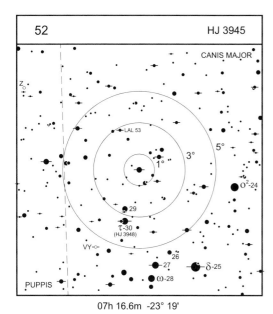

52   HJ 3945

CANIS MAJOR

LAL 53

5°

3°

1°

o²-24

29

τ-30
(HJ 3948)

VY-

26

27

δ-25

ω-28

PUPPIS

07h 16.6m -23° 19'

[270] suggested that there is a near 100% certainty that the two stars are physically connected. Both B and C are metallic-lined late A stars.

Observing and Neighbourhood

HJ 3945 sits in Canis Major, about 2°.5 WNW of the magnitude 3.0 blue supergiant star o² CMa. It is also known as the Winter Albireo. It is a rich area for the telescopic observer. Two degrees S is the luminous star τ CMa (HJ 3948) (4.4, 8.2, 77°, 85″) with additional stars of magnitudes 10.2 and 11.2 much closer, at 9″ and 14″. HJ 3945 is often called the 'Southern Albireo' and indeed there are similarities between the two pairs. Australian observers have also labelled x Vel with the same moniker. Greg Stone [268], however, sees orange going to red with a blue companion and finds this pair more redolent of α Her. Still surviving with its original discoverer's name intact, Lalande 53 (LAL 53) is 1°.5 NNE of HJ 3945. Its stars are 7.6, 7.7 at 165°, 4″.0, 2010.

Measures

Early measure (DOO)	60°.3	27″.12	1898.20
(Linear:	61°.2	27″.40)	
Recent measure (ANT)	50°.8	26″.8	2008.75
(Linear:	50°.7	26″.7)	

53. δ GEM = STF 1066 = WDS J07201+2159AB

Table 9.53 Physical parameters for δ Gem

STF 1066	RA: 07 20 07.38	Dec: +21 58 56.34	WDS: 152(258)		
V magnitudes	Aa: 3.70	Ab: 5.50	B: 8.18		
(*B* − *V*) magnitudes	A: +0.36	B:			
μ(A)	−31.20 mas yr^{-1}	± 1.34	−15.51 mas yr^{-1}	± 1.08 (DR2)	
μ(B)	−28.23 mas yr^{-1}	± 0.11	13.00 mas yr^{-1}	± 0.09 (DR2)	
π(A)	57.74 mas	± 0.81	56.5 light yr	± 0.8 (DR2)	
π(B)	53.98 mas	± 0.07	60.42 light yr	± 0.08 (DR2)	
Spectra	Aab: F2IV	B: K3V			
Masses (M$_\odot$)	Aa: 1.65	Mb: 1.10	B: 0.75		
Luminosities (L$_\odot$)	A: 8	B: 0.15			
Radii (R$_\odot$)	A: 1.1	B: 2.0			
Catalogues	55 Gem	HR 2777	HD 56986	SAO 79294	HIP 35550
DS catalogues (AB)	H 2 27	STF 1066	BDS 3970	ADS 5983	
Radial velocity	+4.1 km s^{-1}	± 2			
Galactic coordinates	195°.985	+15°.885			

History

The duplicity of δ Geminorum, also known as 55 Gem, was found by the elder Herschel on 13 March 1781, the same night as that when he discovered Uranus. Intriguingly, δ Gem was also on the discovery plate of Pluto which was taken by Tombaugh at Lowell Observatory in 1930. Herschel's notes say 'Double. Extremely unequal. L(arge). w(hite). inclining to r(ed).; S(mall). r(ed). With 227, about 2.5 full diameters of L., with 460, 4 or 5 diameters. Position 85° 51′ s(outh). preceding'.

The Modern Era

Abt [469] in 1965 gave a list of 17 spectroscopic binaries in which the primary is an A star. For δ Gem he combined early radial velocities from Lick Observatory with later ones

Finder Chart

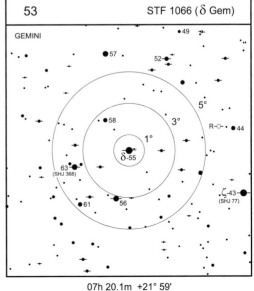

07h 20.1m +21° 59′

Orbit

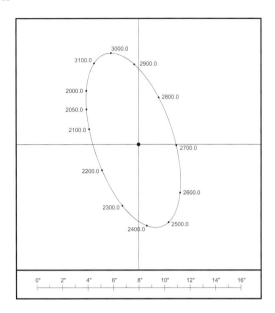

Ephemeris for STF 1066 (2010 to 2100)

Orbit by KSC (2017) Period: 1200 years, Grade: 4

Year	PA(°)	Sep(")	Year	PA(°)	Sep(")
2010.0	227.1	5.59	2060.0	239.5	4.70
2020.0	229.3	5.42	2070.0	242.6	4.53
2030.0	231.6	5.24	2080.0	245.9	4.36
2040.0	234.0	5.06	2090.0	249.5	4.19
2050.0	236.7	4.88	2100.0	253.3	4.04

from Vienna and his own measurements and found a period of 2238 days. The star was single-lined but the derived mass function was 3.8 M$_\odot$, which implied that the secondary was much more massive than the primary star. To try to confirm this, Tremaine *et al.* [273] observed a lunar occultation of δ Gem on 21 November 1974 but found no sign of a secondary several magnitudes fainter within the orientation of the occultation track. Further attempts to resolve A between 1976 and 1993 were also unsuccessful, but in 1997 Hipparcos found an astrometric perturbation and published an orbit [274] which uses some of Abt's spectroscopic elements. The diameters of both components were published by Pasinelli-Fracassini *et al.* [515]. Coryn Bailer-Jones [271] evaluated the astrometric

history of some Hipparcos stars with large parallax on the basis of the derived proper motions obtained by the satellite. In the case of δ Gem he found that the stars had made a close approach to the Solar System about 3 million years ago, coming within 18.5 light years. Data from Gaia will enable much more precise estimates of approaches such as this to be made. The visual orbit is very preliminary but gives a reasonable approximation to the visual measures over almost two centuries.

Observing and Neighbourhood

The colour of the primary is generally agreed to be yellow (although estimates range from pale yellow to very bright yellow) whilst that of the secondary is more problematic. Struve called it blue, whilst Smyth (with a 5.9-inch refractor) noted that it was purple and Webb (3.9-inch refractor) noted 'reddish' in 1852. More recently John Nanson [276] used this star to express doubts about the appearance of colour in faint stars close to very bright primaries. His own experience with a 90-mm aperture telescope indicated that just seeing the companion is not easy and detecting its colour should not be expected. Four degrees WSW is ζ Gem (H 6 9), more commonly called SHJ 77 in WDS. This is an easy pair for binoculars (4.1, 7.7, 347°, 100″, 2015). The primary is a 0″.1 occultation pair not seen double since 1973, and it is a Cepheid. The object 63 Gem is a complex system. SHJ 368 (5.3, 10.9, 324°, 43″, 2006) also has a 10.7 magnitude star at 221°, 133″. H. McAlister found the primary to be a very close binary (MCA 30, P = 46.4 years) whilst Hipparcos added a magnitude 9.2 companion to A (HDS 1050) at 98°, 3″.9, 1991.

Measures

Early (STF)	196°.1	7″.15	1829.72
(Orbit	200°.8	7″.34)	
Recent (ARY)	229°.5	5″.49	2016.22
(Orbit	228°.4	5″.49)	

54. Δ 47 CMA = WDS J07247–3149AB

Table 9.54 Physical parameters for Δ 47 CMa

DUN 47	RA: 07 24 43.87	Dec: −31 48 32.1	WDS: 9144(16) (AC)	
V magnitudes	A: 5.40	B: 7.48	C: 7.58	D: 10.8
$(B − V)$ magnitudes	A: +1.26	B: −0.11		
μ(A)	−21.18 mas yr^{-1}	± 0.17	9.19 mas yr^{-1}	± 0.20 (DR2)
μ(B)	−7.72 mas yr^{-1}	± 0.11	5.42 mas yr^{-1}	± 0.11 (DR2)
μ(C)	−20.17 mas yr^{-1}	± 0.33	14.74 mas yr^{-1}	± 0.72 (DR2)
π(A)	5.14 mas	± 0.11	635 light yr	± 14 (DR2)
π(B)	2.68 mas	± 0.07	1217 light yr	± 32 (DR2)
π(C)	5.45 mas	± 0.21	598 light yr	± 23 (DR2)
Spectra	A: K1+III	B: B	C: B8V	
Luminosities (L$_\odot$)	A: 240	B: 115	C: 25	
Catalogues	HR 2834	HD 58535	SAO 197964	HIP 35957
DS catalogues	DUN 47 (AB-CD)	DAW 129 (AB)	B 1540 (CD)	
Radial velocity	19.90 km s^{-1}	± 0.8		
Galactic coordinates	244°.921	−7°.519		

History

The wide field view of this pair contains the cluster Collinder 140. It was first observed by Lacaille in 1751. In 1922 B. H. Dawson [277] found A was double and seven years later van den Bos [278] doubled the C component. Bernhard Hildebrande Dawson was an Argentine astronomer who was born in Kansas City in 1891 and who worked initially at La Plata Observatory in Argentina. He was perhaps best known as the discoverer of Nova Puppis in 1942. His double star discoveries were made using the 17-inch refractor at La Plata Observatory and were originally denoted by a small greek delta; number 31 is perhaps his most interesting find – a visual pair with a period of 4.56 years.

Finder Chart

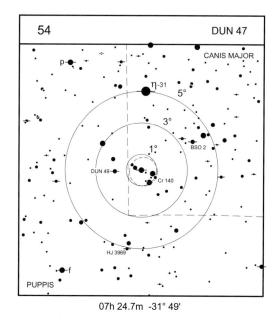

07h 24.7m -31° 49'

The Modern Era

SIMBAD says 'Star in cluster', Collinder 140 – distance 410 ± 30 pc = 1337 light years. Dr Floor van Leeuwen, using Hipparcos data, reported [279] that the cluster distance was 376 pc (1276 light years), so DUN 47 A and C, which appear to be physically connected, happen to lie in the line of sight of the cluster but significantly in front of it, whilst the B star appears to be just a line-of-sight coincidence and actually belongs to the cluster.

Observing and Neighbourhood

This pair can be found 2°.5 S of η CMa, and is visible in 8 × 50 binoculars. Ross Gould reports that both components are white. This is a difficult quadruple for 20-cm, and probably needs more aperture than that. Andrew James noted that the CD pair was easier to see than AB even though it has only half the separation. One degree following is DUN 49 (6.3, 7.0, 54°, 9″, 1999). Two degrees WNW is BSO 2 (6.6, 7.8, 183°, 38″, 2015). Two point five degrees S, a little following, is HJ 3969 (7.1, 8.1, 227°, 17″.5, 2010) – a pair of stars which is moving at about 0″.3 per year together and which may constitute a quadruple system.

Measures

AB			
Early measure (DWS)	309°	2″.17	1922.58
Recent measure (HIP)	311°	2″.12	1991.25
CD			
Early measure (B)	206°	1″.04	1929.22
Recent measure (B)	205°	0″.85	1965.16
AB-CD			
Early measure (BSS)	341°.3	100″.35	1898.20
Recent measure (TMA)	342°.9	98″.49	1999.09

55. α GEM = STF 1110 = WDS J07346+3153AB

Table 9.55 Physical parameters for α Gem

STF 1110	RA: 07 34 36.00	Dec: +31 53 19.1	WDS: 6(1405)	
V magnitudes	Aab: 1.93	Bab: 2.97	Cab: 9.27	
$(B - V)$ magnitudes	Aab:	Bab:	Cab: +1.29	
μ	-191.45 mas yr^{-1}	± 3.95	-145.19 mas yr^{-1}	± 2.95
π (AB)	64.12 mas	± 3.78	51 light yr	± 3
μ(C)	-201.49 mas yr^{-1}	± 0.09	-97.11 mas yr^{-1}	± 0.07 (DR2)
π(C)	66.23 mas	± 0.05	49.24 light yr	± 0.04 (DR2)
Spectra	Aa: A1V	Ab: dM1e	Ba: A4Vm	Bb: dM1e
	Ca: dM1e	Cb: dM1e		
Masses (M$_\odot$)	Aa: 2.74	Ab: 0.37	Ba: 2.26	Bb: 0.53
	Ca: 0.60	Cb: 0.60		
Luminosities (L$_\odot$)	A: 35	B: 13	C: 0.03	
Radii (R$_\odot$)	Ca: 0.60	Cb: 0.60		
Catalogues (A/B)	66 Gem	HR 2891/0	HD 60178	SAO 60198	HIP 36850
DS catalogues	Mayer 21 (AB)	H 2 1 (AB)	STF 1110 (AB)	BDS 4122	ADS 6175
Radial velocity	5.40 km s^{-1}	± 0.5		
Galactic coordinates	187°.441	+22°.479		

History

This is a star with an important place in the history of binary-star astronomy. It was used by William Herschel in 1803 to demonstrate that pairs of stars were genuinely connected by a mutual force of gravity. In the *Histoire de l'Academie* (French Academy of Sciences) publication for 1733 there is a report of observations by Cassini. The translated note reads: 'The night of 3 to 4 May [1678], the Moon being very close to the highest of the three stars of the front of the Scorpion, M. Cassini noticed when observing this part of the sky that this star was double as well as alpha Arietis and the preceding head star of Gemini [Castor].' Thus it seems that Cassini discovered three double stars, including Castor, but it is not known when he actually discovered the duplicity of Castor. It was first measured by Bradley and Pound in 1719 [287]. On 30 March of that year they noted that 'the direction of the double star α

Finder Chart

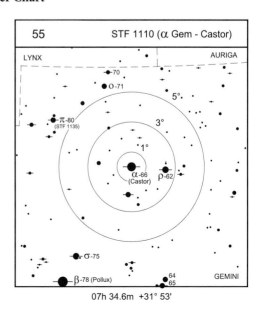

55 STF 1110 (α Gem - Castor)

LYNX   AURIGA

70
O-71
5°.
π-80
(STF 1135)   3°
1°
α-66   ρ-62
(Castor)

σ-75

64
β-78 (Pollux)   65   GEMINI

07h 34.6m +31° 53'

Orbit

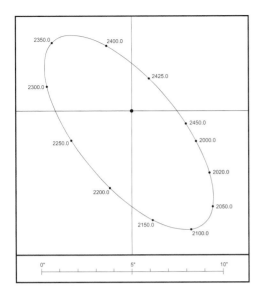

Ephemeris for STF 1110 AB (2015 to 2060)

Orbit by Doc (2014g) Period: 459.8 years, Grade: 3

Year	PA(°)	Sep(")	Year	PA(°)	Sep(")
2015.0	54.6	5.04	2040.0	44.1	6.43
2020.0	52.1	5.38	2045.0	42.4	6.62
2025.0	49.8	5.69	2050.0	40.9	6.78
2030.0	47.7	5.96	2055.0	39.4	6.92
2035.0	45.8	6.21	2060.0	37.9	7.03

of Gemini was so nearly parallel to a line through κ and σ of Gemini that, after many trials, we could scarce determine on which side of σ the line from κ parallel to the line of direction tended; if on either, it was towards β.' Unfortunately they did not attempt to measure or estimate the distance. In 1896 Bélopolsky [282], using a prism spectrograph on the 30-inch Pulkovo refractor, took some spectra of Castor four nights apart. He noted that the radial velocity on each occasion was significantly different and made some further observations to try to determine the cause of this effect. He found that the variation in velocity arose in the fainter component B and that there was a periodic variation over 2.91 days. In 1895 H. D. Curtis [283] determined that the brighter component was also a spectroscopic binary, with a period of nine days. Star C too is a spectroscopic binary. Its periodic variability was found by Adams & Joy [285] in 1917 whilst the spectroscopic orbit was first determined in 1926 by Joy & Sanford [286]. The orbit of the star is close to the line of sight and leads to regular eclipses, resulting in the allocation of the variable star name YY Gem.

The Modern Era

Measures of AB now cover almost 300 years but it will be some years yet before the orbit of this pair can be regarded as

definitive, as the apastron end of the apparent ellipse has not yet been observed. Nevertheless, a number of orbits published in recent years give periods in the range 445 to 467 years. The distance to the system was rather poorly defined by Hipparcos possibly because of the combination of star brightness and orbital motion. Torres & Ribas [280] reworked the distance of C, and hence AB, as 66.90 ± 0.63 milli-arcseconds. Recent photometry of YY Gem by Butler *et al.* [281] put the period at 19.54 hours. A fine light curve can be found on Bruce Gary's website [289]. Recent measurements appear to show that the secondary eclipse is now deeper than the primary, leading to suggestions that there has been a change in star spot activity. Castor is a member of a loose group of bright stars which are moving through space with a similar velocity and direction. The group also includes Vega and Fomalhaut and the visual binaries ψ Vel and μ Dra. The age of this group has been estimated at 400 million years. If the total mass of the Castor system is taken to be 7.1 $M_\odot$, and the separation between AB and C is 70″, then the projected rotation period of Castor C around AB is about 13,200 years.

Observing and Neighbourhood

Castor and Pollux are the bright leaders of the constellation of Gemini. They are easily distinguished, as Castor appears brilliant white whilst Pollux glows orange red. Castor is one of the most spectacular and visited pairs in the sky (more than 1400 measures to date). The companion, having passed within 2″ of A around 1970, is now heading out to the point in the orbit where the two stars are at maximum separation (7″.2 in 2080 or so). The pair is thus visible in small telescopes at all times. Castor C, a ninth magnitude red star 70″ away in PA 166° is also visible in 7.5-cm. Author RWA has measured Castor AB every year since 1990. In spring 2016, the positions of the stars were as given below. Three point five degrees ENE is the 5th magnitude star π Gem, which the elder Herschel found to be double (H 4 53) and which is also catalogued at STF 1135. There are two magnitude 11 stars (11.4 at 214°, 19″.6, 2016, and 11.2 at 343°, 92″, 2016).

Measures

Early measure (STF)	262°.6	4″.42	1826.26
(Orbit	261°.7	4″.52)	
Recent measure (ARY)	52°.8	5″.38	2018.30
(Orbit	53°.0	5″.27)	

56. STF 1126 CMI = WDS J07401+0514AB

Table 9.56 Physical parameters for STF 1126 CMi

STF 1126	RA: 07 40 06.994	Dec: +05 13 51.89	WDS: 144(268)	
V magnitudes	A: 6.55	B: 6.96		
$(B - V)$ magnitudes	A: −0.02	B: −0.07		
μ(A)	−0.15 mas yr^{-1}	± 0.29	−17.76 mas yr^{-1}	± 0.18 (DR2)
μ(B)	−3.16 mas yr^{-1}	± 0.61	−16.55 mas yr^{-1}	± 0.43 (DR2)
π(A)	4.96 mas	± 0.22	658 light yr	± 29 (DR2)
π(B)	2.33 mas	± 0.37	1400 light yr	± 222 (DR2)
Spectra	A0III			
Luminosities (L$_\odot$)	A: 80	B: 250		
Catalogues	HR 2950	HD 61563	SAO 115773	
DS catalogues	H 1 23	STF 1126	BDS 4193	ADS 6263
Radial velocity (A/B)	17.0 km s^{-1}	± 5		
Galactic coordinates	213°.790	+13°.203		

History

William Herschel discovered this pair on 21 November 1781, and he noted that it was 'a most minute double star'. He measured the position angle but simply estimated the distance in terms of the diameter of the stars. Its likely that the stars were closer then than they are today, and a measure of separation would have helped to define the modern orbit by Zirm [288], which gives a predicted separation of 1″.33. Herschel thought the stars closer than those in η CrB, which at the time were 1″.03 apart, but he recognised that the separation estimated in this way depended upon the brightness of the stars. Herschel added a fainter star some 120″ away, preceding the pair. This is not star C (magnitude 11.99) as given in the WDS at a distance of 43″ for 2009. F. G. W. Struve measured the pair in 1820 and it has been followed regularly until today, and, whilst the observed motion in angle is but 50°, with little change in separation, Lewis [194] believed that a maximum distance was reached around 1840.

Finder Chart

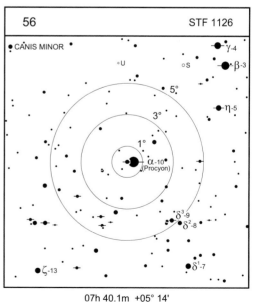

07h 40.1m +05° 14'

Orbit

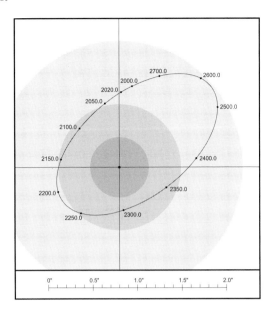

Ephemeris for STF 1126 AB (2015 to 2060)

Orbit by Zir (2015a) Period: 752 years, Grade: 4

Year	PA(°)	Sep(")	Year	PA(°)	Sep(")
2015.0	176.3	0.84	2040.0	187.7	0.75
2020.0	178.4	0.82	2045.0	190.4	0.73
2025.0	180.6	0.80	2050.0	193.1	0.71
2030.0	182.9	0.78	2055.0	195.9	0.70
2035.0	185.2	0.77	2060.0	198.9	0.69

The Modern Era

The pair STF 1126 was not observed by Hipparcos but the parallax given in the table was determined using the Multi-channel Astrometric Photometer on the 30-inch refractor at Allegheny Observatory [291]. This was carried out as part of an investigation of the masses of nearby Procyon A and B. The nature of the pair STF 1126 is thrown into doubt by the Gaia DR2 result. The parallaxes of the two stars are significantly different, and the observations made so far do not seem to show unmistakeable curvature.

Observing and Neighbourhood

This close pair is easy to find as it lies just 12 arcminutes east of Procyon, and for that reason it was too close on the sky to the bright star for the Hipparcos satellite to make meaningful measurements. At its current separation it represents a good resolving test for 12-cm aperture. RWA measured 175° and 1″ in 2015, a movement of 10° since a previous measure in 1992, with the separation unchanged. Procyon is close to the Sun and has an annual proper motion of 1″.3 per year. Like Sirius, it has a white dwarf companion, which was found in the same way – by the regular 'wobble' of Procyon against the background stars as it moved across the sky. The white dwarf was finally seen by J. M. Schaeberle at Lick Observatory in 1896 using the 36-inch refractor [293]. It is outstandingly difficult to see. The visual magnitudes of the two stars are 0.3 and 10.8 (2002) and the maximum separation is 5″. The pair is currently widening; in 2017.0 the white dwarf was found at 304° and 4″.32 whilst the widest separation will occur in 2029 when the position is 8° and 5″.14. The orbit is well defined thanks to direct imaging using WFPC2 on the *HST* [290]. As luck would have it, the lowest possible exposure time (0.11 seconds) on the camera just showed the white dwarf without saturating the detector with the overwhelmingly bright primary. The resulting dynamical parallax (0″.285) is almost identical to that of Hipparcos (0″.28); this represents a distance of 11.44 light years.

Measures

Early measure (STF)	132°.0	1″.46	1829.43
(Orbit	132°.9	1″.35)	
Recent measure (ARY)	177°.3	1″.00	2016.46
(Orbit	177°.0	0″.83)	

57. $\gamma^{1,2}$ VEL = Δ 65 = WDS J08095−4720

Table 9.57 Physical parameters for γ Vel

DUN 65	RA: 08 09 31.95	Dec: −47 20 11.7	WDS: 2543(39)	
V magnitudes	A: 1.79	B: 4.14	C: 7.76	D: 9.4
$(B - V)$ magnitudes	A: −0.25	B: −0.18	C: 0.0	
μ(A)	−6.71 mas yr^{-1}	± 0.58	10.78 mas yr^{-1}	± 0.59 (DR2)
μ(B)	0.50 mas yr^{-1}	± 0.22	3.69 mas yr^{-1}	± 0.26 (DR2)
π(A)	3.56 mas	± 0.33	916 light yr	± 85 (DR2)
π(B)	3.69 mas	± 0.26	884 light yr	± 63 (DR2)
π(A)	2.97 mas	± 0.07	1100 light yr	± 20 (dyn.)
π(A)	2.81 mas	± 0.03	1160 light yr	± 11 (MSF)
Spectra	Aa: WC8	Ab: O7.5e	B: B2III	C: F0
Masses (M$_\odot$)	Aa: 9	Ab: 28.5	B: 14	
Radii (R$_\odot$)	Aa: 6	Ab: 17	B:	
Luminosities (L$_\odot$)	A: 12500	B: 1340	C: 50	D: 10
Catalogues (A/B)	HR 3207/6	HD 68273/243	SAO 219504/1	HIP 39953
DS catalogues	TOK 2 (Aa)	DUN 65 (AB)	WSI 55 (BaBb)	I 1175 (DE)
Radial velocity (A/B)	15.00 km s^{-1}	± 3.1	20 km s^{-1}	± 10
Galactic coordinates	262°.803	−7°.686		

History

James Dunlop observed the three bright components of γ Velorum at Parramatta in 1826. In 1911 Moore [295], using the Mills expedition equipment to Chile, found that the brightest star had a radial velocity which varied by 75 km s^{-1}. Using the 26.5-inch refractor at Johannesburg, R. T. A. Innes [302] found that a fourth star of magnitude 9 which is 34″ distant from C in PA 123° had a companion star, of $V = 12.8$, 1″.5 distant, which has shown some sign of relative motion between 1926 and 1949 but which has not been measured since. Robert Rossiter also independently found this star at Bloemfontein [303].

The Modern Era

Jeffries *et al.* [300] used main sequence fitting to find the distance to the cluster of stars in the association Vel OB2 which surrounds the brilliant pair γ Velorum. This agrees with distances of the close binary AaAb derived from interferometric observations and with the trigonometrical parallax derived by Hipparcos. The primary star is a double-lined spectroscopic binary with a period of 78.5 days and consists of a Wolf–Rayet star (the nearest such object to the Sun) and a late O star; between them they radiate with the energy of about 400,000 Suns. The third star (B), 41″ distant, is probably connected although the proper motions are small and the uncertainties

Finder Chart

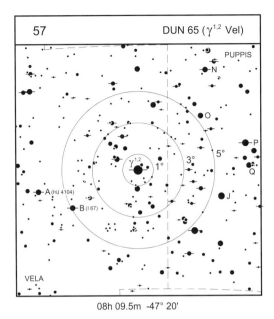

57   DUN 65 ($\gamma^{1,2}$ Vel)

PUPPIS

N

O

P

5°

$\gamma^{1,2}$ 1° 3° Q

A (HJ 4104)

J

B (I 67)

VELA

08h 09.5m -47° 20'

are relatively large. The ground-based interferometer SUSI at Narrabri in Australia has optically defined the orbit of the bright star and, together with the spectroscopic information, allows significant detail to be revealed. The B star is also a spectroscopic binary with a period of 1.482 days [296] and an interferometric pair, as found by the USNO observers [410]. The masses and radii of the close binary come from the work by North *et al.* [299]. Tokovinin *et al.* [297], using the 3.6-metre reflector at ESO, found what appears to be a K4 dwarf only 4″.7 away from A but 14.8 magnitudes fainter in the visual. It is not yet clear whether it actually has an immediate physical connection to the system or whether it is

a pre-main-sequence star in the association. Gaia DR2 shows a significant increase in the parallax of the two bright visual components. It should be taken as a provisional result as the stars are very bright.

Observing and Neighbourhood

The small-telescope user will see a brilliant pair of white stars with a third star of magnitude 6.7 at 152° and a fourth of magnitude 9.3 at 142°, 93″. The ASNSW observers saw a hint of blue in the bright star and recorded white (with a hint of yellow) and orangey-white for stars C and D. The surrounding stars on the finder chart show many more pairs, two of which are well worth seeking out. RWA measured them both in 2008. These lie about 3° E of γ; B Vel (5.1, 6.1, 139°, 0″.9, 2013) was one of Robert Innes' first discoveries (I 67) and is a beautiful pair for 15-cm. Smaller apertures will be able to see HJ 4104 well. This is also known as A Vel (5.5, 7.2, 244°, 3″.5, 2013) and has widened from 2″.3 in 1835. There is a magnitude 9.2 star at 40° and 20″.

Measures

AB			
Early measure (HJ)	220°.2	41″.16	1835.11
Recent measure (ANT)	218°.9	41″.0	2000.01
AC			
Early measure (JC)	151°.4	61″.80	1851.19
Recent measure (TMA)	152°.2	62″.60	2000.01

58. ζ CNC = STF 1196 = WDS J08122+1739AB,C

Table 9.58 Physical parameters for ζ Cnc

STF 1196	RA: 08 12 12.71	Dec: +17 38 53.3	WDS: 7(1181) (AB)	WDS: 30(613) (AC)
V magnitudes	A: 5.30	B: 6.25	C: 5.85	D: ~10
$(B - V)$ magnitudes	A: +0.55	B: +0.69	C: +0.63	
$\mu(A)$	116.98 mas yr^{-1}	± 0.31	−148.89 mas yr^{-1}	± 0.21 (DR2)
$\mu(B)$	58.77 mas yr^{-1}	± 0.24	−120.89 mas yr^{-1}	± 0.14 (DR2)
$\mu(C)$	21.90 mas yr^{-1}	± 0.87	−139.43 mas yr^{-1}	± 0.72 (DR2)
$\pi(A)$	41.30 mas	± 0.17	79.0 light yr	± 0.3 (DR2)
$\pi(B)$	40.96 mas	± 0.16	79.6 light yr	± 0.3 (DR2)
$\pi(C)$	42.13 mas	± 0.48	77.4 light yr	± 0.9 (DR2)
μ	A: +27.61 mas yr^{-1}	± 1.11	B: −151.73 mas yr^{-1}	± 0.99
Spectral types	A: F7V	B: F9V	C: G0V	D: M + M (?)
Masses (M$_\odot$)	A: 1.11	B: 1.00	C: 0.99	D: 0.93
Luminosities (L$_\odot$)	A: 3.5	B: 1.5	C 2.0 :	D: 0.05
Catalogues (A/B/C)	HR 3208/9/10	HD 68257/5/6	SAO 97645/6	HIC 40167
DS catalogues	Mayer 22 (AB,C)	H 1 24(AB)	STF 1196	HUT 1 (DaDb)
Radial velocity (AB,C)	−7.93 km s^{-1}	± 0.9		
Galactic coordinates	205°.301	+25°.531		

History

The wide components of this pair were first noted, according to Roger Griffin [306], by Flamsteed in the course of his regular observations from Greenwich. Flamsteed recorded only the brighter component on 22 March 1680. Subsequent measures have shown that the two components revolve around each other once every 1115 years or so, according to W. D. Heintz [205]. As the ephemeris below shows, the movement in angle amounts to about 4° in 10 years and the separation hardly changes.

About a century later, when William Herschel was conducting a survey for double stars, he came across zeta Cancri and noted on 21 November 1781 'If I do not see extremely ill this morning (4 am) the large star consists of two'. When including the pair in his first published list of new double stars he noted 'Minute treble ...' and it became H 1 24 (see the table for the explanation of Herschel's categories). The pair was not noted by Herschel during a follow-up set of observations in 1802, and it was not until 1825 that Admiral Smyth, observing from Passy near Paris, saw it again at a separation similar to that in 1781.

From then on, the significant orbital motion of AB recommended itself to the observers of the day and it was taken up enthusiastically by F. G. W. Struve at Dorpat and by Dawes, O. Struve, and others. By the middle of the nineteenth century the period was generally agreed to be about 60 years. Observers would make measurements of star C with respect to

Finder Chart

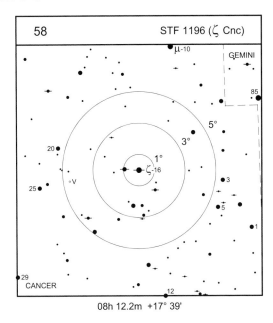

08h 12.2m +17° 39'

Orbit for STF 1196 AB,C

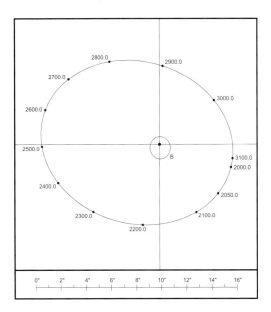

Ephemeris for STF 1196 AB,C (2000 to 2180)

Orbit by KSC (2017) Period: 1115 years, Grade: 4

Year	PA(°)	Sep(")	Year	PA(°)	Sep(")
2000.0	72.8	5.90	2100.0	29.1	5.94
2020.0	64.0	5.93	2120.0	20.4	5.96
2040.0	55.3	5.95	2140.0	11.8	6.01
2060.0	46.6	5.95	2160.0	3.4	6.08
2080.0	37.8	5.94	2180.0	355.2	6.19

came to the conclusion that there another star (D) rotating around C and disrupting its smooth motion around AB.

Surprisingly, Burnham [644] was not convinced of the variable motion of C and went out of his way to compare it to a seventh magnitude star 108″ away. From measures over a period of 14 years by Burnham and Barnard with the Lick 36-inch, he concluded that there was no real evidence for a perturbation.

The Modern Era

Speculation about the nature of D abounded for many years. It was reported to have been observed in the infrared in 1983 but details were never made public. Another speculation involved the idea that D was a white dwarf, and it was not until 2000 that a direct detection of D in the K band was made by J. B. Hutchings & Griffin [308] using the CFHT telescope on Mauna Kea. This showed that the star was actually a late-type dwarf, around spectral type M2. However, it appeared too luminous for its type and Roger Griffin believes that it

Orbit for STF 1196 AB

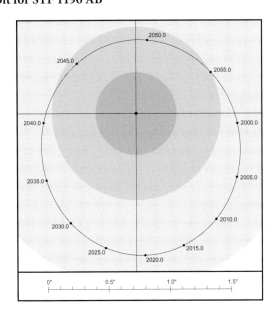

Ephemeris for STF 1196 AB (2016 to 2034)

Orbit by WSI (2006b) Period: 59.582 years, Grade: 1

Year	PA(°)	Sep(")	Year	PA(°)	Sep(")
2016.0	16.8	1.12	2026.0	343.3	1.09
2018.0	10.2	1.13	2028.0	336.0	1.06
2020.0	3.6	1.13	2030.0	328.3	1.03
2022.0	357.0	1.12	2032.0	319.9	0.98
2024.0	350.2	1.11	2034.0	310.8	0.93

the centre of light of stars A and B but also, if the seeing was good, with respect to A and B separately. Otto Struve [305] in 1874 discussed the motion of AB and of C around AB and

Measure of STF 1196 AB with 8″ Refractor

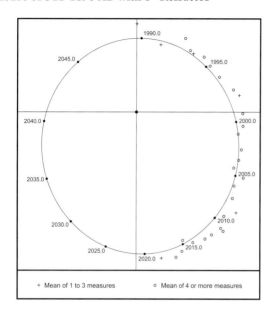

+ Mean of 1 to 3 measures   ○ Mean of 4 or more measures

Year	PA(°)	Sep(″)	No.	Year	PA(°)	Sep(″)	No.
1990.17	179.8	0.70	1	2004.21	61.1	0.92	8
1991.23	159.7	0.57	3	2005.23	59.4	1.03	7
1992.23	145.6	0.71	5	2006.25	54.0	1.00	5
1993.25	137.7	0.66	7	2007.23	49.2	1.05	6
1993.28	134.6	0.66	3	2008.24	45.9	1.15	3
1994.21	128.0	0.71	6	2009.29	41.2	1.10	4
1995.18	121.0	0.71	7	2010.27	37.2	1.19	4
1996.23	112.1	0.74	7	2011.27	37.1	1.16	4
1997.24	102.4	0.82	7	2012.23	32.6	1.16	5
1998.21	98.8	0.86	1	2013.22	29.5	1.16	5
1999.26	89.2	0.88	5	2014.22	24.4	1.14	6
2000.27	82.8	0.89	7	2015.19	20.6	1.09	6
2001.21	78.3	0.86	8	2016.19	18.8	1.17	5
2002.23	71.0	0.92	7	2017.21	15.8	1.20	5
2003.25	65.7	0.93	8	2018.27	10.1	1.18	3

is a close pair of M dwarfs. A confirming observation of D was made by A. Richichi [307] using infrared observations of lunar occultations. As luck would have it, the four stars appeared approximately in a straight line on the sky and as ζ Cnc emerged from the lunar limb the individual signals from stars A, B, C, and D were seen and also, tantalisingly,

another signature corresponding to a star some 0″.06 from C. It is not clear how this component would fit into the system dynamically. No confirming observations of it have been made, and it could be a background object.

Observing

The object ζ Cancri is a visual triple star which is most easily found by extending a line from Castor through Pollux by 2.5 times the distance between the two bright stars. It is an interesting system on several levels. For the small telescope user the star is an attractive fairly close pair which can be readily split in a 7.5-cm aperture. With this aperture the duplicity of the brighter component should be apparent even if the images are not entirely resolved. At the time of writing (2018), AB is at maximum width and will remain above 1″ for the next decade at least, so this is an excellent opportunity to resolve it – providing an aperture of at least 10-cm is used. If the 1115-year period proves correct then AB-C never gets closer than 5″.4 and thus is always visible in a small aperture.

Measures

AB			
Early measure (STF)	35°.4	1″.04	1828.80
(Orbit	47°.3	1″.02)	
Recent measure (ARY)	10°.1	1″.18	2018.27
(Orbit	9°.4	1″.13)	

AB-C			
Early measure (STF)	158°.0	5″.53	1828.99
(Orbit	156°.8	5″.57)	
Recent measure (ARY)	66°.0	5″.86	2018.28
(Orbit	64°.8	5″.93)	

59. $\kappa^{1,2}$ VOL = BSO 17 = WDS J08198−7131AB

Table 9.59 Physical parameters for κ Vol

BSO 17	RA: 08 19 49.00	Dec: −71 30 54.0	WDS: 4508(26)	
V magnitudes	A: 5.33	B: 5.61	C: 7.90	
$(B - V)$ magnitudes	A: −0.07	B: −0.11	C: +0.91	
$\mu(A)$	−17.65 mas yr^{-1}	± 0.14	34.61 mas yr^{-1}	± 0.19 (DR2)
$\mu(B)$	−17.03 mas yr^{-1}	± 0.27	36.96 mas yr^{-1}	± 0.30 (DR2)
$\mu(C)$	−18.24 mas yr^{-1}	± 0.04	35.79 mas yr^{-1}	± 0.06 (DR2)
$\pi(A)$	7.80 mas	± 0.09	418 light yr	± 5 (DR2)
$\pi(B)$	8.00 mas	± 0.15	408 light yr	± 8 (DR2)
$\pi(C)$	2.97 mas	± 0.03	1098 light yr	± 11 (DR2)
Spectra	A: B9III-IV	B: A0IVMn	C:	
Luminosities (L$_\odot$)	A: 100	B: 70	C: 65	
Catalogues (A/B)	HR 3301/2	HD 71046/66	SAO 256497/9	HIP 40817/34
DS catalogues	BSO 17			
Radial velocity (A/B)	+36.00 km s^{-1}	± 7.4	−6.5 km s^{-1}	
Radial velocity (C)	+30.16 km s^{-1}	± 0.14 (DR2)		
Galactic coordinates	284°.825	−19°.046		

History

This object was noted by John Herschel on 21 February 1835. He gave the stars' IDs as Brisbane 2018+2022+2023. The brightest star is also known as LAC 3355. The Brisbane catalogue was instigated by Sir Thomas Brisbane, the Governor of New South Wales, who in 1822 set up an observatory at Parramatta near Sydney and equipped it with with several instruments, including a five-and-a-half foot transit instrument by Troughton and a two-foot mural circle by the same maker. Brisbane employed Carl Rümker and James Dunlop to do the observing and the latter actually did the greater part of the practical side of this project. The result was a catalogue of 7935 stars; the work of reducing the observations and preparing the lists for publication was done by William Richardson, a member of staff at the Royal Observatory, Greenwich.

The Modern Era

The A and B stars have similar proper motions and parallaxes and must be regarded as being physically linked. Star C has a similar proper motion but appears to be about three times further away than the bright pair, according to DR2. The derived parallax is 2.97 ± 0.03 mas whilst DR1 gives the parallax of C as 6.38 ± 0.69 mas. The radial velocity of A has a considerable scatter, which might indicate higher multiplicity.

Finder Chart

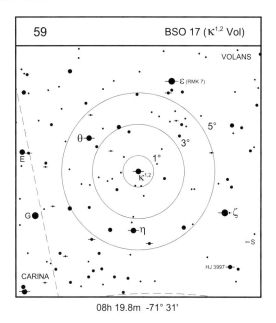

| 59 | BSO 17 ($\kappa^{1,2}$ Vol) |

VOLANS

ϵ (RMK 7)

5°

θ

3°

1°

E

$\kappa^{1,2}$

G

ζ

η

S

HJ 3997

CARINA

08h 19.8m -71° 31'

Observing and Neighbourhood

This coarse triple can be found 3° SSE of ϵ Vol (RMK 7, 4.4, 7.3, 23°, 6″.0, 2010). Volans is a kite-shaped group of mostly fourth and fifth magnitude stars. A fine John Herschel pair (HJ 3997) can be found 4° SW of κ. It is 7.0, 7.1, 306°, 1″.9, 1996 and has increased in position angle by 25°, with little evident change in separation, since it was found in 1836.

Measures

AB			
Early measure (HJ)	55°	64″.97	1835.92
Recent measure (ANT)	58°.1	64″.66	2010.28
AC			
Early measure (GLI)	45°.5	101″.0	1850.24
Recent measure (ANT)	48°.0	99″.42	2010.28

60. δ VEL = I 10 = WDS J08447–5443AaAbB

Table 9.60 Physical parameters for δ Vel

I 10	RA: 08 44 42.23	Dec: −54 42 31.8	WDS: 1283(65)		
V magnitudes	A: 1.99	B:	C: 11	D: 13.5	F: 5.70
(B − V) magnitudes	A: +0.05	B: 5.5	F: +0.47		
μ(A)	28.99 mas yr^{-1}	± 0.50	−103.35 mas yr^{-1}	± 0.54	
π(A)	40.49 mas	± 0.39	80.6 light yr	± 0.8 (Hipp.)	
π(A)	39.8 mas	± 0.4	81.9 light yr	± 0.8 (dyn.)	
μ(F)	24.32 mas yr^{-1}	± 0.18	−91.71 yr^{-1}	± 0.15 (DR2)	
π(F)	41.22 mas	± 0.09	79.1 light yr	± 0.2 (DR2)	
Spectra	Aa: A1V	Ab: A4V	B: F8V	F: F6V	
Masses (M$_\odot$)	Aa: 2.43	Bb: 2.27	C: 1.35		
Radii (R$_\odot$)	Aa: 31.7	Bb: 2.02	B: 1.39		
Luminosities (L$_\odot$)	A: 80	B: 3	F: 2.5		
Catalogues (A/F)	HR 3485/3570	HD 74956/76653	SAO 236232/236405	HIP 42913/43797	
DS catalogues	KEL 1 (AaAb)	I 10 (AB)	HJ 4136 (AC)	SHY 49 (AF)	
Radial velocity	+2.2 km s^{-1}	± 1.78			
Galactic coordinates	272°.079	−7°.371			

History

In 1835 John Herschel noted a faint companion (now called C) to δ Vel at a distance of about 80″, and he estimated its magnitude on two occasions as 10 and 12. About a century later, W. H. van den Bos added a magnitude 13.5 companion to C at a distance of 5″. In 1894, Solon Bailey [248], using the 13-inch Harvard refractor which had been set up at Arequipa in Peru, discovered a close companion of magnitude 5 at a distance of 3″. Robert Innes came across it a year later during one of his early sweeps for new double stars whilst living in Sydney and, as he published his observation first, the pair was given the catalogue number I 10. Owing to the relative difficulty of measuring the system, early measures were scattered and estimates of the brightness of B varied wildly. In 1897 See made it 12 whilst in 1912 Innes estimated 4. The apparently slow motion resulted in only an occasional measure being made of the two stars over the next 60 years or so.

The Modern Era

No measures were made at all between 1953 and 1978, when Tango et al. [311] announced the discovery of a third star in the system using speckle on the Anglo-Australian Telescope. It later transpired that this was not an extra component; rather it was the Innes companion, which had closed in considerably. Using Hipparcos observations and a speckle measure from 1999 Andreas Alzner [310] calculated an apparent orbit with

Finder Chart

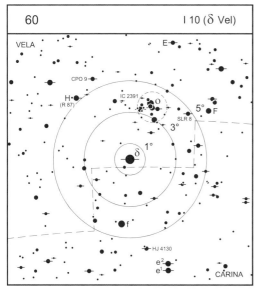

60   I 10 (δ Vel)

08h 44.7m -54° 43'

Orbit

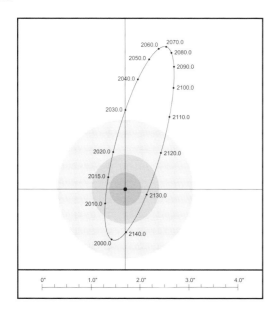

Ephemeris for I 10 AB (2016 to 2034)

Orbit by Msn (2011b) Period: 146.97 years, Grade: 4

Year	PA(°)	Sep(")	Year	PA(°)	Sep(")
2016.0	224.4	0.48	2026.0	184.5	1.27
2018.0	208.3	0.61	2028.0	181.9	1.42
2020.0	198.6	0.77	2030.0	179.9	1.57
2022.0	192.2	0.94	2032.0	178.1	1.71
2024.0	187.8	1.11	2034.0	176.7	1.84

a period of 142 years. In 1997 Sebastian Otero *et al.* [312] noticed visually a series of drops in magnitude of up to 0.3 magnitude, whilst Paul Fieseler had independently found similar behaviour whilst examining data from the Galileo satellite star tracker. They found that the star underwent primary and secondary eclipses with a period of 45.15 days. The primary of δ Vel is thus the brightest known eclipsing variable. More recently, a group of astronomers, including Pribulla, Kervella, Kellerer *et al.* [314], dedicated four papers to the system and, using the NACO adaptive optics instrument on the VLT, determined an astrometric orbit for the eclipsing pair. Using star B as a reference they measured the wobbles in the motion of A with an accuracy of 50 microarcseconds, and found that the orbit of AaAb is 2 mas in diameter. They also recalculated the AB orbit, obtaining values of the elements similar to those of Alzner but with much smaller errors. More recently, Shaya & Olling [270] showed that the nearby bright star HIP 43797, which is 90′ to the SE, has a similar proper motion to δ, and they report that it is almost 100% certain that the connection is physical.

Observing and Neighbourhood

The resolution of δ Vel requires considerable aperture owing to the brightness of the primary, the difference in magnitude between A and B, and the current small separation. No visual sightings of B have been recorded for more than 60 years, but the stars are now separating quite quickly. The open cluster IC 2391 can be found 2° NNW. There are a number of fine pairs in this field. The brightest is H Vel = R 87 (4.7, 7.7, 333°, 2″.6, 2008). Just 1° NW is CPO 9 (6.6, 8.2, 83°, 2″.9, 1991). A good test for 15-cm is SLR 8 (6.1, 7.1, 281°, 0″.8, 2014). Finally, 3° S and slightly W is HJ 4130 (6.5, 8.0, 240°, 4″.0, 1998).

Measures

Early (I)	173°.2	2″.45	1900.19
(Orbit	169°.5	2″.52)	
Recent (TOK)	262°.9	0″.37	2013.13
(Orbit	263°.8	0″.38)	

61. ι CNC = STF 1268 = WDS J08467+2846AB

Table 9.61 Physical parameters for ι Cnc

STF 1268	RA: 08 46 41.82	Dec: +28 45 35.6	WDS: 517(122)		
V magnitudes	A: 4.13	B: 5.99			
(*B − V*) magnitudes	A: +1.12	B: +0.37			
μ(A)	−23.45 mas yr^{-1}	± 0.45	−43.87 mas yr^{-1}	± 0.37 (DR2)	
μ(B)	−24.49 mas yr^{-1}	± 0.09	−44.55 mas yr^{-1}	± 0.06 (DR2)	
π(A)	10.03 mas	± 0.03	325 light yr	± 1 (DR2)	
π(B)	9.41 mas	± 0.06	347 light yr	± 2.2 (DR2)	
Spectra	A: G8IIIa	B: A2V			
Luminosities (L$_\odot$)	A: 180	B: 35			
Radius (R$_\odot$)	A: 24	± 2			
Catalogues (A/B)	48 Cnc	HR 3475/4	HD 74739/8	SAO 80416/5	HIP 43103/0
DS catalogues	Mayer 25 (AB)	H 4 52 (AB)	STF 1268 (AB)	ADS 6988	BDS 3970
Radial velocity (A/B)	15.74 km s^{-1}	± 0.13	25.00 km s^{-1}	± 1.5	
Galactic coordinates	195°.883	+36°.544			

History

Noted by Christian Mayer in 1777, ι Cnc was later swept up by William Herschel on 8 February 1782 when he noted that the stars were considerably unequal and the colours were reddish white and dusky garnet.

The Modern Era

DR2 seems to indicate a significant difference in distance, 22 light years, between the two components, casting some doubt on the physical relationship even though the proper motions are very similar.

Finder Chart

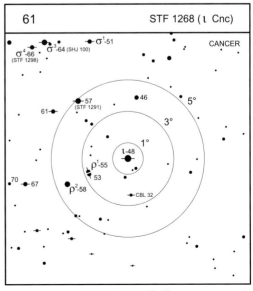

08h 46.7m +28° 46'

Observing and Neighbourhood

ι Cnc is one of the brightest and best double stars in the winter sky. Greg Stone sees the colours as pale yellow and deep sky blue and therefore he considers that it is not a rival to Albireo. Easily a naked-eye object, it is located 14° N and slightly following Praesepe. Alternatively, it is 15° following Pollux. The object 57 Cnc = STF 1291 is a fine close pair of stars with magnitudes 6.09 and 6.37 just 2°.5 NE of ι. One of William Herschel's discoveries, it has slowly widened from 1″ to the current position of 310°.0, 1″.53 (Prieur 2012.24). STF 1298, also known as σ^4 Cnc, magnitudes 5.95, 8.96 is a fairly tough test for the small aperture. RWA found it at 137°.8, 4″.66 on 2016.37. Less than half a degree W in the same low-power field is σ^3 Cnc, also known as SHJ 100, with a deep yellow primary and faint distant companion (5.32, 8.97 115°.7, 88″.26, 2016.37). Just over 1° due S is CBL 32 (7.4, 10.7, 174°, 41″.0, 2010).

Measures

Early measure (STF)	307°.1	30″.46	1828.04
Recent measure (WSI)	307°.6	30″.40	2014.25

62. ε HYDRAE = STF 1273
= WDS J08468+0625AB,C

Table 9.62 Physical parameters for ε Hydrae

STF 1273	RA: 08 46 46.65	Dec: +06 25 08.1	WDS: 66(428)	
V magnitudes	A: 3.49	B: 5.0	C: 6.66	D: ~12.5
μ	A: −228.11 mas yr^{-1}	± 0.98	−43.82 mas yr^{-1}	± 0.64
π	25.23 mas	± 0.98	129.3 light yr	± 5.0
μ(C)	−183.00 mas yr^{-1}	± 0.21	−23.35 mas yr^{-1}	± 0.15 (DR2)
μ(D)	−209.50 mas yr^{-1}	± 0.08	−35.95 mas yr^{-1}	± 0.05 (DR2)
π(C)	25.71 mas	± 0.14	126.9 light yr	± 0.7 (DR2)
π(D)	26.23 mas	± 0.05	124.4 light yr	± 0.2 (DR2)
Spectral types	A: G8III	B: A8V	Ca: F7	D: M2?
Masses	A: 2.5	B: 1.9	Ca: 1.2	Cb: 0.4	D: 0.6
Luminosities (L$_\odot$)	A: 50	B: 15	C: 2.5	D: 0.01
DS Catalogues	SP 1 (AB)	STF 1273 (AB,C)	BDS 4771	ADS 8850
Catalogues	11 Hya	HIP 43109	HD 74874	SAO 117112
Radial velocity	+36.4 km s^{-1}	± 0.9		
Galactic coordinates	220°.718	+28°.527		

History

The duplicity of ε Hya was noted by F. G. W. Struve at Dorpat, having escaped the attention of William Herschel. In 1860, Otto Struve reported that he thought he had seen an elongation in the vertical in the bright star and in 1864 an elongation in PA 190°. On the evening of 6 April 1888, using the 50-cm Merz–Repsold refractor at the Observatory of Brera in Milan, Giovanni Schiaparelli noted that A was a close pair ('elongated without doubt'). He observed it regularly and it was soon evident that the angular motion was rapid. Over the next 12 years he made regular measurements, during which time the position angle increased from 141° to 294°. Less than two years later, the pair, SP1, was single, even through the Lick 36-inch in the hands of R. G. Aitken. In 1876 Joseph Gledhill [316] noted a fourth star (D), which was found at 190°, 12″. This star shares the

Finder Chart

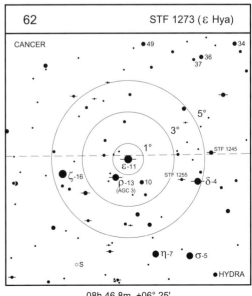

08h 46.8m +06° 25'

Orbit

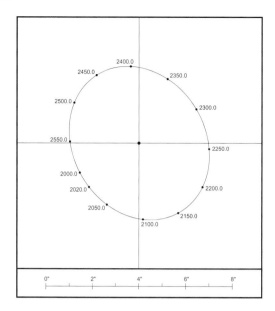

Ephemeris for STF 1273 AB,C (2000 to 2180)

Orbit by Dru (2014) Period: 589 years, Grade: 4

Year	PA(°)	Sep(")	Year	PA(°)	Sep(")
2000.0	295.9	2.83	2100.0	3.2	3.19
2020.0	310.5	2.83	2120.0	14.2	3.29
2040.0	324.9	2.87	2140.0	24.7	3.35
2060.0	338.6	2.96	2160.0	35.0	3.37
2080.0	351.3	3.08	2180.0	45.3	3.35

proper motion and the parallax of the bright triple and hence is physically connected to the system. In 1926 star C was announced to be a single-line spectroscopic binary [319] with a period of 9.905 days, making the group a physical quintuple. This may not be the end of the story. Whilst measuring radial velocities, Underhill [317] found a slight variation in the RV of A with a period of 70 days. This has not yet been confirmed but Parsons [467] suggests that a main sequence companion some 2.5 magnitudes fainter than A would explain certain aspects of the system geometry.

The Modern Era

One of the more powerful facilities for imaging binary stars is located near Albuquerque in New Mexico. This is the 3.5 metre *Starfire* telescope, which is operated under the auspices of the USAF to facilitate research into atmospheric compensation using adaptive optics. Binary stars are used to calibrate the image scale of the detector – the one in current use has a pixel size of $0''.01611 \pm 0''.00003$. This camera was recently used [403] to redefine the apparent orbit of AB,C. The orbit by Heintz, with period 990 years, had stood until 1996, but

the new determination considerably reduces the period to 589 years and changes the eccentricity of the orbit from 0.30 to zero.

Observing and Neighbourhood

Smyth found the colours (of AB and C) to be pale yellow and purple, whilst Crossley *et al.* [320] give yellow and blue. Lewis reports that Struve's colours were yellow and ruddy. Although the separation of the close pair never exceeds $0''.3$ it will be worth looking at the system in the next year or two with apertures of more than 20-cm. With a 20-cm refractor RWA thought that A was elongated a few years ago and Jean-François Courtot, using a 10-inch reflector from France, has also reported seeing an elongation. The AB,C group remains the main attraction in this system. The orbit is open and thus the distance does not change much during the cycle. The difficulty is the relative faintness of star C, nestling about $3''$ next to its third magnitude neighbour, but 10-cm should show them. Indeed, Christopher Taylor, observing from Oxfordshire with a 4-inch refractor, recorded this impression: 'Epsilon Hydrae, showing a tiny, needle-point, companion threaded diamond-ring-like on the outside of the first ring of A, i.e. just in external contact with the second dark interspace, at 320° or so PA. A very dramatic difference of star-disk sizes, A versus B, another (although more difficult) classic instance of this, to use as an illustrative example, like epsilon Bootis. This double star a very difficult "gapper" because of the great inequality of A and B, but: B $= \frac{1}{2}$A or very slightly less, gap $= \left(1 - 1\frac{1}{2}\right) \times$ A. Frequently a truly superb view in seeing II – I, the disks and rings defined to perfection – and it is probably only in such conditions that there would be any chance of the formal "gapping" made tonight. A = yellow.' One degree SE is ρ Hya, which is a severe test for 20-cm. The companion was found by Alvan G. Clark (AGC 3, 4.4, 11.9, 146°, $12''.1$, 2000). Three degrees due W is STF 1245 (6.0, 7.2, 25°, $10''.0$, 2018). There are three more distant stars brighter than $V = 12$. The star STF 1255 is $30'$ following δ – 7.3, 8.6, 33°, $26''.2$, 2012. There are two further stars with $V > 12$.

Measures

Early (STF)	195°.6	3″.14	1831.29
(Orbit	197°.3	3″.31)	
Recent (ARY)	307°.6	2″.83	2016.01
(Orbit	307°.6	2″.82)	

63. STF 1321 UMA = WDS J09144+5241AB

Table 9.63 Physical parameters for STF 1321 UMa

STF 1321	RA: 09 14 22.79	Dec: +52 41 11.8	WDS: 58(451)		
V magnitudes	A: 7.79	B: 7.88			
$(B - V)$ magnitudes	A: +1.55	B: +1.67			
μ(A)	−1546.10 mas yr^{-1}	± 0.06	−569.13 mas yr^{-1}	± 0.06 (DR2)	
μ(B)	−1573.12 mas yr^{-1}	± 0.06	−660.12 mas yr^{-1}	± 0.06 (DR2)	
π(A)	157.88 mas	± 0.04	20.659 light yr	± 0.005 (DR2)	
π(B)	157.89 mas	± 0.04	20.657 light yr	± 0.005 (DR2)	
Spectral types	A: M0V	B: M0V			
Radii (R$_\odot$)	A: 0.60	± 0.01	B: 0.59	± 0.01	
Luminosities (L$_\odot$)	A: 0.025	B: 0.025			
Catalogues (A/B)	HIP45343/120005	HD 79210/1	SAO 27178/9		
DS catalogues	STF 1321	BDS 4972	ADS 7251		
Radial velocity (A/B)	+11.14 km s^{-1}	± 0.1	+12.50 km s^{-1}	± 0.1	
Rotational velocity (A/B)	2.9 km s^{-1}	± 1.2	2.8 km s^{-1}	± 1.2	
Galactic coordinates:	164°.935	+42°.668			

History

Burnham in his Catalogue [50] summarised this pair as belonging to the '61 Cygni' class, containing wide pairs of stars undergoing significant common proper motion but with relatively little apparent orbital motion. Subsequent observations of both 61 Cyg and STF 1321 showed the characteristic curvature exhibited by binary systems. Ball [326] found a star of magnitude 11.9 at 208°, 189″ in 1879 whilst Espin [327] added a fainter object, which the WDS notes was of magnitude 14.5 at 282°.7 28″.2 in 1907.

The Modern Era

K. Chang [325] published a visual orbit for the pair based on observations over about 150 years during which time

Finder Chart

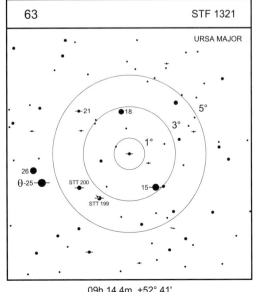

09h 14.4m +52° 41'

Orbit

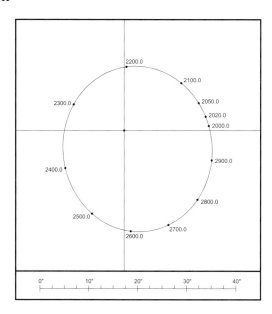

Ephemeris for STF 1321 AB (2000 to 2180)

Orbit by Chg (1972) Period: 975 years, Grade: 4

Year	PA(°)	Sep(")	Year	PA(°)	Sep(")
2000.0	93.0	17.21	2100.0	128.9	14.94
2020.0	99.4	16.76	2120.0	137.5	14.48
2040.0	106.1	16.31	2140.0	146.6	14.02
2060.0	113.3	15.86	2160.0	156.3	13.57
2080.0	120.9	15.40	2180.0	166.8	13.11

the companion moved some 40° in position angle. The resulting elements are therefore uncertain but, as of 2015, they still seem to reflect the motion of the two stars fairly well. The ephemeris for this orbit is given above. In 1973 Abt & Levy [321] published a paper in which they reported that the radial velocities of both components of STF 1321 had led them to the conclusion that each star was itself a spectroscopic binary of small amplitude. Star A had a period of 44.11 days and an eccentricity of 0.3 whilst B had a period of 16.47 days and an eccentricity of 0.76. Some years later Morbey and Griffin [322] re-observed both stars using the radial velocity spectrometer

at Cambridge and reached different conclusions. They were not able to see any significant variation in either star at a level greater than about 0.1 km per second. K. Ward-Duong *et al.* [323] used the VLT and an adaptive optics system, in combination with archive astrometric plates, to look for companions of both stars in the range 1 AU to 10,000 AU and found no evidence for further bodies. By the time of writing, Gaia had measured the fainter component. The difference in accuracy in parallax and proper motion determination between Gaia and Hipparcos is clearly shown in the table.

Exoplanet Host?

More recently, Gagné *et al.* [324] observed each component in a cell in a search for exoplanets and found a scatter in radial velocity of about 50 metres per second.

Observing and Neighbourhood

To be found 4° WNW of θ UMa, this is an easy pair for most telescopes but a larger aperture will show the colours better. Using a 60-mm refractor Sissy Haas found both stars 'peach-white' whilst Webb noted yellow and Struve thought both stars 'yellowish'. Within 2° are two STT systems: STT 199 is 2° SE (6.2, 10.0, 140°, 5".7, 2008) and a magnitude 7.9 at 230" forms ARN 71 AD. Forty minutes of arc NE is STT 200 (6.5, 8.6, 337°, 1".2, fixed).

Measures

Early measure (STF)	48°.5	20".03	1833.54
(Orbit	48°.3	20".20)	
Recent measure (CTT)	97°.2	17".22	2015.28
(Orbit	97°.9	16".87)	

64. 38 LYN = STF 1334 = WDS J09188+3648AB

Table 9.64 Physical parameters for 38 Lyn

STF 1334	RA: 09 18 50.64	Dec: +36 48 09.3	WDS: 259(192)		
V magnitudes	A: 3.92	B: 6.09			
(*B* − *V*) magnitudes	A: +0.05	B: +0.49			
μ(A)	−36.29 mas yr^{-1}	± 0.96	−121.77 mas yr^{-1}	± 0.95 (DR2)	
μ(B)	−26.51 mas yr^{-1}	± 0.19	−127.26 mas yr^{-1}	± 0.23 (DR2)	
π(A)	27.80 mas	± 0.79	117.3 light yr	± 3.3 (DR2)	
π(B)	24.47 mas	± 0.12	133.3 light yr	± 0.7 (DR2)	
Spectra	A1V				
Luminosities (L$_\odot$)	A: 30	B: 5			
Catalogues	HR 3690	HIP 45688	HD 80081	SAO 61391	
DS catalogues	H 1 9 (AB)	STF 1334 (AB)	BDS 5014	ADS 7292	CHR173 (BaBb)
Radial velocity	4.0 km s^{-1}	± 2 km s^{-1}			
Galactic coordinates	186°.810	+44°.471			

History

William Herschel resolved 38 Lyn on 24 November 1780 and he noted that the colours were white for the primary and inclining to red for the secondary. He also called it 'a very fine object' but added that 'a proper motion is suspected in one of the stars'.

The Modern Era

In 1988 McAlister *et al.* [482] found that star B was itself a close double star (CHR 173BaBb). The apparent separation in 1988.17 was 0″.06. No further interferometric observations were made until 2004, when Hartkopf *et al.* [332] found the companion double at 266° and 0″.24. The difference in separation alone indicates a pair in fairly rapid motion and, if the quadrants are correct, then the period cannot be more than about 30 years. De Rosa *et al.* [330], in an investigation

Finder Chart

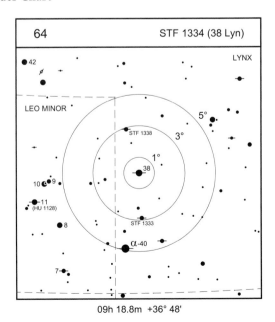

09h 18.8m +36° 48'

of X-ray-emitting A star systems within 75 parsecs of the Sun, found that the radiation was emanating from a magnitude 14 companion star (E), which they found 98″.7 (or 3778 AU) away from the bright star.

Observing and Neighbourhood

The double star 38 Lyncis lies at the extreme eastern edge of Lynx and can be found about 2.5° N of the V = 3.1 magnitude α Lyncis. Smyth noted that the colours were silvery white and lilac, an observation that Webb repeated in the first edition of *Celestial Objects* (1859). In his revision of this work, Espin notes yellow and tawny in 1850 and 1852, presumably reporting Webb's own observations, and adds greenish-white and blue, presumably Espin's own impression. The significant proper motion of the A component appears to be mirrored by the B star, so this is a binary star of long period. Just 1°.5 S of 38 is the fine pair of white stars STF 1333 (magnitudes 6.63 and 6.69, and measured by RWA on 2015.37 at 50°.2, 1″.78) whilst the same distance N is the beautiful binary STF

1338, which is described elsewhere (see Star 65). About 3°.5 E and slightly S is 11 LMi (HU 1128), a visual binary but also a very stiff visual test. Hussey gave the magnitudes as 5.5 and 14.0 but B, whilst very faint, is not as faint as magnitude 14: The stars have magnitudes 4.8 and 12.5 (WDS) and, according to the orbit, derived by Heintz [540], giving a period of 201 years with an eccentricity of 0.88, the companion lies at 49° and 6″.6, 2020, but recent observations disagree with the predictions. John Nanson [126] glimpsed B at ×152 with a 6-inch f/10 refractor, whilst ×253 gave a slightly better view. The star 11 LMi was for many years used as a photometric standard but Skiff & Gatewood [328] concluded that it was a low-amplitude (0.033 magnitude) variable, the cause of the variation being proposed as star spots. It is known as SV LMi.

Measures

Early measure (STF)	240°.2	2″.79	1829.17
Recent measure (ARY)	224°.4	2″.61	2013.39

65. STF 1338 LYN = WDS J09210+3811AB

Table 9.65 Physical parameters for STF 1338 Lyn

STF 1338	RA: 09 20 59.37	Dec: +38 11 17.7	WDS: 65(428)	
V magnitudes	A: 6.72	B: 7.08	C: 12.59	
$(B - V)$ magnitudes	A: +0.42	B: +0.44		
μ	-44.5 mas yr^{-1}	± 0.97	-23.9 mas yr^{-1}	± 0.45
π	23.44 mas	± 1.08	139 light yr	± 6.4
μ(A)	-51.60 mas yr^{-1}	± 0.79	-26.45 mas yr^{-1}	± 0.73 (DR2)
π(A)	14.90 mas	± 0.59	219 light yr	± 8.7 (DR2)
Spectra	F2V	F4V		
Luminosities (L$_\odot$)	A: 8	B: 5		
Catalogues	HR 3701	HD 80441	SAO 61411	HIP 45858
DS catalogues	STF 1338	BDS 5030	ADS 7307	
Radial velocity (A/B)	0.6 km s^{-1}	± 2	-1.9 km s^{-1}	± 2
Galactic coordinates	$184°.892$	$+44°.964$		

History

This object was found at Dorpat by F. G. W. Struve. Component C (magnitude 12.6) at 167° and 143″ was added by Robert Ball in 1879 [333]. Ball was observing stars in certain star fields with the hope of identifying those with a significant or measurable parallax. His argument [326] was based on Schiaparelli's assertion that stars near the Sun may be moving in a stream parallel to the Sun and that the proper motion would not necessarily be an indicator of distance. He also included some red stars and variable stars, in the belief that both were (at the time) held to be small and therefore must be nearby to be seen at all. Neither his initial paper (42 objects) nor the more substantial paper later on (368 stars) showed any sign of a significant parallax, but in the course of this work he found several faint, distant, companions to double stars. As well as holding the Lowndean Chair of Astronomy and Geometry at the Institute of Astronomy, Cambridge, from 1892, Ball (1840–1913) was a great popularizer of astronomy and wrote copiously on astronomical topics.

Finder Chart

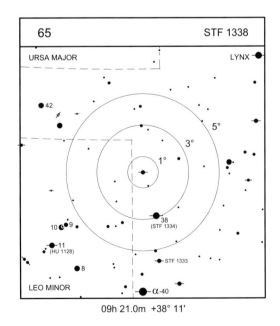

09h 21.0m +38° 11'

Orbit

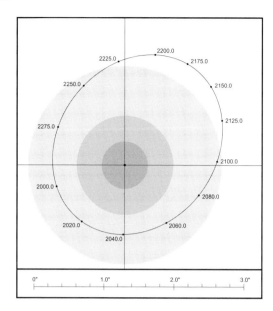

Ephemeris for STF 1338 AB (2010 to 2100)

Orbit by Sca (2002b) Period: 303.27 years, Grade: 3

Year	PA(°)	Sep(")	Year	PA(°)	Sep(")
2010.0	304.2	1.01	2060.0	36.6	1.00
2020.0	321.8	1.00	2070.0	53.5	1.06
2030.0	340.0	0.98	2080.0	69.4	1.13
2040.0	358.9	0.96	2090.0	81.2	1.23
2050.0	18.0	0.97	2100.0	92.2	1.32

The Modern Era

The components of STF 1338 have now completed almost half a revolution since discovery. In 1953 an orbit was calculated by S. Arend [334] based on an observed arc of 90°, and he obtained a period of 389 years and an eccentricity of 0.29. The current orbit is by M. Scardia in 2002 and shortens the period to 303 years.

Observing and Neighbourhood

About 3° S and slightly preceding is 38 Lyn (see Star 64). Continuing another 3° in the same direction brings you to STF 1333, described in our entry for 38 Lyn. John Nanson [126] resolved AB with ×253 in a 6-inch refractor in September 2015, with both main component stars appearing white. On that occasion he noted that he saw the C component (magnitude 12.6 at 166°, 144″) better at ×152.

Measures

First measure (STF)	121°.1	1″.76	1829.53
(Orbit	119°.9	1″.57)	
Recent measure (ARY)	314°.0	1″.08	2018.28
(Orbit	318°.2	1″.00)	

66. ω LEO = STF 1356 = WDS J09285+0903AB

Table 9.66 Physical parameters for ω Leo

STF 1356	RA: 09 28 27.41	Dec: +09 03 24.4	WDS: 24(656)	
V magnitudes	A: 5.69	B: 7.28		
$(B - V)$ magnitudes	A: +0.75	B: +0.52		
μ	36.98 mas yr^{-1}	$\pm$ 2.30	5.78 mas yr^{-1}	$\pm$ 0.70
π	30.15 mas	$\pm$ 1.45	108 light yr	$\pm$ 5
Spectra	F9IV			
Luminosities (L$_\odot$)	A: 5	B: 1		
Catalogues	HR 3754	HD 81858	SAO 117717	HIP 46454
DS catalogues	H 1 26	STF 1338	BDS 5103	ADS 7390
Radial velocity (A/B)	0.6 km s^{-1}	$\pm$ 2	-1.9 km s^{-1}	$\pm$ 2
Galactic coordinates	223°.657	+38°.896		

History

Found by William Herschel on 8 February 1782. He records 'With 227 there is not the least suspicion of its being double; with 460 it appears oblong, and, when perfectly distinct, we see $\frac{3}{4}$ of the diameter of a small star as it were emerge from behind a larger star; with 932 they are more clear of each other, but not separated; . . . ' Herschel also thought that, between the beginning of 1782 and the end of that year, he was able to discern that the two stars were getting further apart. The modern orbit predicts that the motion in distance during this period would have been only 0''.02.

The Modern Era

The current orbit for ω Leo was derived by Muterspaugh et al. [620] and incorporates some very accurate astrometry, derived from observations of the star with the PHASES

Finder Chart

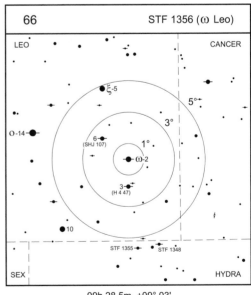

09h 28.5m +09° 03'

Orbit

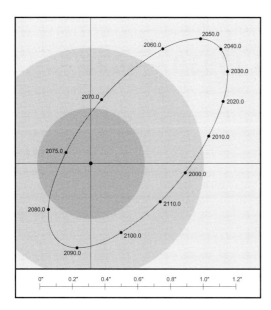

Ephemeris for STF 1356 (2015 to 2060)

Orbit by Mut (2010b) Period: 117.974 years, Grade: 2

Year	PA(°)	Sep(")	Year	PA(°)	Sep(")
2015.0	109.3	0.82	2040.0	130.7	1.05
2020.0	114.6	0.89	2045.0	134.3	1.04
2025.0	119.2	0.95	2050.0	138.0	1.00
2030.0	123.3	1.00	2055.0	142.1	0.93
2035.0	127.1	1.03	2060.0	147.2	0.81

(Palomar High-Precision Astrometric Search for Exoplanet Systems) instrument. This ground-based interferometer ceased operation in late 2008 but when in use utilised three 40-cm mirrors, with one 110-metre and two 87-metre baselines, and was capable of determining the positions of stars to 35 microarcsecond. The pair does not appear in DR2.

Observing and Neighbourhood

Smyth thought that the colours of the stars were pale yellow and greenish at the time the system was below his telescope resolution limit (1843). Hartung considered the pair to be bright yellow, but again the separation was such that he only saw the stars apart occasionally in his 30-cm reflector (1962). RWA could not resolve the pair until the mid-1990s with the 20-cm at Cambridge. At present, they are a real test for apertures of 15- to 20-cm. The stars are currently still below 1″ but the substantial magnitude difference makes resolution more difficult. In the Cambridge 20-cm only the best nights show the stars separated at ×450. About 1° ENE of ω Leo is 6 Leonis (SHJ 107). This is a wide unequal pair with a fifth magnitude K3 giant primary star and a ninth magnitude companion. Greg Stone [126] noted colours of deep orange and violet in a 80-mm f/5 refractor at ×132, agreeing with W. S. Franks, who in 1877 recorded pale orange and purple. Two degrees due S of ω is 3 Leo (H 4 47) – a William Herschel discovery, which escaped Webb but appears in Smyth, where it is described as 'delicate' and the magnitudes as 6.5 (pale yellow) and 13 (blue). The WDS has 5.8, 11.1 at PA 80° and separation 24″.8. Proceed a further 2° SSE and the two rather similar pairs STF 1348 and 1355 (1° apart) can be found: STF 1348 (7.5, 7.6, 313°, 1″.9, 2017); STF 1355 (7.1, 7.8, 354°, 1″.8, 2017).

Measures

Early measure (STF)	153°.9	0″.97	1825.21
(Orbit	148°.7	0″.78)	
Recent measure(ARY)	108°.9	0″.85	2014.94
(Orbit	109°.1	0″.82)	

67. ψ VEL = COP 1 = WDS J09307−4028

Table 9.67 Physical parameters for ψ Vel

COP 1	RA: 09 30 42.00	Dec: −40 28 00.3	WDS: 368(152)	
V magnitudes	A: 3.91	B: 5.12		
(*B* − *V*) magnitudes	A: +0.34	B: +0.53		
μ(A)	−178.49 mas yr^{-1}	± 0.66	93.50 mas yr^{-1}	± 0.81 (DR2))
μ(B)	−203.95 mas yr^{-1}	± 0.57	44.27 mas yr^{-1}	± 0.60 (DR2)
π(A)	54.61 mas	± 0.45	59.7 light yr	± 0.5 (DR2)
π(B)	54.31 mas	± 0.33	60.1 light yr	± 0.4 (DR2)
Spectra	A: F0IV	B: F3IV		
Masses (M$_\odot$)	A: 1.56	B: 1.26		
Radii (R$_\odot$)	A: 1.6	B: 1.2		
Luminosities (L$_\odot$)	A: 7.5	B: 2.5		
Catalogues	HR 3786	HD 82434	SAO 221234	HIP 46651
DS catalogues	COP 1			
Radial velocity	8.80 km s^{-1}	± 1.8		
Galactic coordinates	266°.777	+7°.910		

History

Ralph Copeland (1837–1905) was the third Astronomer Royal for Scotland. He worked at Dun Echt Observatory in Ireland for the 26th Earl of Crawford and was a frequent traveller on worldwide astronomical expeditions, which included seeing transits of Venus, in 1874 from Mauritius, and in 1882 from Jamaica. After the latter event, he made his way through Ecuador and into Peru in March 1883, where he made observations with a 6-inch refractor from sites higher than 12,500 feet [336]. Whilst at Puno from March to June 1883 he recorded the observation of 13 double stars, some of which he was able to attribute to earlier discoverers. The third star on his list was ψ Vel and his journal noted 'ψ Argûs = Stone 5124, 45° 0″.8, 4 and 6 magnitudes; both white; nova?' The last comment is the Latin word for new, as indeed it was and thus became COP 1. On the same night he also resolved the brighter component of the wide pair ε Lupi (COP 2).

Finder Chart

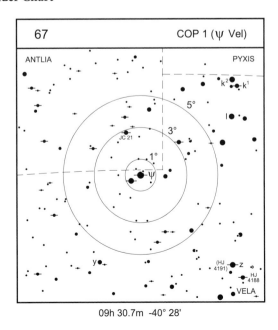

09h 30.7m -40° 28'

Orbit

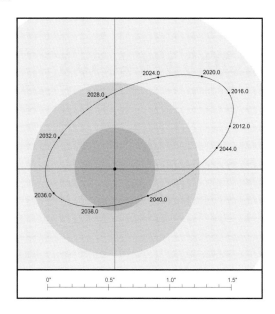

Ephemeris for COP 1 (2016 to 2034)

Orbit by Msn (2001c) Period: 33.95 years, Grade: 2

Year	PA(°)	Sep(")	Year	PA(°)	Sep(")
2016.0	123.1	1.11	2026.0	167.6	0.69
2018.0	129.3	1.08	2028.0	186.8	0.58
2020.0	136.1	1.02	2030.0	212.8	0.51
2022.0	144.0	0.93	2032.0	241.5	0.52
2024.0	154.0	0.81	2034.0	267.0	0.56

The Modern Era

The object ψ Vel is a binary of relatively short period, for a visual system, which can be seen with medium aperture when at its widest. The WDS notes that the primary is a spectroscopic binary but gives no details. The primary is a fast rotator ($v \sin i = 160$ km s^{-1}) and Fuhrmann & Chini [337] have derived the masses given in the table above. The current orbit by Mason has a period of 33.95 years and is assigned Grade 2 in the USNO *Sixth Orbital Catalog*. Recent observations show that the star is falling behind the predicted position angle, indicating that the period is perhaps too short.

Observing and Neighbourhood

This pair should be comfortably within range of 20-cm for the next few years, and is probably accessible to 15-cm on a good night. It is one of the shortest-period visual binaries that can be seen with small apertures. About 4° SW of ψ lie a couple of John Herschel pairs – z Vel (HJ 4191) is an unequal couple (5.1, 9.1, 13°, 6″, 2008) whilst 30′ SW again lies HJ 4188, 6.0, 6.8, 281°, 2″.9, 2015. One of the entries in the small catalogue of Captain Jacob can be found at 1°.5 NNE of ψ Vel (JC 21, 6.5, 8.2, 205°, 56″, 2010).

Measures

Early measure (A)	275°.0	0″.72	1899.3
(Orbit	279°.5	0″.56)	
Recent measure (ANT)	121°.6	1″.02	2016.34
(Orbit	124°.2	1″.11)	

68. υ CAR = RMK 11 = WDS J09471–6504

Table 9.68 Physical parameters for υ Car

RMK 11	RA: 09 47 06.12	Dec: −65 04 19.2	WDS: 3885(29)	
V magnitudes	A: 2.96	B: 6.00		
(B − V) magnitudes	A: +0.28	B:		
μ(A)	−14.52 mas yr^{-1}	± 0.80	7.19 mas yr^{-1}	± 0.70 (DR2)
μ(B)	−11.91 mas yr^{-1}	± 0.11	5.92 mas yr^{-1}	± 0.10 (DR2)
π(A)	1.50 mas	± 0.50	2170 light yr	± 720 (DR2)
π(B)	2.37 mas	± 0.07	1380 light yr	± 41 (DR2)
Spectra	A: A8Ib	B: B3-4IV		
Masses (M$_\odot$)	A: 13	B: 8		
Luminosities (L$_\odot$)	A: 24000	B: 600		
Catalogues (A/B)	HR 3890/1	HD 85123/4	SAO 250695/6	HIP 48002
DS catalogues	RMK 11			
Radial velocity	13.60 km s^{-1}	± 0.5		
Galactic coordinates	285°.038	−8°.820		

History

In 1977, Brian Warner [340] published a list of double stars which had been found by Fearon Fallows during a rather short observing career at the Royal Observatory at the Cape of Good Hope. The list contains 12 objects and Fallows appears to have been the first to notice ten of these objects, the exceptions being Rigel and Acrux. One star on the list is given as μ Arg by Warner, but that already exists in the list and is now known as μ Vel, so it is assumed that Fallows meant υ Carinae and indeed Warner gives the catalogue name RMK 11 to this entry, as it was subsequently found by Carl Rümker at Parramatta. In 1832 he published a *Preliminary Catalogue of Fixed Stars* (short title), which included a list of 28 bright and wide double stars. A recent re-examination by Letchford *et al.* [377] shows that numbers 22 and 23 are the same pair and number 24 does not exist. It is, however, exactly one hour of RA away from number 25 at the same declination and may

Finder Chart

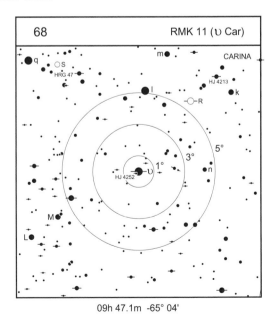

09h 47.1m -65° 04'

be an inadvertent duplicate entry. Of the 26 real systems, the WDS gives credit to Dunlop for five of them, but as Dunlop and Rümker were contemporaneous it is not clear who had precedence in time for the first observation of each pair. One further object appears to have been independently found later at Cordoba Observatory.

The Modern Era

A study by Mandrini & Niemela in 1986 [339] gives a distance of 1333 light years and from this they derive a linear distance of 2000 AU between the stars. With masses of 13 and 8 $M_\odot$ this implies a rotation period of 19,500 years. Gaia (DR2) has announced astrometry results for both components and the proper motions and parallaxes are given above. The uncertainty on the primary's distance is no doubt due to its brightness. If the pair is physical, the best estimate of the distance to the system is that of the secondary star.

Observing and Neighbourhood

'Very fine pair' (Hartung) but the significant magnitude difference and small separation requires sufficient magnification to separate the two stars and move the companion away from the glare of the primary. Gould notes that the double star 'dominates a quite starry field'. An 18-cm refractor shows it as an easy and beautiful pair at ×180. There has been virtually no apparent motion since discovery, supporting the great distance to these stars implied by the DR2 parallaxes. Ten arcminutes to the SE is HJ 4252 (9.3, 9.5, 303°, 12″, 2010). To the NE at 3°.6 distance is HRG 47 (6.3, 7.9, 353°, 1″.1, 2014). Four degrees NW is HJ 4213 (5.8, 9.6, 330°, 8″.8, 2000).

Measures

Early measure (JC)	126°.6	5″.08	1851.21
Recent measure (ANT)	126°.1	4″.98	2010.28

69. γ SEX = AC 5 = WDS J09525–0806AB

Table 9.69 Physical parameters for γ Sex

AC 5	RA: 09 52 30.437	Dec: −08 06 18.13	WDS: 111(309)		
V magnitudes	A: 5.63	B: 6.11	C: 12.3		
(B − V) magnitudes	A: +0.02	B: +0.11			
μ	−57.28 mas yr^{-1}	± 1.38	−49.26 mas yr^{-1}	± 0.36	
π	11.75 mas	± 0.63 mas	278 light yr	± 15	
Spectra	A: A0/A1	B:			
Luminosities (L$_\odot$)	A: 35	B: 20			
Catalogues	8 Sex	HR 3909	HD 5558	SAO 137199	HIP 48437
DS catalogues	H N 49 (AB,C)	HJ 4256 (AB,C)	AC 5 (AB)	BDS 5235	ADS 7555
Radial velocity	12.20 km s^{-1}	± 2.4			
Galactic coordinates	245°.565	+34°.170			

History

The star γ Sex was observed by William Herschel and, on 22 February 1787, he noted that it had a very faint companion. During his work at the Cape in South Africa, John Herschel re-observed γ and in late 1835 was able to glimpse the B star between clouds. He did not refer to his father's previous observation, and gave the stars the number HJ 4256. In April 1852, Alvan Clark [342] was testing a 4.75-inch objective and noticed that the primary star was double. This observation intrigued W. R. Dawes, who wondered why Struve had not seen it at Dorpat. He suspected that there had been orbital motion and the star was then too close for the 9.3-inch Fraunhofer telescope. (The current orbit gives a separation of 0″.48 for 1825.) The stars were close to maximum separation (0″.56) when John Herschel looked at them in admittedly less than perfect conditions. The discovery by Clark is remarkable and demonstrates the quality of even the smaller objectives. The *Aitken Double Star Catalogue* (*ADS*) gives AB-C as HJ 4246, which is incorrect.

Finder Chart

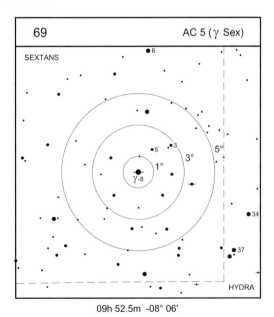

09h 52.5m -08° 06'

Orbit

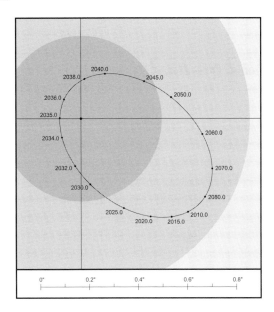

Ephemeris for AC 5 AB (2016 to 2034)

Orbit by USN (2007a) Period: 77.82 years, Grade: 2

Year	PA(°)	Sep(")	Year	PA(°)	Sep(")
2016.0	42.1	0.53	2026.0	23.5	0.37
2018.0	39.2	0.51	2028.0	17.1	0.32
2020.0	36.0	0.48	2030.0	8.1	0.26
2022.0	32.5	0.45	2032.0	352.8	0.19
2024.0	28.4	0.42	2034.0	314.0	0.11

The Modern Era

Gaia DR2 does not include this system, no doubt owing to the close proximity of the two stars at the current time.

Observing and Neighbourhood

The close pair is a severe test of atmosphere and aperture. RWA managed to measure it in 2005 when it was $0''.6$. It is now closing by about $0''.02$ per year and by 2035 will be just $0''.09$ apart before widening to $0''.59$ in 2080. The faint star C is at $333°$, $36''.9$, 2000. The system γ Sex can be found $12°$ following α Hya ($V = 2.0$), which is itself one of William Herschel's widest pairs. The K3 giant primary has a magnitude 9.7 companion at $282°$, $154''$ and another at $85°$ and $257°$.

Measures

Early measure (DA)	38°.2	0″.5	1860.34
(Orbit	41°.7	0″.53)	
Recent measure (WSI)	45°.8	0″.53	2014.31
(Orbit	44°.4	0″.54)	

70. I 173 VEL = WDS J10062–4722

Table 9.70 Physical parameters for I 173 Vel

I 173	RA: 10 06 11.22	Dec: −47 22 11.9	WDS: 2282(42)	
V magnitudes	A: 5.38	B: 7.11		
$(B - V)$ magnitudes	A: +1.17	B: +0.45		
μ(A)	−10.36 mas yr^{-1}	± 0.36	−49.99 mas yr^{-1}	± 0.29 (DR2)
μ(B)	−9.02 mas yr^{-1}	± 1.89	−53.47 mas yr^{-1}	± 1.93 (DR2)
π(A)	12.19 mas	± 0.15	268 light yr	± 3 (DR2)
π(B)	8.34 mas	± 1.06	391 light yr	± 50 (DR2)
Spectra	A: K1III	B: G		
Luminosities (L$_\odot$)	A: 40	B:15		
Catalogues	HR 3976	HD 87783	SAO 221773	HIP 49485
DS catalogues	I 173			
Radial velocity	+21.20 km s^{-1}	± 0.8		
Radial velocity (A)	25.89 km s^{-1}	± 0.28		
Galactic coordinates	276°.175	+6°.734		

History

Innes found this star using a 7-inch refractor at the Royal Observatory at the Cape and observed it occasionally until 1925, when van den Bos found that it had closed in considerably from its discovery separation. During the 1930s the pair was essentially unresolved in the large refractors at Johannesburg and Bloemfontein. Since then it has slowly widened and is currently close to maximum separation, which will be reached in 2034 when the stars will be 1″.0 apart. The separation will then remain above 0″.9 until around 2065, when the stars close again.

The Modern Era

Although DR2 has observed this pair, the results are unsatisfactory; the fainter star gets a very poor parallax

Finder Chart

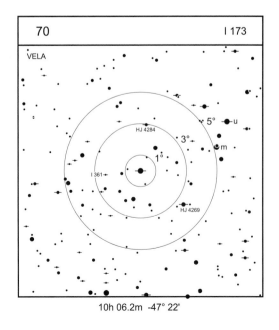

10h 06.2m -47° 22'

Orbit

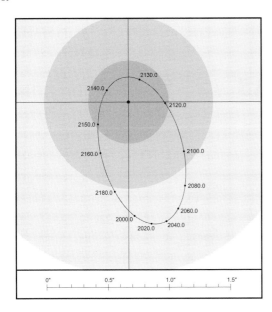

Ephemeris for I 173 (2010 to 2100)

Orbit by Sca (2008e) Period: 202.7 years, Grade: 4

Year	PA(°)	Sep(")	Year	PA(°)	Sep(")
2010.0	7.1	0.95	2060.0	25.5	0.94
2020.0	10.9	0.99	2070.0	29.8	0.88
2030.0	14.5	1.00	2080.0	34.7	0.81
2040.0	18.0	1.00	2090.0	40.8	0.71
2050.0	21.6	0.98	2100.0	49.1	0.60

determination. The pair is now close to widest separation (1″.0 in 2032) and then closes, reaching 0″.2 a century later.

Observing and Neighbourhood

At least 20-cm will be needed to separate the stars at present because the significant difference in magnitude is important. I 173 is located 7° preceding and 2° N of the bright binary μ Vel). There are three pairs in *CDSA2* within 2°: I 361 (8.4, 11.0, 125°, 5″.6, 2001); HJ 4284 (7.4, 9.5, 66°, 6″.6, 2000); and HJ 4269 (6.0, 10.1, 321°, 14″.1, 2010).

Measures

Early measure (I)	57°.4	0″.53	1901.70
(Orbit	53°.6	0″.54)	
Recent measure (ANT)	9°.1	0″.97	2016.34
(Orbit	9°.5	0″.98)	

71. $\gamma^{1,2}$ LEO = STF 1424
= WDS J10200+1950AB

Table 9.71 Physical parameters for γ Leo

STF 1424	RA: 10 19 58.16	Dec: +19 50 30.7	WDS: 13(839)		
V magnitudes	A: 2.37	B: 3.64			
μ	304.30 mas yr^{-1}	$\pm\,0.71$	-154.28 mas yr^{-1}	$\pm\,0.36$	
π	25.07 mas	$\pm\,0.52$	130 light yr	$\pm\,2.7$	
Spectra	K1III	G7III			
Luminosities (L$_\odot$)	A: 145	B: 50			
Catalogues	HD 89484/5 (A/B)	HIP 50583	SAO 81298/9 (A/B)		
DS catalogues	H 1 28	STF 1424	BDS 5388	ADS 7724	BAG 32CaCb
Radial velocity	-36.24 km s^{-1}	$\pm\,0.18$			
Galactic coordinates	$216°.552$	$+54°.652$			

History

This object was discovered by William Herschel on 11 February 1782. 'A beautiful double star. Pretty unequal' is how he describes it. He went to great trouble to examine the star with a large range of eyepieces, with magnifications up to $\times 6652$, when 'I had but a single glimpse of the star quite disfigured; however I ascribe it chiefly to the foulness of the glass, which, on account of its smallness, is extremely difficult to be cleaned.'

The Modern Era

Even after 234 years, the orbit of γ Leo (also called Algieba) is very poorly known. If the current calculation is correct, close approach took place in 1749 when the stars were $0''.22$ apart, beyond the telescopic capabilities of the time. The stars will start to close again around 2065, but in the meantime they remain a glorious sight and easy for small apertures. There are two distant stars, which are unrelated to γ Leo. Star C, magnitude 9.64, is currently $341''$ distant in PA $288°$ and the distance is increasing rapidly. This is a star of considerable

Finder Chart

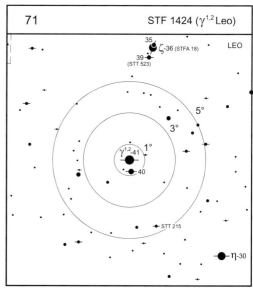

10h 20.0m +19° 50'

Orbit

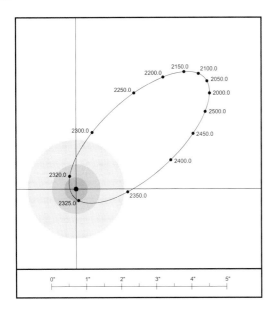

Ephemeris for STF 1424 AB (2000 to 2180)

Orbit by PkO (2014c) Period: 554 years, Grade: 4

Year	PA(°)	Sep(")	Year	PA(°)	Sep(")
2000.0	125.0	4.65	2100.0	132.6	4.75
2020.0	126.6	4.73	2120.0	134.2	4.67
2040.0	128.1	4.78	2140.0	135.8	4.56
2060.0	129.6	4.80	2160.0	137.5	4.41
2080.0	131.1	4.79	2180.0	139.4	4.22

interest. It is an M0 flare star, known as AD Leo, and a parallax by DR2 (0″.20137) puts it 16.197 light years away, with an error of 0.006 light years). It has its own large proper motion of 0″.5 per year. In 1943, Dirk Reuyl [538] analysed a series of photographic plates of AD Leo covering a baseline of 27 years and found a perturbation amounting to about 0″.1 with an apparent period of 26.5 years. In 1968, S. L. Lippincott [346], using plate material taken with the Sproul 24-inch refractor between 1938 to 1966, found that 'no definite analysis of the variable proper motion in terms of Keplerian motion is possible at this time'. In 1981, using the 6-metre reflector at the Special Astrophysical Observatory, Yuri Balega [347] detected a companion at 750 nm at a distance of 0″.07. Four

years later, at two closely spaced epochs, he noted that the pair were 'resolved but uncertain'. Since then no further direct observations of B have been made, so the question of duplicity is still uncertain.

Exoplanet Host?

An exoplanet accompanying star A, period = 428 days, M_J = 8.8, was discovered in 2009 [344].

Observing and Neighbourhood

The object γ Leonis is observable in the smallest apertures. Smyth noted bright orange and greenish-yellow, whilst Struve called them both golden. The writer found both stars to be orange using a 21-cm reflector at ×96, whilst from Australia the ASNSW group called them pure yellow and yellow or yellowish. Just 2°.2 SSW is the binary STT 215. The stars of magnitudes 7.3 and 7.5 rotate around each other in 670 years, according to Zaera [343] but a glimpse at the astrometric history shows that they are beginning to deviate from this orbit. The stars are currently 1″.5 apart and, on the basis of the current orbit, are expected to widen until they reach 2″, two centuries hence. Star D, magnitude 10.62, is 371″ away and the distance is reducing. Due N 3°.5 can be found the wide binocular pair ζ Leonis (STFA 18) (3.5, 6.0, 338°, 335″). The star 39 Leo is a good light-gathering test. The stars in STT 523 are 5.8, 11.3, 300°, 7″.9, 2017. Both have a large and common proper motion and the faint companion is an M dwarf star. DR2 confirms both stars are 73 light years away.

Measures

Early measure (STF)	102°.0	2″.46	1828.24
(Orbit	103°.0	2″.54)	
Recent measure (ARY)	128°.3	4″.78	2018.33
(Orbit	127°.4	4″.74)	

72. J VEL = RMK 13 = WDS J10209−5603

Table 9.72 Physical parameters for J Vel

RMK 13	RA: 10 20 54.90	Dec: −56 02 35.9	WDS: 4064(28)	
V magnitudes	A: 4.50	B: 7.18	C: 9.2	
$(B - V)$ magnitudes	A: −0.12	B: +0.10		
μ(A)	−10.83 mas yr^{-1}	± 0.38	3.67 mas yr^{-1}	± 0.37 (DR2)
μ(B)	−14.31 mas yr^{-1}	± 0.92	2.83 mas yr^{-1}	± 0.09 (DR2)
μ(C)	−19.65 mas yr^{-1}	± 0.07	10.00 mas yr^{-1}	± 0.05 (DR2)
π(A)	2.66 mas	± 0.23	1226 light yr	± 106 (DR2)
π(B)	2.18 mas	± 0.05	1496 light yr	± 34 (DR2)
π(C)	1.61 mas	± 0.04	2025 light yr	± 50 (DR2)
Spectra	A: B3III	B:		
Masses (M$_\odot$)	A: 11.0	B: 2.9	C: 1.9	
Luminosities (L$_\odot$)	A: 1850	B: 230	C: 65	
Catalogues	HR 4074	HD 89890	SAO 237959/60	HIP 50676
DS catalogues	RMK 13			
Radial velocity	10.40 km s^{-1}	± 2.8		
Galactic coordinates	282°.989	+0°.887		

History

This was found by Carl Rümker at Parramatta. There is a magnitude 9.2 star at 191° and 36″ which has a similar proper motion to A and B, and which was found by John Herschel in 1834. He said 'Most beautifully defined. The small star conspicuous in a fully illuminated field.'

The Modern Era

Andrei Tokovinin, in his *Multiple Star Catalogue* [177], notes that a projected period for AB is 21.8 kiloyears whilst that of AB and C is 220.8 kiloyears. Veramendi and González [462] find that both stars have a varying radial velocity and that the B star has a double-line spectrum, but no period has been determined. In addition, star A has a variable line profile, with a period of 2.318 days, but this could be explained by the Be nature of this object. The radial velocities of B also show a long-term variation in the systemic variation, which indicates the presence of another star whose period is estimated to be less than 10,000 days. They conclude that the similarity between the mean velocity of component A, the barycentric velocities obtained for Bab, and the mean velocity found for component C supports the idea that they are gravitationally bound. With the detection of two new hierarchical levels this system would be, at least, quintuple.

Finder Chart

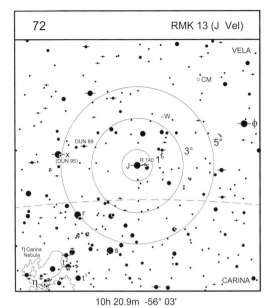

72   RMK 13 (J Vel)

VELA

∘CM

∘W

φ

DUN 89

3°

5°

x
(DUN 95)

R 140 1°

J

r

η Carina
Nebula

S

t²

t¹

η   CARINA

10h 20.9m -56° 03'

Observing and Neighbourhood

The object J Vel sits in a rich part of the sky just N of the border with Carina. Twenty arc minutes W is a binocular pair, the fainter component of which is R 140 (7.5, 8.2, 281°, 3″.2, 1998). Both stars have the same proper motion. Two degrees ENE is DUN 89 (6.8, 7.8, 31°, 26″, 2014), the fainter component of which is a fine test for 10-cm (HLD 106, 7.8, 8.1, 255°, 1″.4, 2014); it is a slow-moving binary which is gradually closing. Move a further 1° E and you come to x Vel (DUN 95). Five degrees to the SE is the η Carinae complex and *CDSA2* shows a number of bright pairs in this area.

Measures

AB			
Early measure (JC)	103°.4	7″.26	1848.12
Recent measure (TMA)	102°.0	7″.14	2000.05

AC			
Early measure (JH)	191°.4	36″.71	1835.91
Recent measure (TMA)	190°.9	36″.16	2000.05

73. x VEL = Δ 95 = WDS J10393–5536AB

Table 9.73 Physical parameters for x Vel

DUN 95	RA: 10 39 18.39	Dec: −55 36 11.7	WDS: 6235(21)	
V magnitudes	A: 4.38	B: 6.06	C: 11.6	D: 11.2
(*B − V*) magnitudes	A: +1.17	B: −0.01		
μ(A)	−18.41 mas yr^{-1}	± 0.36	4.88 mas yr^{-1}	± 0.34 (DR2)
μ(B)	−19.18 mas yr^{-1}	± 0.10	5.53 mas yr^{-1}	± 0.10 (DR2)
π(A)	5.06 mas	± 0.23	645 light yr	± 29 (DR2)
π(B)	4.39 mas	± 0.07	743 light yr	± 12 (DR2)
Spectra	A: G5II	B: B8V		
Radii (R$_\odot$)	A: 1.9 mas	B:		
Luminosities (L$_\odot$)	A: 575	B: 160		
Catalogues (A/B)	HR 4180	HD 92449/63	SAO 238309/13	HIP 52154
DS catalogues	DUN 95 (AB)	HJ 4341 (BC)	CRU9007 (AD)	
Radial velocity	20.10 km s^{-1}	± 0.7		
Galactic coordinates	284°.977	+2°.594		

History

This double was noted by James Dunlop at Parramatta, it was re-examined by John Herschel at the Cape. Herschel added a faint star (Herschel called it 16), some 15″ from B which had been missed by Dunlop. Another star (magnitude 11.2) at 33″ was recorded by Cruls at Rio de Janeiro on 1879.512; it is now 43″ distant in PA 232°. Cruls was using the 10-inch (25-cm) equatorial refractor with a filar micrometer. In 1882, Cruls, who later became Director of the National Observatory in Brazil, co-discovered the Great Comet of 1882, and also observed the transit of Venus of that year from Punta Arenas in Chile. H. C. Russell at Sydney recorded the colours straw yellow and greenish blue.

Finder Chart

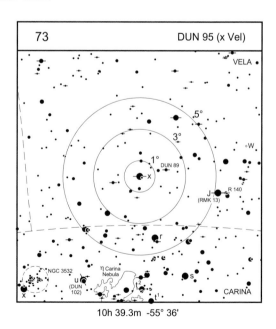

10h 39.3m -55° 36′

The Modern Era

The differences in distance derived for the two components by Hipparcos (A) and Gaia (B) indicate that the stars are formally at different distances but the proper motions are so similar that this seems to point to a systematic error in the satellite astrometry, but which one? The A star is too bright to be amongst the early results from Gaia.

Observing and Neighbourhood

Although x Vel is one of the finest doubles in the southern sky, the riches of Vela are such that E. J. Hartung did not find space for this pair in the pages of his book. With the 350-mm aperture at ×120, Ross Gould from Australia records the colours yellow and deep yellow for the bright components and notes that there are two asterisms in the 'good starry field'. The Astronomical Society of New South Wales observers came to the conclusion that the colours were orange yellow and pale blue.

Measures

Early measure (HJ)	105°.1	51″.85	1838.09
Recent measure (TMA)	105°.4	51″.74	2000.06

74. η CAR = Δ 98 = WDS J10451–5941

Table 9.74 Physical parameters for η Car

DUN 98	RA: 10 45 03.54	Dec: −59 41 03.9	WDS: 10166(15)	
V magnitudes	A: 4.85 (2017)	H: 8.17	G: 9.46	
$(B - V)$ magnitudes	A: +0.75 (2017)	H: +0.13	G: +0.11	
μ(A)	−11.0 mas yr^{-1}	± 0.8	4.1 mas yr^{-1}	± 0.7
μ(H)	−7.16 mas yr^{-1}	± 0.06	2.69 mas yr^{-1}	± 0.05 (DR2)
π(A)	0.4 mas		7500 light yr	
π(H)	0.37 mas	± 0.03	8800 light yr	± 760 (DR2)
Spectra	A: 05III	H: 04.5V	G: O9.5V + B0.5V	
Masses (M$_\odot$)	A: 90 + 30	H:	G: 14 + 6	
Radii (R$_\odot$)	A: 1.9 mas	B:		
Luminosities (L$_\odot$)	A: 60,000 to $> 6 \times 10^6$	H: 3300		
Catalogues (A/H)	HR 4210	HD 93308/303308	SAO 238429/31	
DS catalogues	DUN 98 (AH)	B 2256	I 1092	HJ 4366	WGT 2
Radial velocity (A/H)	−25.0 km s^{-1}	± 2	−1 km s^{-1}	± 10
Galactic coordinates	287°.597	−0°.630		

History

James Dunlop at Parramatta noticed that η Argûs, as it was then called, was a wide and easy double star. John Herschel observed η in 1835 and found two faint companions at 47°, 12″ and 30°, 15″. He gave the brightness of A as magnitude 2 and about two weeks later he stated that the A star was magnitude 1. In 1843 the bright component flared up to magnitude −1, dropped to magnitude 8 or so by 1930, and since then has been slowly increasing in brightness. Today it has reached $V = 4.5$. The star sits in the cluster Trumpler 16 so there are a number of fainter stars close by. Robert Innes at Johannesburg, using the 9-inch refractor, observed the star a number of times. He found a close companion but noted that 'it was impossible to focus η Argûs sharply'. Observations the next year revealed a closer

Finder Chart

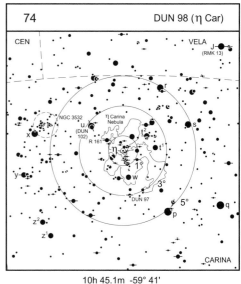

10h 45.1m -59° 41'

companion which he believed accounted for the fuzziness. In all he found four companions to the variable, all between magnitudes 11 and 14 and within 5″. In the 1930s and using the 26-inch at Johannesburg, W. H. van den Bos found another three, one of which was at 0″.2 and which he observed micrometrically. This component appeared single to large-aperture speckle instruments in 1994.

The Modern Era

The system η Carinae, as it is now known, is a unique object – an example of a massive, extremely luminous binary, wrapped in an area of gas and dust and surrounded by the results of earlier periods of mass expulsion from the outer layers of the brighter component. The force of the stellar winds emitted from both cause the surrounding gas to glow and emit X-rays, but the bow shock formed at the boundary of the respective stellar wind bubbles is so energetic that it accelerates particles to very high velocities such that gamma-rays are emitted. The binary period is thought to be 5.54 years and the orbital eccentricity is such that the stars are 20 times closer together at periastron than at apastron. The energetic outflow of gamma-rays occurs at periastron. The companion star has never been seen – its angular separation from the primary is thought to be about 0″.004. Conservative estimates for the luminosity of the five-year system are that of, respectively, 5 million and 1 million Suns, but the variable has a nine magnitude range, and disentangling all the very close visual components is no doubt difficult. The next periastron passage is expected in February 2020. Wiegelt & Ebersberger [353] used infrared speckle to see three close companions at distances of 0″.2 and below. It is possible that one of these is the van den Bos star but the infrequency of high-resolution observations does not enable this identification to be made with certainty. Modern high-resolution techniques have revealed three very close components to A at distances less than 0″.1. Star G is a massive Algol eclipsing binary (V573 Car) with a period of

1.47 days whilst H is a very close visual binary near the limit of resolution of a 4-metre telescope. There is no value for the parallax of η by Gaia (DR2) but H has been measured and found to be almost 8800 light years distant with an uncertainty of 760 light years. Trumpler 16 contains some of the most luminous stars in the galaxy.

Observing and Neighbourhood

This most complex system does, at least, appear relatively simple to the small telescope user. The Dunlop pair is readily seen in small apertures and with a 15-cm the fainter Herschel stars may be glimpsed. The enveloping Homunculus Nebula is about 15″ in extent. The star η itself seems to be slowly brightening after its minimum in 1930 or so and so there is an added reason for observing this system – might another sudden brightening be imminent? A recent paper by Kiminki *et al.* [352] finds evidence for previous outbursts around 1350 and 1550 and again around 1840. They measured the proper motions of knots of material in the outer region of the Homunculus Nebula using narrow-band images from HST. Component R 161 is 0°.7 NE – 6.1, 7.4, 295°, 1″.1, 2008. This pair has moved 40° prograde since discovery; meanwhile 1° S is DUN 97 (6.6, 7.9, 175°, 12″.7, 1998). There are four other stars further out with *V* between 8 and 10.

Measures

AH			
Early measure (JC)	16°.4	61″.32	1845.49
Recent measure (TMA)	16°.9	60″.60	2000.06

AG			
Early measure (PWL)	66°.2	38″.9	1859.92
Recent measure (TMA)	67°.4	37″.95	2000.06

75. μ VEL = R 155 = WDS J10468–4925

Table 9.75 Physical parameters for μ Vel

R 155	RA: 10 46 46.20	Dec: −49 25 12.8	WDS: 1797(51)		
V magnitudes	A: 2.82	B: 5.65			
$(B - V)$ magnitudes	A: +1.09	B: +0.55			
μ	63.22 mas yr^{-1}	±0.29	−54.21 mas yr^{-1}	±0.34	
π	27.84 mas	± 0.38 mas	117 light yr	± 1.6	
μ(A)	67.38 mas yr^{-1}	± 0.90	−49.86 mas yr^{-1}	± 0.85 (DR2)	
π(A)	27.95 mas	± 0.66	117 light yr	± 3 (DR2)	
Spectra	A: G6III	B: F4/5V			
Masses (M$_\odot$)	A: 3.0				
Luminosities (L$_\odot$)	A: 80	B: 6			
Catalogues	μ Vel	HR 4216	HD 93497	SAO 222321	HIP 52727
DS catalogues	R 155				
Radial velocity	6.20 km s^{-1}	± 0.2			
Galactic coordinates	283°.027	8°.572			

History

Brian Warner [340] gave a small list of double stars which had been noted by Fearon Fallows using the mural and transit circles at the Cape of Good Hope in 1829/30, and ten of them predate other discoverers, although the list was not published so Fallows never got any credit for them. One of the stars was μ Argûs, now μ Velorum. With the current orbit the separation at that time would have been 1″.7, so it is surprising that John Herschel did not see it with his considerably bigger aperture, and this must cast doubt on the identification by Fallows. It was re-discovered by Henry Chamberlain Russell, who was born in New South Wales, Australia, in 1836 and attended the University of Sydney, where he graduated with a BA degree in 1859. He was then appointed as a 'computer' at

Finder Chart

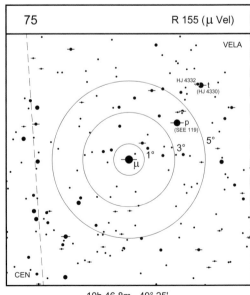

10h 46.8m -49° 25'

Orbit

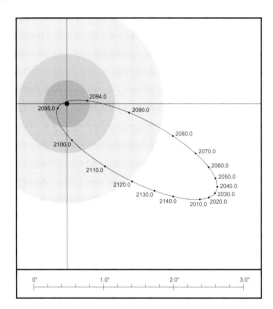

Ephemeris for R 155 (2010 to 2100)

Orbit by Tok (2014a) Period: 149.3 years, Grade: 4

Year	PA(°)	Sep(")	Year	PA(°)	Sep(")
2010.0	55.0	2.32	2060.0	66.5	2.21
2020.0	57.3	2.41	2070.0	69.4	1.97
2030.0	59.5	2.45	2080.0	73.6	1.58
2040.0	61.7	2.44	2090.0	82.1	0.90
2050.0	64.0	2.36	2100.0	8.0	0.51

Sydney Observatory, then under the leadership of its founding director, William Scott. Russell rose to become Government Astronomer in New South Wales in 1870 and in the following year was appointed Director at Sydney. At that time the Observatory had a fine 7.25-inch Merz refractor with micrometer, which was used to re-measure some of John Herschel's double stars. In 1874 an 11.5-inch Schroder refractor of 12 ft 6 in focal length was added. It was fitted with a micrometer and powers ranging from ×100 to ×1500 and used for the discovery of new double stars; Russell himself accumulated a list of 350 objects. Number 155 in this list is μ Vel, which was found in 1880. Russell noted that the colours were pale yellow and pale green and that the magnitudes were 3 and 9.

The Modern Era

The modern value for the visual magnitude of B is 5.65, as derived by Hipparcos. Observation of the star was part of a monitoring programme carried out by Ayres *et al.* [354] with the Extreme Ultraviolet Explorer, and they observed a significantly intense flare from the primary star which lasted 1.5 days. Yellow giants are not expected to produce such outbursts unless they happen to be in an RS CVn binary system. The WDS catalogue indicates that A is a spectroscopic binary but there are no references to the source of this information.

Observing and Neighbourhood

The apparent orbit of the visual pair has an eccentricity of almost 0.97 – one of the largest known. The two stars are 80 times closer together at periastron (0″.03 in 1945) than at apastron (2″.45 in 1880). The large difference in magnitude suggests that a night of good seeing would be advantageous and then 15-cm should show it well. The stars will remain near widest separation for several decades. The system μ Velorum lies 10° due N of the η Carinae nebula, and the sky between the two is filled with a range of visual double stars including DUN 95 (see Star 73) and DUN 102 (see Star 76). For a real challenge there is p Vel (SEE 119), one of T. J. See's most interesting discoveries. This pair has a period of only 16.65 years and is therefore correspondingly close. The minimum distance will be 0″.14 in 2020, the widest separation of 0″.44 occurring in 2030. The stars are magnitudes 4.1 and 5.8. The star t Vel is also HJ 4330 (5.2, 8.6, 163°, 40″, 1999). Star A is again a close pair, also called YSJ 1 (0″.4). Close by in the field is HJ 4332 (7.0, 9.8, 163, 25″, 2010).

Measures

Early measure (I)	61°.7	2″.72	1900.36
(Orbit	63°.6	2″.37)	
Recent measure (TOK)	56°.9	2″.28	2016.34
(Orbit	56°.5	2″.39)	

76. U CAR = Δ 102/3 VEL = WDS J10535–5851AB

Table 9.76 Physical parameters for u Car

DUN 102	RA: 10 53 29.57	Dec: −58 51 11.8	WDS: 16912(11)		
V magnitudes	A: 3.88	B: 6.23	C: 7.75		
(*B − V*) magnitudes	A: +1.07	B: −0.11	C: +0.86		
μ(A)	77.61 mas yr^{-1}	± 0.45	39.43 mas yr^{-1}	± 0.41 (DR2)	
μ(B)	−15.28 mas yr^{-1}	± 0.17	2.84 mas yr^{-1}	± 0.14 (DR2)	
μ(C)	19.49 mas yr^{-1}	± 0.07	−14.07 mas yr^{-1}	± 0.05 (DR2)	
π(A)	33.88 mas	± 0.25	96.3 light yr	± 0.7 (DR2)	
π(B)	1.38 mas	± 0.09	2360 light yr	± 154 (DR2)	
π(C)	4.51 mas	± 0.04	723 light yr	± 6 (DR2)	
Spectra	A: K0IV	B: B3IV/V			
Luminosities (L$_\odot$)	A: 20	B: 1400	C: 30		
Catalogues (A/B)	HR 4257	HD 94510/491	SAO 238574/0	HIP 53253/31	SAO 238574/70
DS catalogues	DUN 102 (AB)	DUN 103 (AC)			
Radial velocity	8.10 km s^{-1}	± 0.7			
Galactic coordinates	288°.182	+0°.599			

History

James Dunlop observed this coarse triple and gave it two entries in his double star catalogue. 'Very nearly in a line', recorded Dunlop. He did not record a distance for AB. John Herschel only estimated the distance of B, as $2\frac{1}{2}$ times the distance of C, but since then B and C have been interchanged in the catalogues.

The Modern Era

The bright star u Carinae is one of 65 stars brighter than *V* = 4.5 which do not have either a Bayer letter or Flamsteed number. It is not to be confused with U Carinae, a long-period and very luminous Cepheid variable, which lies 1°.5 SE. In fact u Car has been used as a photometric

Finder Chart

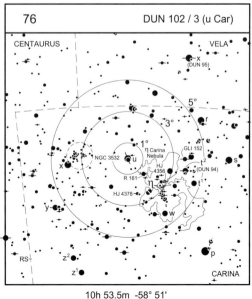

10h 53.5m -58° 51'

comparison star [355] for the photometry of U Car when it has been suspected of variability itself. It is number 5011 in the New Suspected Variables (NSV) catalogue, with a reported range of 3.75 to 3.80. Gaia DR2 shows that the three stars appear to be completely unassociated, having significantly different parallaxes and proper motions.

Observing and Neighbourhood

This is a bright, easy binocular triple located in a gloriously rich area of sky for the deep-sky and double star observer. Ross Gould notes that the three stars are almost in a line, with the bright orange primary in the middle. The fainter stars are unassociated, component A being considerably further from the Sun. Both B and C have been observed by Gaia. One point five degrees SW is the η Carinae Nebula and in the 3° quadrant preceding and south of u Car there are a number of double stars: HJ 4356 (7.3, 10.4, 150°, 2″.9, 2010); HJ 4378 (6.8, 10.2, 347°, 30″.9, 2010); GLI 152 (6.2, 8.0, 80°, 26″.2, 2001). The brightest is DUN 94 (4.9, 7.5, 20°, 14″.7, 2008), also known at t^2 Car, but perhaps the pick is R 161 (6.1, 7.4, 295°, 1″.1, 2008, now slowly widening).

Measures

AB			
Early measure (WFC)	201°.6	153″.02	1893.16
Recent measure (UC)	204°.4	159″.38	2000.06
AC			
Early measure (WFC)	10°.2	61″.86	1893.16
Recent measure (TMA)	7°.1	56″.43	2000.06

77. 54 LEO = STF 1487 = WDS J10556+2445

Table 9.77 Physical parameters for 54 Leo

STF 1487	RA: 10 55 36.80	Dec: +24 44 59.0	WDS: 161(252)	
V magnitudes	A: 4.48	B: 6.30		
(*B* − *V*) magnitudes	A: 0.00	B: +0.08		
μ(A)	−78.06 mas yr^{-1}	± 0.67	−16.52 mas yr^{-1}	± 0.69 (DR2)
μ(B)	−75.37 mas yr^{-1}	± 0.15	−18.60 mas yr^{-1}	± 0.09 (DR2)
π(A)	9.83 mas	± 0.35	332 light yr	± 12 (DR2)
π(B)	10.18 mas	± 0.06	320 light yr	± 2 (DR2)
Spectra	A: A1V	B: A2Vn		
Radii (R$_\odot$)	A: 2.88	B: 2.59		
Luminosities (L$_\odot$)	A: 135	B: 25		
Catalogues	HR 4259/60 (A/B)	HD 94601/2 (A/B)	SAO 81583/4 (A/B)	HIP 53417
DS catalogues	Mayer 28	H 3 30	STF 1487	BDS 5603	ADS 9797
Radial velocity	A: −0.49 km s^{-1}	± 0.98	B: +1.30 km s^{-1}	± 0.92
Galactic coordinates	211°.595	+63°.910		

History

Whilst Thomas Lewis records that 54 Leo was found to be double by William Herschel, it was actually known to Christian Mayer and is number 28 in his catalogue of 1784. Herschel's own observation was made on 21 February 1781. 'Double. Considerably unequal. L(arge). brilliant w(hite); S(mall). Ash-coloured or greyish w(hite). Distance 7″ 6″ (sic) mean measure. Position 9° 14″s(outh). f(ollowing).'

The Modern Era

The two stars are both fast rotators. The A component is rotating at 180 km s^{-1} whilst the companion is spinning even more rapidly, at 250 km s^{-1}. The WDS notes file says that B is a spectroscopic binary; however, recent echelle spectrograph measurements of the radial velocity of both components using

Finder Chart

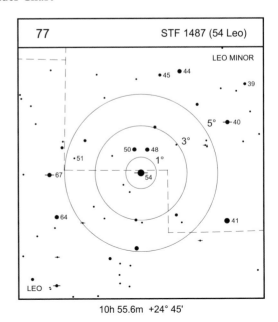

10h 55.6m +24° 45′

the HIRES spectrograph on *Keck I* by Becker [356] show no clear sign of radial velocity scatter. The radius of each star was calculated by Slettebak [357]. The proper motion of A is largely shared by B and so the pair form a visual binary of very long period.

Observing and Neighbourhood

The star can be found about midway between ξ UMa and γ Leo and is 2° SE of the line joining the two. It is a fine sight in a small aperture and, although both components are early A stars, there are reports of various hues being observed, although William Herschel found the stars to be brilliant white and ash-white or greyish white. John Nanson [126] looked at the pair on 8 March 2013 with a 6-inch f/10 refractor at ×109. He found that prolonged gazing at the stars tended to bring out a yellow hue in the primary and a blue hue in the companion. RWA has recorded only white for both components.

Measures

Early measure (STF)	102°.8	6″.14	1830.35
Recent measure (WSI)	113°.3	6″.49	2014.30

78. ξ UMA = STF 1523 = WDS J11182+3132AB

Table 9.78 Physical parameters for ξ UMa

STF 1523	RA: 11 18 10.90	Dec: +31 31 45.0	WDS: 3(1617)	
V magnitudes	Aab: 4.33	Bab: 4.80		
(*B − V*) magnitudes	Aab: +0.59	Bab: +0.73		
μ	−453.7 mas yr^{-1}	± 2.0	−591.4 mas yr^{-1}	± 2.0
π	119.7	± 0.8	27.2 light yr	± 0.2
μ(B)	−339.40 mas yr^{-1}	± 0.78	−607.89 mas yr^{-1}	± 0.76 (DR2)
π(B)	114.49 mas	± 0.43	28.5 light yr	± 0.1 (DR2)
Spectra	F9V	G9V		
Luminosities (L$_\odot$)	A: 1.2	B: 0.8		
Catalogues	HD 98230	SAO 62484	HIP 55203	
DS catalogues (AB)	H 1 2	STF 1523	BDS 5734	ADS 8119
Radial velocity (A/B)	−18.2 km s^{-1}	± 2.7		
Galactic coordinates	195°.107	+69°.246		

Introduction

The second entry in Sir William Herschel's catalogue of 1782, ξ UMa, has attracted much attention from double observers since its discovery on 2 May 1780. Herschel recorded 'Double. A little unequal. Both w(hite). and very bright. The interval with 227 is $\frac{2}{3}$ diameters of L(arge).; with 222, 1 diameter of L(arge).; with 275, near $1\frac{1}{2}$ diameter of L(arge). Position 53° 47′ s(outh). following.' In 1804, Herschel [363] used it, along with a number of other stars, as an example of the existence of gravitational attraction outside the Solar System and as confirming Herschel's belief in the existence of binary systems (double stars that are gravitationally connected). In 1827 ξ UMa was also the first visual binary to have its orbit computed, by Felix Savary [364]. The title of Savary's paper is 'Sur la détermination des orbites que découvrent autour de leur centre de gravité deux étoiles très rapprochées l'une de l'autre'. Lick Observatory spectra taken with the 36-inch refractor by Campbell & Wright [360] showed radial velocity

Finder Chart

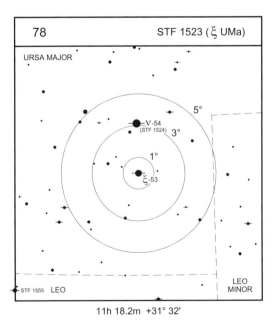

11h 18.2m +31° 32′

Orbit

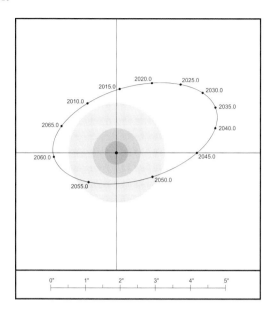

Ephemeris for STF 1523 AB (2018 to 2036)

Orbit by Msn (1995) Period: 59.878 years, Grade: 1

Year	PA(°)	Sep(")	Year	PA(°)	Sep(")
2018.0	161.0	2.01	2028.0	128.7	2.86
2020.0	152.4	2.19	2030.0	124.3	2.97
2022.0	145.2	2.38	2032.0	120.1	3.05
2024.0	139.0	2.55	2034.0	116.1	3.09
2026.0	133.6	2.72	2036.0	112.1	3.08

Measure with 8″ Refractor

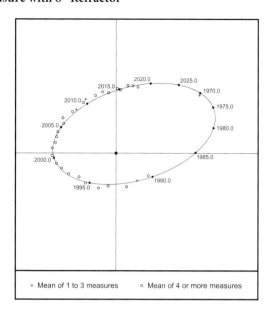

+ Mean of 1 to 3 measures   ○ Mean of 4 or more measures

Year	PA(°)	Sep(")	No.	Year	PA(°)	Sep(")	No.
1970.35	123.7	2.91	1	2004.30	250.5	1.79	7
1990.28	56.5	1.13	4	2005.34	241.4	1.73	7
1991.27	38.7	0.97	1	2006.34	237.4	1.83	6
1992.42	18.0	0.98	5	2007.34	227.5	1.71	5
1993.48	346.2	0.94	5	2008.43	223.5	1.68	3
1994.29	332.7	1.09	5	2009.37	213.8	1.70	6
1995.35	313.2	1.21	6	2010.30	207.0	1.74	3
1996.35	301.6	1.26	5	2011.41	200.1	1.71	4
1997.37	292.8	1.49	5	2012.40	192.6	1.75	4
1998.38	284.8	1.63	5	2013.35	187.3	1.70	6
1999.31	279.3	1.75	5	2014.42	178.8	1.80	6
2000.35	272.1	1.85	6	2015.38	175.6	1.77	5
2001.32	265.5	1.85	6	2016.43	169.7	1.92	5
2002.31	260.4	1.78	5	2017.40	165.5	1.94	6
2003.42	254.9	1.78	6	2018.44	161.0	1.95	4

variations in the system. In 1905, Norlund [359], in calculating a new visual orbit for AB, discovered a systematic run of residuals amounting to an amplitude of 0″.05 and a period of 1.8 years. Abetti, at Lick, confirmed that the radial velocity of star A varied periodically in 1.8 years, thus identifying the source of the effect found by Norlund. Radial velocity measurements of star B began in 1902, but it was not until 1918 that it was shown by Lick observers that this star too was a spectroscopic binary. A comprehensive study of the astrometry and radial velocities of the system was carried out by van den Bos in 1928 [366], and there the matter largely rested until Roger Griffin's [358] thorough analysis.

The Modern Era

The object ξ UMa is one of the most observed visual binary systems known. The WDS lists more than 1600 observations of position angle and separation, and the pair with its period of just under 60 years has now been observed through more than three revolutions. The current orbit (the ephemeris of which appears above) is classified as Grade 1.

According to Griffin, the period of Aa is 670.24 ± 0.09 days with a significant eccentricity (0.532) whilst B has a circular orbit of period 3.980507 days (known to an accuracy of half a second). In 1995 Mason *et al.* [361] noted an additional companion star to B during speckle interferometric observations, but a positive detection was noted in only one out of 27 attempts. Since no confirming observations have been made, the reality of this star must be in some doubt. In addition, Daniel Bonneau [365] questioned its existence on dynamical grounds. In 2013, Wright *et al.* [362] announced the discovery of a common proper motion companion to the quadruple system, located 511″ away. This star is a T8.5 dwarf which appears to be a single object. It is some 16 magnitudes fainter than AB in the *K* band. Gaia DR2 gives only the position and *G* magnitude of the A component.

Observing and Neighbourhood

The system is easily found by starting at δ Leo. Move 11° in declination: ξ is the southerly of two bright stars, the more northerly of which is the V = 3.5 mag ν. It will be easy to resolve ξ in 7.5-cm apertures for several decades to come. At the last periastron in 1995, the stars closed in to just under 1″. As the two components widen the angular motion will reduce – at present it is about 5° per year. The nearby ν (STF 1524), one of William Herschel's last discoveries, is also known as H N 53, and is much more of a challenge. The primary is a K3 giant and the companion is 7″.8 away in PA 145° but it is of magnitude 10.1. For a resolution test, try STF 1555, which lies 5° SE of ξ and is described at greater length elsewhere (see Star 82).

Measures

Early measure (STF)	238°.8	1″.75	1826.20
(Orbit	240°.3	1″.71)	
Recent measure (ARY)	161°.0	1″.95	2018.44
(Orbit	159°.0	2″.05)	

79. STF 1527 LEO = WDS J11190+1416

Table 9.79 Physical parameters for STF 1527 Leo

STF 1527	RA: 11 18 59.91	Dec: +14 16 06.9	WDS: 138(270)	
V magnitudes	A: 7.01	B: 7.99		
$(B - V)$ magnitudes	A: +0.57	B: +0.88		
μ	-453.7 mas yr^{-1}	± 2.0	-591.4 mas yr^{-1}	± 2.0
π	32.53 mas	± 1.39	100 light yr	± 4
Spectra	F9V			
Luminosities (L$_\odot$)	A: 1.2	B: 0.5		
Catalogues	HD 98354	HIP 55254	SAO 99551	
DS catalogues	H N 142	STF 1527	BDS 5739	ADS 8128
Radial velocity (A/B)	21.80 km s^{-1}	± 0.2		
Galactic coordinates	239°.123	+64°.918		

History

Herschel noted the separation as 4± whilst Thomas Lewis [194] stated that the star was a binary, although at that time the apparent motion since discovery had amounted to just +10° in PA and −0″.6 in separation. The significant proper motion would have separated the stars rapidly had they been unassociated.

The Modern Era

This pair has attracted some attention from observers with large apertures and a number of orbits have been computed. The form of the apparent orbit will be much clearer once the stars begin to separate, as they will do from 2015 or so. If the current orbit is correct, they will reach apastron in around 2200, when the separation will again be 4″. In 2012 Prieur *et al.* [504] obtained a dynamical parallax of 31.2 mas, in good agreement with the revised Hipparcos value. This paper also concluded that the spectral type of B was around G3V on the basis of the observed magnitude difference.

Finder Chart

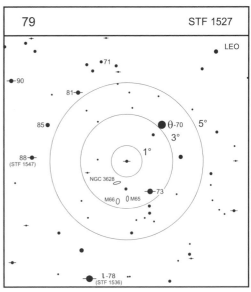

11h 19.0m +14° 16'

Orbit

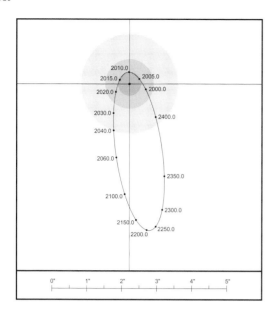

Ephemeris for STF 1527 (2015 to 2060)

Orbit by Tok (2012b) Period: 415.0 years, Grade: 4

Year	PA(°)	Sep(")	Year	PA(°)	Sep(")
2015.0	247.1	0.30	2040.0	340.9	1.36
2020.0	299.9	0.45	2045.0	343.9	1.56
2025.0	320.9	0.69	2050.0	346.2	1.74
2030.0	330.9	0.93	2055.0	348.1	1.91
2035.0	336.8	1.15	2060.0	349.7	2.07

Observing and Neighbourhood

In 1992 and 1994 the writer was able to measure this pair with the Cambridge 8-inch when the separation was near $0''.9$. Since then it has closed to $0''.3$ and only now is beginning to widen again. At present the pair is a real challenge for the larger aperture. A 30-cm should show that the image is not round, although to actually divide the stars will probably take 40-cm. By 2025 it will be back within the scope of 20-cm. It is easily found, being $2°$ SE of the bright star θ Leo. Three degrees further E is 88 Leo (STF 1547). RWA measured this pair of stars, whose magnitudes, are 6.3 and 9.1, and found $332°$, $15''.6$ in 2017. According to W. Hartkopf [367] they form a binary pair with period 3453 years. The stars' separation is predicted to vary between $7''$ and $16''.5$ and both are yellow. Four degrees S and slightly E is ι Leo.

Measures

Early measure (STF)	10°.0	3″.83	1826.20
(Orbit	9°.9	4″.08)	
Recent measure (ARY)	54°.5	0″.89	1994.27
(Orbit	52°.0	0″.76)	
Latest measure (SCA)	237°.2	0″.29	2014.26
(Orbit	237°.9	0″.30)	

80. ι LEO = STF 1536 = WDS J11239+1032AB

Table 9.80 Physical parameters for ι Leo

STF 1536	RA: 11 23 55.37	Dec: +10 31 46.9	WDS: 39(541)	
V magnitudes	A: 4.06	B: 6.71		
(B − V) magnitudes	A: +0.44	B: +0.91		
μ(A)	149.41 mas yr^{-1}	± 0.72	−90.70 mas yr^{-1}	± 0.69 (DR2)
μ(B)	177.27 mas yr^{-1}	± 0.50	−52.97 mas yr^{-1}	± 0.84 (DR2)
π(A)	42.36 mas	± 0.40	77.0 light yr	± 0.7 (DR2)
π(B)	41.50 mas	± 0.36	78.6 light yr	± 0.7 (DR2)
Spectra	F3	F3s		
Luminosities (L$_\odot$)	A: 11	B: 1		
Catalogues	HR 4399	HD 99028	SAO 99587	HIP 55642
DS catalogues	STF 1536	BDS 5765	ADS 8148	
Radial velocity (A/B)	−11.80 km s^{-1}	± 0.2		
Galactic coordinates	247°.600	+63°.548		

History

Found by F. G. W. Struve in 1827, this pair showed little motion during the remainder of the century, with maximum distance occurring around 1874. From 1910 or so the stars began to close up quickly, and a close approach occurred in 1943 when they were 0″.63 apart.

The Modern Era

In 1976 Abt & Levy [369] used radial velocities of the A star in combination with visual measures to derive a period of 191 years. The period was reduced by five years in the analysis by Staffan Söderhjelm [370]. The stars are now almost back at the discovery position.

Finder Chart

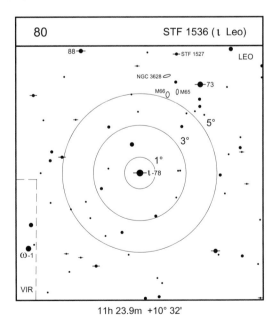

11h 23.9m +10° 32'

Orbit

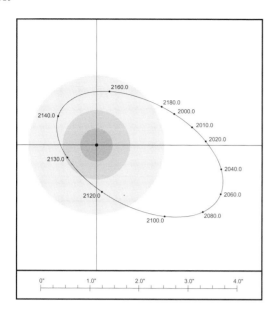

Ephemeris for STF 1536 AB (2010 to 2100)

Orbit by Sod (1999) Period: 186 years, Grade: 2

Year	PA(°)	Sep(")	Year	PA(°)	Sep(")
2010.0	100.3	1.97	2060.0	69.0	2.71
2020.0	92.0	2.22	2070.0	63.9	2.67
2030.0	85.2	2.43	2080.0	58.5	2.55
2040.0	79.3	2.59	2090.0	52.3	2.32
2050.0	74.0	2.68	2100.0	44.4	1.99

Observing and Neighbourhood

The pair ι Leo is 5°.5 SSE of θ Leo, and exactly half-way between the two stars are the bright galaxies M65 and M66. It is not particularly easy for the small aperture owing to the significant difference in magnitude between the components, and for the northern observer there is also the relatively low declination; however, the stars will continue to separate for the next 30 years or so. F. G. W. Struve found the colours to be yellow and blue, whilst Smyth notes 'pale yellow and light blue'. About 7°.5 S of ι is the bright binocular pair τ Leo (STFA 19), 5.1, 7.5, 182°, 88″, 2016, with the bright pair 83 Leo = STF 1540 (6.6, 7.5, 149°, 28″, 2017) nestling about 20′ W and visible in the same low-power field.

Measures

Early measure (STF)	90°.1	2″.41	1837.39
(Orbit	89°.5	2″.30)	
Recent measure RWA	96°.1	2″.19	2018.32
(Orbit	93°.2	2″.18)	

81. STT 235 UMA = WDS J11323+6105AB

Table 9.81 Physical parameters for STT 235 UMa

STT 235	RA: 11 32 20.76	Dec: +61 04 57.9	WDS: 96(334)	
V magnitudes	A = 5.69	B = 7.75		
$(B - V)$ magnitudes	A: +0.51	B: +0.75		
μ	-16.48 mas yr^{-1}	± 0.51	-96.10 mas yr^{-1}	± 0.43
π	35.73 mas	± 0.54	91.3 light yr	± 1.4
μ(A)	-24.23 mas yr^{-1}	± 0.52	-67.23 mas yr^{-1}	± 0.62 (DR2)
π(A)	34.84 mas	± 0.38	94 light yr	± 1 (DR2)
Spectra	F7V			
Masses (M$_\odot$)	2.05	± 1.10 (dyn.)	2.34 (phot.)	1.10 (spec.)
Luminosities (L$_\odot$)	A: 3.5	B: 0.5		
Catalogues	HD 100203	SAO 15542	HIP 56290	
DS catalogues	STT 235	BDS 5811	ADS 8197	
Radial velocity (A/B)	0.6 km s^{-1}	± 2	-1.9 km s^{-1}	± 2
Galactic coordinates	$138°.904$	$+53°.524$		

History

Found by O. Struve in 1843, this pair is now well into its third orbital cycle since discovery. Smyth has not observed it and Haas suggested only that a 300-mm aperture should be able to resolve this pair in 2006, when the separation was 0″.7.

The Modern Era

The pair is amongst those predicted to pass close to the solar system (Bailer-Jones [271]). The orbit is tolerably well known and the period of 72.7 years assigned to it by Söderhjelm has been graded 2 in the USNO *Sixth Orbit Catalog*.

Finder Chart

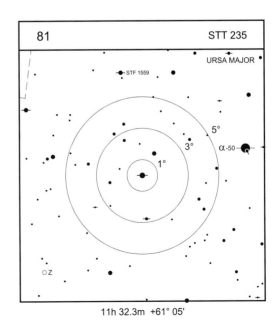

11h 32.3m +61° 05'

Orbit

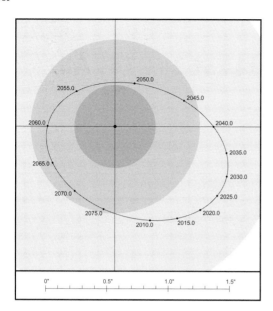

Ephemeris for STT 235 AB (2015 to 2042)

Orbit by Sod (1999) Period: 72.7 years, Grade: 2

Year	PA(°)	Sep(")	Year	PA(°)	Sep(")
2015.0	34.9	0.89	2030.0	66.5	0.99
2018.0	42.0	0.93	2033.0	72.7	0.96
2021.0	48.5	0.97	2036.0	79.3	0.91
2024.0	54.7	0.99	2039.0	87.1	0.84
2027.0	60.6	1.00	2042.0	96.7	0.73

Observing and Neighbourhood

The object STT 235 can be found 4° ESE of α UMa and for many years this pair has been too close for resolution with the apertures available to the writer, but in the spring of 2017 the companion appeared very closely preceding and slightly S of the primary using the 8-inch refractor at Cambridge. It was difficult owing to the significant difference in brightness, but it was certainly there. The next 10 years now offers the best chance to resolve the system but even then it will require at least 15-cm and a good night, and the significant difference in magnitude makes splitting the pair harder. From the end of the 2020s the stars will close quickly, reaching a minimum separation of 0″.35 in 2052 before widening again to 1″ in 2100. Three point five degrees N and slightly E is STF 1559 (6.8, 8.0, 323°, 2″.0, 2017).

Measures

Early measure (STT)	153°.9	0″.97	1825.21
(Orbit	148°.7	0″.78)	
Recent measure (SCA)	108°.9	0″.85	2014.94
(Orbit	109°.1	0″.82)	

82. STF 1555 LEO = WDS J11363+2747AB

Table 9.82 Physical parameters for STF 1555 Leo

STF 1555	RA: 11 36 17.94	Dec: +27 46 52.7	WDS: 122(294)	
V magnitudes	A: 6.41	B: 6.78	C: 11.7	
$(B - V)$	A: +0.25	B: +0.29		
μ	24.11 mas yr^{-1}	± 1.31	15.83 mas yr^{-1}	± 1.63
π	13.97 mas	± 0.75	233 light yr	± 13
Spectra	F0V			
Luminosities (L$_\odot$)	A: 13	B: 8		
Catalogues	HR 4465	HD 100808	SAO 81893	HIP 56601
DS catalogues	STF 1555 (AB)	HJ 503 (AB-C)	BDS 5841	ADS 8231
Radial velocity	8.00 km s^{-1}	± 3.7		
Galactic coordinates	206°.347	+73°.334		

History

Discovered at Dorpat by F. G. W. Struve, this was a relatively easy pair to resolve. It was, nonetheless, missed by John Herschel with his 20-foot reflector at Slough in 1827 when he noted a faint companion (C) of magnitude 10 some 18″ distant in PA 50° and assigned it number 503 in his catalogue (WDS – magnitude 11.7 at 158°, 22″.5 – distance increasing). There is also some confusion over this star as the next entry, number 504, appears to be the same object – it is at slightly different RA and the magnitudes and position are 6 and 12 at 15″ and 50°.

The Modern Era

Some doubt still lingers over the exact nature of the close pair. The early measures were discussed by Burnham in his 1906 catalogue and he said '… at this time the motion is practically

Finder Chart

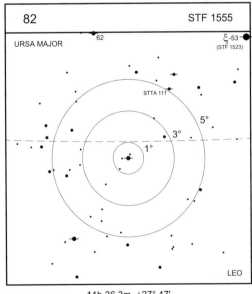

11h 36.3m +27° 47'

Orbit

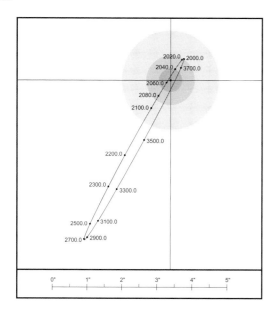

Ephemeris for STF 1555 AB (2010 to 2100)

Orbit by Doc (2017a) Period: 1730 years, Grade: 4

Year	PA(°)	Sep(")	Year	PA(°)	Sep(")
2010.0	149.1	0.72	2060.0	298.7	0.13
2020.0	150.8	0.66	2070.0	317.5	0.34
2030.0	153.1	0.53	2080.0	321.8	0.56
2040.0	157.6	0.34	2090.0	323.8	0.76
2050.0	176.3	0.13	2100.0	324.9	0.95

rectilinear and uniform. It will soon be a difficult pair with a minimum distance of about 0″.1.' Between 1920 and 1945 the distance remained below 0″.2 but measures were scattered. Since then the separation has slowly increased and as at 2015 stands at about 0″.75. This is certainly significantly closer than the distance predicted by the linear ephemeris, whilst the orbit predicts a smaller separation. In the next five or ten years it should be possible to come down on the side of one prediction or the other. One component of the system is a spectroscopic binary.

Observing and Neighbourhood

This is an ideal system for testing the resolution of a 15-cm telescope. The relative brightness and equality of the two stars fit Dawes' criterion well. The pair appears on Map 9 of the *CDSA2*, but it is not labelled, reflecting the linear nature of the system in the opinion of the cartographers. RWA measured this pair in 1995, 2003, and 2014 and has recorded a slow increase in separation with little change in position angle. This is a fairly sparse area of the sky but about 3° NNW is the wide pair STTA 111 of magnitudes 6.95 and 9.49, measured by RWA in 1996 at 33°.2, 67″.52. There is very little motion in this system. The primary star is the W-UMa-type binary AW UMa, which has a period of 0.44 days and an amplitude of 0.25 magnitudes in *V*. Pribulla *et al.* [371] indicated the likelihood of two further components with periods of 398 and 6250 days respectively. Moving another 3° in approximately the same direction will bring you to the beautiful binary ξ UMa (see Star 78).

Measures

Early measure (STF)	339°.4	1″.25	1829.12
(Orbit	337°.4	1″.20)	
(Linear	341°.8	1″.17)	
Recent measure (ARY)	148°.2	0″.81	2014.17
(Orbit	150°.1	0″.68)	
(Linear	148°.6	0″.91)	

83. 65 UMA = STF 1579
= WDS J11551+4629AB,D

Table 9.83 Physical parameters for STF 1579

STF 1579	RA: 11 55 05.75	Dec: +46 28 36.6	WDS: 981(78)		
V magnitudes	A = 6.5	B = 9.0	C = 8.72	D = 6.97	
(B − V) magnitudes	A: +0.11	B:			
μ	+10.41 mas yr^{-1}	± 0.49	+2.38 mas yr^{-1}	± 0.43	
π	4.72 mas	± 0.58	690 light yr	± 85	
μ(A)	22.71 mas yr^{-1}	± 0.49	−19.43 mas yr^{-1}	± 0.63 (DR2)	
μ(D)	12.05 mas yr^{-1}	± 0.28	−3.07 mas yr^{-1}	± 0.41 (DR2)	
π(A)	11.52 mas	± 0.48	283 light yr	± 12 (DR2)	
π(D)	6.23 mas	± 0.27	524 light yr	± 23 (DR2)	
Spectra	A3Vn				
Masses	Aa: 3.28	Ab: 2.52	B: 1.81	C: 1.9	D: 2.9
					Db: 1.6
Luminosities (L$_\odot$)	A: 15	B: 4	C: 2	D: 35	
Catalogues	HR 4560	HD 203483	SAO 43495	HIP 58112	
DS catalogues	H 1 72 (AC)	STF 1579 (AC)	A1777 (AB)	BDS 5962	ADS 8347
	BAG 46 (Da)				
Radial velocity (A)	−3.90 km s^{-1}	± 4.4			
Galactic coordinates	149°.122	+67°.682			

History

This pair, now known as AC, was found by William Herschel at Slough on 20 November 1782, although with some difficulty. He could not see it with ×227 'unless it has been seen first with a higher power'. He also noted a further star at 60″, which appeared as bright as component A; it is now known as D. When Robert Aitken observed the pair in 1908 he found that A itself was a close pair (A1777), which was also a rapid binary. The stars were separated by 0″.26 at discovery.

The Modern Era

In 1979, Giménez & Quesada [372] found that star A was a light variable and in 1982 they confirmed [376] that it was an eclipsing binary. The two stars (called Aa1 and Aa2) form DN UMa, which has a period of 1.7304 days with an amplitude of about 0.09 magnitude. These observations were carried out with a 254-mm reflector with a CCD and BVRI filters over four years, as a consequence of which they found another variation of 640 days, representing another component, Ab, around which the Aa1–Aa2 pair were revolving. In 2012, Zasche *et al.* [373] published a comprehensive survey of the astrometry, photometry, and radial velocity of this complex

Finder Chart

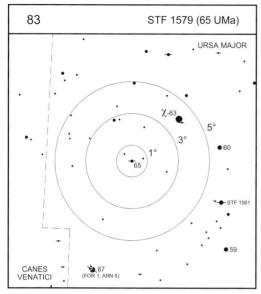

83   STF 1579 (65 UMa)

URSA MAJOR

χ-63

5°

3°

1°

65

● 60

●— STF 1561

● 59

CANES
VENATICI   67
(FOR 1, ARN 5)

11h 55.1m +46° 29'

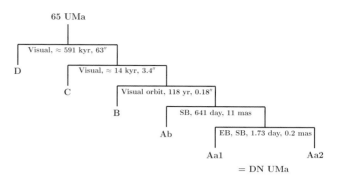

Figure 9.3 shows the hierarchical structure:

65 UMa

Visual, ≈ 591 kyr, 63″

D

Visual, ≈ 14 kyr, 3.4″

C

Visual orbit, 118 yr, 0.18″

B

SB, 641 day, 11 mas

Ab

EB, SB, 1.73 day, 0.2 mas

Aa1   Aa2

= DN UMa

Figure 9.3 The hierarchical structure of STF 1579 as presented by Zasche *et al.* [373].

Observing and Neighbourhood

The system 65 UMa can be found about 8° S of γ UMa. In this group a small aperture will show only the stars A, C, and D. The close visual pair AB is also unequal so any attempt to resolve it visually at present would require a large aperture and good seeing. Unfortunately it will remain below 0″.2 for another 10 years or so, and the significant difference in magnitude (2.5) only adds to the difficulty. Three degrees WSW of 65 is STF 1561 – 6.5, 8.2, 245°, 89″.2, 2020, an orbital pair with period 2050 years. Three point five degrees SSE is 67 UMa, a binocular triple. The components AB are FOR 1 (5.2, 6.7, 62°, 274″, 2012) whilst C is 8.5 at 25°, 395″ to A (ARN 5).

Measures

AC			
Early measure (STF)	1832.43	36°.5	3″.71
Recent measure (PRI)	2012.38	41°.9	3″.90

AD			
Early measure (STF)	1833.45	113°.8	62″.91
Recent measure (GAT)	2012	113°.9	63″.05

system. They used recent interferometric measures of AB, which have helped to pin down the period to 118 years and the maximum orbital separation to 0″.31. The same study also found a dynamical parallax for 65 UMa of 234 ± 29 parsecs (763 ± 95 light years), placing the system somewhat further away than that found by Hipparcos (see the table). Figure 9.3 shows the hierarchical structure of this complex system. Both the distant Herschel stars are physically connected but estimates of the periods of rotation around AB must be speculative at present. The WDS catalogue shows a further component. Balega *et al.* [374] found component D to be a very close pair (0″.1) in 2009 but there have been no confirming observations as yet. Septuple systems are extremely rare. Tokovinin [557] listed only two – AR Cas and ν Sco. They were not aware of the duplicity of D when they drew up their table. Gaia DR2 places the connection between D and the rest of the group in some doubt.

84. D CEN = RMK 14 = WDS J12140−4543

Table 9.84 Physical parameters for D Cen

RMK 14	RA: 12 14 02.70	Dec: −45 43 26.1	WDS: 2215(43)	
V magnitudes	A: 5.78	B: 6.98		
$(B - V)$ magnitudes	A: +1.84	B: +1.44		
μ(A)	−37.19 mas yr^{-1}	± 0.21	6.60 mas yr^{-1}	± 0.13 (DR2)
μ(B)	−33.60 mas yr^{-1}	± 0.07	5.43 mas yr^{-1}	± 0.05 (DR2)
π(A)	5.34 mas	± 0.14	611 light yr	± 16 (DR2)
π(B)	4.93 mas	± 0.06	662 light yr	± 8 (DR2)
Spectra	A: K3III	B:		
Radii (R$_\odot$)	A: 1.19 ±	0.13 mas		
Luminosities (L$_\odot$)	A: 140	B: 55		
Catalogues	HR 4652	HD 106321	SAO 223297	HIP 59654
DS catalogues	RMK 14			
Radial velocity	6.8 km s^{-1}	± 2		
Galactic coordinates	296°.134	+16°.654		

History

Such a bright and relatively easy pair should not have escaped the notice of James Dunlop at Parramatta, but Andrew James notes that the separation would have been close to the separation limit of the pairs found by Dunlop, probably as a result of the quality of his home-made reflecting telescope. Rümker, also at Parramatta, did however find this pair. In 1832 he published a *Preliminary Catalogue of Fixed Stars* (short title), which included a list of 28 bright and wide double stars. A recent re-examination by Letchford *et al.* [377] shows that numbers 22 and 23 are the same pair and number 24 does not exist. It is, however, exactly one hour of RA away from number 25 at the same declination and may be a slip of the pen. Of the 26 real systems, the WDS gives credit to Dunlop for five of them but, as Dunlop and Rümker were contemporaneous and did not record the dates of their observations, it is not clear who had

Finder Chart

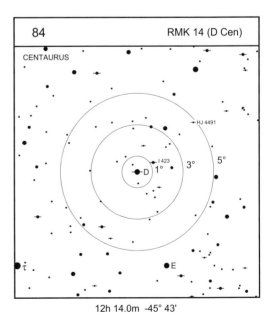

12h 14.0m -45° 43'

precedence in time to the first observation of each pair. One further object appears to have been independently found later at Cordoba Observatory. It did not escape John Herschel: 'Both stars yellow; a very fine object' was his conclusion.

The Modern Era

If A and B were unrelated the separation between them would have widened by about $1''$ per hundred years. However, A is now known to be a spectroscopic binary and so this is a physical triple star. The system D Cen was one of thousands of stars monitored by the Diffuse Infrared Background Explorer experiment on the COBE satellite. The star was observed in four infrared wavelengths from 1.25 to 4.9 μm but no sign of variation was found.

Observing and Neighbourhood

This is a lovely, bright pair that is attractive in small apertures but better seen with 10-cm or more. It can be found in mid-Centaurus above the Southern Cross, forming a rough equilateral triangle with γ and δ^1 Cen. Andrew James notes colours of deep yellow or orange-ish and orangey-red. Hartung, with 30-cm, noted orange-red and white in a 'well-sprinkled star field', whilst Richard Jaworski [378] using a 100-mm refractor at $\times 80$ found orange and red and E. J. Hartung noted orange and white. HJ 4491 ($2°.5$ NW) is a physical pair (8.1, 8.8, $41°$, $23''.2$, 2010). DR2 shows both stars at a distance of 173 light years. A more difficult target is I 423, which is just $30'$ NW of D Cen. The stars have magnitudes 6.8 and 10.6 but are only separated by $2''.7$ in PA $166°$ (1991).

Measures

Early measure (SLR)	$243°.4$	$2''.90$	1891.34
Recent measure (ARY)	$242°.8$	$2''.70$	2008.40

85. STF 1639 COM = WDS J12244+2535AB

Table 9.85 Physical parameters for STF 1639 Com

STF 1639	RA: 12 24 26.81	Dec: +25 34 56.7	WDS: 68(422)	
V magnitudes	A: 6.74	B: 7.83	C: 11.42	
$(B - V)$ magnitudes	A: +0.27	B: +0.45		
μ(A)	-9.41 mas yr^{-1}	± 0.10	-10.87 mas yr^{-1}	± 0.06 (DR2)
μ(B)	-15.47 mas yr^{-1}	± 0.12	-6.43 mas yr^{-1}	± 0.07 (DR2)
π(A)	11.64 mas	± 0.05	280 light yr	± 1 (DR2)
π(B)	11.70 mas	± 0.06	279 light yr	± 1 (DR2)
Spectra	A7V	F4V		
Luminosities (L$_\odot$)	A: 12	B: 4	C: 0.15	
Catalogues	HR 4719	HD 108007	SAO 82293	HIP 60525
DS catalogues	STF 1639	BDS 6158	ADS 8539	
Radial velocity (A/B)	-1.10 km s^{-1}	± 1.8		
Galactic coordinates	225°.758	+83°.760		

History

Found by F. G. W. Struve in 1828, the pair closed slowly throughout the nineteenth century until in 1892 it appeared single even to Burnham with the Lick 36-inch refractor (when the current orbit predicts a separation of 0″.09 for that epoch). It reappeared several years later and has been widening ever since. Burnham thought that the period would be more than 400 years, and the current orbit shows a period of 575 years, although it is still far from certain how long it will be before the companion slowly starts to turn inwards again.

The Modern Era

According to Olevic's orbit [380] the stars will reach their widest separation around 2175, when they will be separated by 2″.35. At present it is a fine object for 15-cm although there is a considerable difference in brightness between the stars. A third star of magnitude 11.4 can be found 91″ distant in PA 159°. DR2 does not include this star but does show that

Finder Chart

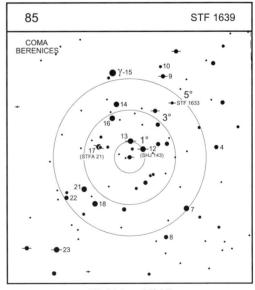

Orbit

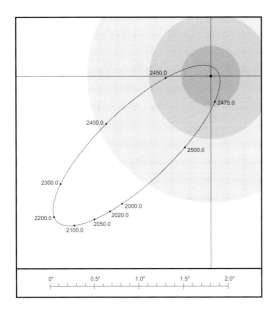

Ephemeris for STF 1639 AB (2000 to 2180)

Orbit by Ole (2000b) Period: 575.44 years, Grade: 4

Year	PA(°)	Sep(")	Year	PA(°)	Sep(")
2000.0	324.6	1.71	2100.0	316.9	2.24
2020.0	322.7	1.86	2120.0	315.7	2.29
2040.0	321.0	1.98	2140.0	314.6	2.33
2060.0	319.5	2.09	2160.0	313.4	2.34
2080.0	318.2	2.17	2180.0	312.3	2.35

the RS CVn variable IL Com (V = 8.1), which is 480″ distant to the E) also has a very similar parallax to the components of STF 1639.

Observing and Neighbourhood

The pair STF1639 is located in the heart of the Coma cluster (Melotte 111) and is a member of the group. A recent paper by Guerrero *et al.* [379] finds that the 25% of the stars in the Coma cluster which they observed were found to be double or multiple, whilst only 5% of the surrounding field stars were not alone in space. STF 1639 forms a tight equilateral triangle with 12 and 13 Com and the finding chart will help to locate it. There are a number of pairs within a few degrees of STF 1639. The nearest is 12 Com, which was catalogued by Sir James South and Sir John Herschel (SHJ 143). This is a wide and very unequal pair, visible in binoculars. The stars are of magnitudes 4.9 and 8.9 and the separation is 58″. Admiral Smyth noted the colours as straw yellow and rose red. About 1°.5 NW of 12 Com is STF 1633 (magnitudes 7.0, 7.1, 245°, 8″.9, 2017), the yellow components of which are moving through space at more than 0″.1 per year. Webb says 'Very pretty. Solitary'. The object 17 Com is a fine binocular pair – STFA 21 (5.2, 6.6, 250°, 145″, 2012).

Measures

Early measure (STF)	90°.1	2″.41	1837.39
(Orbit	89°.5	2″.30)	
Recent measure (CTT)	324°.2	1″.79	2014.37
(Orbit	323°.2	1″.82)	

86. $\alpha^{1,2}$ CRU = Δ 252 = WDS J12266–6306AB

Table 9.86 Physical parameters for α Cru

DUN 252	RA: 12 26 35.90	Dec: −63 05 56.7	WDS: 694(99) (AB)		
V magnitudes	A: 1.32	B: 1.55	C: 4.82		
$(B - V)$ magnitudes	A: +0.08	B: +0.75	C: −0.14		
μ(A)	−35.83 mas yr^{-1}	± 0.47	−14.86 mas yr^{-1}	± 0.43	
μ(B)	−35 mas yr^{-1}	± 18	25 mas yr^{-1}	± 25	
μ(C)	−38.0 mas yr^{-1}	± 1.0	−15.3 mas yr^{-1}	± 0.9	
π(A)	10.13 mas	± 0.50	320 light yr	± 15	
μ(C)	−39.59 mas yr^{-1}	± 0.33	−14.54 mas yr^{-1}	± 0.37 (DR2)	
π(C)	10.56 mas	± 0.23	309 light yr	± 7 (DR2)	
Spectra	A: B0.5IV + B1V	B: B0.5IV	C: B3/5V		
Masses (M$_\odot$)	Aa: 17.8	Ab: 6.1	B: 15.5		
Luminosities (L$_\odot$)	A: 2400	B: 1950	C: 90		
Catalogues (A/B/C)	26/27 G Cru	HR 4730/1/29	HD 108248/9/50	SAO 251904/—/3	HIP 60718
DS catalogues	DUN 252				
Radial velocity	11.90 km s^{-1}	± 2.4			
Galactic coordinates	300°.127	−0°.363			

History

Discovered by Fr Fontenay whilst in South Africa on 2 June 1685. Using an excellent 12-foot telescope, Fontenay notes that the star in the head of the Cross which is marked 3 by Bayer is actually two beautiful stars separated by their own diameter, and he compares it with Castor in the northern hemisphere. He then mentions a third star (C), which is more distant. James Dunlop observed the three stars in 1826 and allocated number 122 in his catalogue to AB whilst AC became 123; he was the first to measure them. For some unknown reason the system is now known as DUN 252 whilst Dunlop's original number 252 is a very wide pair 2°.5 NW of ϵ Tucanae. John Herschel comments on Rümker's notes about α Crucis: 'Mr Rümker describes it as involved in a milk-white nebulosity (*Preliminary Catalogue*, pp. 15, 17), a description which I am at a loss to understand, unless it refers to the bright surrounding light of the Milky Way.' Moore at the Lick Observatory [385] found a velocity amplitude of 60 km s^{-1} in the C component and in 1932 Neubauer [384] announced that both A and B were spectroscopic binaries with almost identical periods of just under one day.

The Modern Era

Thackeray & Hill [383] presented the results of a radial velocity study of α^1 and α^2 Crucis. They found no sensible variation of velocity in α^2, and the new measurements of the radial velocity of α^1 did not corroborate Neubauer's data.

Finder Chart

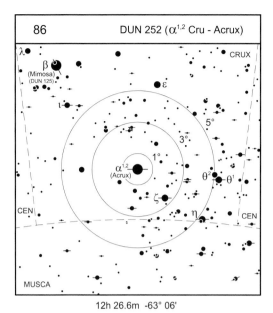

| 86 | DUN 252 ($\alpha^{1,2}$ Cru - Acrux) |

12h 26.6m -63° 06'

A period of 75.769 days was found with an eccentricity of 0.48 (later refined to 75.779 days and $e = 0.46$), with the conclusion that α^1 consisted of a B1IV primary with a main sequence companion. In 1979 Hernández & Hernández [387] conducted a new spectroscopic study of the C component and determined a period of 1.225 days and a velocity amplitude of 80 km s^{-1}. The systemic velocity was less than 1 km s^{-1} away from that of the A component binary system, thus linking C to AB as a physical member of the system, although Jim Kaler [390] has some doubts about this association. Shatsky & Tokovinin [417] added a faint infrared object 2″.4 distant

from C at PA 53°. The two brightest stars are not included in DR2 but the distant magnitudes 4.8 star is.

Observing and Neighbourhood

The object α Crucis, also known as Acrux, is the brightest triple star that is accessible to the small telescope. In addition to the two first magnitude components there is the third star, with $V = 4.8$ (HD 108250) at 202° and 91″.8, making it an easy binocular object. The three bright stars are all hot B stars and all appear bluish-white. E. J. Hartung has seen this pair in bright sunshine using a 7.5-cm aperture. More recently Rainer Anton [382] imaged the system and found five stars between magnitudes 12 and 13 which were within 30″ to 125″ of star A. Several other bright stars of Crux are all wide and unequal doubles. The star β (DUN 125) (1.3, 7.2, 23°, 373″, 2000) is an easy binocular target whilst 145″ distant in PA 260 is a carbon star called Espin-Birmingham 365.

Measures

AB			
Early measure (HJ)	120°.6	5″.65	1835.33
Recent measure (ANT)	112°.3	4″.00	2014.23
AC			
Early measure (JC)	202°.0	88″.93	1847.31
Recent measure (ANT)	202°.0	91″.80	2014.24

87. γ CRU = Δ 124 = WDS J12312–5707AB

Table 9.87 Physical parameters for γ Cru

DUN 124	RA: 12 31 09.95	Dec: −57 06 47.5	WDS: 7263(19)	
V magnitudes	A: 1.83	B: 3.65	C: 9.74	
$(B − V)$ magnitudes	A: +1.82	B: +0.14		
μ	−28.23 mas yr^{-1}	± 0.14	− 265.08 mas yr^{-1}	± 0.12
π	36.83 mas	± 0.18	88.6 light yr	± 0.4
Spectra	A: M3.5III	B: A3V		
Luminosities (L$_\odot$)	A: 115	B: 25		
Catalogues	HR 4763/4	HD 108903/25	SAO 240019/22	HIP 61084
DS catalogues	DUN 124			
Radial velocity	21.00 km s^{-1}	± 0.1		
Galactic coordinates	300°.159	+5°.650		

History

A common alternative name for this star is Gacrux. The name seems to have been derived by Elijah Hinsdale Burritt (1794–1893), who produced *The Atlas of the Heavens*. The wide companions were seen by Dunlop.

The Modern Era

The object γ Cru is the brightest M giant in the sky and has been investigated for variability. Petit [389] found the range was a few hundredths. It has certainly been investigated for radial velocity variations by Murdoch & Clark [388]. Over 900 days they found a total range of 1200 metres per second; although there was no obvious periodicity, several small peaks were seen in frequency space at 90 days or below. The atmosphere of Gacrux contains barium, and Kaler [390] believes that this indicates the existence of a white dwarf companion which is seeding the primary. The Dunlop stars are both optical. None of the three stars listed above appear in Gaia DR2.

Finder Chart

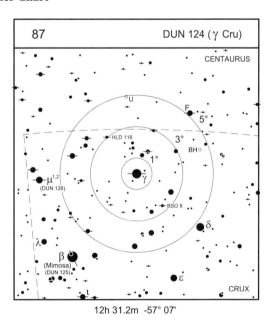

12h 31.2m -57° 07'

Observing and Neighbourhood

Easily found, γ Crucis is the top star of the Cross. It is also a binocular target: the primary is an M3 giant and there is a significant colour difference between the reddish-orange primary and the white companion 130″ distant, which is bright and far enough away to be seen in binoculars. Four degrees SE is Mimosa (β Cru, DUN 125). The Dunlop star has magnitude 7.2 and is 23° and 373″ distant from the magnitude 1.3 primary. Robert Innes found an 11th magnitudes star at 326°, 42″.6 and more recently Rainer Anton, working in Namibia, has imaged two more distant magnitude 11 stars. There is a neat pair (BSO 8) at 1°.5 SW of γ. Its stars are magnitudes 7.8 and 8.0 at 335° and 5″.2, 2016. Head due E to find μ (DUN 126) (see Star 93), which is possibly the finest pair for the small aperture besides α. Its white components of magnitudes 3.9 and 5.0 are 35″ apart in PA 17°, 2016. One degree N and a little E is HLD 116 (7.1, 8.9, 182°, 1″.9, 2000).

Measures

AB			
Early measure (PWL)	36°.5	99″.17	1860.11
(Linear	36°.4	99″.11)	
Recent measure (ANT)	25°.0	128″.9	2010.27
(Linear	25°.0	129″.42)	
AC			
Early measure (CRU)	86°.2	155″.2	1879.50
Recent measure (ANT)	70°.0	165″	2010.27

88. 24 COM = STF 1657 = WDS J12351+1823

Table 9.88 Physical parameters for 24 Com

STF 1657	RA: 12 35 07.76	Dec: +18 22 37.40	WDS: 404(143)	
V magnitudes	A: 5.11	B: 6.53		
$(B - V)$ magnitudes	A: +1.34	B: +0.31		
μ(A)	-3.72 mas yr^{-1}	± 0.38	23.11 mas yr^{-1}	± 0.24 (DR2)
μ(B)	-2.32 mas yr^{-1}	± 0.12	21.24 mas yr^{-1}	± 0.08 (DR2)
π(A)	8.58 mas	± 0.20	380 light yr	± 9 (DR2)
π(B)	8.84 mas	± 0.07	369 light yr	± 3 (DR2)
Spectra	A: K0II-III	B: A9V		
Luminosities (L$_\odot$)	A: 100	B: 25		
Catalogues (A/B)	HR 4792/1	HD 109511/0	SAO 100160/59	HIP 61418/5
DS catalogues	H 4 27	STF 1657	BDS 6212	ADS 8600
Radial velocity (A/B)	3.03 km s^{-1}	± 0.1	4.90 km s^{-1}	± 0.5
Galactic coordinates	278°.862	+80°.478		

History

Found by William Herschel on 28 February 1781. 'Double. Considerably unequal. L(arge). whitish r(ed).; S(mall). blueish r(red). Mean distance 18″24‴. Position 3° 25′ n(orth) preceding.'

The Modern Era

Star B is a double-lined spectroscopic binary. Adams [391] first noted this on a spectrum which showed two sets of spectral lines separated by an amount equivalent to 127 km s^{-1}. Petrie [392] later derived the correct period, 7.3366 days. Star A is a suspected variable (NSV 5748). Observations were by Petit [389]. The quoted errors on the Hipparcos astrometry of the A component are larger than might be expected, but DR2 has nailed the parallaxes with a much higher precision, and the stars are at the same distance within the given errors, so it

Finder Chart

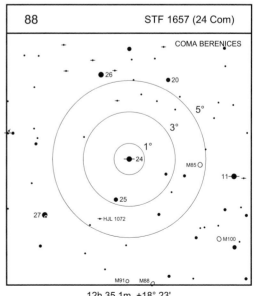

12h 35.1m +18° 23′

seems certain that A and B are physical and it is likely that this is a triple star if not quadruple. The pair can be used as a scale calibrator. At 2015.5, Gaia DR2 measured a position angle of 270°.2 and a separation of 20″.13. The similar proper motions mean that this value will not change significantly for a few years.

Observing and Neighbourhood

F. G. W. Struve gave yellow and blue for the colours whilst Smyth commented that they were 'very brilliant' and found 'orange colour and emerald tint' for the two stars. Greg Stone [126] was impressed with 24 Com but noted that it was not particularly easy to find, and he gave two different methods which can be found on the Star Splitters website. He noted colours of yellow going to orange and summer-sky blue which were most intense in 30-cm, becoming less so with smaller apertures. Two degrees SSE is HJL 1072, a rather wide pair but which appears to have common proper motion (7.6, 9.3, 23°, 118″, 2010).

Measures

Early measure (STF)	271°.9	20″.42	1830.03
Recent measure (HSW)	269°.9	20″.06	2015.36

89. γ CEN = HJ 4539 = WDS J12415−4858AB

Table 9.89 Physical parameters for γ Cen

HJ 4539	RA: 12 41 31.04	Dec: −48 57 35.5	WDS: 282(181)	
V magnitudes	A: 2.82	B: 2.88	C: 14.4	
(B − V) magnitudes	A: +0.02	B: +0.03		
μ	A: −185.72 mas yr⁻¹	± 0.20	5.79 mas yr⁻¹	± 0.16
π	25.06 mas	± 0.28	130 light yr	± 1.5
Spectra	A: A0IV	B: A1IV		
Luminosities (L$_\odot$):	A: 100	B: 90		
Catalogues	HR 4819	HD 110304	SAO 223603	HIP 61932
DS catalogues	HJ 4539 (AB)	SEE 159 (AB,C)		
Radial velocity	−5.50 km s⁻¹	± 1.78		
Galactic coordinates	301°.255	+13°.880		

History

John Herschel, 1 March 1835: 'A star 4 m, which I am very much inclined to believe a close double, but could not verify owing to bad definition. Tried 320 but it will not bear that power.' Then 30 days later: '180 with triangular aperture shows it elongated; 320 fairly double and almost divided.' He observed the pair 10 times in all, recording PAs between 345° and 360° and estimated separations ranging between 2/3″ and 1″. T. J. J. See found a very faint companion of magnitude 14.4 at 114° and 58″.2. The distance is increasing rapidly owing to the proper motion of γ.

The Modern Era

The number of measures of this pair (181 to date) compares unfavourably with its more northerly twin γ Vir (1580), reflecting the inequality of double star research between northern and southern hemispheres, even allowing for the fact that γ Virginis has been observed for much longer. Even during the current periastron passage, when the stars have

Finder Chart

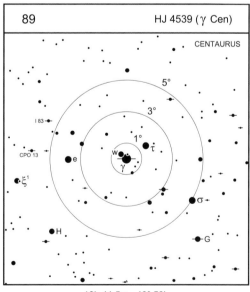

12h 41.5m -48° 58'

Orbit

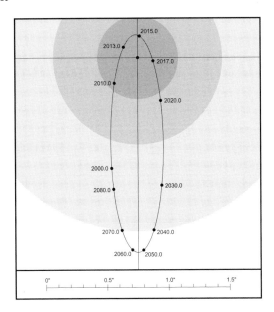

Ephemeris for HJ 4539 AB (2016 to 2034)

Orbit by Ary (2015b) Period: 83.57 years, Grade: 2

Year	PA(°)	Sep(")	Year	PA(°)	Sep(")
2016.0	137.3	0.12	2026.0	14.4	0.83
2018.0	47.9	0.21	2028.0	12.4	0.94
2020.0	28.5	0.39	2030.0	10.7	1.04
2022.0	21.3	0.57	2032.0	9.4	1.12
2024.0	17.2	0.70	2034.0	8.2	1.20

been comfortably within the range of a 4-metre telescope, only one observation has been made. With an apparent orbit such as this, the close approaches need to be sufficiently monitored to nail down the orbital period. Even an error of one year in the period can result in angular residuals of tens of degrees.

Observing and Neighbourhood

This is a brilliant pair, which is almost a carbon copy of γ Virginis in the northern sky (which is only 30 seconds of time away in RA), but it is a more difficult proposition. At the time of going to press, the stars are now widening, having reached 0″.12 in 2016, but they will not be within range of a 20-cm until about 2022/3 if the current orbit is correct. Like γ Virginis, the stars are equally bright and the apparent orbit is a very eccentric one, maximum separation being reached around 2050 when they will be 1″.53 apart. When the stars were considerably wider, around 1980, the ASNSW group made the colours pure white and white (or possibly yellowish). About 3° ENE of γ is I 83, whose magnitudes are 7.4 and 7.7. This is an orbital pair with a period of 173.4 years and is currently at its widest point (237°, 0″.86, 2020.0). Three degrees due E is CPO 13 (7.2, 9.2, 70°, 5″.1, 2016).

Measures

Early measure (JC)	13°.7	1″.11	1857.97
(Orbit	15°.8	0″.77)	
Recent measure (TOK)	67°.4	0″.15	2017.28
(Orbit	68°.3	0″.15)	

90. γ VIR = STF 1670 = WDS J12416−0127AB

Table 9.90 Physical parameters for γ Vir

STF 1670	RA: 12 41 39.60	Dec: −01 26 57.9	WDS: 4(1578)		
V magnitudes	A: 3.48	B: 3.53			
(B − V) magnitudes	A: +0.39	B: +0.39			
μ	−614.76 mas yr⁻¹	± 0.88	+61.34 mas yr⁻¹	± 0.47	
π	85.58 mas	± 0.60	38.1 light yr	± 0.3	
Spectra	F0V	F0V			
Masses (M⊙)	A: 1.45		B: 1.45		
Radii (R⊙)	A: 1.59 mas		B: 1.59 mas		
Luminosities (L⊙)	A: 4.5	B: 4.3			
Catalogues (A/B)	HR 4825/6	HD 110379/80	SAO 138917	HIP 61941	
DS catalogues	Mayer 33	H 3 18	STF 1670	BDS 6243	ADS 8630
Radial velocity (A/B)	−20.42 km s⁻¹	± 0.4			
Galactic coordinates	297°.834	+61°.326			

History

Thomas Lewis records that the first mention of duplicity for γ Vir was made by Fr Richaud from the Cape in 1689 [397]. The entry reads: 'La nuit du trois au quatre May, la Lune étant fort proche de la plus haute des trois Etoiles du front du Scorpion, M. Cassini remarqua en observant cet endroit du Ciel, que cette Etoile toit double comme la première d'Aries, et la Tête précédente des Gemeaux.' The translation, kindly provided by Professor James Lequeux [396] says 'the night of 3 to 4 May [1678], the Moon being very close to the highest of the three stars of the front of the Scorpion, M. Cassini noticed when observing this part of the sky that this star was double as well as alpha Arietis and the preceding head star of Gemini [Castor]'. Lewis later on then refutes this possibility by arguing that, a period of 150 years being the accepted value when he wrote his book (1906), projecting back to the late 1680s would mean the stars were then less than 1″.5 apart and probably not visible to Richaud's early telescope.

Finder Chart

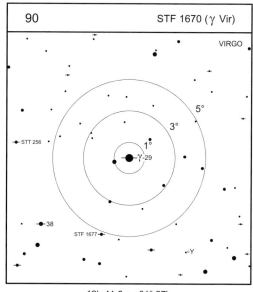

12h 41.6m -01° 27'

Orbit

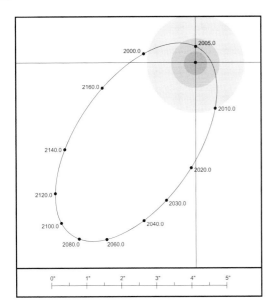

Ephemeris for STF 1670 AB (2010 to 2055)

Orbit by Sca (2007c) Period: 169.104 years, Grade: 2

Year	PA(°)	Sep(")	Year	PA(°)	Sep(")
2010.0	23.7	1.39	2035.0	344.3	4.41
2015.0	6.0	2.27	2040.0	341.4	4.65
2020.0	357.5	2.93	2045.0	338.9	4.93
2025.0	351.9	3.47	2050.0	336.7	5.17
2030.0	347.7	3.93	2055.0	334.6	5.38

In fact the current orbit gives a projected separation of 3″.7 for Richaud's epoch of observation, making it certainly feasible. The first measures of γ were made by Bradley and Pound whilst making circle observations at Wanstead in March 1718. The system was one of the earliest targets for orbit computers, and the increasingly rapid orbital motion in the 1820s drew the attention of double star observers to it. F. G. W. Struve followed the pair using the 9.3-inch Fraunhofer refractor at Dorpat and measured the closest separation at 0″.25 in early 1836. John Herschel, observing from South Africa with his 20-foot reflector, was unable to get accordant measures.

The Modern Era

One of the most observed pairs in the WDS *Observations Catalog*, this is also one of the best pairs for the small aperture anywhere in the sky. The orbit computed by Scardia [447] gives the period to an accuracy of four days. During the last periastron passage in 2005, a series of measurements were made using a speckle camera on the 1-metre Zeiss reflector at Merate in Italy. These helped to define that part of the orbit

Measures with 8″ Refractor

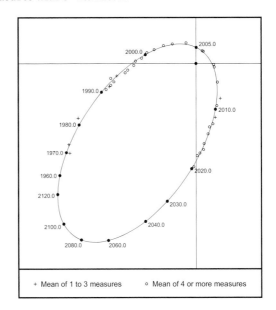

+ Mean of 1 to 3 measures   ○ Mean of 4 or more measures

Year	PA(°)	Sep(")	No.	Year	PA(°)	Sep(")	No.
1970.40	304.3	4.43	1	2004.31	206.5	0.64	7
1973.41	301.4	4.32	3	2005.31	153.0	0.40	5
1982.40	294.0	3.74	3	2005.43	148.0	0.40	4
1990.34	285.6	2.70	5	2006.27	87.1	0.51	4
1991.35	284.5	2.59	2	2006.40	79.6	0.54	4
1992.41	283.8	2.54	5	2007.36	49.8	0.81	8
1993.40	282.1	2.43	7	2008.28	35.8	1.20	1
1994.30	278.5	2.32	2	2009.38	27.1	1.32	7
1994.31	279.4	2.50	6	2010.37	20.3	1.60	2
1995.33	277.1	2.07	6	2011.33	16.1	1.75	6
1996.32	275.2	1.99	4	2012.38	12.4	1.81	5
1997.31	271.6	1.81	5	2013.31	10.0	1.99	7
1998.38	267.8	1.76	5	2014.34	5.9	2.24	9
1999.35	264.6	1.59	6	2015.36	5.1	2.41	6
2000.36	258.1	1.46	8	2016.35	2.5	2.50	5
2001.31	252.3	1.27	8	2017.41	0.7	2.57	6
2002.31	242.1	1.06	8	2018.37	358.6	2.88	6
2003.35	231.5	0.87	7				

which had previously been observed only once, in 1836, and for which the measures were marginal at best. At the closest point, the two stars were separated by 0″.37 in early August, 2005. The precision of this orbit and knowledge of the parallax from Hipparcos combined to give the total mass of the system at 2.9 $M_\odot$, which seems to be divided evenly between the two components. The similarity of the two stars is also reflected in the determination of the angular diameters as recorded by Richichi [393]. Scardia *et al.* [447] also tackled another possibility which had been pointed out by Sir John Herschel in 1847. In computing the orbit of γ he noted alternating runs of positive and negative residuals from the orbit and speculated that this might be due to an invisible, perturbing mass. Infrared imaging of γ with a 4-metre telescope in the infrared by Law [398] at the time of periastron in 2005 showed no sign of any star as faint as M0V at a separation of 0″.4, and there appear to have been no perturbations that might have been induced by an extra component with a mass of 0.3 $M_\odot$. Reports of the variability of one or both components have been made but neither star is in the GCVS. Although many observers of this system see no difference in magnitude, in recent years a Δm value of 0.69 at V was recorded by Horch et al. [395] whilst Riddle et al. [394] found Δm to be as much as 0.9 at 745 nm.

Observing and Neighbourhood

The components are both very bright (each around $V = 3.5$), in fact almost equally bright, and well separated even for the small aperture for the next century and a half. RWA has measured this pair every year, on average for seven nights each time, for the last 28 years. In 1990 it was at 286° and 2″.70, and it closed to around 0″.4 in 2005 (when it was just elongated with the 8-inch refractor at Cambridge at a power of ×600 in PA 150°). At this time the stars were moving in position angle at the rate of 1° every five days. Currently the pair is near 0° and 2″.75 – a total arc of about 285° over the 28 years). There are very few systems in the sky which afford such a good view of orbital motion in action using only the eye. Christopher Taylor, observing with his 12.5-inch Calver reflector at Hanwell, demonstrates this in a series of sketches [13] which he made during the recent periastron passage. Others might be mentioned – α Centauri, which is currently only 4″ apart and the angular motion of which is almost 14° per year, although that southern pair is now widening with increasing (and decelerating) position angle. More impressive is MLO 4 (see Star 129). In 2018 the stars moved through almost 60° in the year although the separation is a more challenging 0″.5 and the declination an equally challenging −35°! There are two other pairs in the vicinity: STF 1677 is 2°.5 SSE (7.3, 8.1, 348°, 15″.9, 2015) and, more difficult, is STT 256, 3°.75 E (7.3, 7.6, 103°, 1″.1, 2017).

Measures

Early measure (DA)	249°.9	1″.30	1832.20
(Orbit	251°.8	1″.24)	
Recent measure (ARY)	358°.6	2″.88	2018.37
(Orbit	359°.8	2″.73)	

91. β MUS = R 207 = WDS J12463–6806AB

Table 9.91 Physical parameters for β Mus

R 207	RA: 12 46 16.87	Dec: −68 06 29.1	WDS: 849(86)		
V magnitudes	Aa: 3.55	Ab: 6.6	B: 4.03		
$(B - V)$ magnitudes	Aab: −0.18	B: −0.16			
μ	A: −41.97 mas yr^{-1}	± 0.43	B: −8.89 mas yr^{-1}	±0.31	
π	9.55 mas	± 0.41	340 light yr	± 15	
Spectra	A: B2.5V	B:			
Luminosities (L$_\odot$)	A: 330	B: 35			
Catalogues	β Mus	HR 4844	HD 110879	SAO 252019	HIP 62322
DS catalogues	R 207(AB)	RIZ 5 (AaAb)	RIZ 5 (AC)		
Radial velocity	42.00 km s^{-1}	± 7.4			
Galactic coordinates	302°.449	−5°.241			

History

The object β Mus was found to be double by H. C. Russell [525] in the course of his work with the 11.5-inch refractor at Sydney Observatory. He notes 'First seen (18)78.284; one of the closest doubles I know'. The appellation β Mus was given to R 203 in error.

The Modern Era

The pair β Mus is a member of the Sco-Cen OB2 association, which is the nearest site of recent massive star formation to the Sun. It includes three subgroups – Upper Scorpius, Upper Centaurus Lupus, and Lower Centaurus Crux – and its members range from OB stars down to brown dwarfs. The brightest member is Antares. Also included are molecular cloud complexes in Lup, Cha, Cr, and Oph. In 2013 Rizzuto *et al.* [475] published the results of a survey of some of the brighter members of the association using the SUSI interferometer to assess the frequency of multiplicity amongst these stars. Using a 15-metre baseline they were sensitive to double

Finder Chart

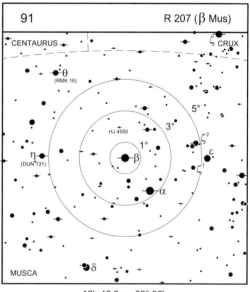

12h 46.3m -68° 06'

Orbit

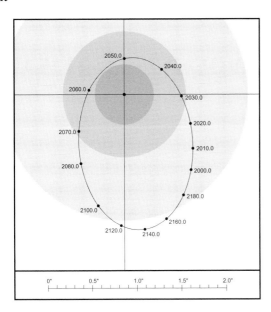

Ephemeris for R 207 AB (2015 to 2060)

Orbit by FMR (2012g) Period: 194.276 years, Grade: 5

Year	PA(°)	Sep(")	Year	PA(°)	Sep(")
2015.0	58.9	0.89	2040.0	123.2	0.50
2020.0	66.7	0.81	2045.0	148.0	0.45
2025.0	76.3	0.72	2050.0	179.7	0.39
2030.0	88.3	0.64	2055.0	221.0	0.35
2035.0	103.6	0.57	2060.0	263.1	0.39

stars with separations between 7 and 200 mas. One of the new binary systems to emerge from this survey was β Mus Aab. It has a close companion at a distance of about 0″.05 but there is no information on the nature of any binary motion, so it is not yet clear whether the suspected radial velocity variation in star A can be directly attributable to the newly discovered component Ab. To be complete, Rizzuto *et al.* also looked at the space up to 10,000 AU from each star and from 2MASS data they found another faint star with the same proper motion as β. This object has magnitude 15.9 and is 95″ distant.

Observing and Neighbourhood

The β Mus system can be found 8° due S of β Cru (Mimosa). Recent micrometrical and CCD measures have indicated that the orbit is predicting position angles which are more than 5° in advance of those observed – a fact that suggests that the current orbital period may be too short. Both stars are blue-white according to the ASNSW observers. Three point five degrees NE is the fine pair RMK 16 (see Star 95), and 2°.8 due E is η Mus (DUN 131), whose stars of magnitudes 4.8 and 7.2 are 58″ apart in PA 332° and therefore accessible to binoculars. One degree N of μ and slightly E is HJ 4550 (7.6, 8.7, 97°, 13″.6, 2016).

Measures

Early measure (I)	341°.0	1″.33	1900.36
(Orbit	342°.6	1″.18)	
Recent measure (ANT)	53°.2	1″.01	2016.35
(Orbit	60°.5	0″.87)	

92. 35 COM = STF 1687 = WDS J12533+2115AB

Table 9.92 Physical parameters for 35 Com

STF 1687	RA: 12 53 17.77	Dec: +21 14 42.1	WDS: 78(388)	
V magnitudes	A: 5.15	B: 7.08	C: 9.76	
(*B − V*) magnitudes	A: +1.18	B: +0.43		
μ(A)	−49.40 mas yr^{-1}	± 0.85	−35.37 mas yr^{-1}	± 0.55 (DR2)
μ(B)	−53.99 mas yr^{-1}	± 0.11	−27.53 mas yr^{-1}	± 0.06 (DR2)
π(A)	11.56 mas	± 0.34	282 light yr	± 8 (DR2)
π(B)	11.54 mas	± 0.06	283 light yr	± 1 (DR2)
Spectra	G7III			
Luminosities (L$_\odot$)	A: 50	B: 9	C: 0.8	
Catalogues	HR 4894	HD 112033	SAO 82550	HIP 62886
DS catalogues	H 5 130 (AC)	STF 1639 (AB)	BDS 6296	ADS 8695
Radial velocity (A)	−6.091	± 0.075 km s^{-1}		
Galactic coordinates	307°.148	+84°.102		

History

The star first entered the double star catalogues when Sir William Herschel found a faint third component, C, on 26 February 1783. He noted that the primary star was red and the companion dusky. At this time the binary companion would have been at a distance of around 1″.4, assuming the current orbit is correct, and perhaps the difference in magnitude was the deciding factor in Herschel's not seeing the close star. It was left to Struve at Dorpat to find B.

The Modern Era

The current orbit is by Drummond [403]. The eccentricity is 0.206 and the stars were closest around 1960 at a separation of 0″.99 whilst the widest separation comes in 2155 when it reaches 1″.56. Star C is physically attached. The French double star astronomer Jacques Le Beau [402] was of the opinion that

Finder Chart

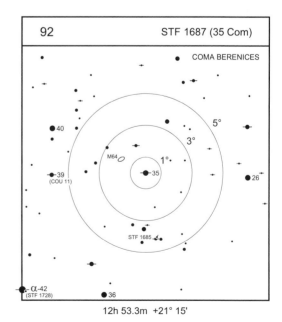

Orbit

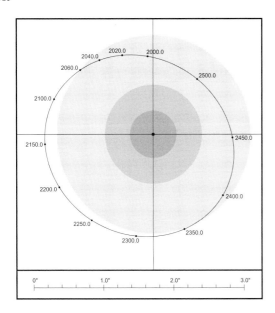

Ephemeris for STF 1687 AB (2010 to 2100)

Orbit by Dru (2014) Period: 539.4 years, Grade: 4

Year	PA(°)	Sep(")	Year	PA(°)	Sep(")
2010.0	193.3	1.13	2060.0	229.3	1.38
2020.0	201.7	1.19	2070.0	235.1	1.42
2030.0	209.4	1.24	2080.0	240.6	1.45
2040.0	216.6	1.29	2090.0	245.9	1.48
2050.0	223.2	1.33	2100.0	251.0	1.50

C was in very slow direct motion around AB. In 2016 C was at 127°, 28″.5. Halbwachs *et al.* [401] in 2012 found that the radial velocity of A varies. They derived a period of 2905 days with an amplitude of 5.5 km s^{-1} and an eccentricity of 0.63.

Observing and Neighbourhood

One degree to the NNE is the bright galaxy M64. Two degrees S and slightly W is STF 1685 (7.3, 7.8, 200°, 15″.8, 2015). The star 39 Com, 3° E, is one of Paul Couteau's early discoveries (COU 11), although still a formidable challenge for 20–25-cm (6.1, 8.8, 314°, 1″.8, 2015). Star A is double again (CHR 150) – an interferometric discovery by the CHARA group.

Measures

Early measure (STF)	25°.3	1″.43	1829.99
(Orbit	29°.9	1″.37)	
Recent measure (TOK)	193°.6	1″.17	2015.35
(Orbit	195°.1	1″.14)	

93. $\mu^{1,2}$ CRU = Δ 126 = WDS J12546–5711AB,C

Table 9.93 Physical parameters for μ Cru

DUN 126	RA: 12 54 35.66	Dec: −57 10 40.4	WDS: 2866(36)	
V magnitudes	A: 4.04	B: 5.20		
$(B - V)$ magnitudes	A: −0.17	B: −0.11		
μ (A)	−30.69 mas yr^{-1}	± 0.13	−13.08 mas yr^{-1}	± 0.11
μ (B)	−32.49 mas yr^{-1}	± 0.23	−10.92 mas yr^{-1}	± 0.19
π (A)	7.87 mas	± 0.17	414 light yr	± 9
π (B)	8.01 mas	± 0.29	407 light yr	± 15
μ (A)	−28.16 mas yr^{-1}	± 0.22	−10.34 mas yr^{-1}	± 0.34 (DR2)
π (A)	8.95 mas	± 0.23	364 light yr	± 9 (DR2)
Spectra	A: B2IV-V	B: B5Vne		
Luminosities (L$_\odot$)	A: 250	B: 100		
Catalogues (A/B)	HR 4898/9	HD 112092/1	SAO 240366/7	HIP 63003/5
DS catalogues	DUN 126			
Radial velocity (A/B)	13.90 km s^{-1}	± 0.7	13.00 km s^{-1}	± 3.7
Galactic coordinates	303°.362	+5°.691		

History

The constellation of Crux, because of the density of bright stars within such a small area of sky, has long been known in history to the peoples of the southern hemisphere. It was not located correctly on celestial globes until 1598. Crux could be used as a guide by navigators to locate due south. A line drawn through γ and α when extended passes close to the South Celestial pole. If a line is drawn through the midpoint of, and at right angles to, the line joining α and β Centauri then that line when extended south crosses the Crux marker close to the SCP.

Finder Chart

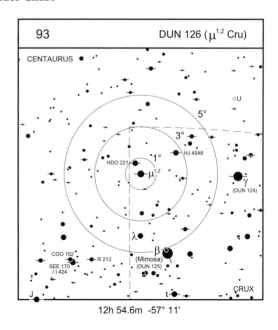

12h 54.6m -57° 11'

The Modern Era

Some of the brightest stars in Crux are members of the Scorpius–Centaurus OB association, the nearest such grouping to the Sun. The members include α, β, δ and also both members of μ Crucis. Both bright components were examined by variability by Jakate [405] and colleagues in a search for variability of the β CMa type, i.e. a low-amplitude variation with a period of less than a day. The B star was found to vary but the period found was too long for that class of star. The Hipparcos parallaxes and proper motions clearly indicate that the two bright components are close together in space and physically connected.

Observing and Neighbourhood

Jeremy Perez [404] was able to see the stars from the summit of Haleakala in Maui with 15 × 70 binoculars on a tripod. He made the colours bright blue for A and an elusive orange colour for B. Andrew James notes that this is a beautiful blue duo, in a star-studded field, which is easy to resolve in smaller telescopes and is a good test for 7 × 50 binoculars. Richard Jaworski [13], using 80-mm, finds it a wide, easy pair at ×80.

The stars appear to him to be white and yellowish. The two naked-eye stars nearer to μ plotted in *CDSA2* are both wide, unequal doubles. Northeast by about half a degree is HDO 221 (5.3, 11.8, 322°, 30″, 2011), the distance increasing owing to the proper motion of B, and 1° NW is HJ 4548 (4.6, 8.9, 166°, 52″, 2010), closing quickly owing to the proper motion of A. Four degrees SSE is a group of three bright stars, each of which is double or multiple. The object R 213 is a fine pair, which has widened since Russell found it at 0.3″, −6.6, 7.9, 21°, 0″.8, 2016. Just following 45′ distant are COO 152 and I424; COO 152 is 6.2, 9.4, 146°, 25″.3, 2016 but Hipparcos found A to be a close binary, one component of which is an SB2. The wide companion is physical, making this a quadruple system. The star I 424 is 4.6, 8.4, 15°, 1″.9, 2015 and so not easy in anything under 15-cm. Star A is the close pair SEE 170, currently 0″.3 apart, and one of these stars is itself a β Lyr type variable.

Measures

Early measure (JC)	17°.2	35″.7	1846.28
Recent measure (ARY)	16°.3	34″.75	2010.64

94. $\alpha^{1,2}$ CVN = STF 1692
= WDS J12560+3819AB

Table 9.94 Physical parameters for α CVn

STF 1692	RA: 12 56 01.67	Dec: +38 19 06.2	WDS: 362(163)		
V magnitudes	A(α^2): 2.85	B(α^1): 5.52	C: 16.2		
$(B - V)$ magnitudes	A: -0.06	B: $+0.34$			
μ(A)	-224.68 mas yr^{-1}	± 1.45	59.32 mas yr^{-1}	± 1.14 (DR2)	
μ(B)	-233.72 mas yr^{-1}	± 0.18	61.55 mas yr^{-1}	± 0.13 (DR2)	
π(A)	26.57 mas	± 0.59	123 light yr	± 3 (DR2)	
π(B)	30.39 mas	± 0.09	107.3 light yr	± 0.3 (DR2)	
Spectra	A: B9p	B: F0V			
Luminosities (L$_\odot$)	A: 80	B: 5	C: 0.0002	D:	
Catalogues (A/B)	HR 4915/4	HD 112413/2	SAO 63257/6	HIP 63125/1	
DS catalogues	Mayer 34 (AB)	STF 1692 (AB)	ADS 8706	BDS 6313	TOK 562 (BC)
Radial velocity	-4.10 km s^{-1}	± 0.2	-0.60 km s^{-1}	± 0.9	
Galactic coordinates	118°.305	+78°.769			

History

This star is one of Christian Mayer's discoveries and is number 34 in Juerg Schlimmer's version of Mayer's catalogue [29]. Sir William Herschel noted it on 7 August 1780 and listed it as H 4 17. The star was included in Piazzi's Palermo catalogue of 1814 and a proper motion in RA of 0″.34 per year towards the west was found. The modern value, given above, is about two-thirds of this value.

The Modern Era

Although the brighter of the two, HR 4915, is called α^2 whereas the fainter companion, HR 4914, is α^1. The former, α^2 CVn, is a star of surpassing interest to astronomers. SIMBAD contains no less than 600 references to papers which discuss or mention it. This primary is the prototype of the chemically peculiar A stars (called Ap stars, although a recent

Finder Chart

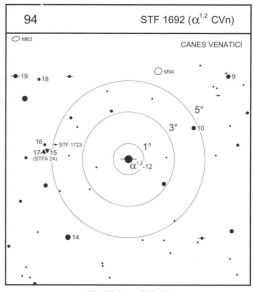

classification of the spectrum makes it slightly earlier than A0). The spectrum and brightness of α^2 are both variable. The surface of the star is contaminated by unusual abundances of the metals Sr, Eu, Hg, and Cr. There is also a strong magnetic field, amounting to several thousand gauss, distributed as a dipole field with the axis inclined to the line of sight by 120° and to the rotational axis of the star by 80°. The star is thus called an oblique rotator. The rare earth and other elements found at the surface are churned up from lower levels in the star by convective columns. Whilst many Ap stars are X-ray emitters, α^2 unusually is not but X-rays have been detected from the companion star B. Andrei Tokovinin [406], in an astrometric study of Hipparcos stars closer than 67 pc, found a common proper motion companion (magnitude 16) at 205°, 18″.6, 1980. The WDS catalogue notes that both bright stars are spectroscopic binaries. The DR2 results should be treated with caution. The G magnitude for A is given as 5.53, which is 0.02 magnitudes fainter than that for B, even though the V magnitudes differ by 2.4.

Observing and Neighbourhood

A fairly bright naked-eye star, α CVn sits in an empty area of sky some 18° due S of ϵ UMa (Alioth). The smallest apertures will show the companion star. The spectral types suggest colours of white and pale yellow. In 1830–1831 Sir John Herschel could '… perceive no contrast of colours'. In *Sidereal Chromatics* Admiral Smyth conducted an experiment whereupon six ladies and five gentlemen were invited to view the star through a 5-inch Gregorian reflector and asked to give their opinions of the colours of each component. All noted that A was white or yellowish, whilst for B the hues included lilac, purple, plum colour, and darker blue. Dembowski noted white and pale olive blue in 1856, whilst T. W. Webb found pale yellow and pale copper with his 9-inch mirror in 1870. RWA found both stars white with a 21-cm at ×96 on 5 April 1968. Just 3° preceding α is a group of sixth magnitude stars including the binocular pair 15 and 17 CVn (STFA 24) (6.0, 6.3, 296°, 276″, 2010). Thirty minutes of arc to the NW of 15 CVn is STF1723 (8.7, 10.1, 11°, 6″.1, 2014).

Measures

Early measure (STF)	226°.8	19″.85	1833.46
Recent measure (WSI)	229°.0	19″.20	2010.07

95. θ MUS = RMK 16 = WDS J13081−6518

Table 9.95 Physical parameters for θ Mus

RMK 16	RA: 13 08 07.15	Dec: −65 18 26.6	WDS: 3723(30)	
V magnitudes	Aa: 5.9	Ab: 6.6	B: 7.55	C: 12.59 (K band)
(B − V) magnitudes	A: −0.05	B:		
μ	A: −4.26 mas yr^{-1}	± 0.39	−2.18 mas yr^{-1}	± 0.48
π	0.26 mas	± 0.41	(> 8000) light yr	
μ(B)	−3.93 mas yr^{-1}	± 0.06	−1.94 mas yr^{-1}	± 0.06 (DR2)
π(B)	0.34 mas	± 0.04	9600 light yr	± 1150 (DR2)
Spectra	A: WC8 + WN8 + O I	B: O9III + ?		
Masses (M$_\odot$)	Aa: 16.7 + 12	Ab:		
Luminosities (L$_\odot$)	A: 45000	B: 6500	C: -	D:
Catalogues	HR 4952	HD 113904	SAO 252162	HIP 64094
DS catalogues	RMK 16 (AB)	CHR 247 (AaAb)	SNA 30 (AC)	
Radial velocity	−28.4 km s^{-1}	± 0.2		
Galactic coordinates	304°.675	−2°.491		

History

Noted by Rümker at Parramatta, this object was measured four times between 1833 and 1836 by John Herschel, who noted 'Fine *'.

The Modern Era

The primary star of RMK 16 is the second brightest Wolf–Rayet star in the sky (WR 48). Such stars are usually accompanied by shells of gas, which indicates previous episodes of violent mass ejection. However, a paper by Stupar *et al.* [407] proposes that the filamentary structure around θ Mus is not a WR shell but possibly a supernova remnant (SNR) and a series of complex HII regions. Star A is a spectroscopic binary of period 18 days [412] but is also accompanied by an O supergiant whose spectral lines show either little or slow variation

Finder Chart

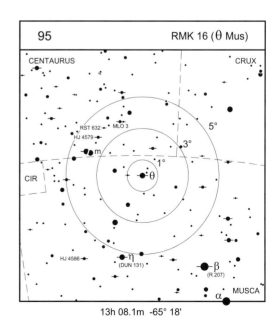

13h 08.1m -65° 18'

in velocity with time. The A star has also been recorded as a pair, using speckle interferometry, with a distance of around 0″.04 [410] and what appears to be rapidly varying position angles with time. This is not the spectroscopic pair but probably the O star, making the A component a triple star. There are two recent reports that B is also a spectroscopic binary – one that the star is an SB2 (Chini *et al.* [409]) and the other, by Sota *et al.* [408], that it is an SB1. An examination of B by Mason *et al.* [410] showed that it was unresolved in a 4-metre aperture. The stars are much too distant to give a parallax measurable by Hipparcos, and an estimate from spectroscopy gives a distance of 2500 pc. Zhekov *et al.* [413] conclude that component A consists of two WR stars and an O supergiant, and they tie the system in to the nearby cluster Danks 2, which is 7500 parsecs away. Sana *et al.* [414] found another component (C) 3″.5 from B in PA 207°. Unravelling the complex nature of the A component will have to wait for DR3. If B is physical, the DR2 parallax confirms the extreme distance of the star, even an approximate value of which defeated Hipparcos.

Observing and Neighbourhood

The system θ Mus sits on the opposite side of the Coalsack Nebula about 5° SE of α Crucis. It is a splendid sight for the small telescope – two white stars, unequal and rather close. There has been very little motion since discovery, so it will remain a fine object for a long time to come. The fine binary star β Mus (see Star 91) can be found 3°.5 to the NW, whilst 3°.1 SSE is a beautiful binocular pair η Mus (DUN 131) (4.8, 7.2, 332°, 58″, 2002). A further 1° E is HJ 4586 (7.3, 9.1, 141°, 2″.9, 1991), which is slowly closing. In a curve radiating away from the star m Mus are three visual pairs, the nearest of which is HJ 4579 (7.9, 8.6, 98°, 5″.1, 2010). Next is RST 632 (7.7, 10.6, 302°, 4″.9, 2000), which should be accessible to 20-cm. It is one of the easier pairs in Rossiter's predominantly difficult-to-observe catalogue. Finally MLO 3 (7.0, 9.1, 39°, 1″.7, 2014) has an 09 giant primary; *K* band imaging studies add four stars within 5″ and B is a very close pair.

Measures

Early measure (HJ)	187°.5	5″.82	1836.38
Recent measure (ANT)	186°.6	5″.36	2007.37

96. α COM = STF 1728 = WDS J13100+1732AB

Table 9.96 Physical parameters for α Com

STF 1728	RA: 13 09 59 29	Dec: +17 31 46.0	WDS: 25(642)	
V magnitudes	A: 4.85	B: 5.53		
$(B - V)$ magnitudes	A: +0.59	B: +0.47		
μ	-433.13 mas yr^{-1}	± 0.70	141.24 mas yr^{-1}	± 0.51
π	56.10 mas	± 0.89	58.1 light yr	± 0.9
Spectra	F5V	F6V		
Luminosities (L$_\odot$)	A: 3	B: 1.5		
Catalogues (A/B)	HR 4968/9	HD 114378/9	SAO 100443	HIP 64241
DS catalogues	STF 1728	BDS 6406	ADS 8804	
Radial velocity (A)	-16.05 km s^{-1}	± 0.21		
Galactic coordinates	327°.933	+79°.489		

History

F. G. W. Struve found this star, also known as α or 42 Com, in autumn 1827 at a distance of 0″.57 (according to Lewis, or 0″.64 according to Smyth) but by 1833 it was single in his 9.3-inch Fraunhofer refractor and remained that way until 1836, when he measured a distance of 0″.30. It attracted much attention with the observers of the time but it was always a challenging object as it never exceeded 0″.6 in separation and was frequently much less than this. After 20 years or so of observing it was clear that the position angle remained fixed near 192° or 12°; because the stars were ostensibly equally bright it was never clear which quadrant the pair was in. After the conjunction had taken place, did the companion continue to widen in the opposite quadrant, or did it speed around the primary and head back to the same quadrant? In 1866 Otto Struve assumed that the former condition was true and calculated an orbit for the star which gave the period as 25.5 years, which agrees fairly closely with the most recent value and which was based on measures made with telescopes of around 10-inch aperture, with the exception of Otto's own 15-inch refractor at Pulkovo. Otto gave the inclination as 90°

Finder Chart

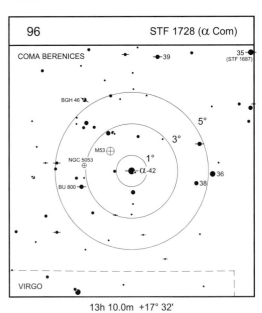

13h 10.0m +17° 32'

Orbit

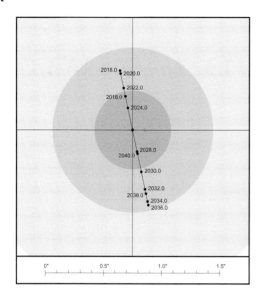

Ephemeris for STF 1728 AB (2018 to 2036)

Orbit by Mut (2015) Period: 25.8524 years, Grade: 1

Year	PA(°)	Sep(")	Year	PA(°)	Sep(")
2018.0	192.2	0.51	2028.0	12.5	0.18
2020.0	192.2	0.48	2030.0	12.4	0.36
2022.0	192.1	0.36	2032.0	12.3	0.51
2024.0	192.0	0.19	2034.0	12.3	0.61
2026.0	175.2	0.00	2036.0	12.3	0.64

and Burnham notes that an occultation takes place every 13 years and that the orbit is almost in the plane of sight but, without any knowledge of the apparent angular diameters of stars, it appears that no one tried to predict the epoch of a possible eclipse.

The Modern Era

Interest was re-awakened in this system in a paper by Hartkopf & McAlister [415], who produced a new orbit for the pair, based on speckle interferometry, which gave much higher astrometric precision. Their paper, which was received by the *Astronomical Journal* on 28 February 1989, predicted that an eclipse would occur in February 1989 with a depth of about 0.1 mag in *V* and lasting 1.3 days. In 2014, Muterspaugh *et al.* [416], with the benefit of 20 years more observations of the orbit, predicted that the next eclipse would occur around 25 January 2015 (but hopefully on Muterspaugh's sister's birthday, 23 January!) and would last between 28 and 44 hours. As the event approached Huib Henrichs and Marcella Wijngaarden from the University of Amsterdam had planned to observe the eclipse from their home institute and began to take an interest in the logistics of the event. They began to do the photometry on 25 November 2014 and received Muterspaugh's and Henry's paper on 15 December. On 21

December Henrichs calculated the orbit from WDS data, but using separations only. This gave an eclipse date of 28 October 2014. On 16 January 2015 CHARA interferometry measures confirmed that the eclipse had passed. Henrichs also confirmed that the eclipse would have been observable in Amsterdam for at least two hours.

In December 2014 Muterspaugh *et al.* [416] published a paper on astro-ph. They admitted that the eclipse had been missed because three visual observations of position angle out of a total of 609 used in the orbital analysis were 180° out. This was enough to skew the time of conjunction by several months and in fact the eclipse had occurred on 18 November (RWA's birthday!). Henry had spent 22 years monitoring the brightness of α Com using the Automated Photometric Telescope at Fairborn Observatory at the University of Tennessee. He had noted that there was a slight sinusoidal variation in brightness, with a period of about a decade.

A secondary eclipse is due in 2026 and, although favourable for northern observers, there is only a 5% chance of observing it, and the next favourable primary eclipse will not be until 2066.

Observing and Neighbourhood

Admiral Smyth thought that 42 Com '... is placed fortunately for the out-door gazer, being midway between Arcturus and Denebola on the parallel, and vertically half way down from Spica to Cor Caroli'. Another way is to start from ε Vir, head 7°.5 due N to pick up the stars 36 and 38 Com and then head 2.5° due W. There is an 11.5 magnitude star (349°, 85″, 2013); the distance is decreasing owing to the proper motion of AB. The current ephemeris shows that the stars are rapidly widening and by Spring 2018 they should be resolvable in 25-cm and be certainly elongated in 20-cm. Three degrees NE is BGH 46 a binocular pair (6.5, 7.6, 58°, 203″, 2011). DR2 shows that the stars are at similar distances but not the same within the quoted errors (302 light years). A more interesting system is BU 800 (6.7, 9.5, 105°, 7.8, 2014), which was once thought to have a hyperbolic orbit. This is a nearby system; DR2 puts it at 35.8 light years. There are two faint field stars, magnitudes 12.6 and 13.3.

Measures

Early measure (STF)	189°.5	0″.57	1827.83
(Orbit	12°.3	0″.63)	
Recent measure (NPI)	192°.62	0″.03874	2015.02
(Orbit	192°.66	0″.038)	

97. J CEN = Δ 133 = WDS J13226−6059AB,C

Table 9.97 Physical parameters for J Cen

DUN 133	RA: 13 22 37.94	Dec: −60 59 18.3	WDS: 6726(20)	
V magnitudes	AB: 4.51	C: 6.15		
$(B − V)$ magnitudes	AB: −0.14	C: +0.01		
μ(AB)	−28.75 mas yr^{-1}	± 0.43	−23.84 mas yr^{-1}	± 0.47 (DR2)
μ(C)	−10.26 mas yr^{-1}	± 0.08	10.00 mas yr^{-1}	± 0.09 (DR2)
π(AB)	9.70 mas	± 0.36	336 light yr	± 12 (DR2)
π(C)	2.43 mas	± 0.07	1342 light yr	± 39 (DR2)
Spectra	A: B2.5V	B: B3/5		
Luminosities (L$_\odot$)	A: 140	B: 480		
Catalogues (AB/C)	HR 5035/4	HD 116087/72	SAO 252284/3	HIP 65271
DS catalogues	FIN 208 (AB)	DUN 133 (AB,C)		
Radial velocity	6.00 km s^{-1}	± 3.7		
Galactic coordinates	306°.707	+1°.655		

History

In 1930 W. S. Finsen [421] found the primary star of DUN 133 to be a very close pair (FIN 208) using the Johannesburg 26.5-inch refractor. W. H. van den Bos also observed it in 1934 but recorded only an upper limit to the separation, 0″.100. In each case the distance between the stars was below the telescope resolution limit of 0″.17. Finsen made some measurements in the 1960s with his eyepiece interferometer but the stars have remained mostly at or below the 0″.1 level.

The Modern Era

Direct imaging [417] around the star in the K band has revealed seven objects whose V magnitudes are assumed to range from 190 to 15 within 11″. Even the use of 4-metre-class telescopes and speckle techniques has not contributed

Finder Chart

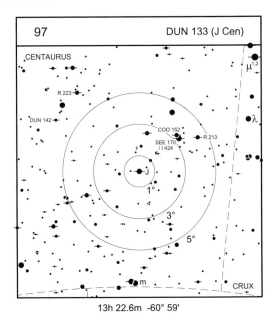

13h 22.6m -60° 59'

much to our knowledge of the motion of the close pair. In 2009 [419] and 2013 [420], Andrei Tokovinin observed the star with the 4.1-metre SOAR telescope in Chile, and since then there have been no further positive measures. It is likely that the stars form a binary with a period of maybe ten years, and the apparent orbit is highly inclined to the line of sight, restricting the position angle changes to a small range. One of the components of FIN 208 is a variable and goes under the name V790 Cen. The Gaia astrometry of star C shows that there is no physical connection with AB, but Veramendi & González [462] have recently shown that C is itself a short-period spectroscopic binary with a period of 3.85 days.

Observing and Neighbourhood

The system J Cen can be seen almost halfway between β Cen and β Crucis. The Dunlop pair is a fine sight in binoculars and small apertures. Both bright components are B stars. Another Dunlop pair (DUN 142) is 3° NE (6.5, 7.6, 90°, 33″.2, 2010). The pair R 223 (6.6, 9.9, 13°, 2″.6, 2008) was almost 6″ apart when first found by Russell.

Measures

FIN 208			
Early measure (FIN)	160°	0″.14	1930.47
Recent measure (TOK)	160°.0	0″.051	2013.24

DUN 133			
Early measure (R)	343°.1	60″.19	1871.42
Recent measure (ARY)	345°.0	60″.56	2010.62

98. ζ UMA = STF 1744 = WDS J13239+5456AB,C

Table 9.98 Physical parameters for ζ UMa

STF 1744	RA: 13 23 55.42	Dec: +54 55 31.5	WDS: 48(485)	
V magnitudes	AaAb: 2.23	BaBb: 3.88	C: 4.01	
(B − V)	A: +0.02	B: +0.17	C: +0.22	
μ (Mizar)	−120.21 mas yr^{-1}	± 0.12	−16.04 mas yr^{-1}	± 0.14
μ (Alcor)	−121.2 mas yr^{-1}	± 0.5	−22.0 mas yr^{-1}	± 0.5
π (Mizar)	39.91	± 0.13	81.7 light yr	± 0.3
π (Alcor)	38.01	± 1.71	85.8 light yr	± 3.9
μ (Mizar B)	114.01 mas yr^{-1}	± 0.54	−26.45 mas yr^{-1}	± 0.50 (DR2)
μ (Alcor)	113.31 mas yr^{-1}	± 0.38	−28.56 mas yr^{-1}	± 0.32 (DR2)
π (Mizar B)	40.50 mas	± 0.35	80.5 light yr	± 0.7 (DR2)
π (Alcor)	40.47 mas	± 0.22	80.6 light yr	± 0.4 (DR2)
Spectra	A: A2V + A2V	B: A1m + A5V	C: A5V + M3-4V	
Masses	Aa: 2.5	Ab: 2.5	Ba: 1.9	Bb: 0.3
	Ca: 1.8	Cb: 0.3		
Luminosities (L$_\odot$)	A: 65	B: 14	C: 13	
Catalogues (Mizar)	HR 5054/5	HD 116656/7	SAO 28737	HIP 65378
Catalogues (Alcor)	80 UMa	HR 5062	HD 116842	SAO 28751	HIP 65477
DS catalogues	Mayer 37 (AB)	PEA 1 (AaAb)	STF 1744 (AB)	BDS 6406	ADS 8891
	PCF 1 (CaCb)			
Radial velocity (Mizar):	−9.6 km s^{-1}	± 1.0		
Radial velocity (Alcor):	−6.31 km s^{-1}	± 0.38		
Gal. coordinates (Mizar)	113°.111	+61°.579		
Gal. coordinates (Alcor)	112°.769	+61°.469		

History

As one of the components of the tail of the Great Bear, ζ UMa appears to be two stars, which can be seen easily with the naked eye. The brighter of the two is also known as Mizar (V = 2.3), whilst 11′.8 distant is Alcor (C) which is 80 UMa (V = 4.0). Mizar occupies a unique place in the pantheon of visual double stars. It was the first telescopic pair ever noted, it was the first star system to be photographed, and it contains the first spectroscopic binary to be recognised. Benedetto Castelli, using one of his friend Galileo's telescopes, found Mizar to be a close double star on 7 January 1617 [422] and

Finder Chart

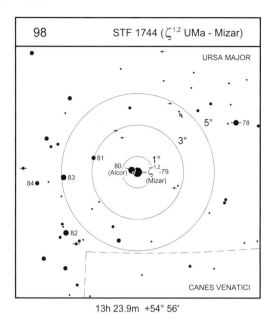

98   STF 1744 (ζ¹,² UMa - Mizar)

URSA MAJOR

5°

3°

1°

ζ¹,²-79
(Mizar)

80
(Alcor)

81

83

84

82

CANES VENATICI

13h 23.9m +54° 56'

this was confirmed by Galileo eight days later. Mizar A and B were approximately 15″ apart at that time. On 21 April 1857 at Harvard, J. L. Whipple took a photographic image of Mizar and Alcor, and in 1889 Antonia Maury noted the cyclically changing and splitting spectral lines in Mizar A. Writing to *The Observatory* on 12 November 1889, E. C. Pickering [423] noted 'The only satisfactory explanation of this phenomenon as yet proposed is that the brighter component of this star is itself a double star having components nearly equal in brightness and too close to have been separated as yet visually'. He goes on to say that the period of revolution is 104 days. In 1908 Mizar B was also revealed to be a spectroscopic binary but with only one spectrum visible. The next major advance in our knowledge of the system was in 1920 when Pease [424] observed ζ^1 with the stellar interferometer on the 100-inch at Mount Wilson and found that the relative position angle and separation of the two components could be interpreted from the interference fringes obtained. The period was then found to be 20.3 days.

The Modern Era

An improved orbit for ζ^2 was calculated in 1965 by Gutmann [425] but since then little new knowledge has been added about the system. Modern ground-based multiple aperture

interferometry uses a technique called phase closure to reconstruct stellar images in close binary systems. Hummel *et al.* [426] demonstrated this with Mizar A, and a movie of the orbital motion of the two stars is available at [684]. The size of the orbit is approximately 10 milliarcseconds. The great accuracy of this technique yields a dynamical parallax which gives a distance for Mizar A of 81.7 light years. This can be compared with the Hipparcos distance of 85.8 light years for Alcor, and it seems that the two multiple stars are physically related. The linear separation is 74.0 ± 39.0 thousand astronomical units. In 2007 Mamajek *et al.* [427] found a faint, close companion to Alcor, in the *M* band (4.8 μm), at a distance of 1″.1, which they interpret as a low-mass dwarf. It was independently found by Zimmerman *et al.* [428], who were able to observe it on two occasions and confirmed its physical relationship to Alcor. Alcor and Mizar therefore form a physical sextuple system, the second closest such group to the Sun. Another possible component has been suggested by Gontcharov & Kiyaeva [429]. From consideration of the linear motion of Aa with time and using both historical and Hipparcos observations, they proposed a new component to Aa which would have a putative mass of 1.5 M$_\odot$ and be approximately 0″.5 from A. There are no confirming observations.

Observing and Neighbourhood

It is an unmistakeable object. The second star from the end of the 'tail' of Ursa Major, Alcor can be seen with the naked eye on any decent night. The two bright stars which make up ζ UMa are a beautiful sight in almost any aperture. Smyth made them brilliant white and pale emerald, whilst Sissy Haas sees both as green-white. A wide-field view which includes Alcor will show other bright stars, some of which are moving through space with Mizar and Alcor as part of the Ursa Major Group, a system which is about 400 million years old.

Measures

Early measure (STF)	1830.63	147°.6	14″.37
Recent measure (ARY)	2015.15	153°.6	14″.43

99. 25 CVN = STF 1768 = WDS J13375+3618AB

Table 9.99 Physical parameters for 25 CVn

STF 1768	RA: 13 37 27.70	Dec: +36 17 41.4	WDS: 44(498)	
V magnitudes	A: 4.98	B: 6.95		
(*B* − *V*) magnitudes	+0.24	B: +0.41		
μ(A)	−93.74 mas yr^{-1}	± 0.21	23.47 mas yr^{-1}	± 0.18 (DR2)
μ(B)	−97.96 mas yr^{-1}	± 0.12	30.48 mas yr^{-1}	± 0.11 (DR2)
π(A)	17.96 mas	± 0.18	181.6 light yr	± 1.8 (DR2)
π(B)	18.32 mas	± 0.12	178.0 light yr	± 1.2 (DR2)
Spectra	A: A7IV	B:		
Masses (M$_\odot$)	A: 2.23	B: 1.58		
Luminosities (L$_\odot$)	A: 25	B: 4		
Catalogues	HD 118623	HR 5127	SAO 63638	HIP 66458
DS catalogues	STF 1768	BDS 6566	ADS 8974	
Radial velocity	−10.40 km s^{-1}	± 2.1		
Galactic coordinates	78°.969	+76°.610		

History

Found by F. G. W. Struve in 1827, the stars then closed up and by the 1860s the system was single to all intents and purposes. With a 7-inch aperture, Dawes recorded the distance as 0″.15 in spring 1860. Bearing in mind the significant difference in magnitude, this observation should be regarded with some scepticism. In fact, the current orbit for 25 CVn gives an angle of about 355° and a separation of 0″.23 for the date in question. Since then the pair has widened and the stars reached a maximum separation of 1″.80 in 1983; they are now beginning to close slowly, with the next close conjunction due in around 2091.

Finder Chart

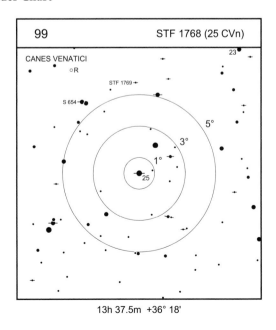

13h 37.5m +36° 18'

Orbit

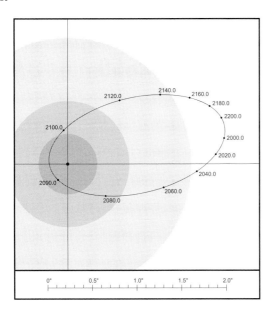

Ephemeris for STF 1768 AB (2015 to 2060)

Orbit by Sod (1999) Period: 228 years, Grade: 3

Year	PA(°)	Sep(")	Year	PA(°)	Sep(")
2015.0	95.1	1.70	2040.0	86.8	1.44
2020.0	93.6	1.66	2045.0	84.7	1.37
2025.0	92.1	1.61	2050.0	82.4	1.29
2030.0	90.4	1.57	2055.0	79.7	1.20
2035.0	88.7	1.51	2060.0	76.6	1.10

The Modern Era

This star was examined by De Rosa *et al.* [514], who were looking for higher multiplicity in a sample of A stars within 75 pc of the Sun – known as the VAST project (Volume limited A STars). They did not find any further components either close in (via observations with adaptive optics on several large telescopes) or distant (using wide-field plate archives) but derived the masses of A and B as given above.

Observing and Neighbourhood

The relatively wide separation at which the pair currently resides is offset by the significant difference in magnitude, and a good night is needed with a 15-cm to get a satisfactory view of the companion. F. G. W. Struve noted colours of white and blue. Smyth did not include the pair in the *Bedford Catalogue*, owing, no doubt, to its very small separation during the early Victorian epoch. The star STF 1769, 3° N, is rather faint and wide but is actually a physical quintuple: AC is 7.9, 9.3, 259°, 56″.5, 2013 whilst Struve found a closer component B of magnitude 10.4, now at 46°, 1″.6, 2013. A 16.1 magnitude star found by Tokovinin has common proper motion whilst A is also an SB1. The whole group moves through space with a proper motion > 0″.2 per year. The object S 654 is an easy telescopic pair (5.6, 8.9, 239°, 71″.8, 2013). It too also has common proper motion and indeed a third very distant star (SHY 633 AC) also shares it.

Measures

Early measure (STF)	78°.9	1″.04	1830.54
(Orbit	77°.5	1″.13)	
Recent measure (ARY)	97°.7	1″.65	2013.30
(Orbit	95°.6	1″.71	

100. STF 1785 BOO = WDS J13491+2659

Table 9.100 Physical parameters for STF 1785 Boo

STF 1785	RA: 13 49 04.00	Dec: +26 58 47.6	WDS: 12(875)	
V magnitudes	A: 7.36	B: 8.15		
$(B - V)$ magnitudes	A: +1.2	B: +1.23		
μ(A)	-416.52 mas yr^{-1}	± 0.19	-112.02 mas yr^{-1}	± 0.12 (DR2)
μ(B)	-462.74 mas yr^{-1}	± 0.07	-71.47 mas yr^{-1}	± 0.05 (DR2)
π(A)	73.92 mas	± 0.07	44.12 light yr	± 0.04 (DR2)
π(B)	74.20 mas	± 0.05	43.96 light yr	± 0.03 (DR2)
Spectra	A: K4V	B: K6V		
Luminosities (L$_\odot$)	A: 0.2	B: 0.08		
Catalogues	HD 120476	SAO 83011	HIP 67422	
DS catalogues	SHJ 168	STF 1785	BDS 6641	ADS 9031
Radial velocity	-20.46 km s^{-1}	± 0.09		
Galactic coordinates	$78°.969$	$+76°.610$		

History

The pair was measured by Sir James South in 1823, who derived a distance of 5″.66. This, when plotted with observations taken over the next few decades, seemed to show that the motion of the companion was linear. However, by the turn of the nineteenth century, the characteristic curvature of an orbit became clear and it was supposed that South had made an error in reducing his measurement. Burnham discussed the measurements up to about 1900 and showed that it was perfectly possible to fit the two orbits which were extant then, with periods of 125 years and 199 years, equally well.

The Modern Era

This pair of stars is moving through space at almost 0.5 arcseconds per year and is only 43.7 light years distant. The orbit is reasonably eccentric and at the time of writing the

Finder Chart

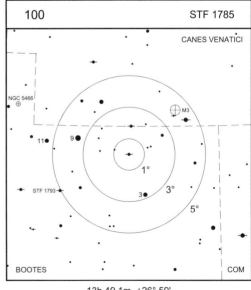

13h 49.1m +26° 59'

Orbit

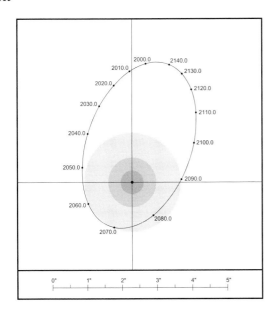

Ephemeris for STF 1785 (2015 to 2060)

Orbit by Hei (1988d) Period: 155.75 years, Grade: 2

Year	PA(°)	Sep(")	Year	PA(°)	Sep(")
2015.0	185.8	2.93	2040.0	223.4	1.85
2020.0	190.9	2.74	2045.0	237.0	1.65
2025.0	196.9	2.54	2050.0	254.1	1.49
2030.0	203.9	2.31	2055.0	274.0	1.41
2035.0	212.6	2.08	2060.0	295.3	1.40

stars are beginning to close slowly and will reach their minimum distance of $1''.09$ in 2081. Thereafter they widen again, reaching $3''.45$ in 2142. At present a 10-cm should show them separated but one should bear in mind that both components are fairly faint. DR2 measures both stars and increases the accuracy of the parallax and proper motion determinations by a factor of almost ten over Hipparcos.

Observing and Neighbourhood

F. G. W. Struve noted that both stars were white in the Dorpat survey, but he was using a fairly small aperture for this work. T. W. Webb reports that George Knott saw colours of very pale yellow and bluish in 1871, as did W. S. Franks in 1876. This rather faint pair can be found $6°$ due N of 6 Boo. It is also only $2°.5$ S following the globular cluster M 3. About $2.5°$ S following is STF 1793, a neat pair of stars magnitudes 7.5 and 8.4 at $241°$, $4''.6$, 2015.

Measures

Early measure (STF)	$164°.4$	$3''.49$	1830.54
(Orbit	$163°.1$	$3''.45$)	
Recent measure (ARY)	$189°.0$	$3''.15$	2016.51
(Orbit	$187°.2$	$2''.88$)	

101. 3 CEN = H 3 101 = WDS J13518−3300

Table 9.101 Physical parameters for 3 Cen

H 3 101	RA: 13 51 49.60	Dec: −32 59 38.7	WDS: 1288(65)	
V magnitudes	A: 4.50	B: 6.01		
(B − V) magnitudes	A: −0.16	B: +0.01		
μ(A)	−34.70 mas yr^{-1}	± 0.86	−27.91 mas yr^{-1}	± 0.79 (DR2)
μ(B)	−36.74 mas yr^{-1}	± 0.27	−23.77 mas yr^{-1}	± 0.19 (DR2)
π(A)	11.10 mas	± 0.43 mas	294 light yr	± 11 (DR2)
π(B)	10.27 mas	± 0.14 mas	318 light yr	± 4 (DR2)
Spectra	A: B5III	B: B8V		
Luminosities (L$_\odot$)	A: 105	B: 30		
Catalogues (A/B)	HR 5210	HD 120709/10	SAO 204916/7	HIP 67669
DS catalogues	H 3 101	DUN 148		
Radial velocity	7.50 km s^{-1}	± 1.6		
Galactic coordinates	317°.282	+28°.189		

History

Herschel observed 3 Cen and in his report of 31 January 1783 also called it k Cen. His report says 'Double. Considerably unequal. L(arge). d(usky)w(hite).; S(mall). d(usky)p(ale)r(ose). Distance 11″35‴. Position 22° 0′ s(outh) following.' John Herschel also observed it in South Africa and called the star Δ 148.

The Modern Era

Hipparcos photometry of star A reports a range of variability between H_p = 4.27 and 4.32. There are two obvious dips in the data which suggest the star is an eclipsing binary, but there are not enough data to derive a period.

Finder Chart

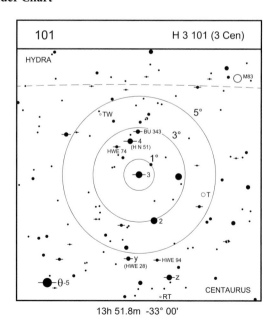

13h 51.8m -33° 00'

Observing and Neighbourhood

CDSA2 shows eight pairs within a radius of 4° of 3 Cen. To the north 1°.5 and slightly following is 4 Cen (H N 51). The stars are 4.7, 8.5 at 185°, 15″, 2013. In the same low-power field (20′ to the NW) is BU 343 (6.3, 8.9, 199°, 0″.7, closing) with a 9.5 magnitude star at 95°, 94″. Twenty-five arcminutes to the SE is HWE 74 (7.2, 9.8, 117°, 5″.8, 2013), a common proper motion pair. Two point five degrees S is HWE 28, an orbital pair with a period of 373 years. It is currently at its widest separation (317°, 1″.0, 2020). One degree due W is HWE 94 (6.6, 10.2, 0°, 11″.9, 2016).

Measures

Early measure (JC)	110°.8	8″.33	1846.30
Recent measure (ARY)	104°.2	7″.87	2013.68

102. $\kappa^{1,2}$ BOO = STF 1821
= WDS J14135+5147AB

Table 9.102 Physical parameters for κ Boo

STF 1821	RA: 14 31 29.00	Dec: +51 47 23.8	WDS: 256(193)	
V magnitudes	A (κ^2): 4.54	B (κ^1): 7.08		
$(B-V)$ magnitudes	A (κ^2): +0.21	B (κ^1): +0.39		
μ(A)	60.96 mas yr^{-1}	± 0.34	−9.27 mas yr^{-1}	± 0.33 (DR2)
μ(B)	44.09 mas yr^{-1}	± 0.65	−39.33 mas yr^{-1}	± 0.61 (DR2)
π(A)	20.31 mas	± 0.22	160.6 light yr	± 1.7 (DR2)
π(B)	21.20 mas	± 0.41	153.8 light yr	± 3.0 (DR2)
Spectra	A: A7V	BaBb: F1V+ ?		
Luminosities (L$_\odot$)	A: 30	B: 2.5		
Catalogues	HR 5329/8	HD 124675/-	SAO 29046/5	HIP 69483/1
DS catalogues	H 3 11	STF 1821	BDS 6778	ADS 9173
Radial velocity (A/B)	−15.60 km s^{-1}	± 0.9	−20.4 km s^{-1}	± 2
Galactic coordinates	96°.455	+60°.914		

History

The object κ Bootis was observed by Sir William Herschel on 27 September 1779. He noted that the stars were very unequal and that the colours were red and dusky. Measurements of the separation in 1779, 1780, and 1781 gave a mean distance of 12″.503 and the position angle was about 240°. Plaskett *et al.* [431] noted that component B had variable radial velocity.

The Modern Era

Bakos [432] confirmed that the B component was a single-line spectroscopic binary. He produced an orbit which showed that the period was 1791.2 days. The primary spectral type was given as F2 with corresponding mass 1.5 M$_\odot$. The secondary was at least one magnitude fainter and therefore it could be an F8 star. If a statistical argument is used to derive a possible inclination for the orbital plane then the mass drops to

Finder Chart

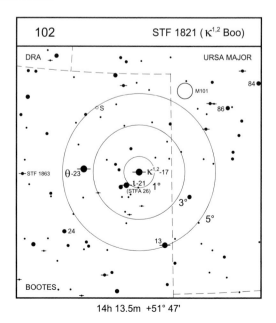

14h 13.5m +51° 47'

0.5 $M_\odot$, in which case the companion to B is as late as K7. In 2006 O. Kiyaeva [433] did an extensive astrometric study of the triple system based on photographic plates taken with the 26-inch refractor at Pulkovo between 1982 and 2004. Despite the angular motion over the last two centuries, amounting to just 3° in position angle, Kiyaeva nevertheless produced two orbits for AB with periods of 6101 and 6675 years. These imply systemic masses of 4.2 and 3.6 $M_\odot$. Kiyaeva also produced four astrometric orbits for the BaBb system, and concluded that there was evidence for the duplicity of star A, which was originally inferred by examining the radial velocity of the star as presented by Abt. He produced two large papers [434], [435] of stellar radial velocity determinations taken from the early days at Lick Observatory, and over a 20-year timespan the radial velocity of component A varied by up to 22 km s^{-1}, which was significantly larger than the quoted errors.

Observing and Neighbourhood

The star κ Boo is the easternmost star in a group of three naked-eye stars located 5° NE of the end star in the tail of the Great Bear (η). Sue French finds both stars yellow-white with 130-mm at ×23. Another star in the loose group is ι (STFA 26), which is a wide and easy binocular double (magnitudes 4.8, 7.4, 33°.1, 38″.82, 2016) and which Smyth describes as pale yellow and creamy white, whereas Sue French records it as yellow-white and golden. A close companion to A noted by Struve in 1836 was also tentatively seen by Smyth in 1838, but has been seen by no one else since. This is presumably the source of the appellation STF 4026AB which appears in the SIMBAD entry for this star. W. H. Smyth noted that the stars were 'pale white and bluish', whilst W. S. Franks in 1915 made them yellow and purple. Continue a line from κ through θ Boo and you come to two pairs which are resolution tests for 10-cm and 20-cm: STF 1863 is a binary with period 534 years which is moving very slowly at present (7.7, 7.8, 59°, 0″.65, 2020); STF 1871, just off the chart, 35′ E of STF 1863, is 8.0, 8.1, 313°, 1″.9, 2017.

Measures

Early measure (STF)	237°.7	12″.60	1832.50
Recent measure (WSI)	235°.8	13″.49	2012.66

103. α CEN = RHD 1 = WDS J14396−6050AB

Table 9.103 Physical parameters for α Cen

RHD 1	RA: 14 39 36.49	Dec: −60 50 02.4	WDS: 56(452)	
V magnitudes	A: −0.01	B: +1.33	C: 11.13	
(*B* − *V*) magnitudes	A: +0.71	B: +0.88	C: +1.82	
μ(AB)	−3619.9 mas yr^{-1}	± 3.9	639.8 mas yr^{-1}	± 3.9 (Kervella)
μ(C)	−3773.8 mas yr^{-1}	± 0.4	770.5 mas yr^{-1}	± 2.0 (Benedict)
μ(C)	−3781.31 mas yr^{-1}	± 0.08	769.77 mas yr^{-1}	± 0.19 (DR2)
π(AB)	747.17 mas	± 0.61	4.365 light yr	± 0.004 (Kervella)
π(C)	768.13 mas	± 1.04	4.246 light yr	± 0.006 (Benedict)
π(C)	768.50 mas	± 0.21	4.244 light yr	± 0.001 (DR2)
Spectra	A: G2V	B: K1V	C: M5.5Ve	
Masses (M$_\odot$)	A: 1.1055	± 0.0039	B: 0.9373	± 0.0033
Masses (M$_\odot$)	C: 0.1221	± 0.0037		
Luminosities (L$_\odot$)	A: 1.5	B: 0.4	C: 0.00005	
Radii (R$_\odot$)	A: 1.2234	± 0.0053	B: 0.8632	± 0.0037
Radii (R$_\odot$)	C: 0.1542	± 0.0045		
Catalogues (A/B)	HR 5459/60	HD 128620/1	SAO 202538	HIP 71683/1
DS catalogues	RHD 1 (AB)	SCZ 1 (CaCb)		
Radial velocity (AB)	−22.332 km s^{-1}	± 0.005 (Kervella)		
Radial velocity (C)	−22.204 km s^{-1}	± 0.032 (Benedict)		
Galactic coordinates	315°.733	−0°.681		

History

The history of α Centauri as a double star goes back to 1689 when Fr Richaud, a French Jesuit priest, who was in Pondicherry and heading for Siam as part of a six-man mission for King Louis XIV, made a telescopic observation of a comet ('je remarquai que le pied le plus oriental & le plus brillant etoit un double etoile aussi bien que le pied de la Croissade – I noted that the brightest star at the easternmost foot (of Centaurus) was a double star as good as that at the foot of the Cross'), i.e. α Crucis. Proxima Centauri was discovered by R. T. Innes [443] in 1915, while working at the Union Observatory in Johannesburg. Innes was blinking two plates taken with the Franklin Adams camera in 1910 and 1915, and after a considerable amount of work he noticed a faint star which showed considerable movement between the two dates. The amount of movement and direction convinced Innes that his star was connected to α Centauri, but its distance was not established until 1928, when H. L. Alden [438] confirmed that it was marginally closer to the Solar System than α.

Finder Chart

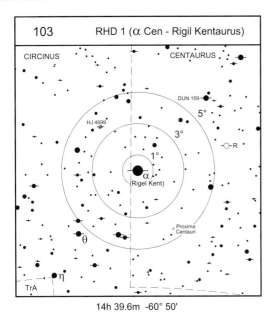

14h 39.6m -60° 50'

Orbit

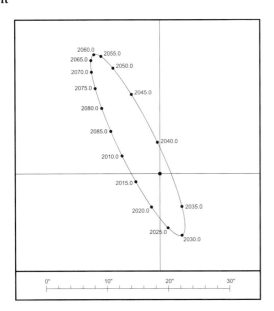

Ephemeris for RHD 1 AB (2015 to 2042)

Orbit by Pbx (2016) Period: 79.91 years, Grade: 2

Year	PA(°)	Sep(")	Year	PA(°)	Sep(")
2015.0	288.2	4.13	2030.0	20.1	10.44
2018.0	327.1	4.46	2033.0	26.9	9.01
2021.0	352.1	6.12	2036.0	42.3	4.48
2024.0	5.5	8.11	2039.0	169.8	3.04
2027.0	13.8	9.78	2042.0	194.5	8.78

The Modern Era

In 1998 Schultz *et al.* [436] announced that they had found a possible substellar companion to Proxima as a result of observations with the Faint Object Spectrograph fitted to the HST and used as a coronographic camera. They tentatively concluded that it is co-moving with Proxima but the WDS clearly states 'Not a binary' and no confirming observations have yet been made. The refinement of radial velocity methods which allows astronomers to detect shifts as small as several metres per second in stars accompanied by planets has led to the discovery of a planet accompanying Proxima Cen. A recent study by Kervella *et al.* [444] has produced precise values for the masses and radii of the three stars, along with the proper motion, parallax, and radial velocity of the centre of mass of the AB binary system as given above. Whilst it was assumed that Proxima was gravitationally bound to α, a critical piece of evidence had been missing and that was an accurate value for the radial velocity of Proxima. With this established, the authors pronounced on the physical relationship of Proxima Centauri to its bright neighbours. They found Proxima to be gravitationally bound with a high degree of confidence. The period is 550,000 years and the apparent orbit takes Proxima from about 13,000 AU from α to 4300 AU. The star is currently near apastron. Whilst the bright stars are not in DR2, there is a measurement of Proxima, although the G magnitude given (8.95) seems at odds with the WDS V magnitude of 11.13. The data is given in the table and should be regarded as provisional. The astrometry of Proxima given above is by Benedict *et al.* [439]. Dedicated books on α and Proxima have been written by Martin Beech [441] and Ian Glass [442].

Exoplanet Host

In 2016 Anglada-Escudé *et al.* [440] announced that Proxima was accompanied by a planetary body. The mass was 1.3 times that of the Earth and the period was 11.2 days. The reflex motion that this body produced on Proxima Centauri was only 3 m s^{-1} peak to peak.

Observing and Neighbourhood

The pair α Centauri is quite simply the best visual double star in the sky. The combination of brightness, rapid orbital motion, and the knowledge that this is the nearest binary star to the Sun make this a 'must-see' for all apertures. Even in the smallest telescopes the stars are brilliant but in large telescopes they are almost too bright to bear. Both stars appear deep yellow to RWA, whilst the ASNSW group make them yellowish and yellowish but darker. The pair is currently widening towards the shorter end of the apparent ellipse but around 2030 it will begin to close again rapidly, passing periastron in 2035 and minimum separation of $1''.68$ in 2037. The maximum separation of $21''.8$ is reached in 2060. Two degrees NE is HJ 4699 (6.8, 9.1, 125°, 37″, 2013), whilst 3°.2 NW is DUN 159 (5.0, 7.6, 157°, 10″.7, 2016) with two more distant stars of magnitude 10.7.

Measures

Early measure (JC)	237°.8	7″.97	1848.02
(Orbit	239°.9	7″.94)	
Recent measure (RWA)	312°.9	4″.11	2016.70
(Orbit	311°.0	4″.09)	

104. ζ BOO = STF 1865
= WDS J14411+1344AB

Table 9.104 Physical parameters for ζ Boo

STF 1865	RA: 14 41 08.95	Dec: +13 43 41.9	WDS: 18(746)	
V magnitudes	A: 4.46	B: 4.55	C: 10.98	
$(B - V)$	A: +0.07	B: +0.09		
μ	51.95 mas yr^{-1}	± 0.81	−11.08 mas yr^{-1}	± 0.53
π	18.56 mas	± 0.76	176 light yr	± 7
Spectra	A: A0V	B: A0V		
Luminosities (L$_\odot$)	A: 38	B: 35		
Catalogues	HR 5477	HD 129246J	SAO 101145	HIP 71795
DS catalogues	H N 114 (AB)	STF 1865 (AB)	H 6 104 (AC)	BDS 6955 ADS 9343
Radial velocity	−8.50 km s^{-1}	± 0.6 km s^{-1}		
Galactic coordinates	10°.804	+61°.115		

History

On 29 November 1782 William Herschel noted that ζ Boo had 'a very obscure star in view. Extremely unequal … Distance about $1\frac{1}{2}$ minutes. Position almost directly preceding' but, whilst re-observing this system on 5 April 1796, he found the bright star was a close double 'very nearly in contact. I can however see a small division.' The ephemeris from the orbit by M. Scardia gives a distance of 0″.89 for that epoch. South & Herschel [518] measured it in 1823 but got a distance about twice that expected, and from 1830 onwards, when Struve began to observe the pair, the stars closed and the trajectory appeared almost linear. A close approach in 1897 was followed by a reverse of direction around 1950, since when the stars have been closing more rapidly.

The Modern Era

This system is closely related to γ Vir in terms of orbital dynamics. Two stars of similar mass, brightness, and spectral type are involved, and in each case the apparent orbit is a

Finder Chart

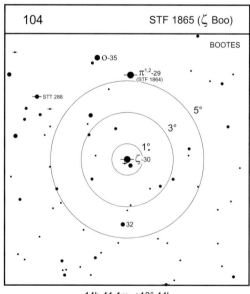

14h 41.1m +13° 44'

Orbit

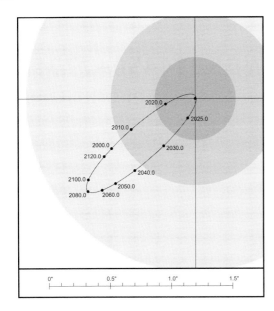

Ephemeris for STF 1865 AB (2016 to 2052)

Orbit by Sca (2007f) Period: 125.24 years, Grade: 2

Year	PA(°)	Sep(")	Year	PA(°)	Sep(")
2016.0	289.3	0.40	2036.0	321.0	0.66
2020.0	280.2	0.25	2040.0	319.1	0.76
2024.0	3.8	0.04	2044.0	317.7	0.84
2028.0	328.2	0.37	2048.0	316.5	0.91
2032.0	323.6	0.53	2052.0	315.5	0.97

highly eccentric ellipse. In the case of γ Vir, e = 0.88 whilst that of ζ Boo is 0.98 according to the orbit of M. Scardia [447]. In fact, the eccentricity is the highest known of any visual system and means that the stars are 82 times closer to each other at periastron than at apastron. In 2010, Muterspaugh *et al.* [620] re-calculated the orbit, giving it an eccentricity of 0.9977 which, if true, means that the stars approach each other to 0.3 AU at periastron and are almost 2000 times further away at apastron. The formal error on the eccentricity of this orbit allows a small possibility that the orbit may even be parabolic. The *Sixth Double Star Orbital Catalog* prefers the

slightly less eccentric Scardia orbit, and the ephemeris from that work is given here. Either way the next close approach, due in late 2023, opens up the possibility of observing a change in orbital elements due to mutual tidal effects. The star is included here because the components are both bright and equal and will form a stiff test for larger apertures from 2020 or so.

Observing and Neighbourhood

This pair is closing very quickly and by 2020 the separation will be pushing the resolution of a 50-cm telescope. The shorter orbital ephemeris given above shows that the motion will be extremely swift around autumn 2023, when the stars will be moving at almost 6° per day. At that point the separation will be only 5 mas but in the range of ground-based interferometers. By the spring of 2030 the pair should be split in a 20-cm aperture once more. The wider Herschel companion C is being left behind by the proper motion of AB and is currently 105″ away. There are some splendid pairs in the vicinity. The star ζ forms an equilateral triangle 3° on the side with both π Bootis (STF 1864) (due N) and STT 288 (NE). The object π is a beautiful pair, whose stars are of magnitudes 4.9 and 5.8 and they are currently separated by 5″.4 at PA 113°. STT 288 is a binary pair of 313-year period. The stars are of magnitudes 6.9 and 7.6 and are currently 1″.1, apart although now slowly closing, passing 0″.8 around 2030 before reaching 0″.5 in 2060 or so. A further 4° NE of π is ξ Bootis (STF 1888; see Star 106).

Measures

Early measure (STF)	309°.2	1″.19	1830.47
(Orbit	309°.8	1″.15)	
Recent measure (ARY)	290°.9	0″.50	2013.21
(Orbit	292°.5	0″.49)	

105. ϵ BOO = STF 1877 = WDS J14450+2704AB

Table 9.105 Physical parameters for ϵ Boo

STF 1877	RA: 14 45 59.25	Dec: +27 04 27.0	WDS: 54(461)		
V magnitudes	A: 2.58	B: 4.81			
$(B-V)$ magnitudes	A: +1.36	B: +0.06			
μ	-50.95 mas yr^{-1}	0.59	$+21.07$ mas yr^{-1}	$\pm\,0.58$	
π	16.10 mas	$\pm\,0.66$	203 light yr	$\pm\,8$	
$\mu(B)$	-43.53 mas yr^{-1}	$\pm\,0.73$	22.50 mas yr^{-1}	$\pm\,1.16\ (DR2)$	
$\pi(B)$	14.88 mas	$\pm\,0.41$	219 light yr	$\pm\,8\ (DR2)$	
Spectra	A: K0II-III	B: A0V			
Luminosities (L$_\odot$)	A: 340	B: 45			
Catalogues (A/B)	36 Boo	HR 5506/5	HD 129989/8	SAO 83500	HIP 72105
DS catalogues	H 1 1	STF 1877	ADS 9372	BDS 6993	
Radial velocity	-16.31	$\pm\,0.19$			
Galactic coordinates	$39°.385$	$+64°.784$			

History

The binary system ϵ Bootis was observed by Sir William Herschel in September 1779 and has the distinction of being the first star in his first class. He noted that the stars were very unequal and the bright star was reddish whilst the companion was blue, 'or rather a faint lilac'. F. G. W. Struve was so taken with the colour of the stars that he referred to ϵ by the soubriquet of *pulcherrima* – 'most beautiful'. In his famous paper in *Philosophical Transactions* read to the Royal Society on 9 June 1803, Sir William Herschel [363] said 'This beautiful double star, on account of the different colours of the stars of which it is composed, has much the appearance of a planet and its satellite, both shining with innate but differently coloured light.' His son John, writing in 1826 [448], observed that 'After a long and obstinate contest with ϵ Bootis, which is certainly one of the most difficult double stars to measure correctly, Rigel itself excepted, I remain unconvinced of its motion.'

Finder Chart

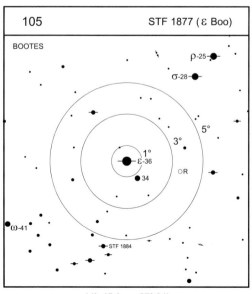

14h 45.0m +27° 04'

The Modern Era

Component B is known to be a spectroscopic binary but no orbit has been published. The two bright stars form a physical system. Over the last two centuries there has been little, if any, movement in separation, whilst the position angle has increased about 30°. Herschel's angles, measured in the 1780s, all seem to be about 10° out since the annual angular motion over the last two centuries has been about 0°.14 per year. Projecting this angular motion through 360° leads to an orbital period of about 2850 years; however, assuming that the masses of the primary and secondary are 4.6 and 2.4 M$_\odot$, on the basis of the spectral types, then a projected period of under 600 years results if the orbit is circular. This period reduces if the B component consists of two A dwarfs, so there is an inconsistency between the assumed mass and projected period, which could be resolved if the apparent orbit is highly inclined and/or eccentric.

Observing and Neighbourhood

Struve: the colours are a decided yellow and a decided green. Webb: light yellow and greenish (1850). Smyth found the stars pale orange and sea green and assigned magnitudes of 3 and 7 to them. Hartung, observing from Victoria where the stars are never more than about 25° in altitude, says 'deep yellow and white (or bluish by contrast)'. RWA found orange-yellow and light blue with a 21-cm mirror in 1970. The binary ε is the middle star on the eastern side of the kite-shape formed by the bright stars of Boötes and is about 13° due W of Alphekka (α CrB). Three degrees S and slightly following is the fine close pair STF 1884 (6.6, 7.5, 57°, 2″.2, 2016), slowly widening since discovery and certainly a long-period binary.

Measures

Early measure (STF)	321°.0	2″.64	1829.39
Recent measure (WSI)	344°.5	2″.92	2015.31

106. ξ BOO = STF 1888 = WDS J14514+1906AB

Table 9.106 Physical parameters for ξ Boo

STF 1888	RA: 14 51 23.28	Dec: +19 06 02.3	WDS: 5(1430)		
V magnitudes	A: 4.76	B: 6.82			
$(B - V)$ magnitudes	A: +0.82	B: +1.16			
μ(A)	128.21 mas yr^{-1}	± 0.40	−42.06 mas yr^{-1}	± 0.52 (DR2)	
μ(B)	132.68 mas yr^{-1}	± 0.07	−181.62 mas yr^{-1}	± 0.10 (DR2)	
π(A)	148.52 mas	± 0.24	21.96 light yr	± 0.04 (DR2)	
π(B)	148.21 mas	± 0.05	22.01 light yr	± 0.01 (DR2)	
Spectra	G7Ve	K5Ve			
Masses (M$_\odot$)	A: 0.860		B: 0.70		
Luminosities (L$_\odot$)	A: 0.45		B: 0.07		
Radii (R$_\odot$)	A: 0.90	± 0.04	B: 0.66	± 0.07	
Catalogues	36 Boo	HR 5544	HD 131156	SAO 101250	HIP 72659
DS catalogues	H 2 18	STF 1888	BDS 7034	ADS 9413	
Radial velocity	1.59 km s^{-1}	± 0.3			
Galactic coordinates	23°.086	+61°.356			

History

Observed by William Herschel on 9 April 1780. He noted the colours as 'Pale r(ed) or nearly r(ed). Garnet or deeper r(ed) than the other'.

The Modern Era

In 1943 K. Strand [611], using plates from Sproul and Johannesburg, announced that he had detected a perturbation in the ξ Boo system with a period of 2.2 years and an amplitude of 0″.02, and postulated an invisible body with a mass of about 0.1 M$_\odot$. This has not been confirmed. The GCVS notes that A is a small amplitude variable (magnitudes 4.52 to 4.67) of the BY Dra variety with a period of 10.137 days. The Chandra X-ray satellite [449] resolved the two stars, showing that the visually brighter A component produced 89% of the observed X-ray flux from the binary, indicating a very active corona.

Finder Chart

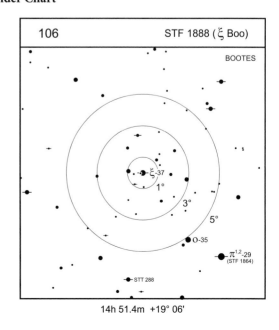

14h 51.4m +19° 06′

Orbit

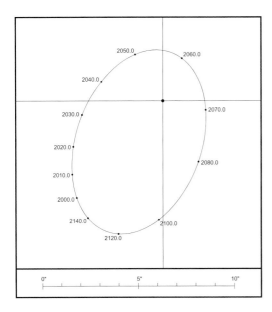

Ephemeris for STF 1888 AB (2015 to 2060)

Orbit by Sod (1999) Period: 151.6 years, Grade: 2

Year	PA(°)	Sep(")	Year	PA(°)	Sep(")
2015.0	302.9	5.64	2040.0	253.0	3.34
2020.0	296.5	5.21	2045.0	234.1	3.00
2025.0	288.9	4.74	2050.0	211.2	2.77
2030.0	279.6	4.26	2055.0	185.2	2.61
2035.0	267.9	3.78	2060.0	155.3	2.39

Exoplanet Host?

As a nearby star, ξ Boo has been the target of several observing campaigns to look for signs of exoplanets, which, so far, have proved fruitless. Astronomers used radial velocities taken at the Lick and Keck observatories over an almost 30 year period [451] to make accurate measures of the radial velocity of the brighter component, but no significant variations were seen apart from a slow linear drift due to the binary companion.

Observing and Neighbourhood

One of the best targets for small telescopes in the northern sky, this much observed system can be seen fully resolved at all times in its 151.5-year period orbit even with small apertures. It is easily found about 9° following α Boo (Arcturus). The pair will be resolvable in 7.5-cm aperture through the whole orbital cycle. The two stars were separated by 7″.3 in 1978 and will close up to a separation of 2″.1 in 2066. The colour contrast is marked in this pair, owing to the difference in spectral types in the two stars of nearly two full spectral classes, although the bluish tinge mentioned by Smyth (orange and purple) and STF (yellow and purple) has not been experienced by Hartung or the writer, for whom yellow and deep orange seems a better description of the colours of A and B. Nearby are ζ Boo (see Star 104) and π Boo.

Measures

Early measures (STF)	334°.2	7″.22	1829.40
Orbit	332°.1	7″.27)	
Recent measure (ARY)	300°.3	5″.26	2018.44
Orbit	298°.5	5″.35)	

107. 44,i BOO = STF 1909 = WDS J15038+4739

Table 9.107 Physical parameters for 44,i Boo

STF 1909	RA: 15 03 47.30	Dec: +47 39 14.6	WDS: 15(821)	
V magnitudes	A: 5.20	B: 6.10		
$(B - V)$ magnitudes	A: +0.65	B: +0.94		
μ	-445.84 mas yr^{-1}	± 1.44	19.86 mas yr^{-1}	± 1.67
μ(A)	-332.06 mas yr^{-1}	± 1.97	87.35 mas yr^{-1}	± 2.02 (DR2)
μ(B)	-463.68 mas yr^{-1}	± 0.67	-7.52 mas yr^{-1}	± 0.80 (DR2)
π(A)	77.25 mas	± 1.20	42.22 light yr	± 0.7 (DR2)
π(B)	78.09 mas	± 0.53	41.77 light yr	± 0.3 (DR2)
Spectra	A: F7V	Ba: G0Vn	Bb: G0Vn	
Masses (M$_\odot$)	A: 1.04	± 0.10	B: 1.28	± 0.02
Radii (R$_\odot$)	A: 0.98	± 0.04	B: 0.87	± 0.02
Luminosities (L$_\odot$)	A: 1.1	B: 0.5		
Catalogues	HD 133640	HR 5618	SAO 45357	HIP 73695
DS catalogues	H 1 15	STF 1909	BDS 7120	ADS 9494
Radial velocity	-10.40 km s^{-1}	± 2.1		
Galactic coordinates	$80°.370$	$+57°.066$		

History

Although William Herschel discovered the duplicity of 44 Bootis on 17 August 1782 the pair were in the process of closing rapidly, and by 1811 they were only 0″.23 apart, but were then easily seen by Struve at Dorpat in 1832 when the distance had increased to almost 3″. For much of the next century the motion appeared to be mainly in separation as the stars widened. Agnes Clerke [452] summarized some of the early impressions of variability. She notes that in 1781 Herschel considered them considerably unequal, yet in 1787 they were perfectly matched. John Percy [453] says that Herschel discovered two variable stars, 44 Boo and α Her. In 1819, F. G. W. Struve found two magnitudes difference between the stars, from 1822–1833 consistently one magnitude, and between 1833 and 1838 only half a magnitude. In 1921, W. S. Adams [454] took a spectrum of the star and found that it resembled W UMa. The period of variation (0.268 days) was found by

Finder Chart

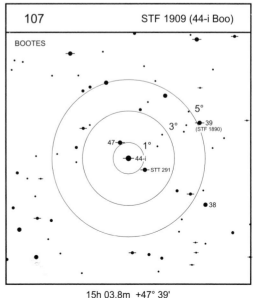

15h 03.8m +47° 39'

Orbit

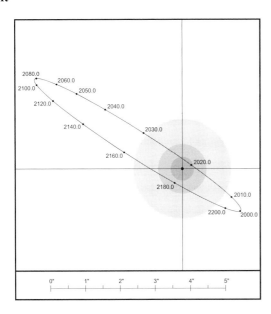

Ephemeris for STF 1909 (2016 to 2052)

Orbit by Zir (2011) Period: 209.8 years, Grade: 2

Year	PA(°)	Sep(")	Year	PA(°)	Sep(")
2016.0	70.0	0.84	2036.0	231.7	2.29
2020.0	112.1	0.27	2040.0	233.2	2.75
2024.0	212.2	0.57	2044.0	234.2	3.16
2028.0	225.3	1.19	2048.0	235.0	3.51
2032.0	229.5	1.76	2052.0	235.6	3.81

Schilt [455] in 1926 using photographic plates exposed on the Mount Wilson 60-inch telescope, with 100 images on each plate. The variable was confirmed to be the B component.

The Modern Era

J. M. Saxton [456], using a 21-cm reflector and a photo-multiplier tube, was able to show by carrying out photo-electric photometry of i Boo that there was a significant period change in its ephemeris. Pustylnik *et al.* [457], how-ever, attribute this effect to increased activity on the stellar surface.

Observing and Neighbourhood

The apparent orbit of 44 Boo is a highly elongated ellipse, the widest separations being in 2005 at 2″.0 and in 2087 at 4″.9 whilst the two closest approaches were in 1970 at 0″.45 and will be in 2020 when the stars are 0″.23 apart. As this volume goes to press, the visual system is closing rapidly, as can be seen from the ephemeris. In 2016, using the Cambridge 8-inch, the stars were resolvable on one night only, and for the next few years the pair will be very difficult with apertures less than 30- or 40-cm. The colours of B have apparently ranged from 'lucid gray' – Smyth 1842 – to blue (Webb and Secchi) and later reddish (Webb and Engelmann). Moving WNW by 2°.5 one comes to 39 Boo (STF 1890), a lovely pair, the colour of which Struve reported as white and purple. The spectral types are F6V and F5V and the magnitudes 6.3 and 6.8. The stars have identical proper motions and are closing (47°, 2″.7, 2016), but the WDS gives a linear solution. Forty arcminutes SW is STT 291 (6.3, 9.6, 170°, 36″, 2015), a Herschel discovery from 1783 (H 5 122).

Measures

Early measure (STF)	234°.0	2″.86	1832.24
(Orbit	233°.7	2″.96)	
Recent measure (ARY)	72°.0	0″.91	2016.54
(Orbit	71°.9	0″.75)	

108. π LUP = HJ 4728 = WDS J15051−4703

Table 9.108 Physical parameters for π Lup

HJ 4728	RA: 15 05 07.16	Dec: −47 03 04.3	WDS: 697(99)	
V magnitudes	A: 4.59	B: 4.67		
$(B - V)$ magnitudes	A: −0.15	B: −0.12		
μ	A: −22.98 mas yr^{-1}	± 0.67	B: −22.81 mas yr^{-1}	± 0.52
π	7.36 mas	± 0.55 mas	490 light yr	± 37
μ(A)	−25.18 mas yr^{-1}	± 1.02	−17.86 mas yr^{-1}	± 0.71 (DR2)
π(A)	6.86 mas	± 0.52	475 light yr	± 36 (DR2)
Spectra	A: B5V	B: B5IV		
Luminosities (L$_\odot$)	A: 250	B: 230		
Catalogues (A/B)	HR 5605/6	HD 133242	SAO 225426	HIP 73807
DS catalogues	HJ 4728			
Radial velocity	4.50 km s^{-1}	± 3.7		
Galactic coordinates	325°.326	+9°.925		

History

'Suspected with 180, plainly seen double with 320 and the whole aperture, in spite of much flare, which assists the division.' This is John Herschel's report on π Lupi in the *Results of Astronomical Observations at the Cape of Good Hope* (1847) and which he observed on 17 February 1835 (sweep 544). Subsequent measures in the latter half of the nineteenth century show that the values of separation are quite scattered for such a bright pair, and those using smaller apertures tend to overestimate the separation. In the 1880s for instance the separations by Tebbutt with an 8-inch OG were close to 1″.5, whilst those by Sellors with 11.5-inches tended to be nearer 1″.1.

The Modern Era

The WDS has a note which indicates that both stars are spectroscopic binaries. A paper by Nitschelm [458] gives a

Finder Chart

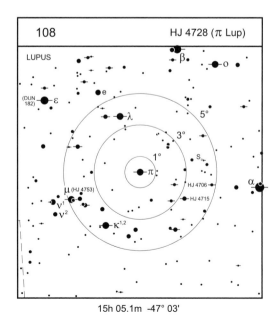

15h 05.1m -47° 03'

period of 517 years for the two visual components although the USNO Sixth Orbital Catalogue contains no such orbit for the star. Nitschelm [459] later reported that the spectra of π Lupi show three distinct sets of lines and argued therefore that the system is quadruple. It seems possible that one star is an SB2 and the other an SB1. Orbital motion is very slow and it would seem that the companion is reaching one end of the apparent ellipse. With just 20° or so of position angle covered it is impossible to calculate an orbital period, but it is likely to be at least 1000 years.

Observing and Neighbourhood

A very bright pair, which should be resolved in 10-cm for the foreseeable future. Two degrees WSW is HJ 4715 (6.0, 6.8, 281°, 2″.1, 2016), a beautiful pair which is slowly closing. Go another 1° to the WNW to reach HJ 4706 (7.7, 9.0, 220°, 7″.0, 2016).

Measures

Early measure (SLR)	85°.8	0″.86	1891.56
Recent measure (ANT)	64°.5	1″.82	2014.66

109. $\mu^{1,2}$ LUP = HJ 4753
= WDS J15185–4753AB

Table 9.109 Physical parameters for μ Lup

HJ 4753	RA: 15 18 32.02	Dec: −47 52 31.0	WDS: 1347(63)		
V magnitudes	A: 4.95	B: 5.05	C: 6.76	D: 14.70	
$(B - V)$ magnitudes	A: −0.10	B: −0.08	C: +0.11	D:	
μ(AB)	−29.59 mas yr^{-1}	± 0.78	−35.07 mas yr^{-1}	± 0.57	
μ(C)	−25.50 mas yr^{-1}	± 0.71	−34.75 mas yr^{-1}	± 0.51 (DR1)	
π(AB)	9.72 mas	± 0.71	336 light yr	± 25	
μ(C)	−18.60 mas yr^{-1}	± 0.34	−24.32 mas yr^{-1}	± 0.51 (DR2)	
π(C)	4.01 mas	± 0.18	813 light yr	± 36 (DR2)	
Spectra	A: B8V	B:	C: A2V		
Masses (M$_\odot$)	A: 3.1	B: 3.2	Ca: 1.8	Cb: 0.04	D: 0.05
Luminosities (L$_\odot$)	A: 90	B: 80	C: 100	D: 0.02	
Catalogues	HR 5683	HD 135734	SAO 225638	HIP 74911	
DS catalogues	HJ 4753 (AB)	DUN 180 (AB,C)	HUB 15 (AD)		
Radial velocity (AB)	14.90 km s^{-1}	± 1.78			
Radial velocity (C)	−6.9 km s^{-1}				
Galactic coordinates	326°.857	+8°.046			

History

James Dunlop had his paper 'Approximate Places of Double Stars in the Southern Hemisphere observed at Parramatta in New South Wales in 1829' published in the *Memoirs of the RAS*. He measured the difference in RA and declination to yield position angle and separation. Dunlop listed μ Lupi as entry number 180 and measured the difference in RA and declination between the two stars, which can be converted into 21″.00 and 45°.9. John Herschel mentions merely 'Close double' and gives a distance of 2″ and a PA of 173°.5 when he first observed the duplicity of the brighter component in 1836.

The Modern Era

Modern high-resolution techniques are demonstrating that B stars often form systems of higher multiplicity, and

Finder Chart

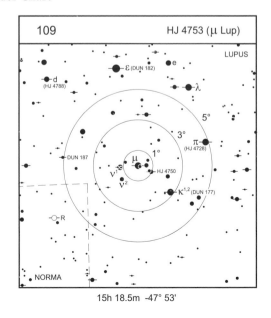

15h 18.5m -47° 53'

Orbit

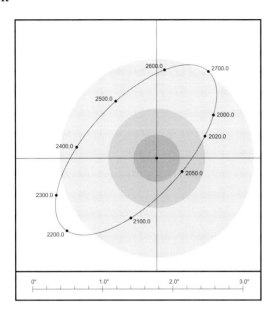

Ephemeris for HJ 4753 AB (2010 to 2100)

Orbit by Zir (2015a) Period: 772 years, Grade: 5

Year	PA(°)	Sep(")	Year	PA(°)	Sep(")
2010.0	121.3	0.89	2060.0	32.5	0.41
2020.0	114.1	0.76	2070.0	8.2	0.50
2030.0	103.8	0.62	2080.0	352.8	0.63
2040.0	87.7	0.49	2090.0	342.8	0.77
2050.0	62.7	0.41	2100.0	336.0	0.91

μ Lupi is no exception. Hubrig *et al.* [461] found an infrared component 6″.15 distant from A which they conclude is a pre-main sequence star with a mass of 0.05 M$_\odot$. The spectra of C (Veramendi & Genzález [462]) showed that C was a spectral type A2V but that it was also a single-lined spectroscopic binary of period 12.53 days. When they took a spectrum of the A component they found that it showed two spectra of similar spectral type but radically different rotation velocities

(278 km s^{-1} and 50 km s^{-1}), the latter being similar to that of C. If the visual orbit is correct, the stars will close to 0″.5 around 2055 then widen to a maximum separation of 1″.7 in 2230 or so. Gaia has found a proper motion for C very close to that of AB but C gives a negative parallax, although as it is a spectroscopic binary this may affect this result. The radial velocity of C is significantly different from that of AB but Veramendi does not rule out a physical connection, given that the orbit of C is preliminary. If the star found by Hubrig is connected then this is potentially a quintuple system.

Observing and Neighbourhood

A beautiful pair of bright, white stars with the Dunlop companion adding to the beauty of the view. The close binary is slowly getting more difficult to resolve and will need at least 15-cm from now on. This is a rich region for the double star observer. Within 5° are κ^1 Lup (DUN 177) (3.8, 5.5, 143°, 27″, 2010), π Lup (see Star 108), DUN187 (7.1, 9.2, 217°, 24″, 2010), and HJ 4788 (4.7, 6.5, 12°, 2″.1, 2016) whilst just 30′ away from μ is HJ 4750 (6.0, 10.4, 15°, 12″.5, 2013).

Measures

HJ4753AB

Early measure (JC)	170°.1	1″.5	1851.49
(Orbit	168°.7	1″.31)	
Recent measure (ARY)	119°.0	0″.83	2016.66
(Orbit	116°.7	0″.80)	

DUN 180AC

Early measure (SEE)	130°.2	23″.44	1897.13
Recent measure (ARY)	128°.1	23″.22	2016.66

110. ε LUP = Δ182 = WDS J15227−4441AB,C

Table 9.110 Physical parameters for ε Lup

DUN 182	RA: 15 22 40.87	Dec: −44 41 22.6	WDS: 5872(22)		
V magnitudes	Aa: 3.6	Ab: 5.1	B: 5.1	C: 9.1	D: 16.2
(B − V) magnitudes	Ab: −0.15	B:	C: −0.12		
μ(AB)	A: −22.86 mas yr^{-1}	± 0.67	B: +8.87 mas yr^{-1}	± 0.59	
μ(C)	A: −12.78 mas yr^{-1}	± 2.5	B: −16.4 mas yr^{-1}	± 2.5	
π	6.37 mas	± 0.70 mas	512 light yr	± 56	
μ(A)	−16.64 mas yr^{-1}	± 1.79	−24.14 mas yr^{-1}	± 1.59 (DR2)	
μ(C)	−6.46 mas yr^{-1}	± 0.09	−13.87 mas yr^{-1}	± 0.06 (DR2)	
π(A)	8.24 mas	± 0.85	396 light yr	± 41 (DR2)	
π(C)	3.69 mas	± 0.52	884 light yr	± 125 (DR2)	
Spectra	A: B5V + ? + ?	B: ?	C: A5V		
Masses (M$_\odot$)	Aa1: 8.5	Aa2: 7.4	Ab: 4.6	B: 4.8	
Luminosities (L$_\odot$)	Aa: 440	Ab: 110	B: 110	C: 12	
Catalogues	HR 5708	HD 136504	SAO 225712	HIP 75264	
DS catalogues	RIZ 13 Aa,Ab	COP 2 (AB)	DUN 182 (AB,C)	RIZ 13 (AD)	
Radial velocity (AB)	7.90 km s^{-1}	± 4			
Radial velocity (C)	−26 km s^{-1}	± 5			
Galactic coordinates	329°.228	+10°.323			

History

At Parramatta John Dunlop measured a faint star 19″ away from ε Lupi at a position angle of 174°. He used an achromatic telescope of 46-inches focal length in conjunction with a parallel wire micrometer which he had made himself. The close pair was discovered by Copeland in 1883 (see the entry on ψ Velorum, Star 67). In the 1900s Lick Observatory astronomers were in Chile using a two-prism spectrograph on bright southern hemisphere stars; during this expedition they discovered the spectroscopic binary nature of ε, and a period of 4.55 days was quickly established.

The Modern Era

The system ε Lupi is the first known double magnetic massive binary. In a recent survey for duplicity using the SUSI interferometer, Rizzuto et al. [475] detected a close companion star at a distance of 0″.049, which probably does not correspond to the spectroscopic component but may be related to the star found by Uytterhoeven et al. [465], who used 106 high-resolution spectra to find a period of 4.55970 years for the primary; in doing so, however, they also detected apsidal motion with a period of 430 years. This makes ε a physical quadruple, and the physical connection of Dunlop's

Finder Chart

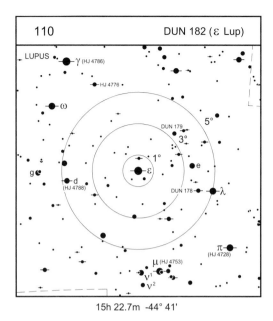

15h 22.7m −44° 41'

Orbit

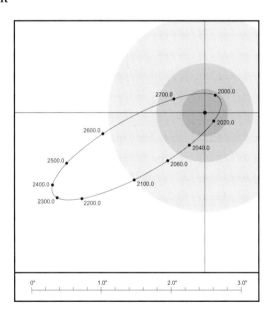

Ephemeris for COP 2 AB (2010 to 2100)

Orbit by Tok (2015c) Period: 736.69 years, Grade: 5

Year	PA(°)	Sep(")	Year	PA(°)	Sep(")
2010.0	115.1	0.26	2060.0	321.6	0.85
2020.0	47.7	0.17	2070.0	318.5	1.00
2030.0	350.8	0.31	2080.0	316.2	1.13
2040.0	333.8	0.50	2090.0	314.3	1.26
2050.0	326.2	0.69	2100.0	312.8	1.37

component C is questionable. Its spectral type and radial velocity led Thackeray [464] to doubt its connection to the system. The proper motion is also significantly different. DR2 confirms that C is twice as far as A and hence a field star. Another star, with $V = 16$, was found by Rizzuto using wide-field databases.

Observing and Neighbourhood

The Dunlop pair will remain a relatively easy object for 10-cm apertures. The distance is slowly increasing. The star found by Copeland, however, has closed up and is currently near minimum separation, placing it beyond the range of most telescopes for at least 20 years. The system ε Lupi lies in a rich field of interesting visual double stars. Within 5° are γ, μ, and two DUN pairs, DUN 178 (6.5, 7.3, 256°, 31″, 2016) – the primary is a close pair (B 1267, 6.5, 9.6, 314°, 1″.2, 1991) – and DUN 179 (7.3, 8.5, 45°, 10″.5, 2016). The star HJ 4788 (magnitudes 4.7, 6.5) was found by John Herschel during sweep 718 (10 July 1836); it is new called d Lup. A spectacular sight in the Johannesburg telescope, RWA measured it at 12°, 2″.1 in 2016, confirming that the pair has closed since discovery (3″.1).

Measures

COP 2			
Early measure (SEE)	280°.8	0″.96	1897.12
(Orbit	275°.4	0″.70)	
Recent measure (TOK)	95°.3	0″.21	2014.18
(Orbit	93°.4	0″.21)	
DUN 182			
Early measure (DUN)	173°.6	19″.12	1826
Recent measure (ARY)	168°.9	26″.14	2016.66

111. η CRB = STF 1937 = WDS J15232+3017AB

Table 9.111 Physical parameters for η CrB

STF 1937	RA: 15 23 12.31	Dec: +30 17 16.2	WDS: 10(1074)	
V magnitudes	A: 5.70	B: 5.93	C: 17	
$(B - V)$ magnitudes	A: +0.60	B: +0.71		
μ	116.83 mas yr^{-1}	± 0.40	-171.37 mas yr^{-1}	± 0.49
π	55.98 mas	± 0.78	58.2 light yr	± 0.8 (Hipp.)
π	54.1 mas	± 0.64	60.3 light yr	± 0.7 (dyn.)
Spectra	A: G2V	B: G2V		
Masses (M$_\odot$)	A: 1.243	± 0.054	B: 1.100	± 0.039
Luminosities (L$_\odot$)	A: 1.5	B: 1.3	C: 0.00005	
Catalogues	HD 137107	HR 5727	SAO 64673	HIP 75312
DS catalogues	H 1 16	STF 1937	BDS 7251	ADS 9617
Radial velocity	7.26 km s^{-1}	± 0.05		
Galactic coordinates	47°.536	+56°.725		

History

This pair was noted by Sir William Herschel on 9 September 1781 (when the current orbit puts them at a distance of just over 1″). He said that he was able to see them as double at ×227 but commented that he would not have been able to discover them at that power. The stars spend only nine years in every 41-year cycle at a separation of 1 arcsecond or greater, so it was a fortuitous observation. Lewis lists 15 orbits which had been calculated by 1906 including the first, by John Herschel in 1833, when very few reliable observations had been made. Herschel obtained a period of 44 years but by 1850 the elements of the orbit were more or less those which characterize the current orbit.

The Modern Era

The *WDS* gives the spectra as F8V and G0V – the types in the table above are from Muterspaugh [621].

Finder Chart

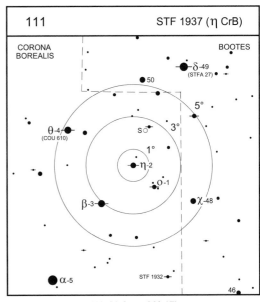

15h 23.2m +30° 17'

Orbit

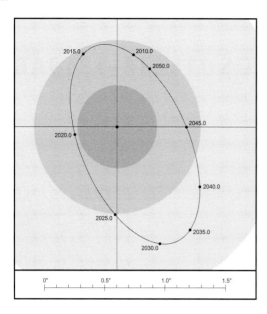

Ephemeris for STF 1937 AB (2018 to 2036)

Orbit by Mut (2010b) Period: 41.6296 years, Grade: 1

Year	PA(°)	Sep(")	Year	PA(°)	Sep(")
2018.0	237.9	0.46	2028.0	13.8	0.91
2020.0	280.0	0.37	2030.0	21.0	1.00
2022.0	326.0	0.43	2032.0	27.2	1.04
2024.0	351.1	0.61	2034.0	33.3	1.03
2026.0	4.7	0.78	2036.0	39.6	0.99

Kirkpatrick *et al.* [466] found a very faint companion of spectral type L8 194″ SE of the bright pair (equivalent to a linear separation of 3600 AU) and they concluded that it is physically connected to AB. The derived *V* magnitude was 17.0. A search in DR2 around η to 500″ shows no companion stars with a parallax near 55 mas. The WDS notes two faint and distant stars which are unrelated: C is 13.4 at 359°, 74″ and D is 11.0 at 41°, 218″; in each case the separation from AB is increasing owing to the proper motion of the binary pair.

Observing and Neighbourhood

The binary η CrB is easily found nestling close to the bright stars of Corona Borealis. It is one-third of the way between β CrB and δ Boo. This is a wonderful pair for 30-cm as they can be followed throughout their entire orbital cycle and the annual change can be seen without instrumentation. The stars are currently closing quickly and will reach the minimum separation of 0″.36 in 2020. By about 2024 they should again be resolvable in 20-cm. Four degrees NNW is the bright, wide binocular pair δ Boo (STFA 27) (3.6, 7.9, 78°, 105″, 2015),

Measures with 8″ Refractor

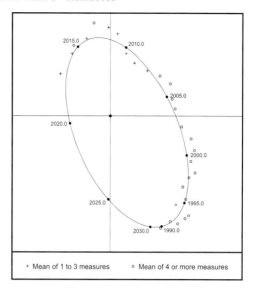

| | + Mean of 1 to 3 measures | ○ Mean of 4 or more measures | |

Year	PA(°)	Sep(")	No.	Year	PA(°)	Sep(")	No.
1990.42	25.3	1.04	5	2003.55	96.0	0.57	4
1991.46	30.8	1.03	5	2004.52	104.6	0.56	5
1992.58	33.2	1.08	5	2005.47	117.1	0.59	5
1993.54	37.3	1.08	5	2006.43	127.6	0.55	5
1993.59	37.8	0.94	3	2007.43	139.2	0.50	2
1994.48	39.4	1.08	5	2008.53	154.5	0.49	2
1995.52	43.5	0.96	5	2009.58	165.2	0.54	3
1996.49	46.9	0.87	5	2010.56	174.8	0.69	3
1997.52	53.1	0.86	6	2011.58	181.3	0.74	2
1998.54	57.2	0.88	8	2012.55	190.3	0.79	6
1999.52	61.6	0.80	7	2013.56	197.5	0.68	3
2000.48	68.5	0.79	6	2014.59	208.4	0.67	6
2001.48	74.9	0.73	7	2015.66	212.5	0.62	2
2002.52	81.7	0.63	6	2017.59	231.0	0.56	2

which, despite its large apparent angular separation, appears to be a genuine binary pair, although an orbit of 760 centuries in the catalogue might seem to be a leap of faith. Three point six degrees SSW of η is the binary pair STF 1932 (7.3, 7.4, 267°, 1″.6, 2020), which orbits much more swiftly – every 203 years. RWA has seen the companion to θ CrB = COU 610 a couple of times with the Cambridge 20-cm but always with difficulty. The WDS seems to show a gradual widening, so it's worth trying now and again (4.3, 6.3, 199°, 0″.8, 2016).

Measures

Early measure (STF)	43°.3	0″96	1829.55
(Orbit	45°.6	0″.94)	
Recent measure (ARY)	212°.5	0″.62	2015.66
(Orbit	211°.4	0″.62)	

112. γ CIR = HJ 4757 = WDS J15234–5919AB

Table 9.112 Physical parameters for γ Cir

HJ 4757	RA: 15 23 22.66	Dec: −59 19 14.5	WDS: 1234(67)	
V magnitudes	A: 4.83	B: 6.93		
(*B* − *V*) magnitudes	A: −0.11	B: +1.22		
μ	A: −12.97 mas yr^{-1}	± 0.77	B: −34.24 mas yr^{-1}	± 0.70
π	7.27 mas	± 0.81	449 light yr	± 50 (Hipp.)
Spectra	A: B5V + B5V?	B: F8V		
Luminosities (L$_\odot$)	A: 190	B: 25		
Catalogues (A/B)	HR 5704	HD 136415/6	SAO 242463	HIP 75323
DS catalogues	HJ 4757			
Radial velocity	−16.90 km s^{-1}	± 1.78		
Galactic coordinates	321°.245	−1°.963		

History

John Herschel, using his 20-foot (18-inch aperture) reflector, notes on sweep 575 (22 April 1835): 'Pos 109.8, 110.4. A momentary glimpse with 180 induced me to apply 320 with 12-inches aperture. It was then cleanly divided, and these are good measures.' He gives the separation as 3/4″ and later on during sweep 717 (9 July 1836): 'A beautiful object. Pos 104.4 101.3. The first with 480, which separates the stars by $\frac{1}{4}$ diameter; the other with the same power and an aperture reduced to 6 inches.' On this occasion the separation is given as 1″. He also makes measures on five nights over a two-year period with the 7-foot (6-inch aperture) reflector, which show the angle scattered between 102° and 109°.

The Modern Era

The parallax derived from the orbit by Hartkopf [468] is substantially different to that from Hipparcos. This led

Finder Chart

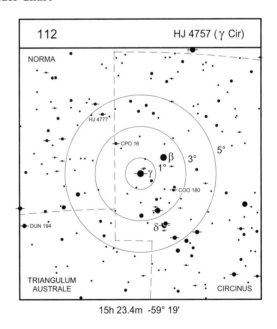

15h 23.4m -59° 19′

Orbit

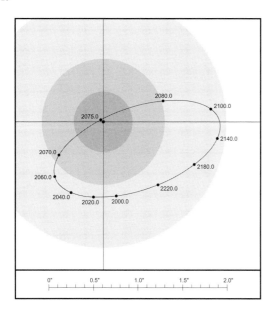

Ephemeris for HJ 4757 (2015 to 2060)

Orbit by Hrt (2010a) Period: 258.00 years, Grade: 4

Year	PA(°)	Sep(")	Year	PA(°)	Sep(")
2015.0	356.8	0.82	2040.0	335.0	0.85
2020.0	352.3	0.83	2045.0	330.8	0.85
2025.0	347.9	0.83	2050.0	326.6	0.85
2030.0	343.5	0.84	2055.0	322.3	0.84
2035.0	339.2	0.85	2060.0	317.8	0.81

Parsons [467] to speculate that the primary is a close pair of B5 dwarfs, and he uses the photometry from Tycho to add weight to his conclusion.

Observing and Neighbourhood

This was found about 6° following and 1°.5 N of α Cen. Although 15-cm might suffice to divide the close pair, the significant difference in magnitude might require 20-cm for a good view. The distance has remained unchanged for several decades with the position angle reducing by about 1° per year over the same time interval. *CDSA2* shows a number of available target pairs within 3° or 4°. One point five degrees WSW of γ is COO 180 (7.6, 9.2, 288°, 12″.4, 2004). About 1° NE will bring you to CPO 16, a multiple star with five components for the medium aperture. The central binary has magnitudes of 7.0 and 8.0 and is at 34° and 2″.6 (2016) and is slowly widening. A magnitude 10.2 can be found at 67″ and two further stars of magnitudes 11.7 and 12.9 are also nearby. A further 1° NE and you alight on HJ 4777 (7.5, 9.1, 295°, 5″.6, 2013).

Measures

Early measure (I)	79°.1	1″.11	1900.36
(Orbit	75°.8	1″.23)	
Recent measure (ANT)	359°.4	0″.79	2014.25
(Orbit	357°.3	0″.82)	

113. $\mu^{1,2}$ BOO = STFA 28
= WDS J15245+3723AB

Table 9.113 Physical parameters for μ Boo

STFA 28	RA: 15 24 29.43	Dec: +37 22 37.756	WDS: 19(735) (BC)		
V magnitudes	AaAb(μ^1): 4.33	B(μ^2): 7.09	C: 7.63		
$(B - V)$ magnitudes	AaAb: +0.31	B: +0.63	C: +0.65		
μ(A)	−149.42 mas yr^{-1}	± 0.99	93.96 mas yr^{-1}	± 1.04 (DR2)	
μ(B)	−139.14 mas yr^{-1}	± 0.05	90.26 mas yr^{-1}	± 0.04 (DR2)	
μ(C)	−152.39 mas yr^{-1}	± 0.04	89.78 mas yr^{-1}	± 0.05 (DR2)	
π(A)	28.08 mas	± 0.59	116.2 light yr	± 2.4 (DR2)	
π(B)	27.15 mas	± 0.03	120.1 light yr	± 0.1 (DR2)	
π(C)	27.23 mas	± 0.27	119.8 light yr	± 1.2 (DR2)	
Spectra	Aa: F0IV + ?	B: G0V	C: ?		
Masses (M$_\odot$)	Aa: 2.7	Ab: 1.5	B: 1.1	C: 1.0	
Luminosities (L$_\odot$)	Aab: 20	B: 1.5	C: 1		
Catalogues (A/B)	HD 137391/2	HR 5733/4	SAO 64686/7	HIP 75411/5	
DS catalogues	CHR 181 (AaAb)	H 6 17 (A-BC)	STFA 28 (A-BC)	BDS 7258	ADS 9626
DS catalogues	H 1 17 (BC)	STF1938 (BaBb)			
Radial velocity (Aa)	8.60 km s^{-1}	± 0.3			
Radial velocity (Bb)	8.50 km s^{-1}	± 0.3			
Galactic coordinates	60°.394	+56°.316			

History

On 30 July 1780, Sir William Herschel first noted that μ Boo consists of two bright stars. He described the colours as 'L(arge) reddish w(hite). S(mall) pale r(ed)' The distance was given as 2′8″, which Herschel regarded as an exact estimate. This appears to be an error as the stars have remained fixed at 108″ since the mid 1820s. Returning to the area on 10 September 1781, he then noticed that the fainter of the two was also double and they were 'Both dusky w(hite) inclined to r(ed)'. As it transpired later, the stars in this closer pair were then at their greatest separation.

The Modern Era

Between 1850 and 1880 the stars that make up component B remained at about 0″.6 and moved 160° retrograde in position angle. Since then they have been widening and reached widest separation in 2010. In a large radial velocity project, using spectra taken at McDonald Observatory around 1960, H. A. Abt [469] and colleagues in 1965 took 13 spectra of μ^1 Boo. Although the spectra showed very broad lines due to rapid rotation, they concluded that the derived velocities fitted a periodic variation with a period of 299 days and an eccentricity of 0.54. Niehaus & Scarfe [470] re-observed

Finder Chart

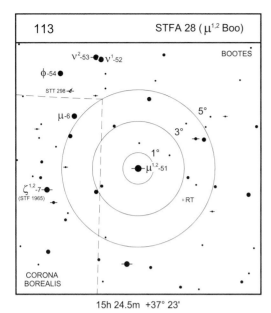

113   STFA 28 ($\mu^{1,2}$ Boo)

BOOTES

ν^2-53 ν^1-52

ϕ-54

STT 298

μ-6

5°

3°

1°

$\mu^{1,2}$-51

$\zeta^{1,2}$-7
(STF 1965)

RT

CORONA
BOREALIS

15h 24.5m +37° 23'

Orbit

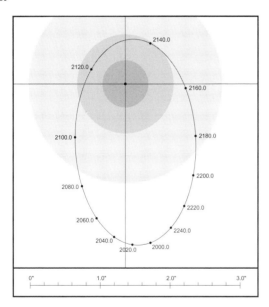

2140.0
2120.0
2160.0
2100.0
2180.0
2200.0
2080.0
2220.0
2060.0
2240.0
2040.0
2020.0 2000.0

0"   1.0"   2.0"   3.0"

Ephemeris for STF 1938 BC (2010 to 2100)

Orbit by Kiy (2014) Period: 265 years, Grade: 2

Year	PA(°)	Sep(")	Year	PA(°)	Sep(")
2010.0	5.9	2.25	2060.0	347.5	1.91
2020.0	2.6	2.24	2070.0	342.6	1.75
2030.0	359.2	2.20	2080.0	336.4	1.55
2040.0	355.6	2.14	2090.0	328.2	1.31
2050.0	351.8	2.04	2100.0	315.9	1.03

μ^1 Boo and found no significant variation; the errors in the velocities amounted to 3 or 4 km s^{-1} due to the width of the lines. In a postscript to this paper, however, they said that recent new observations received from Dr Abt, combined

with their own spectra, could be interpreted as representing a spectroscopic binary with a period of 4500 days. In 1988, using the 3.6-metre CFHT on Hawaii, McAlister et al. [482] and colleagues found that μ^1 was a close pair, and they recorded a separation of 0″.069 with a position angle of 350°.3. In 2010 Muterspaugh et al. [620] published a series of very precise measurements of this pair (giving errors in separation in the region of 0″.0002) and found a period of 1368.02 days or 3.75 years. The separation of the stars ranges from 0″.05 to 0″.11 during this cycle. The note in Webb [209] says that μ^1 and μ^2 together 'form one vast system' but this idea is disputed in a recent astrometric study of the group made by Kiyaeva et al. [473]. They derived orbits for the interferometric pair AaAb (CHR 181) and the visual binary BC, the ephemeris from which appears above. Kiyaeva also suggested that the two systems μ^1 and μ^2 are not related but happen to be passing in space, because there are significant differences in element abundances between the two pairs.

Observing and Neighbourhood

The stars μ^1 and μ^2 form an easy binocular double. The BC pair has now almost returned to the discovery position in 1782, is well resolved in 75-mm, and will remain so for most of the rest of the century. Smyth noted that μ^1 was flushed-white whilst the components of μ^2 were both greenish-white. Hartung notes pale yellow for the bright star and deep-yellow for the visual binary companion. Nearby are ζ CrB (STF 1965) (see Star 116), and STT 298 (see Star 115). The components of μ Boo – STFA 28 – although very different in brightness and at a large angular separation are moving through space together and have the same parallax, according to DR2 (equivalent to a distance of 120 light years). The WDS notes that an orbit is extant but this does not appear in the Sixth Orbit Catalogue.

Measures

STF1938

Early measure (STF)	324°.1	1″.25	1829.73
(Orbit	323°.1	1″.18)	
Recent measure (ARY)	3°.7	2″.31	2016.53
(Orbit	3°.7	2″.25)	

STFA 28

Early measure (STF)	172°.6	108″.73	1821.78
Recent measure (ARN)	170°.9	108″.13	2012.33

114. γ LUP = HJ 4786 = WDS J15351−4110AB

Table 9.114 Physical parameters for γ Lup

HJ 4786	RA: 15 35 08.45	Dec: −41 10 00.3	WDS: 666(102)		
V magnitudes	Aab: 3.44	B: 3.54	C: 17.0	D: 16.4	
(*B − V*) magnitudes	Aab: −0.18	B: −0.10			
μ	A: −15.62 mas yr^{-1}	± 0.69	−25.43 mas yr^{-1}	± 0.43	
π	7.75 mas	± 0.50	420 light yr	± 27	
Spectra	Aa: B2IV	Ab: —	B: —	C: F1?	D: K5?
Masses (M$_\odot$)	Aa: 9.7	Ab: 1.1	B: 9.5	C: 0.4	D: 0.5
Luminosities (L$_\odot$)	A: 600	B: 550	C: 0.002	D: 0.002	
Catalogues	HR 5776	HD 138690	SAO 225938	HIP 76297	
DS catalogues	HJ 4786 (AB)	RIZ 17 (AaAb)	RIZ 15 (AC)	RIZ 15 (AD)	
Radial velocity	2.3 km s^{-1}	± 5			
Galactic coordinates	333°.194	+11°.891			

History

John Herschel alighted upon γ Lupi on his sweep 718 (10 July 1836) at Feldhausen. 'Cleanly divided with 480, and the black division well seen. Measures perfectly good and to be depended on.' He gives the distance as 2/3 of an arcsecond. Herschel continues 'With 800 well separated; with 180 and triangular aperture perceived to be double. [This shows the distance to be underrated at two-thirds of a second.]' Herschel observes this star 15 times during his work at the Cape, but concluded that 'there is no evidence for angular motion which can be relied on'.

The Modern Era

The multiplicity of stars on OB associations has been pursued by a number of groups of astronomers in the last few decades. In 1987 Levato *et al.* [474] using the 0.9-metre and

Finder Chart

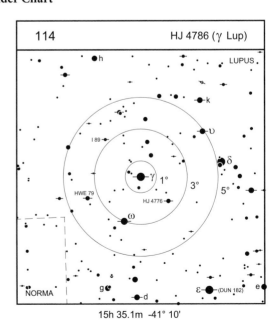

15h 35.1m -41° 10'

Orbit

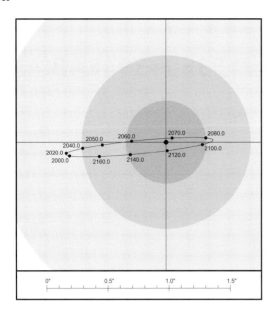

Ephemeris for HJ 4786 AB (2015 to 2060)

Orbit by Hei (1990c) Period: 190 years, Grade: 3

Year	PA(°)	Sep(")	Year	PA(°)	Sep(")
2015.0	276.6	0.83	2040.0	274.0	0.68
2020.0	276.1	0.82	2045.0	273.3	0.61
2025.0	275.7	0.80	2050.0	272.4	0.52
2030.0	275.2	0.78	2055.0	271.0	0.41
2035.0	274.7	0.74	2060.0	268.4	0.28

1.5-metre reflectors at CTIO, and their associated spectrographs in May 1974, found a number of new spectroscopic binaries. Amongst this list was γ Lupi. This was also one of the stars chosen by Rizzuto et al. [475] when they looked at stars in the Sco–Cen OB association using the SUSI optical interferometer at Narrabri in Australia. They found a third star at a PA of 89°.6 and 0″.026. Rizzuto et al. also looked at the wider field around each target and found two very faint stars at 39″ and 53″ from AB, which Andrei Tokovinin in his MSC suggests may be physical. The true orbit of γ Lupi is highly inclined to the line of sight so that the apparent orbital motion is mostly confined to changes in distance. The closest approach is 0″.03 in 2068 if the orbit is correct, but it will only be after that event that the orbital elements can be regarded as reliable.

Observing and Neighbourhood

In terms of brightness the stars of this system are equivalent to those in γ Virginis. However, the closest approach in 2069 brings the components within 0″.04 but the pair remains within reach of 20-cm for the next two decades or so. The star HJ 4776 can be found SW (6.3, 8.4, 228°, 5″.6, 2010), whilst 1°.5 NE is I 89 (6.8, 8.1, 169°, 1″.5, 2016), which has increased in PA by 20° in the century since discovery. The star HWE 79 is 1°.8 ESE (6.1, 7.9, 338°, 3″.7, 2011).

Measures

Early measure (HWE)	274°.0	0″.59	1856.17
(Orbit	275°.4	0″.70)	
Recent measure (ARY)	277°.4	0″.95	2016.65
(Orbit	276°.3	0″.83)	

115. STT 298 BOO = WDS J15360+3948AB

Table 9.115 Physical parameters for STT 298 Boo

STT 298	RA: 15 36 02.22	Dec: +39 48 08.9	WDS: 40(537)	
V magnitudes	A: 6.77	B: 8.44	C: 7.56	
$(B - V)$ magnitudes	A: +1.06	B: +1.30	C: +0.97	
μ(A)	−442.57 mas yr^{-1}	± 0.39	55.83 mas yr^{-1}	± 0.47 (DR2)
μ(B)	−463.88 mas yr^{-1}	± 0.39	45.74 mas yr^{-1}	± 0.61 (DR2)
μ(C)	−449.10 mas yr^{-1}	± 0.04	50.78 mas yr^{-1}	± 0.06 (DR2)
π(A)	44.80 mas	± 0.23	72.8 light yr	± 0.4 (DR2)
π(B)	44.51 mas	± 0.02	73.28 light yr	± 0.03 (DR2)
π(C)	43.99 mas	± 0.21	74.1 light yr	± 0.4 (DR2)
Spectra	A: K1V	B:	C: K3V	
Masses (M$_\odot$)	A: 0.8	B: 0.7	C: 0.9	
Luminosities (L$_\odot$)	A: 0.8	B: 0.2	C: 0.4	
Catalogues (AB/C)	HD 139341/23	SAO 64800/799	HIP 76382/75	
DS catalogues	STT 298	BDS 7732	ADS 9716	
Radial velocity	−66.7 km s^{-1}	± 0.6		
Galactic coordinates	64°.113	+53°.779		

History

Discovered by Otto Struve in 1846 with the 15-inch at Pulkovo when close to its widest separation, this binary was observed infrequently until at the end of the 1860s it suddenly began to close quickly and then attracted the attention of the observers of the time. By 1877 the period was given as 56.653 years with rather unwarranted precision.

The Modern Era

The V magnitudes quoted by Hipparcos and given above seem at odds with the visual estimates of Δm recorded in the WDS, when the stars never seem to be more than about 0.3 magnitudes different, and agree with the experiences of RWA when measuring this system. The magnitude 7.6 star at 121″ distance (HIP 76375) is a physical member of the system with almost 100% probability, according to Shaya & Olling [270].

Finder Chart

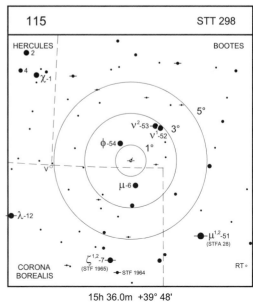

15h 36.0m +39° 48'

Orbit

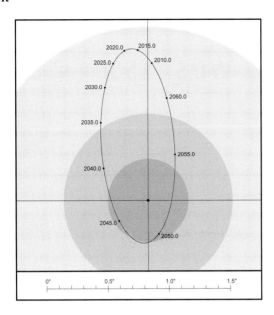

Ephemeris for STT 298 AB (2015 to 2042)

Orbit by Sod (1999) Period: 55.6 years, Grade: 1

Year	PA(°)	Sep(")	Year	PA(°)	Sep(")
2015.0	183.9	1.20	2030.0	201.4	0.96
2018.0	187.0	1.21	2033.0	207.0	0.83
2021.0	190.1	1.19	2036.0	215.0	0.67
2024.0	193.3	1.15	2039.0	228.2	0.50
2027.0	197.0	1.07	2042.0	254.9	0.34

The distance is the same as that to STT 298 AB and the proper motions are very similar. Component D (at 224°, 167″) in the WDS is actually the galaxy NGC 5966.

Observing and Neighbourhood

This binary is quite easily swept up by starting at β Boo at the top of the 'kite' and moving E by 6° to arrive at the wide and bright binocular pair $\nu^{1,2}$ Bootis (magnitudes 5.0, 5.0, separation 10′.5), which sports colours of orange and white. Moving a further 1.5° SE will bring you to a fainter wide pair of stars. This is STT 298AB and C, the binary pair being the northern component. At the time of writing STT 298 is close to its maximum separation and should remain resolvable in 15-cm for a decade or so. Four degrees SW is the fine triple $\mu^{1,2}$ Bootis (see Star 113) and 4° SSE are $\zeta^{1,2}$ CrB (STF 1965) (see Star 116) and STF 1964.

Measures

	PA	Sep	Epoch
Early measure (STF)	183°.8	1″.19	1846.49
(Orbit	183°.0	1″.17)	
Recent measure (ARY)	185°.8	1″.18	2014.65
(Orbit	183°.5	1″.19)	

116. $\zeta^{1,2}$ CRB = STF 1965 = WDS J15394+3638

Table 9.116 Physical parameters for ζ CrB

STF 1965	RA: 15 39 22.68	Dec: +36 38 07.0	WDS: 54(461)		
V magnitudes	A (ζ^2): 4.96	B (ζ^1): 6.00			
$(B-V)$ magnitudes	A: −0.11	B: 0.00			
μ(A)	−15.62 mas yr^{-1}	± 0.27	−7.31 mas yr^{-1}	± 0.34 (DR2)	
μ(B)	−14.20 mas yr^{-1}	± 0.09	−3.87 mas yr^{-1}	± 0.13 (DR2)	
π(A)	6.64 mas	± 0.18	491 light yr	± 13 (DR2)	
π(B)	6.22 mas	± 0.07	524 light yr	± 6 (DR2)	
Spectra	A: B7V + B7V + ?	B: B7V?			
Masses (M$_\odot$)	A: 1.156	± 0.14	B: 1.05	± 0.14	
Luminosities (L$_\odot$)	A: 200	B: 85			
Catalogues	7 CrB	HR 5834	HD 139891	SAO 64834	HIP 76669
DS catalogues	H 2 8	STF 1965	BDS 7352	ADS 9737	
Radial velocity	−0.00 km s^{-1}	± 0.4			
Galactic coordinates	58°.698	+53°.412			

History

Found by Sir William Herschel on 1 October 1779. He noted that the stars were considerably unequal and that the primary was a fine white with the fainter star white inclining to red. In his *Cycle of Celestial Objects*, Admiral Smyth said 'I have obtained a splendid set of measures at Hartwell House, by which it is rendered still more evident, that these stars are optical, and relatively at rest. Indeed, there are few of these objects whose details come out so satisfactorily.' He made the colours bluish white and smalt blue. More recently Sissy Haas finds both stars white. The spectroscopic duplicity of ζ^2 CrB was discovered by W. H. Christie from a plate taken on 4 May 1924 with the 72-inch reflector of the Dominion Astrophysical Observatory at Victoria, British Columbia. More plates were taken and an orbit was calculated by J. S. Plaskett [476], who found a double-line system with a period of 12.60 days and an orbit of very low eccentricity.

Finder Chart

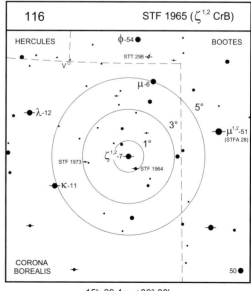

The Modern Era

Although the proper motion of each star is only 15 mas per year, over 200 years that would have separated the pair by about $3''$ so it is clear from the fixed nature of the separation that the stars are physical; in addition, new evidence from Gaia DR2 determines that the parallaxes of the bright two components are almost equal. The spectroscopic orbit of ζ^2 was revisited by Abhyankar & Sarma [477], who, using plates taken with the 74-inch David Dunlap telescope, found essentially the same elements. They speculated about the presence of a third body and an investigation by Gorden & Mulliss [478] did indeed find that the situation was more complex. They found spectroscopic evidence for three stars, a close pair with an orbital period of 1.723 days, which corresponded to Plaskett's system, and a more distant third star orbiting them with a period of 251 days. They also found that the components of the short-period binary were rotating at significantly different speeds (46 and 7.5 km s^{-1} respectively), posing questions about why both stars had not achieved synchronization between their orbital and rotational speeds. Both ζ^1 and ζ^2 have been observed with a large aperture and a speckle camera but no visual companions wider than $0''.05$ have been found.

Observing and Neighbourhood

The star can be found $10°$ N and slightly E of α CrB. It is a bright and easy pair in small apertures, in which it will appear white. In the same finder field lies STF 1964. This contains some good tests for 20- and 30-cm apertures. The bright STF stars are magnitudes 8.1 and they are $14''.9$ apart on PA $86°$ (2014). Star C itself has a companion of magnitude 9 sitting just $1''.7$ away in PA $223°$ (2016), the measuring of which RWA found to be a stiff test for the 8-inch refractor with a bright field. Hussey found another companion to A of magnitude 10 at a distance of $1''.3$, but a larger aperture is needed – certainly it has never been seen in the 8-inch. The WDS also records a single observation of a $0''.1$ companion to C which has never been confirmed. About $1°.5$ E and slightly S is STF 1973 (7.6, 8.8, $320°$, $30''.8$, 2015).

Measures

Early measure	STF	$300°.8$	$6''.02$	1828.79
Recent measure	RWA	$307°.4$	$6''.24$	2016.52

117. Δ 194 TRA = WDS J15549–6045AB,C

Table 9.117 Physical parameters for Δ 194 TrA

DUN 194	RA: 15 54 52.64	Dec: −60 44 37.1	WDS: 11487(24)	
V magnitudes	A: 6.35	B: 8.09	C: 9.97	D: 9.02
$(B − V)$ magnitude	A: +0.12	B: −0.04	C: 0.00	D: +0.01
μ(A)	−0.70 mas yr^{-1}	± 0.11	−3.84 mas yr^{-1}	± 0.13 (DR2)
μ(B)	−1.44 mas yr^{-1}	± 0.41	−1.77 mas yr^{-1}	± 0.54 (DR2)
μ(C)	−3.70 mas yr^{-1}	± 0.06	−6.22 mas yr^{-1}	± 0.06 (DR2)
μ(D)	−0.63 mas yr^{-1}	± 0.06	−4.20 mas yr^{-1}	± 0.06 (DR2)
π(A)	1.16 mas	± 0.09	2811 light yr	± 218 (DR2)
π(B)	2.97 mas	± 0.42	1098 light yr	± 155 (DR2)
π(C)	1.88 mas	± 0.05	1735 light yr	± 46 (DR2)
π(D)	1.16 mas	± 0.04	2811 light yr	± 97 (DR2)
Spectra	A: B9II	B:	C:	D: B5/7III
Luminosities (L$_\odot$)	A: 1750	B: 55	C: 25	D: 150
Catalogues	HR 5898	HD 141913	SAO 253344	HIP 77927
DS catalogues	SLR 11 (AB)	DUN 194 (AB,C)	HJ 4809 (AB,C)	
Radial velocity	−5.00 km s^{-1}	± 4.3		
Galactic coordinates	323°.561	−5°.446		

History

First catalogued by James Dunlop, this system was recovered by John Herschel in 1836 when he listed the two ninth magnitude companions, one of which (D) Dunlop appears to have missed. Herschel does not acknowledge Dunlop's previous observation and allocates HJ 4809 to the three stars. He does not notice the duplicity of A, which was not found until 1891 when Sellors was using the 11.5-inch refractor at Sydney and gave the measured separation as 0″.56. Two years later he found the separation to be 1″.0 and it has remained between 1″ and 1″.2 ever since.

The Modern Era

Sinbad gives A as a variable star; it has been observed to vary between V = 6.11 and 6.15 [479]. The proper motions of the four brightest stars in the system are all similar but very small, reflecting the distance of the group.

Observing and Neighbourhood

The two distant companions are within range of large binoculars whilst the close pair needs at least 10-cm. Star C is

Finder Chart

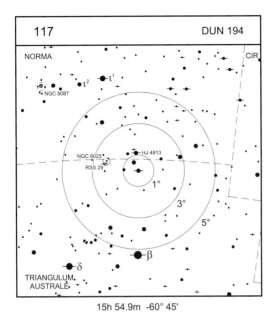

117   DUN 194

NORMA

CIR

ι²   ι¹

NGC 6087

NGC 6025   HJ 4813

RSS 29

1°

3°

5°

β

δ

TRIANGULUM
AUSTRALE

15h 54.9m -60° 45'

45″ from A in PA 47° (2016), whilst D is 49″ away in PA 256° (2016). Richard Jaworski [378] likens the group to the shape of a boomerang, with the bright star at the apex. Hartung reports that the Sellors pair could be divided with 105-mm. Since then (1960) it has slowly widened. The system sits in a rich star field, amongst which can be found HJ 4813 (5.9, 8.4, 103°, 4″.8 (widening), 2016) and RSS 29 (7.3, 8.1, 81°, 53″, 2016), which is embedded in the galactic cluster NGC 6025.

Measures

SLR 11 (AB)			
Early measure (SLR)	91°.4	1″.09	1895.54
Recent measure (ARY)	94°.6	1″.25	2016.70
DUN 194 (AC)			
Early measure (HJ)	50°.9	43″.47	1836.24
Recent measure (ARY)	47°.3	44″.10	2016.70
DUN 194 (AD)			
Early measure (HJ)	257°.0	49″.30	1836.24
Recent measure (ARY)	256°.9	48″.43	2016.70

118. ξ SCO = STF 1998
= WDS J16044−1122AB,C

Table 9.118 Physical parameters for ξ Sco

STF 1998	RA: 16 04 22.13	Dec: −11 22 27.2	WDS(AB): 31(610)		
V magnitudes	A: 5.10	B: 4.87	C: 7.30	D: 7.43	E: 7.97
(B − V) magnitudes	+0.48	B: 0.48	C: -	D: +0.86	E: +0.98
μ(A)	−41.94 mas yr^{-1}	± 0.51	−19.16 mas yr^{-1}	± 0.37 (DR2)	
μ(B)	−74.34 mas yr^{-1}	± 0.45	−32.91 mas yr^{-1}	± 0.34 (DR2)	
μ(C)	−75.04 mas yr^{-1}	± 0.09	−12.03 mas yr^{-1}	± 0.06 (DR2)	
μ(D)	−61.59 mas yr^{-1}	± 0.08	−22.25 mas yr^{-1}	± 0.04 (DR2)	
μ(E)	−56.39 mas yr^{-1}	± 0.09	−20.32 mas yr^{-1}	± 0.05 (DR2)	
π(A)	36.24 mas	± 0.26	90.0 light yr	± 0.6 (DR2)	
π(B)	35.31 mas	± 0.23	92.4 light yr	± 0.6 (DR2)	
π(C)	35.82 mas	± 0.05	91.1 light yr	± 0.1 (DR2)	
π(D)	35.91 mas	± 0.04	90.8 light yr	± 0.1 (DR2)	
π(E)	35.84 mas	± 0.05	91.0 light yr	± 0.1 (DR2)	
Spectra	A: F5IV	B: F6V	C: G1V	D: G8V	E: K0V
Masses (M$_\odot$)	A: 1.5	B: 1.5	C: 1.0	D: 1.0	E: 0.9
Radii (R$_\odot$)	A: 1.5	B: 2.5	C: 0.9		
Luminosities (L$_\odot$)	A: 6	B: 7	C: 0.8	D: 0.7	E: 0.4
Catalogues (A/B/C)	HR 5978/7/—	HD 144070/69/—	SAO 159665/5/6	HIP 78727	
Catalogues (D/E)	HD 144087/8	SAO 159668/70	HIP 78738/9		
DS catalogues (ABC)	H 1 33 (AB)	H 2 20 (AB,C)	STF 1998	BDS 7487	ADS 9909
DS catalogues (DE)	H 2 21	STF 1999	BDS 7488	ADS 9910	
Radial velocity (AB)	−36.37 km s^{-1}	± 0.83			
Radial velocity (C)	−31.5 km s^{-1}	± 5			
Radial velocity (D)	−31.60 km s^{-1}	± 0.2			
Radial velocity (E)	−32.10 km s^{-1}	± 0.4			
Galactic coordinates	0°.020	+29°.424			

History

Bayer called this system ξ Sco but Flamsteed allocated the magnitude 4.2 star to Libra and called it 51 Lib. This was an error, as the star is certainly in Scorpius and was reallocated to that constellation by Argelander. On 23 May 1710 (presumably 1780 is meant) William Herschel noted the wider component of ξ Sco and a nearby fainter and wider

Finder Chart

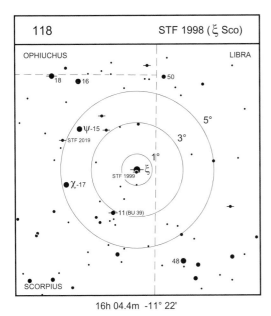

16h 04.4m −11° 22′

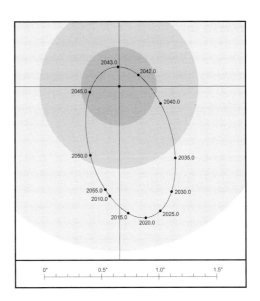

Ephemeris for STF 1998 AB (2018 to 2036)

Orbit by Doc (2009g) Period: 45.90 years, Grade: 1

Year	PA(°)	Sep(″)	Year	PA(°)	Sep(″)
2018.0	8.6	1.11	2028.0	23.7	1.05
2020.0	11.6	1.13	2030.0	27.3	0.99
2022.0	14.5	1.13	2032.0	31.3	0.92
2024.0	17.4	1.11	2034.0	36.2	0.82
2026.0	20.5	1.09	2036.0	42.5	0.71

pair as follows: 'Double double. The first set very unequal. L(arge). fine w(hite). With 227, nearly 2 diameters of L. By the micrometer 6″ 23 ‴, but too large a measure position 1° 23′ n(orth), following. The other set both small and obscure. With 227, perhaps 5 or 6 of their diameters asunder.' There is a footnote to say that the large star is a close double, and the details appear in the second catalogue of 1784. The entry for 12 May 1782 reads 'Treble. Without great attention, and a considerable power, it may be mistaken for a double star; but the largest of them consists of two. Both w(hite). with $460\frac{1}{4}$ or at most $1\frac{1}{3}$ diameter asunder; with 932, full $\frac{1}{3}$ diameter of L(arge). or near $\frac{1}{2}$ diameter of S(mall). Position, with 278, 82° 2′ n(orth). following.' By 1846, Maedler had computed an orbit for AB even though the motion since Struve in 1825 amounted to only about 30°, apart from the discovery measure of Herschel, which taken at face value gave a period of 105 years. The equality of the components meant that the quadrant in which the secondary was located was uncertain. In 1858 Captain Jacob [480] had observed the stars at or below 0″.4 and pointed out that if the circular orbit mentioned by Smyth was correct, this separation could not happen. Jacob considered that the orbit was highly elongated, with a period of 52 years, but no one took any notice. For the next 50 years, orbit computers continued to assume that the stars had a near circular apparent orbit and it was left to Robert Aitken [481] to clarify the situation. By reversing the position angles of the early measures he showed that the orbit was both very eccentric and much shorter, according to him 44.5 years.

Orbit for STF 1998 AB

Orbit for STF 1998 AC

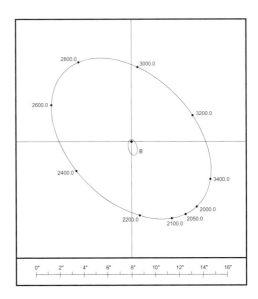

Ephemeris for STF 1998 AC (2000 to 2180)

Orbit by Zir (2008) Period: 1514.43 years, Grade: 5

Year	PA(°)	Sep(″)	Year	PA(°)	Sep(″)
2000.0	46.3	7.57	2100.0	28.8	7.04
2020.0	43.0	7.53	2120.0	24.9	6.86
2040.0	39.6	7.45	2140.0	20.8	6.65
2060.0	36.2	7.34	2160.0	16.4	6.44
2080.0	32.6	7.20	2180.0	11.6	6.21

The Modern Era

The system is now known to be quintuple. Aitken noted that micrometer measures had been made between AB of STF 1998 and star A in STF 1999 but stopped short of concluding that the two pairs were moving through space together, a fact confirmed by the results from Hipparcos and Gaia.

Observing and Neighbourhood

The system ξ Sco is always a challenge from northern Europe or the UK because of the low altitude of the stars in the sky. At present, they are near widest separation and now is a good opportunity to see this short-period visual system. Use preferably 15-cm. The maximum separation of 1″.13 occurs in 2021, ducks beneath 1 arcsecond by 2030 and reaches 0″.16 by 2043 before widening again. One way to find them is to start at Graffias (β Sco) and move due N by 18°. Two degrees SSE is 11 Sco (BU 39, 5.8, 9.8, 263°, 3″.3), one of S. W. Burnham's earliest discoveries with the 6-inch Clark, which he found on 23 June 1872 (the first was BU 40 on 27 April 1870). The faintness and closeness of the companion suggests that 20-cm would be needed for a definite view of the two stars. There is another STF pair nearby – STF 2019 (8.0, 9.7, 143°, 23″.2). The primary was doubled by Robert Rossiter (RST 3936) into two stars of magnitude 8.2 separated by 0″.2.

Measures

AB			
Early measure (STF)	4°.4	1″.21	1831.38
(Orbit	4°.8	1″.07	
Recent measure (ANT)	6°.2	1″.10	2016.33
(Orbit	6°.0	1″.09)	
AC			
Early measure (STF)	81°.3	7″.53	1823.73
(Orbit	76°.6	6″.86)	
Recent measure (ANT)	41°.6	8″.00	2016.33
(Orbit	43°.5	7″.54)	
DE = STF 1999			
Early measure (STF)	101°.3	10″.11	1831.49
Recent measure (ANT)	98°.4	11″.86	2016.33

Table 9.119 Physical parameters for ν Sco

H 5 6	RA: 16 11 59.74	Dec: −19 27 38.3	WDS: 921(82)		
V magnitudes	A: 4.35	B: 5.31	C: 6.60	D: 7.23	
(*B* − *V*) magnitudes	A: −0.02	B: +0.07	C: +0.12	D:	
μ(A)	−6.87 mas yr^{-1}	± 0.64	−28.26 mas yr^{-1}	± 0.48 (DR2)	
μ(B)	−8.58 mas yr^{-1}	± 0.62	−26.54 mas yr^{-1}	± 0.50 (DR2)	
μ(C)	−11.94 mas yr^{-1}	± 0.12	−24.40 mas yr^{-1}	± 0.09 (DR2)	
μ(D)	−8.03 mas yr^{-1}	± 0.37	−24.05 mas yr^{-1}	± 0.31 (DR2)	
π(A)	7.36 mas	± 0.35	443 light yr	± 21 (DR2)	
π(B)	7.42 mas	± 0.28	440 light yr	± 17 (DR2)	
π(C)	7.09 mas	± 0.07	460 light yr	± 5 (DR2)	
π(D)	6.47 mas	± 0.21	504 light yr	± 16 (DR2)	
Spectra	A: B2V +? +?	B: ?	C: B8V	D: ?+?	
Luminosities (L$_\odot$)	A: 280	B: 110	C: 35	D: 30	
Catalogues (A/C)	14 Sco	HR 6027/6	HD 145502/1	SAO 159763	HIP 79374
DS catalogues	Mayer 40 (AC)	H 5 6 AC	BU 120 (AB)	MTL 2 (CD)	CHR 146 (AaAb)
	BDS 7533	ADS 9951			
Radial velocity	2.4 km s^{-1}	± 5			
Galactic coordinates	354°.609	+22°.700			

History

Discovered by Christian Mayer in 1777, this beautiful, wide, and unequal pair was swept up two years later by William Herschel. He noted on 19 September 1779 'Double. Very unequal. Both w(hite). Distance 38″ 20‴, pretty accurate. Position 69° 28′ n(orth), preceding'. Smyth notes the colours as bright white and pale lilac but does not mention any further components. The star culminates in the UK at about 20° altitude but the lack of multiplicity at that time may be significant. In 1846 Ormsby M. Mitchel, using an 11-inch Munich refractor at Cincinnati Observatory, noticed that C was double but his observation was not reported for some time afterwards. Mitchel (1810–1862) studied at West Point, the American military academy, and later taught mathematics there. He joined Cincinnati College in 1836 and by 1845 had raised enough funds to build a dome, in which he placed an 11-inch refractor which was purchased in Germany. It was at the time the second biggest instrument in the world. With it he was able to do some double star micrometry, but when the Civil War broke out he joined the Union Army and was appointed a Brigadier General. He never returned to Cincinnati, dying of yellow fever in South Carolina. This pair was independently found by Jacob in 1847.

Finder Chart

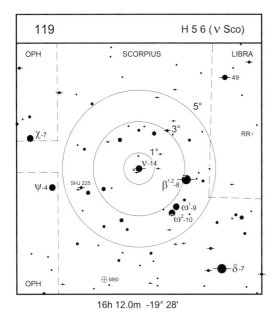

119   H 5 6 (ν Sco)

OPH   SCORPIUS   LIBRA

49

5°

χ-7   3°   RR°

1°

ν-14

ψ-4   SHJ 225   β¹·²-8

ω¹-9

ω²-10

δ-7

OPH   M80

16h 12.0m -19° 28'

The Modern Era

In the late 1980s another component was discovered close to A by H. A. McAlister [482] and colleagues. The new component was found 58 mas distant in PA 172° and it was again resolved in the following year. In 2008 attempts to observe the two stars with the 8.1-metre *Gemini* telescope failed, but it was picked up again in 2009–2011 at roughly the same distance as before but on the other side of the brighter star, thus suggesting a highly inclined apparent orbit. Despite the apparently short period, there is no orbit for this system. The visual pairs BU 120 and MTL 2 also show changes, admittedly quite small but mostly in distance, so are the three orbits of these stars all aligned? Stat A is also an SB, as is the Mitchel companion. D. Grellmann *et al.* [483] suggested there was evidence for another late-type companion in the CD system

but they regarded the AB and CD groups as independent and unrelated.

Observing and Neighbourhood

Easily found just 1°.5 following the $V = 2.6$ magnitude β Sco, ν Sco is probably the best visual quadruple in the sky. It is much closer than ϵ Lyrae and the stars are of comparable brightnesses. It is low from Europe and the northern USA, and Sue French notes that the stars which ought to appear blue-white according to their early spectral type actually look more like white to yellow to her eyes. The system β Sco (H 3 6) is likely to be at least a physical quadruple star. The Herschel number refers to A (2.6, 4.5, 20°, 13″.7, 2017). Burnham doubled A (BU 947) but the companion is magnitude 10 or fainter and at present only 0″.3 away from A. The period is given as 610 years. Star C has common proper motion with B and is also a close pair (MCA 42) with a period of 38.8 years, never wider than about 0″.1. Much easier is SHJ 225 (7.4, 8.1, 333°, 46″.5, 2013).

Measures

BU 120 (AB)			
Early measure (D)	0°.0	0″.73	1876.35
Recent measure (ARY)	1°.8	1″.55	2016.70
MTL 2 (CD)			
Early measure (SLR)	39°.1	1″.11	1846.58
Recent measure (ARY)	56°.2	2″.48	2016.70
H 5 6 (AC)			
Early measure (SHJ)	338°.2	40″.82	1824.37
Recent measure (ARY)	335°.9	41″.31	2013.70

120. $\sigma^{1,2}$ CRB = STF 2032 = WDS J16147+3352AB

Table 9.120 Physical parameters for STF 2032 CrB

STF 2032	RA: 16 14 40.86	Dec: +33 51 31.0	WDS: 9(1094)		
V magnitudes	AaAb: 5.62	B: 6.49	Ea,Eb: 12.31		
$(B - V)$ magnitudes	A: +0.64	B: +0.69			
μ(A)	-268.33 mas yr^{-1}	± 0.10	-86.93 mas yr^{-1}	± 0.15 (DR2)	
μ(B)	-291.12 mas yr^{-1}	± 0.04	-78.65 mas yr^{-1}	± 0.05 (DR2)	
π(A)	44.14 mas	± 0.06	73.89 light yr	± 0.10 (DR2)	
π(B)	44.15 mas	± 0.02	73.87 light yr	± 0.03 (DR2)	
Spectra	Aa: G0V	Ab: G1V	B: G1V	Ea: M2.5	
Masses (M$_\odot$)	Aa: 1.137	Ab: 1.090			
Radii (R$_\odot$)	Aa: 1.244	Ab: 1.244			
Luminosities (L$_\odot$)	A: 2.5	B: 1.0	E: 0.005		
Catalogues	HD 146361/2 (A/B)	HR 6063/4 (A/B)	SAO 65165	HIP 79607	
DS catalogues	H 1 3 (AB)	STF 2032 (AB)	BDS 7563	ADS 9979	YSC152 (EaEb)
Radial velocity	-12.30 km s^{-1}	± 0.06			
Galactic coordinates	$54°.667$	$+46°.141$			

History

Sir William Herschel alighted on this star on 7 August 1780 with the following comment: 'Treble. The two nearest pretty unequal; the third very faint with powers lower than 460. The two nearest both w.(hite), the third d.(usky).' Herschel observed the close pair again in 1804, by which time they had moved 24° prograde in position angle. Unfortunately no distance measure was made at this time and it was left to Sir John Herschel to do this in 1822, when a further 59° had been traced out. After that the pairs steadily separated, and the brightness of the stars combined with the relatively wide separation made the binary a favourite of observers; to date almost 1100 measures have been made.

Finder Chart

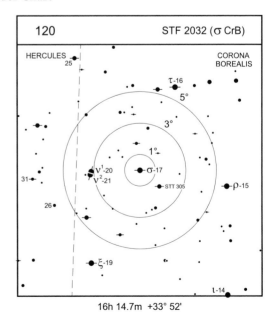

16h 14.7m +33° 52'

Orbit

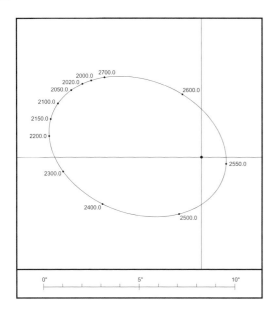

Ephemeris for STF 2032 AB (2000 to 2180)

Orbit by Rag (2009) Period: 726 years, Grade: 4

Year	PA(°)	Sep(")	Year	PA(°)	Sep(")
2000.0	235.8	6.96	2100.0	249.9	7.99
2020.0	239.0	7.26	2120.0	252.4	8.07
2040.0	241.9	7.51	2140.0	254.9	8.11
2060.0	244.7	7.71	2160.0	257.3	8.12
2080.0	247.4	7.87	2180.0	259.8	8.10

The Modern Era

The star σ CrB is a complex system comprising a triple and a binary system separated by about 14,000 AU. The WDS components C (13.1 at 26″) and D (magnitude 10.8 at 93″) appear to be background objects as they do not share the considerable proper motion of AB. However, there is a more distant star of magnitude 12.3 at 10.5′ which is travelling through space at the same rate and which Shaya and Olling consider has an almost 100% chance of being physically connected. What is more, this star is also a binary. It was seen to have an astrometric perturbation by Heintz [485] and was later resolved into a very close pair by Horch *et al.* [484]. It appears to be an M-dwarf binary. The period given by Heintz is 52 years but his orbital elements give position angles and separations for 2014 which disagree with Horch's measurements completely.

Observing and Neighbourhood

The bright pair is observable in all apertures. In 1968, using an 8.3-inch (21-cm) reflector, RWA recorded colours of 'yellow and purple(?)' but, two years later, through the 28-inch refractor at Herstmonceux they appeared white and blue. More recently Sue French [486] sees them as both yellow. Users of binoculars will note the very wide pair ν^1 and ν^2 CrB about 1.5° E of σ. The stars have $V = 5.4$ and $V = 5.6$ and are currently separated by 355″; the separation is slowly decreasing. They are respectively M2III and K5III and consequently appear orange-red in optical aid. DR2 shows that they are not related: ν^1 CrB is 650 light years away whilst ν^2 is only 540. The other nearby pair in *CDSA2* is STT 305 (6.4, 10.2, 265°, 5″.8, 2015). A 13th magnitude star at 264°, 28″, reported by Madler in 1847, has never been confirmed.

Measures

Early measure (STF)	104°.9	1″.22	1830.11
(Orbit	105°.4	1″.29)	
Recent measure (ARY)	238°.6	7″.17	2015.67
(Orbit	238°.3	7″.21)	

121. α SCO = ANTARES = GNT 1 = WDS J16294–2626

Table 9.121 Physical parameters for α Sco

GNT 1	RA: 16 29 24.45	Dec: −26 25 55.2	WDS: 400(144)		
V magnitudes	A: 0.91V	B: 5.2			
$(B − V)$ magnitudes	A: +1.84	B: −0.2			
μ	−12.11 mas yr^{-1}	± 1.22	−23.30 mas yr^{-1}	± 0.76	
π	5.89 mas	± 1.00	554 light yr	± 94	
Spectra	A: M1.5Ia-Ib	B: B2.5V			
Masses (M$_\odot$)	A: 12.5	B: 7.2			
Luminosities (L$_\odot$)	A: 11,400	B: 200			
Catalogues (A/B)	21 Sco	HR 6134	HD 148478/9	SAO 184415	HIP 80763
DS catalogues	GNT 1	BDS 7631	ADS 10074		
Radial velocity	−3.5 km s^{-1}	± 1.8			
Galactic coordinates	351°.947	−15°.064			

History

The earliest observation of Antares as a double appears to be by the Austrian Johann (or Johannes) Tobias Bürg (1766–1835) from Vienna on 13 April 1819 (1819.28), who suspected a companion in PA 270° [491] during a lunar occultation. When he reported this to his peers, they too hastily dismissed it as caused by lunar atmospheric refraction. The next confirmed direct visual discovery was by James W. Grant (1788–1865) on 23 July 1844, made from Bengal, India, and now recognised as GNT 1. Yet another claim appears in Gledhill *et al.* [488], saying that the American Ormsby M. Mitchel (1810–1862) found this while using the newly acquired 11-inch Munich refractor at Cincinnati Observatory. The first measures in the WDS are those of Mitchel in 1846, who, according to Burnham, measured the separation of the pair on 16 nights and the position angle on two. Mitchel was not aware of Grant's observation [487]. It is interesting to note that Dunlop, Rumker, and John Herschel never

Finder Chart

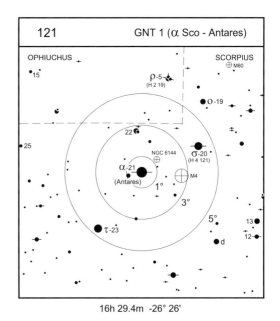

16h 29.4m -26° 26'

Orbit

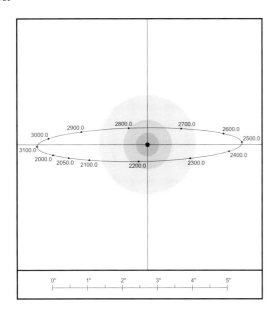

Ephemeris for GNT 1 (2000 to 2180)

Orbit by Pal (2005b) Period: 1217.536 years, Grade: 5

Year	PA(°)	Sep(")	Year	PA(°)	Sep(")
2000.0	276.3	2.72	2100.0	284.7	1.72
2020.0	277.5	2.56	2120.0	287.9	1.48
2040.0	278.8	2.38	2140.0	292.3	1.22
2060.0	280.4	2.18	2160.0	299.2	0.97
2080.0	282.3	1.96	2180.0	310.7	0.73

saw the secondary. W. H. Wright [489] at Lick Observatory noted that Miss Antonia Maury 'finds indications of a faint superimposed spectrum' but this 'probably refers to the telescopic companion'.

The Modern Era

In contrast with our Sun, Antares is a red supergiant, whose diameter is roughly equivalent to the diameter of Jupiter's orbit or about 1.2 billion kilometres.

Measures suggest that these stars either have slowly widened, reaching a maximum of 3″.6, and are now closing, or alternatively are slowly closing. The former possibility suggested to the late French double observer Paul Baize in 1978 a preliminary orbit spanning 878 years, with periastron in AD 1461. Evidence for this is rather scarce, as there has been too little motion (just 3° in position angle) to justify an orbit, and the measures that do exist are quite scattered.

Using both HIP2 parallax and a maximum separation of 3″.6, the projected separation is about 610 AU while the minimum orbital period is between 2130 and 7200 years. Difficulties arise from our poor knowledge of each star. The estimated absolute magnitudes are −5.3 and −0.8, respectively, and the respective effective surface temperatures are 2980 K and 22,500 K. Recent estimates find ages between 10 and 12 million years.

The supergiant primary is an LC-type variable star that unpredictably fluctuates between 0.88 and 1.16 magnitudes, over a period spanning 5.8 years.

Observing and Neighbourhood

Reddish or orangey-red coloured, Antares' popularity is mostly due to its greatly enhanced colour contrast with its fainter blue or green companion, the latter being variously described as blue, turquoise, pale green, greenish, or green. They are split in 10.5-cm but can be glimpsed in 7.5-cm under ideal conditions. Both are more easily seen in full daylight in 20-cm, though smaller apertures might also possible. Whilst in the area, check out the pair σ Sco (H 4 121) (2.9, 8.4, 273°, 20″.3, 2016). In 1976, Morgan et al. [490] reported duplicity of the bright star, noting a magnitude 5.7 star at a distance of 0″.3. RWA felt that he could see this component well enough in 2016 to measure it with the Johannesburg refractor, but it was difficult. Star A is additionally a very close pair (NOR 1) with a period of 33 years and a separation of a few milliseconds of arc. Two degrees N of σ and a little to the E is ρ Oph which is more well known for being the site of a large area of interstellar dust clouds and gas. It is also a pretty double star, found by William Herschel – H 2 19 (5.1, 5.7, 338°, 3″.2, 2016).

Measures

Early measure (DA)	273°.2	3″.46	1848.02
(orbit	279°.5	3″.12)	
Recent measure (ARY)	276°.0	3″.16	2016.67
(orbit	277°.2	2″.58)	

122. λ OPH = STF 2055 = WDS J16309+0159AB

Table 9.122 Physical parameters for λ Oph

STF 2055	RA: 16 30 54.84	Dec: +01 59 02.8	WDS: 17(788)		
V magnitudes	A: 4.15	B: 5.15	C: 11.84	D: 10.9	
(B − V) magnitudes	A: +0.06	B: +0.09			
μ	−30.98 mas yr^{-1}	± 0.61	−73.42 mas yr^{-1}	± 0.58	
π	18.84 mas	± 0.55	173 light yr	± 5	
μ(A)	−32.60 mas yr^{-1}	± 0.71	−66.02 mas yr^{-1}	± 0.54 (DR2)	
μ(C)	−28.24 mas yr^{-1}	± 0.06	−74.50 mas yr^{-1}	± 0.04 (DR2)	
π(A)	20.33 mas	± 0.41	160 light yr	± 3 (DR2)	
π(C)	19.43 mas	± 0.04	167.9 light yr	± 0.3 (DR2)	
Spectra	A: A0V	B: A0V			
Luminosities (L$_\odot$)	A: 40	B: 20	C: 0.02	D:	
Catalogues	10 Oph	HR 6149	HD 148857	SAO 121658	HIP 80883
DS catalogues	H 1 83	STF 2055	BDS 7649	ADS 10087	
Radial velocity	−16.00 km s^{-1}	± 1.5			
Galactic coordinates	17°.123	+31°.845			

History

William Herschel observed λ Oph on 9 March 1783. His note says 'A very beautiful and close double star. L.(arge). w(hite). S.(mall) blue; both fine colours'. On 20 May 1802 he also reports '…with ×527 I saw it well, but with great difficulty. The object is uncommonly beautiful; but it requires a most excellent telescope to see it well…' In 1929, E. B. Frost *et al.* [492] published a large paper listing the radial velocities of 500 A stars. Between 1907 and 1928, 11 plates of λ Oph had been taken with the Bruce spectrograph on the 40-inch Yerkes refractor. The conclusion drawn was that the apparent range of velocity variation 'can be attributed to the uncertainty of measurement on very poor lines'. The stars are both early A type dwarfs and spinning fairly, so that the lines are quite broad.

Finder Chart

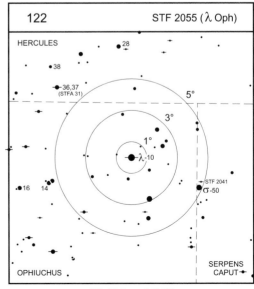

16h 30.9m +01° 59'

Orbit

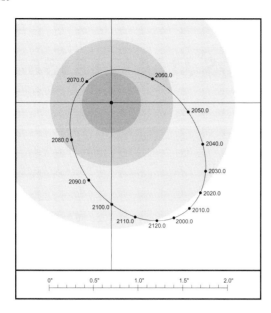

Ephemeris for STF 2055 AB (2015 to 2042)

Orbit by Hei (1993b) Period: 129.0 years, Grade: 2

Year	PA(°)	Sep(")	Year	PA(°)	Sep(")
2015.0	41.1	1.43	2030.0	54.6	1.29
2018.0	43.6	1.41	2033.0	57.7	1.25
2021.0	46.2	1.39	2036.0	61.1	1.20
2024.0	48.9	1.36	2039.0	64.8	1.14
2027.0	51.6	1.33	2042.0	68.9	1.07

The Modern Era

The current orbit, the ephemeris of which is given above, was calculated by Heintz [493] in 1993. In the same short paper he also put forward a strong case for the existence of a third body, evidence for which he found from a series of astrometric plates taken with the 61-cm Sproul refractor. The

WDS mentions only the variable radial velocity, and no third body has yet been confirmed or directly imaged. The WDS indicates that star C (169°, 120″, 2013) has the same proper motion as the bright pair, so it is likely that it is connected, whilst D (247°, 308″, 2002) appears to be a field star. DR2 confirms that C is at a similar distance, and has a similar proper motion, to AB whilst D has a parallax of 15.02 mas.

Observing and Neighbourhood

This star benefits from being at a higher altitude in the sky. In Cambridge, where it culminates 50° from the zenith, it is always a problematic star to measure, and the separations obtained with the 8-inch refractor over a number of years seem to differ from the expected orbital value by being too large by 0″.2 or so. Smyth found the colours yellowish-white and smalt blue. E. J. Hartung found both stars pale yellow in his 30-cm reflector from Victoria, Australia, whilst more recently Sissy Haas found both lemon yellow in 125-mm at ×200. The faint field stars are unconnected: C is at 167°, 120″, 2013, whilst D is at 247°, 308″, 2002. Three degrees NE is the wide pair 37/36 Her (STFA 31) (5.8, 6.9, 229°, 70″, 2015), which is fixed. More difficult is STF 2041, which is about 10′ NW of the nearby σ Ser. Its stars have magnitudes 7.5 and 10.5 and they are separated by 2″.6 in PA 1° (2010).

Measures

Early measure (STF)	342°.1	0″.81	1828.51
(Orbit	334°.9	0″.79)	
Recent measure (ARY)	43°.9	1″.76	2016.50
(Orbit	42°.4	1″.42)	

123. ζ HER = STF 2084 = WDS J16413+3136

Table 9.123 Physical parameters for ζ Her

STF 2084	RA: 16 41 17.16	Dec: +31 36 09.8	WDS: 14(838)	
V magnitudes	A: 2.95	B: 5.40		
(B − V) magnitudes	A: +0.72	B: +0.92		
μ	−461.52 mas yr⁻¹	± 0.38	+342.28 mas yr⁻¹	± 0.48
π	93.32 mas	± 0.47	35.0 light yr	± 0.2
Spectra	Aa: G0IV	B		
Masses (R☉)	A: 1.45	± 0.01	B: 0.98	± 0.02
Radii (R☉)	A: 2.58	B: 0.92		
Luminosities (L☉)	A: 6.3	B: 0.7		
Catalogues	40 Her	HD 150680	HR 6212	SAO 65485	HIP 81693
DS catalogues	H 1 36	STF 2084	BDS 7717	ADS 10157
Radial velocity	−67.80 km s⁻¹	± 0.2		
Galactic coordinates	52°.661	+40°.289		

History

Found by William Herschel on 18 July 1782. His catalogue notes 'A fine double star. Very unequal. L.(arge star) w.(hite); S.(mall star) ash-colour.' By 1802 he was unable to separate the two stars and although he recorded three widely varying angles in 1802 and 1803, our current knowledge of the orbit suggests that the stars would have been about 1″.1 apart at that time and should have been visible. In his compendium on the Struve stars, Lewis [194] took 13 pages to cover ζ Herculis. From discussion of the micrometer measures from 1826 to 1905, Lewis thought that he could see an increase in the orbital period over three revolutions and this he concluded was due to the duplicity of the primary star. Aitken notes that a further review by G. Comstock [499], which took into account the effect of systematic observational bias in the measures, concluded that there was no evidence for a third body from that source but that other evidence did point to a possible third body with a period of 18 years and amplitude 0″.1. Modern theory on the stability of triple systems finds that there are no instances where the orbital periods of the inner and outer systems are so close in value.

Finder Chart

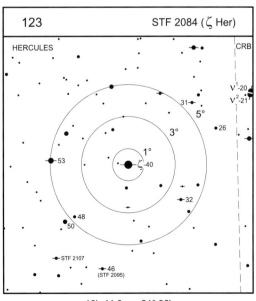

16h 41.3m +31° 36'

Orbit

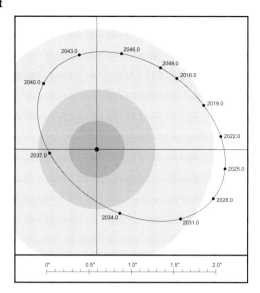

Ephemeris for STF 2084 (2016 to 2034)

Orbit by Sod (1999) Period: 34.45 years, Grade: 1

Year	PA(°)	Sep(")	Year	PA(°)	Sep(")
2016.0	130.9	1.25	2026.0	77.0	1.53
2018.0	117.8	1.32	2028.0	67.6	1.49
2020.0	106.2	1.40	2030.0	57.2	1.37
2022.0	95.8	1.47	2032.0	43.7	1.15
2024.0	86.2	1.52	2034.0	20.4	0.79

The Modern Era

Talk of a third component surfaced again in 1976, when Paul Baize [497] found evidence for a third body with a period of 12 years, and in 1983 McCarthy *et al.* [652] recorded a direct detection using infrared speckle. Nine years later, a paper by X. Pan *et al.* [494] (reported the results of Mark III stellar interferometer measures of the primary of ζ Her. They noted that there was evidence for a close companion to A with a separation of 10 mas and a Δm of 3.5. However, in the discussion after the talk, Professor C. D. Scarfe noted that he had been making radial velocity measurements of ζ Her for 30 years and found that a two-body orbit fitted the observations with a scatter of only 0.25 km s^{-1}. Additionally, the WDS notes that (visual) speckle and Hipparcos data show no sign of the third component. Allen *et al.* [495] searched for common proper motion components of nearby SBs and reported nothing for ζ Her. Hutter *et al.* [496] included ζ in a search for multiplicity in bright stars and did not find any further components. They estimated that the detection limits of Δm and the separation were 3.5 and 6 mas respectively.

Observing and Neighbourhood

Lewis, using the 28-inch refractor at Greenwich, noted that between 1895 and 1898 the companion appeared dark blue

Measures with 8″ Refractor

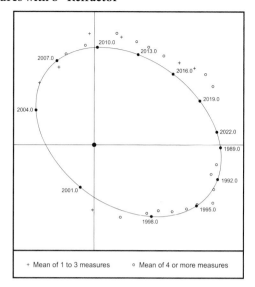

+ Mean of 1 to 3 measures   ○ Mean of 4 or more measures

Year	PA(°)	Sep(")	No.	Year	PA(°)	Sep(")	No.
1990.52	81.2	1.48	5	2007.59	198.1	1.10	6
1991.54	76.4	1.44	4	2008.66	185.4	1.15	4
1992.61	70.3	1.52	5	2009.57	182.6	1.28	3
1993.67	64.8	1.57	6	2010.69	168.4	1.31	4
1994.55	61.0	1.41	6	2011.69	165.6	1.28	3
1995.57	56.4	1.33	5	2012.64	157.0	1.29	6
1996.57	50.6	1.21	6	2013.75	149.9	1.30	5
1997.55	45.5	1.08	5	2014.59	140.9	1.32	5
1998.57	38.8	1.01	5	2015.63	132.9	1.39	3
1999.61	20.3	0.89	8	2016.57	126.4	1.49	2
2000.50	358.5	0.75	2	2017.60	121.4	1.56	5
2005.47	222.7	0.97	1	2018.58	115.4	1.57	5
2006.49	205.5	0.99	2				

whilst in 1900 it was a decided green and in 1904 appeared light green. It must be said that at this time the companion was near its smallest separation from the primary and, even in a large aperture, the pair is a difficult object because of the large difference in magnitudes. The stars are currently opening and can be readily seen with 20-cm, but a night of good seeing is essential as the flare from the primary tends to obscure the secondary. When the companion is seen clearly it does tend to be greenish. The star STF 2107, which is 4° SE is a well-observed visual binary – the WDS records 435 measures – and the period is 268 years. The stars are currently near their widest separation – 6.9, 8.5, 107°, 1″.4, 2020. Nearby 46 Her is also a visual pair – STF 2095 – 7.4, 9.1, 163°, 5″.2, 2013.

Measures

Early measure (STF)	196°.9	1″.09	1835.45
(Orbit	198°.4	1″.09)	
Recent measure (ARY)	133°.9	1″.33	2015.64
(Orbit	133°.4	1″.24)	

124. μ DRA = STF 2130 = WDS J17053+5428AB

Table 9.124 Physical parameters for μ Dra

STF 2130	RA: 17 05 20.20	Dec: +54 28 14.3	WDS: 16(797)		
V magnitudes	A: 5.66	B: 5.69	C: 13.7		
$(B - V)$ magnitudes	A: +0.52	B: +0.54			
μ(A)	-56.63 mas yr^{-1}	$\pm$ 0.15	74.11 mas yr^{-1}	$\pm$ 0.14 (DR2)	
μ(B)	-98.41 mas yr^{-1}	$\pm$ 0.13	94.82 mas yr^{-1}	$\pm$ 0.12 (DR2)	
μ(C)	-55.57 mas yr^{-1}	$\pm$ 0.16	85.39 mas yr^{-1}	$\pm$ 0.06 (DR2)	
π(A)	36.80 mas	$\pm$ 0.10	88.6 light yr	$\pm$ 0.2 (DR2)	
π(B)	36.80 mas	$\pm$ 0.06	88.6 light yr	$\pm$ 0.1 (DR2)	
π(C)	36.67 mas	$\pm$ 0.03	88.94 light yr	$\pm$ 0.07 (DR2)	
Spectra	A: F7V	B: F7V			
Masses (M$_\odot$)	A: 1.31	B: 1.51			
Luminosities (L$_\odot$)	A: 3.4	B: 3.3	C: 0.002	D:	
Catalogues	21 Dra	HD 154906/5	HR 6370/69	SAO 30239	HIP 83608
DS catalogues	H 2 13	STF 2130	BU 1088 (AC)	BDS 7875	ADS 10345
Radial velocity	-17.30 km s^{-1}	$\pm$ 0.5			
Galactic coordinates	82°.299	+37°.018			

History

William Herschel found this pair on 19 October 1781. He noted that both stars were white and equally bright, and his mean distance measure came to 4″.354. In 1889, using the Lick 36-inch, S. W. Burnham found a faint companion at a distance of 12″.25 which showed no change over the next 10 years, leading Thomas Lewis to declare that it was a physical member of the system.

The Modern Era

In 1943 K. Strand [611] found a perturbation with a period of 3.2 years in the motion of AB due to a third body. Using plates taken with the large refractors at Potsdam, Lick, and Sproul, he considered that the amplitude of the perturbation was 'entirely too small to be detected from the visual observations'. In 1966 G. Ishida [503] took a number of spectra of both components of μ Dra but neither conformed to the elements of Strand's orbit, although Ishida did say that he thought the B star was a spectroscopic binary. Heintz [501] was of the opinion that the reported perturbations do not exist. A spectroscopic orbit for the B component is now included in Pourbaix's *Ninth Catalogue of Spectroscopic Binary Orbits* with a period of 2270 days, but the *WDS* notes that it is a low-grade orbit which needs confirmation. In 2012 J.-L. Prieur *et al.* [504] produced an orbit for AB (the ephemeris of which is given below), which gave a dynamical parallax of 36.3 mas, in close agreement with the Gaia DR2 value given above.

Finder Chart

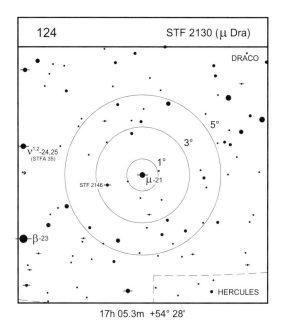

17h 05.3m +54° 28'

Orbit

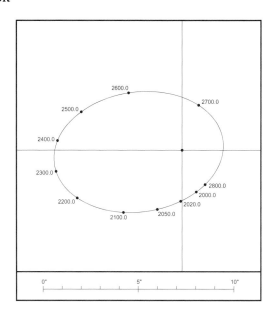

Ephemeris for STF 2130 AB (2010 to 2100)

Orbit by Pru (2012) Period: 812.0 years, Grade: 4

Year	PA(°)	Sep(")	Year	PA(°)	Sep(")
2010.0	8.0	2.41	2060.0	331.6	3.50
2020.0	358.3	2.60	2070.0	327.0	3.74
2030.0	350.1	2.81	2080.0	323.0	3.96
2040.0	343.0	3.04	2090.0	319.3	4.18
2050.0	336.9	3.27	2100.0	316.0	4.40

Star C is at 175°, 13″, 2006 and is confirmed as being physical by Heintz [500].

Observing and Neighbourhood

This is a beautiful pair of white stars, easily seen in 7.5-cm or above. Closest approach (2″.1) was in 1973 and the stars will be separating for many years to come, reaching 6″.7 by around 2330. Four degrees E is the beautiful twin pair of ν Dra (STFA 35) (see Star 133), a splendid sight with any optical aid from binoculars up, and 4°.5 WSW is 16/17 Dra, a fine triple for the small aperture (but just off the chart here). The close pair is 5.4, 6.4, 103°, 3″.1 whilst a third star of magnitude 5.5 is 89″ distant on PA 193° but has the same proper motion as the close pair. Like ν Dra this is also a physical system. Closer to μ is STF 2146 (6.9, 8.8, 224°, 2″.6, 2015) with a third star of magnitude 8.9 at 235°, 89″.

Measures

Early measure (STF)	208°.1	3″.34	1828.52
(Orbit	205°.2	3″.29)	
Recent measure (ARY)	1°.7	2″.30	2016.89
(Orbit	1°.2	2″.53)	

125. η OPH = BU 1118 = WDS J17104–1544AB

Table 9.125 Physical parameters for η Oph

BU 1118	RA: 17 10 22.67	Dec: −15 43 29.7	WDS: 198(225)	
V magnitudes	A: 2.97	B: 3.44		
(B − V) magnitudes	A: +0.13	B: +0.12		
μ	+40.13 mas yr^{-1}	± 1.06	99.17 mas yr^{-1}	± 0.39
π	36.91 mas	± 0.8	88 light yr	± 2
Spectra	A: A2V	B: A3V		
Masses (M$_\odot$)	AB: 7.05 (dyn.)			
Luminosities (L$_\odot$)	A: 40	B: 25		
Catalogues	HR 6378	HD 155125	SAO 160332	HIP 84012
DS catalogues	BU 1118	BDS 7885	ADS 10374	
Radial velocity	−2.40 km s^{-1}	± 0.3		
Galactic coordinates	6°.721	+14°.008		

Finder Chart

History

This bright binary was discovered by S. W. Burnham in May 1889 using the Lick 36-inch refractor. He was of the opinion that the stars were then near maximum separation, whilst in fact they did not reach this point until 1923 when they were about 0″.6 apart.

The Modern Era

The current orbit predicts a period of 87.5 years. The stars are now closing (338°, 0″.49, 2020) reaching a minimum distance five years hence (38°.0, 0″.14, 2025) and then widening to reach 24°, 0″.35 by 2030.

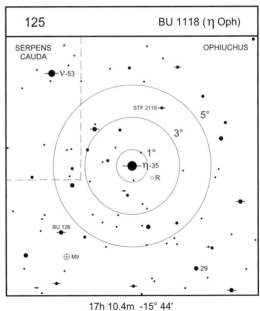

17h 10.4m -15° 44'

Orbit

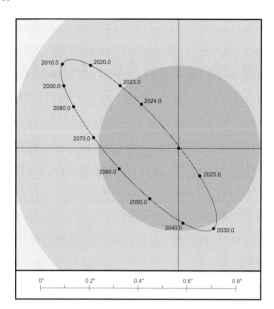

Ephemeris for BU 1118 AB (2016 to 2034)

Orbit by Doc (2007d) Period: 87.58 years, Grade: 2

Year	PA(°)	Sep(")	Year	PA(°)	Sep(")
2016.0	231.0	0.56	2026.0	33.0	0.27
2018.0	229.5	0.53	2028.0	28.1	0.33
2020.0	227.7	0.49	2030.0	24.2	0.35
2022.0	225.5	0.41	2032.0	20.6	0.35
2024.0	221.2	0.23	2034.0	16.8	0.34

Observing and Neighbourhood

Two degrees N and slightly W is STF 2119 (8.2, 8.3, 185°, 2″.4, 2017), whilst 3° SE is BU 126 (6.3, 7.6, 265°, 2″.4, 2016). Just about 1° S of this pair is the globular cluster M9.

Measures

Early measure (BU)	273°.7	0″.35	1889.39
(Orbit	275°.7	0″.30)	
Recent measure (ANT)	232°.3	0″.55	2014.25
(Orbit	232°.1	0″.57)	

126. α HER = STF 2140 = WDS J17146+1423AB

Table 9.126 Physical parameters for α Her

STF 2140	RA: 17 14 38.86	Dec: +14 23 25.2	WDS: 45(491)		
V magnitudes	AaAb = 3.48	Ba = 5.6	Bb: 6.6		
(*B* − *V*) magnitudes	AaAb = +1.45	BaBb = +0.71			
μ(A)	−7.32 mas yr^{-1}	± 0.92	+36.07 mas yr^{-1}	± 0.97	
π(A)	9.07 mas	± 1.32	360 light yr	± 53	
μ(B)	−8.33 mas yr^{-1}	± 0.63	41.39 mas yr^{-1}	± 0.70 (DR2)	
π(B)	9.91 mas	± 0.50	329 light yr	± 17 (DR2)	
Spectra	A: M5Ib-II	Ba: G5III	Bb: A9IV-V		
Masses (M$_\odot$)	A: 2.7	± 0.6	Ba: ∼ 2.5	B: ∼ 2.0	
Radii (R$_\odot$)	A: 400	± 70			
Luminosities (L$_\odot$)	Aab: 400	Ba: 60	Bb: 25		
Catalogues	HD 156014/5	HR 6406/7	SAO 102680/1	HIP 84345	
DS catalogues (AB)	Mayer 43	H 2 2	STF 2140	BDS 7914	ADS 10418
	CHR 139 (AaAb)				
Radial velocity	−32.09 km s^{-1}	± 0.22			
Galactic coordinates	35°.534	+27°.818			

History

The system α Herculis has been known as Rasalgethi for thousands of years. The name means the kneeler's head in Arabic and harks back to a time when the stars referred to a kneeling man, before Hercules became part of the constellation canon. It is unmistakeable and sits about 10° directly S of δ Herculis, the southeastern star in the group of six bright stars which make up the Keystone of Hercules. First seen by Neville Maskelyne in 1777, this pair was swept up by William Herschel on 29 August 1779. He found the primary star to be red and the companion 5 arcseconds distant was 'blue tending to green'. Herschel also observed in 1795 that the primary star varied in brightness and thought the period to be 66 days. In 1921 Sanford [505] found that B was a single-lined spectroscopic binary with a period of 51.59 days moving in

a circular orbit. He noted the difference in radial velocity between A and B and wondered if the two bright stars were actually an optical pair.

The Modern Era

The current picture of α Herculis is still rather unclear. It is known to consist definitely of four stars. Dr Myron Smith has suggested the presence of another component, because of a long-term change in the radial velocity of the primary star, over and above what is known about the pulsational variations and the effect of the close interferometric companion (a) to A found by Harold McAlister *et al.* [665] using the KPNO 4-metre telescope in 1986. This system, known as CHR 139AaAb, was resolved three times between

Finder Chart

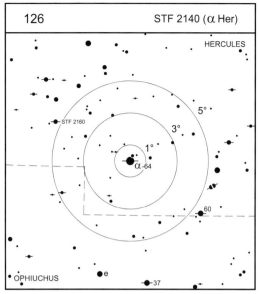

126   STF 2140 (α Her)

HERCULES

5°

3°

1°

α -64

OPHIUCHUS   e
37

17h 14.6m +14° 23'

Orbit

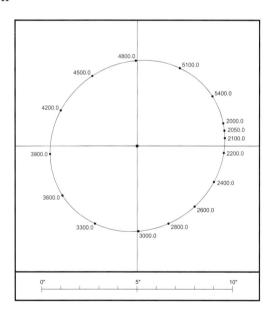

Ephemeris for STF 2140 AB (2000 to 2225)

Orbit by Baz (1978) Period: 3600 years, Grade: 4

Year	PA(°)	Sep(")	Year	PA(°)	Sep(")
2000.0	104.5	4.65	2125.0	92.8	4.59
2025.0	102.2	4.64	2150.0	90.4	4.57
2050.0	99.8	4.63	2175.0	88.0	4.56
2075.0	97.5	4.61	2200.0	85.6	4.54
2100.0	95.1	4.60	2225.0	83.2	4.52

1986 and 1991 but has not been positively reported since. The two main visual components, A and B, certainly form a visual binary with a period of thousands of years, but the observed motion to date is just too scant to predict a meaningful period. Theiring & Reimers [506] found that Ba is a G8 III star whilst the companion is of spectral type A9IV-V with corresponding masses of 2.5 and 2 $M_\odot$. The diameter of Rasalgethi has been measured at 34 $\pm$ 0.8 mas, which equates to about 400 $R_\odot$. Theiring and Reimers, using the International Ultraviolet Explorer (IUE), measured the diameter as changing between 264 and 303$R_\odot$. This pulsation also affects the visual brightness of star A. It has been under constant surveillance by the American Association of Variable Star Observers (AAVSO) since the mid 1930s, and the range of brightness varies between V = 2.5 and 4.0 with a mean of about V = 3.3. In 2010 Moravveji *et al.* [507] reported on the results of a photometric monitoring programme of star A. They found a range in V from 2.76 to 3.62 and detected seven separate modes of pulsation, of which one, with a period of 1343 days, was connected to diameter changes within the star. The estimation of masses has been hindered by the lack of any dynamical information about either the period of Aa or the long-period AB systems. Theiring and Reimers obtained 2.5 and 2.0 $M_\odot$ for components Ba and Bb, with corresponding spectral types of G8III and A9IV-V. Moravveji *et al.* [507] obtained a range between 2.17 and 3.25 $M_\odot$ for the mass of A. The WDS notes two faint field stars.

Observing and Neighbourhood

The star α Herculis is most noted for the colour contrast between the two bright visual components. Admiral Smyth noted colours of orange and emerald or bluish green in his *Cycle of Celestial Objects*. He goes on to note that the tenth and 12th magnitude stars in the field are remarkable for their lilac tinge even in his 5.9-inch refractor. In 1968, using an 8.3-inch reflector, RWA saw it as reddish-orange and deep green at powers of ×96 and ×216. Only E. J. Hartung disagrees about the colour of the B star; from Australia with his 12-inch reflector he made it white. Two point five degrees ENE of α is STF 2160 (6.4, 9.2, 66°, 3".9, 2016).

Measures

Early measure (STF)	118°.5	4".65	1829.67
(Orbit	120°.5	4".68)	
Recent measure (ARY)	104".6	4".69	2014.62
(Orbit	102°.5	4".64)	

127. δ HER = STF 3127 = WDS J17150+2450AB

Table 9.127 Physical parameters for δ Her

STF 3127	RA: 17 15 01.91	Dec: +24 50 21.2	WDS: 210(219)		
V magnitudes	Aa: 3.31	Ab: 4.4	B: 8.3	C: 10.5	D: 10.6
(B − V) magnitudes	A: +0.08	B:			
μ(A)	−22.79 mas yr^{-1}	± 0.57	−156.52 mas yr^{-1}	± 0.62 (DR2)	
μ(B)	−105.30 mas yr^{-1}	± 0.04	+5.29 mas yr^{-1}	± 0.05 (DR2)	
π(A)	42.78 mas	± 0.35	76.2 light yr	± 0.6 (DR2)	
π(B)	10.11 mas	± 0.30	323 light yr	± 10 (DR2)	
Spectra	A: A1IV	B: G4IV-V			
Masses (M$_\odot$)	A: 2.5				
Radii (R$_\odot$)	A: 2.2				
Luminosities (L$_\odot$)	Aa: 22	Ab: 8	B: 4		
Catalogues	65 Her	HR 6410	HD 156164	SAO 84951	HIP 84379
DS catalogues	H 5 1 (AB)	STF 3127 (AB)	BNU 5 (AaAb)		
Radial velocity (A/B)	−40.0 km s^{-1}	± 2	−4 km s^{-1}	± 5	
Radial velocity (B)	4.05	± 2.57 (DR2)			
Galactic coordinates	46°.824	+31°.423			

History

This binary was found by William Herschel on 9 August 1779. 'Double. Extremely unequal. L(arge).w(hite).; S(mall). inclining to r(ed). Distance 33″.75. Position 72° 28′ s(outh). following'. Herschel calls the primary star Flamsteed 11. Merrill [571] observed the bright component with the 100-inch Mount Wilson stellar interferometer and found it single at the 0″.03 level. Abetti [511] added faint companions C (353°, 174″, 2013) and D (93°, 192″, 2010), which appear to be unconnected to δ.

The Modern Era

The primary is a multiple system. The WDS notes that it is a spectroscopic binary and in 1980 Bonneau & Foy [512] resolved A into two stars, with a magnitude difference of about 1.5 in V and a separation of 0″.095 using a speckle camera on the 1.9-metre reflector at Haute Provence. Three further resolutions were reported in 1988/9 by Ismailov et al. [513], which, if accurate, show rapid orbital motion, but no observations of this companion have been made since then. It is possible that this is the spectroscopic companion but no

Finder Chart

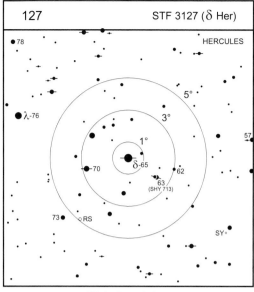

127   STF 3127 (δ Her)

HERCULES

17h 15.0m +24° 50'

Linear Ephemeris for AB (2015 to 2100)

Year	PA (°)	Sep (″)
2015.0	289.2	12.90
2020.0	291.9	13.58
2025.0	294.3	14.25
2030.0	296.5	15.01
2035.0	298.5	15.76
2040.0	299.9	16.53
2045.0	301.3	17.31
2050.0	302.8	18.11
2075.0	308.5	22.22
2100.0	312.3	26.50

reports of double lines have been recorded for A. In the VAST survey looking at duplicity in A stars using the K band, de Rosa *et al.* [514] found a star 12″ distant with a projected period of 676 years and a mass of 0.45 $M_\odot$. This seems to coincide with the position angle and separation of the Struve companion, which is clearly unrelated to star A. The CADARS catalogue [515] is a compilation of the apparent diameters and absolute radii of stars taken from the astronomical literature. It contains data for 7778 stars taken from 13,573 records. The results for δ Her give it a diameter of about 1.5 $R_\odot$, whereas the paper by Malagnini & Morossi [517] suggests it is nearer 2.2 $R_\odot$.

Observing and Neighbourhood

The bright star and the Struve companion form a classic example of an optical double star where the two components are moving in completely different directions in space and are therefore unconnected with each other. According to the elements found by Bill Hartkopf, the nearest approach was in 1964 when the separation was 8″.89. F. G. W. Struve made the colours green and ashy-white. RWA finds this quite a difficult pair to measure and the magnitude of 8.3 for B seems rather optimistic. In 2011, Greg Stone [516], with 60-mm aperture and ×90 saw B only with averted vision and with good dark adaption. The nearest entry in *CDSA2* is SHY 713, also known as 63 Her, a pair of stars with common proper motions, and, according to DR2, a common distance of 260 light years (6.2, 7.0, 74°, 196″, 2015).

Measures

Early measure (STF)	174°.1	25″.53	1832.11
(Linear	173°.9	25″.87)	
Recent measure (ARY)	290°.8	13″.10	2017.65
(Linear	290°.5	13″.25)	

128. 36 OPH = SHJ 243 = WDS J17153−2636AB

Table 9.128 Physical parameters for 36 Oph

SHJ 243	RA: 17 15 20.78	Dec:−26 36 06.1	WDS: 273(135)		
V magnitudes	A: 5.12	B: 5.12	C: 6.34	D: 7.8	E: 12.3
(B − V) magnitudes	A: +0.97	B: +0.98	C: +0.66		
μ(A)	−498.67 mas yr^{-1}	± 0.27	−1148.96 mas yr^{-1}	± 0.18 (DR2)	
μ(B)	−466.54 mas yr^{-1}	± 0.65	−1142.06 mas yr^{-1}	± 0.45 (DR2)	
μ(C)	−479.85 mas yr^{-1}	± 0.10	−1124.55 mas yr^{-1}	± 0.07 (DR2)	
π(A)	167.82 mas	± 0.16	19.44 light yr	± 0.02 (DR2)	
π(B)	167.78 mas	± 0.22	19.44 light yr	± 0.03 (DR2)	
π(C)	168.07 mas	± 0.08	19.41 light yr	± 0.01 (DR2)	
Spectra	A: K0V	B: K1V	C: K5V		
Masses (M$_\odot$)	A: 0.82	B: 0.81	C: 0.71		
Radii (R$_\odot$)	A: 0.813	± 0.016	B: 0.81		
Luminosities (L$_\odot$)	A: 0.3	B: 0.3	C: 0.1	D:	
Catalogues (A/B/C)	HR 6402/1/—	HD 155856/5/6026	SAO 185198/9/213	HIP 84405/—/84478	
DS catalogues	SHJ 243(AB)	HDO 144(AB)	SHY 87(AC)	BDS 7905(7946)	ADS 10417
Radial velocity (AB)	−0.10 km s^{-1}	± 0.5			
Radial velocity (C)	−0.04 km s^{-1}	± 0.22 (DR2)			
Galactic coordinates	358°.279	+6°.878			

History

The WDS has two entries for observations of this system by Christian Mayer (1777.5) and Tobias Mayer (1780), neither of which record the position angle but both give distances around 13″. This must be another pair. The stars were measured by South and Herschel and the result published in 1824 [518]. The Harvard observers discovered a pair catalogued as HDO 144, but Burnham showed that it was identical to 36 Oph, hence the extra BDS catalogue number. In 1941 Luyten [521] showed that the proper motion of C some 732″ distant in PA 74° was almost the same as that of AB and the Hipparcos satellite has confirmed that the distance to C and the AB pair is to all intents and purpose the same, making this a physical triple.

The Modern Era

Between 1980 and 1992 the radial velocity of both components was monitored using CFHT on Hawaii. Irwin et al. [520] reported on the results. Star C is the RS CVn variable V2215 Oph. The two companions D and E given in the WDS are field stars: D is 7.8 at 338°, 277″, distance increasing rapidly, 2000, and E is at 313°, 38″, 2000.

Finder Chart

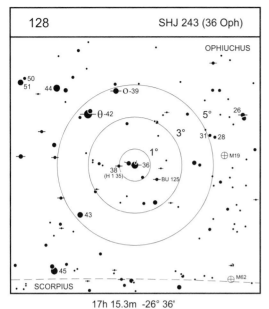

128   SHJ 243 (36 Oph)

OPHIUCHUS

50
51
44
O-39
θ-42
5°
26
3°
31 28
1°
M19
38 36
(H 1 35)
BU 125

43

45
M62
SCORPIUS

17h 15.3m -26° 36'

Orbit

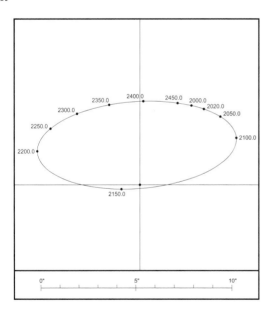

2350.0 2400.0 2450.0 2000.0
2300.0   2020.0
2250.0   2050.0
2200.0   2100.0

2150.0

0"   5"   10"

Ephemeris for SHJ 243 AB (2010 to 2100)

Orbit by Irw (1996) Period: 470.9 years, Grade: 4

Year	PA(°)	Sep(")	Year	PA(°)	Sep(")
2010.0	142.7	4.96	2060.0	126.7	5.52
2020.0	139.2	5.09	2070.0	123.9	5.59
2030.0	135.9	5.21	2080.0	121.1	5.63
2040.0	132.7	5.33	2090.0	118.3	5.62
2050.0	129.7	5.43	2100.0	115.5	5.55

Exoplanet Host?

Research by Wittenmayer *et al.* [519] has ruled out planets with masses between 0.13 and 5.4 Jupiter masses existing between 0.05 and 5.2 AU from the primary stars.

Observing and Neighbourhood

36 Oph lies in the area of sky about 15° E of Antares in an area where there are a number of clusters, including M19. Half a degree E is the pair 38 Sco = H 1 35 (6.9, 9.0, 337°, 5″.8, 2016), whilst 1° WSW is BU 125 (6.9, 9.7, 66°, 1″.8, 1991).

Measures

Early measure (SHJ)	227°.3	5″.55	1822.52
(Orbit	225°.3	4″.99)	
Recent measure (ARY)	141°.4	5″.13	2013.70
(Orbit	140°.7	5″.01)	

129. MLO 4 SCO = WDS J17190–3459AB

Table 9.129 Physical parameters for MLO 4 Sco

MLO 4	RA: 17 18 57.16	Dec: −34 59 23.1	WDS: 165(248)		
V magnitudes	A: 7.38	B: 8.82	C: 10.6	D: 12.5	
$(B - V)$ magnitudes	A: +1.02	B: +1.30			
μ	1129.76 mas yr^{-1}	± 9.72	−77.02 mas yr^{-1}	± 4.67	
π	146.29 mas	± 9.03	22.3 light yr	± 1.4	
$\mu(C)$	1131.61 mas yr^{-1}	± 0.11	−215.55 mas yr^{-1}	± 0.08 (DR2)	
$\pi(C)$	138.02 mas	± 0.09	23.63 light yr	± 0.02 (DR2)	
Spectra	A: K3V	B: K5V	C: M1.5V		
Masses (M$_\odot$)	A: 0.73	B: 0.69	C: 0.37		
Luminosities (L$_\odot$)	A: 0.04	B: 0.01	C: 0.002	D:	
Catalogues	142 G Sco	HR 6426	HD 156384	SAO 208670	HIP 84709
DS catalogues	MLO 4 (AB)	BU 416 (AB)	HJ 4935 (AB,C)	SEE 509 (AD)	BDS 7929
Radial velocity	0.00 km s^{-1}	± 3.7			
Galactic coordinates	351°.842	+1°.423			

History

On sweep 792 at Feldhausen during his survey of the southern heavens, John Herschel observed the star Brisbane 6097 to be double and noted 'position estimated from diagram'. He gives no separation or magnitudes, simply an estimated position angle of ±130°. The pair was given the catalogue number HJ 4935 and the WDS gives the magnitude as 10.27 and the current separation as 33″. What is more interesting is why Herschel did not divide the primary star, which we now know must have been separated by 1″.6 at the epoch at which he observed it. The double nature of the primary was reported at Melbourne Observatory in 1867 [526]. In 1875, S. W. Burnham [523] noted the close pair with the 6-inch Clark, and it became β 416. He also included it in his *General Catalogue* even though it lies 3° S of the southernmost limit of the catalogue (121° from the North Pole). Aitken is more

Finder Chart

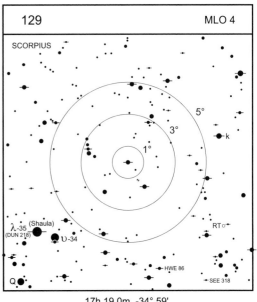

17h 19.0m -34° 59'

Orbit

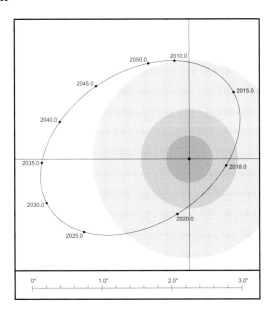

Ephemeris for MLO 4 AB (2016 to 2034)

Orbit by Sod (1999) Period: 42.15 years, Grade: 2

Year	PA(°)	Sep(")	Year	PA(°)	Sep(")
2016.0	133.2	0.98	2026.0	300.0	1.92
2018.0	80.1	0.53	2028.0	292.9	2.07
2020.0	347.4	0.78	2030.0	286.6	2.14
2022.0	320.9	1.32	2032.0	280.5	2.16
2024.0	308.5	1.69	2034.0	274.5	2.14

strict and excludes it. In 1877 Herbert Howe [524] at Cincinnati also observed the star as a new double and called it Melbourne 870, as it occupies that position in the *First Melbourne General Catalogue* of 1864. Only a month or so after Howe's observation, H. C. Russell [525] at Sydney also happened upon the pair and called it R 298. T. J. J. See added another faint star (D) of magnitude 12.5, now at 287°, 65″, 1999, but it appears to be a field star.

Exoplanet Host?

In 2013 it was announced [522] that analysis of radial velocity measurements of the faint red dwarf (C) suggested evidence of six, or even seven, planets including three super-Earths located in the habitable zone. Since then doubt has been cast on this interpretation, and the current `exoplanet.au` catalogue shows two planets with masses of > 5.7 and > 3.7 Earth masses with respective periods of 7.20 and 28.14 days.

The Modern Era

This is a physical triple system and it is of added interest to researchers because of the planetary system attached to the faintest component.

Observing and Neighbourhood

This nice triple is found 3.5° NW of Shaula (λ Sco) and Lesath (υ Sco), the two bright stars in the sting of the Scorpion's tail. The close pair are both orange in colour. In the calendar year 2018 the stars rotated almost 60° in position angle whilst the separation remained virtually unchanged at just over 0″.5. To see this phenomenon properly needed 25-cm but 20-cm would have shown strongly elongated images which will also change obviously in orientation. Shaula (magnitude 1.6) has a distant 9.2 star at PA 330° and 94″ (distance increasing), forming DUN 218. The primary consists of two massive young B stars, one of which is an eclipsing binary and is 29 light years away. Four degrees S and slightly W is HWE 86 (6.9, 9.0, 150°, 2″.6, 2016). One and a half degrees further W and a little S from HWE 86 is SEE 318, a binary with a period of 285 years. The stars are currently near widest separation (8,4, 8.7, 358°, 0″.92, 2020).

Measures

AB			
Early measure (HWE)	226°.3	1″.68	1877.53
(Orbit	226°.8	2″.09)	
Recent measure (ARY)	124°.0	0″.86	2016.67
(Orbit	121°.9	0″.85)	
AB-C			
Early measure (BAR)	129°.4	30″.01	1894.61
Recent measure (ARY)	141°.6	32″.93	2013.69

130. 41G ARA = BSO 13 = WDS J17191−4638AB

Table 9.130 Physical parameters for 41G Ara

BSO 13	RA: 17 19 03.85	Dec: −46 38 10.1	WDS: 455(134)		
V magnitudes	A: 5.61	B: 8.88			
(*B* − *V*) magnitudes	A: +0.89	B:			
μ(A)	1029.64 mas yr^{-1}	± 0.15	107.00 mas yr^{-1}	± 0.16 (DR2)	
μ(B)	952.22 mas yr^{-1}	± 0.67	143.18 mas yr^{-1}	± 0.56 (DR2)	
π(A)	113.82 mas	± 0.13	28.66 light yr	± 0.03 (DR2)	
π(B)	120.18 mas	± 0.48	27.14 light yr	± 0.11 (DR2)	
Spectra	A: G8V	B: M0V			
Masses (M$_\odot$)	A: 0.81	B: 0.52			
Radii (R$_\odot$)	A: 0.79	B: 0.48			
Luminosities (L$_\odot$)	A: 0.4	B: 0.02			
Catalogues (A)	41G Ara	HR 6416	HD 156274	SAO 227816	HIP 84720
DS catalogues	BSO 13				
Radial velocity	25.3 km s^{-1}	± 0.1			
Radial velocity (A/B)	26.04 km s^{-1}	± 0.20	25.06 km s^{-1}	± 1.07 (DR2)	
Galactic coordinates	342°.299	−5°.268			

History

This pair was first noted at Parramatta in 1825 during the observations being made for the Brisbane Catalogue. Innes [528] noted that it was subsequently unobserved for more than half a century, until 1880, when it was included in the Argentine General Catalogue, and a few weeks later picked up independently by Russell [525], who gave it his running number R 297.

The Modern Era

The current orbital period is 953 years but this is a rather preliminary value as the stars have not yet been observed at widest separation.

Exoplanet Host?

The pair BSO 13 has been surveyed [528] by radial velocity methods for signs of a planetary family; there is a linear

Finder Chart

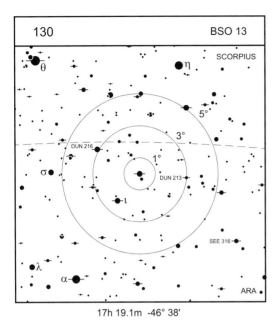

130   BSO 13

SCORPIUS

θ   η

5°

DUN 216   3°

1°

σ   DUN 213

ι

SEE 316

λ

α

ARA

17h 19.1m -46° 38'

Orbit

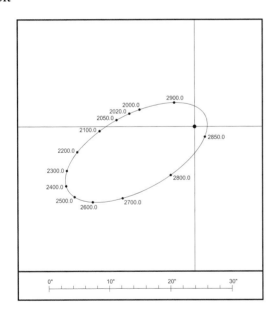

2900.0
2000.0
2020.0
2050.0
2100.0   2850.0
2200.0
2300.0   2800.0
2400.0
2500.0   2700.0
2600.0

0"   10"   20"   30"

Ephemeris for BSO 13 AB (2010 to 2100)

Orbit by Sca (2013d) Period: 953.0 years, Grade: 5

Year	PA(°)	Sep(")	Year	PA(°)	Sep(")
2010.0	256.3	10.05	2060.0	267.0	13.31
2020.0	259.0	10.76	2070.0	268.6	13.89
2030.0	261.3	11.43	2080.0	270.0	14.44
2040.0	263.4	12.08	2090.0	271.4	14.97
2050.0	265.3	12.71	2100.0	272.6	15.49

increase in the systemic radial velocity of the primary star of 9.6 metres per second per year, which indicates the presence of a stellar mass companion with an estimated period of over 500 years.

Observing and Neighbourhood

Found half-way between η Sco and α Ara, this is a beautiful pair with colours of deep yellow and orange. The different spectral types and the size and period of the apparent orbit almost make it a southern equivalent of η Cas. An easy object for the small aperture, the stars of this nearby binary were separated by only 2″ in 1905 but will widen for about four centuries until reaching a maximum separation of 23″.1 around 2420. Two faint field stars, magnitudes 13.4 and 14.4, are being left rapidly behind by the proper motion of AB. As of 1999, they were at 267°, 145″ and 290°, 95″ respectively. Nearby is the fine triple DUN 216+HJ 4949 (see Star 132) and also DUN 213 (6.9, 8.3, 168°, 8″.3, 2016). The fine pair SEE 316 is 4° SW (6.3, 7.7, 173°, 1″.0, 2016). Little orbital motion has been shown since discovery, but the WDS notes that the primary has a composite spectrum.

Measures

Early measure (I)	69°.6	2″.12	1900.64
(Orbit	68°.8	2″.09)	
Recent measure (ARY)	258°.3	10″.78	2014.25
(Orbit	258°.0	10″.52)	

131. $\rho^{1,2}$ HER = STF 2161 = WDS J17237+3709AB

Table 9.131 Physical parameters for $\rho^{1,2}$ Her

STF 2161	RA: 17 23 40.97	Dec: +37 08 45.3	WDS: 80(385)		
V magnitudes	A (ρ^2): 4.96	B (ρ^1): 5.91			
($B - V$) magnitudes	A: −0.03	B: 0.00			
μ(A)	−42.32 mas yr^{-1}	± 0.57	2.67 mas yr^{-1}	± 0.61 (DR2)	
μ(B)	−40.39 mas yr^{-1}	± 0.19	7.14 mas yr^{-1}	± 0.18 (DR2)	
π(A)	8.34 mas	± 0.34	391 light yr	± 16 (DR2)	
π(B)	9.04 mas	± 0.11	361 light yr	± 4 (DR2)	
Spectra	A: B9.5III	B: A0Vn			
Luminosities (L$_\odot$)	A: 120	B: 45	C:	D:	
Catalogues (A/B)	75 Her	HR 6485/4	HD 157779/8 (B/A)	SAO 66001/0	HIP 85112
DS catalogues	Mayer 46	H 2 3	STF 2161	BDS 8003	
	ADS 10526	MCA 48 (AaAb)			
Radial velocity	−21.0 km s^{-1}	± 2			
Galactic coordinates	61°.442	+32°.710			

History

This double star was first noted by Christian Mayer and appears as number 46 in his catalogue, as arranged by Jorge Schlimmer. William Herschel then observed it on 29 August 1779, when he found the stars 'Pretty unequal. Both w.(hite)'.

The Modern Era

The brightness and relatively wide separation of the stars has made this a popular object for measurement, and the *WDS Observations Catalog* currently contains 385 observations of the STF numbered pair. In 1981, during a survey with the Canada–France–Hawaii Telescope (CFHT), H. McAlister [529] and colleagues resolved the primary component of ρ into two stars. The measured separation was 0″.29. In 1985 the pair was unresolved when observed by two independent groups of observers using telescopes of apertures 1.9 and

Finder Chart

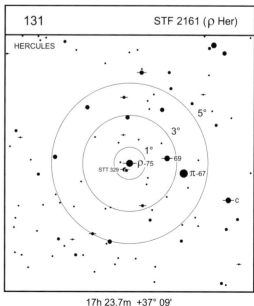

131   STF 2161 (ρ Her)

HERCULES

5°
3°
1°
STT 329
ρ-75
69
π-67
c

17h 23.7m +37° 09'

4 metres. Between 1989 and 2008 there were six further positive resolutions; in each case the position angle was between $0°$ and $30°$. In 2004, M. Scardia *et al.* [530] found the distance to be $0''.476$, and in 2007 B. D. Mason *et al.* [531] published a magnitude difference of 1.0 in the V band, whilst the distance had halved in the intervening three years. Such a pair at a distance close to $0''.5$ should be visible on occasion in relatively small apertures, but there appear to have been no sightings in the historical records. Burnham and Aitken avoided observing this pair, presumably because its very slow motion (see below) precludes any chance of getting an orbit, otherwise they might have seen Aab in one of the large American refractors. It may be worth taking a close scrutiny of ρ Her from time to time. It is likely that the orbital period of the close pair AaAb is short – perhaps 25 years or less.

Observing and Neighbourhood

The two stars are hot blue types and have appeared white or greenish white to various observers. F. G. W. Struve thought them greenish-white and greenish, whilst Admiral Smyth recorded bluish-white and pale emerald. RWA found them both white with an 8.3-inch (21-cm) reflector at ×96. About 15′ to the SE is STT 329 (6.4, 9.9, 12°, 33″.5, 2012).

Measures

Early measure (STF)	307°.2	3″.60	1830.35
Recent measure (ARY)	321°.9	3″.92	2014.67

132. Δ 216 ARA = WDS J17269–4551AB,C

Table 9.132 Physical parameters for Δ 216 Ara

DUN 216	RA: 17 26 51.98	Dec: −45 50 34.7	WDS: 1712(53)	
V magnitudes	A: 5.63	B: 6.46	C: 7.11	
(B − V) magnitudes	A: −0.07	B: −0.05	C: +0.04	
μ(A)	−8.41 mas yr^{-1}	± 0.21	−28.28 mas yr^{-1}	± 0.17 (DR2)
μ(B)	−3.96 mas yr^{-1}	± 0.31	−29.73 mas yr^{-1}	± 0.21 (DR2)
μ(C)	−5.91 mas yr^{-1}	± 0.12	−29.38 mas yr^{-1}	± 0.08 (DR2)
π(A)	5.01 mas	± 0.13	651 light yr	± 17 (DR2)
π(B)	5.23 mas	± 0.18	624 light yr	± 21 (DR2)
π(C)	5.15 mas	± 0.07	633 light yr	± 9 (DR2)
Spectra	A: B8V	B: B9V	C: B9.5V	
Luminosities (L$_\odot$)	A: 190	B: 75	C: 45	
Catalogues (AB/C)	HR 6477	HD 157661/49	SAO 227971/66	HIP 85389
DS catalogues	HJ 4949 (AB)	DUN 216 (AB,C)		
Radial velocity	+8.00 km s^{-1}	± 3.1		
Galactic coordinates	343°.718	−5°.941		

History

For this object, Dunlop gives a place for the epoch of 1826 that precesses (at 2000) to 17 25 54 −45 50 18, which agrees to within 1 minute of RA with the present catalogue position. He gives a distance of 33″.07 between two stars of magnitude 7 and a position angle of 59°56′ N following. He also adds that 'The southernmost star is double'. When John Herschel alighted on the same position in 1835 he found 'a very coarse double star, the position of which agrees with DUN 216, but the other particulars differ completely. The large star is a close double star hitherto undescribed.' Herschel found the difference in RA between A and C to be 7.37 seconds of time whereas Dunlop made it 1.55 seconds.

The Modern Era

It seems likely that this is a physical triple system. The parallax for the distant C star derived by Gaia agrees, within admittedly large errors, with that of AB found by Hipparcos. Veramendi & González [462] studied the stars looking for spectroscopic sub-systems but found none. Their conclusion was that the period of AB was about 2600 years whilst that of C around AB was over 700 centuries.

Observing and Neighbourhood

This is a spectacular object for the small telescope and it sits just at the extreme northern border of Ara with Scorpius,

Finder Chart

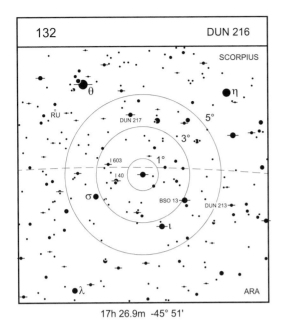

| 132 | DUN 216 |

SCORPIUS

θ

η

RU

DUN 217

5°

3°

I 603

1°

I 40

σ

BSO 13

DUN 213

ι

λ

ARA

17h 26.9m -45° 51'

3.5° SW of θ Sco. The close pair is easy in 10-cm whilst binoculars will show the wide pair. Within a 3° radius there are a number of other systems of interest: BSO 13, a nearby

binary of long period; DUN 213 (7.0, 8.3, 168°, 8″.3, 2016) and DUN 217 (6.3, 8.5, 169°, 13″.3, 2016). Just over 1° to the ESE is I 40 (6.0, 10.5, 210°, 20″, 2016, – the position angle is decreasing and the separation increasing). Moving 40′ NE of I 40, there is another Innes pair (I 603, 7.2, 9.4, 78°, 1″.3, 2010).

Measures

HJ 4949AB			
Early measure (JC)	266°.5	2″.51	1856.39
Recent measure (RWA)	251°.1	2″.12	2016.71
DUN 216AC			
Early measure (JC)	313°.2	103″.18	1856.47
Recent measure (RWA)	311°.6	102″.54	2016.71

133. $\nu^{2,1}$ DRA = STFA 35 = WDS J17322+5511

Table 9.133 Physical parameters for ν Dra

STFA 35	RA: 17 32 15.88	Dec: +55 10 22.1	WDS: 185(277)		
V magnitudes	A(ν^2): 4.87	B(ν^1): 4.90			
$(B - V)$ magnitudes	A: +0.32	B: +0.27			
μ(A)	143.73 mas yr^{-1}	$\pm$ 0.35	61.55 mas yr^{-1}	$\pm$ 0.34 (DR2)	
μ(B)	147.97 mas yr^{-1}	$\pm$ 0.26	54.89 mas yr^{-1}	$\pm$ 0.29 (DR2)	
π(A)	33.36 mas	$\pm$ 0.16	97.8 light yr	$\pm$ 0.5 (DR2)	
π(B)	33.09 mas	$\pm$ 0.13	98.6 light yr	$\pm$ 0.4 (DR2)	
Spectra	A: A4m + ?	B: A5V			
Masses (M$_\odot$)	Aa: 1.61	Ab: 0.24	B: 1.61		
Luminosities (L$_\odot$)	A: 8	B: 8			
Catalogues (A/B)	25/24 Dra	HR 6555/4	HD 159560/41	SAO 30450/47	HIP 85829/819
DS catalogues	H 5 11	BDS 8076	ADS 10628	SHY 299	
Radial velocity (A/B)	-16.00 km s^{-1}	$\pm$ 0.7	-15.20 km s^{-1}	$\pm$ 1.78	
Galactic coordinates	83°.019	33°.122			

History

This system was almost certainly discovered by Hodierna and described in his 1654 volume [533]. He notes a telescopic pair in the 'head of the Dragon which consists of four stars in a rhombus'. It was later recorded by Flamsteed in 1690, and subsequently by William Herschel on 19 October 1779. 'Double. A little unequal. L(arge) pale r(ed).; S(mall) pale r(ed). Distance 54′ 48′′′. Position 44° 19′ n(orth). preceding. From the right ascension and declination of these stars in FLAMSTEAD'S [sic] catalogue we can gather, that in his time the distance was 1′ 11′′.448; the position 41° 23′ preceding; their magnitudes equal or nearly so. The difference in the distance of the two stars is so considerable that we can hardly account for it otherwise than by admitting a proper motion in either one or other of the two stars, or in the solar system; most probably [none] of the three is at rest.' The WDS observations catalogue showed that the PA and separation

Finder Chart

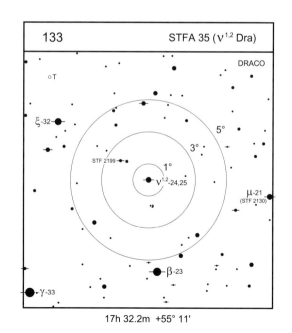

17h 32.2m +55° 11'

have varied little from 312° and 61″.8 over 200 years, so the large differences in separation in the measures of Herschel and Flamsteed are unexplained.

The Modern Era

Both components are spectroscopic binaries, according to WDS, but the MSC catalogue has only orbital parameters for the A component, which has a period of 38 days.

Observing and Neighbourhood

Found in the head of the Dragon, a glorious pair of white stars that can be seen in hand-held binoculars and are brilliant even in small telescopes. Gary Seronik [534] produced a rough guide to deciding whether you can resolve a particular separation in binoculars. Take the power of the binoculars (7 in 7 × 50, say) and divide into 300. In this case you get 43″. Greg Stone has tested this and finds it works quite well but will need to be modified in cases where the stars are faint or unequal. He reports that he can resolve ν Dra in 8 × 42. Moving 4° to the W brings you to μ Draconis (see Star 124), a beautiful binary; continue in the same direction by 5° and you alight on the bright triple star 17/16 Dra (STF2078, 5.4, 6.4, 5.5, 103°, 3″.1, 19°, 89″.3, 2016). Just over 1° to the NNE of ν is the binary STF 2199 (8.0, 8.6, 51°, 1″.9, 2020), which is seen well in 20-cm. This pair has a near 1300-year orbit and will be slowly widening for the next century or so.

Measures

Early measure (STF)	313°.0	61″.72
Recent measure (SMR)	310°.9	61″.85

134. 95 HER = STF 2264 = WDS J18015+2136

Table 9.134 Physical parameters for 95 Her

STF 2264	RA: 18 01 30.41	Dec: +21 35 44.8	WDS: 137(271)		
V magnitudes	A: 4.85	B: 5.29			
$(B - V)$ magnitudes	A: +0.17	B: +0.95			
$\mu(A)$	8.73 mas yr^{-1}	± 0.09	36.23 mas yr^{-1}	± 0.13 (DR2)	
$\mu(B)$	9.27 mas yr^{-1}	± 0.14	39.07 mas yr^{-1}	± 0.19 (DR2)	
$\pi(A)$	7.84 mas	± 0.09	416 light yr	± 5 (DR2)	
$\pi(B)$	7.89 mas	± 0.12	413 light yr	± 6 (DR2)	
Spectra	A: A5IIIn	B: G8III			
Luminosities (L$_\odot$)	A: 150	B: 100			
Catalogues	HR 6730	HD 164669	SAO 85648	HIP 88267	
DS catalogues	Mayer 50	H 3 26	STF 2264	BDS 8302	ADS 10993
Radial velocity	−33.30 km s^{-1}	± 0.5			
Galactic coordinates	47°.500	+20°.285			

History

The pair 95 Herculis was discovered by Christian Mayer and is number 50 in his catalogue [29]. The colours of these stars have caused considerable comment from observers over the past two centuries. Sir William Herschel noted that both components were white. Admiral W. H. Smyth found them to be apple green and cherry red, as did his son Charles Piazzi Smyth until 1856, when he was surprised to find them both white. It should be noted that at this point Piazzi Smyth was observing from altitude (8000 feet) on Tenerife. Admiral Smyth then re-observed them and confirmed his earlier colours and then, in correspondence with a number of the leading observers of the day, he found that all of them had found significant colours in the components of 95 Her. Later in the century, Dúner found bright green and yellow, whilst Flammarion found the hues to be gold and azure. In his 1906 book on the Struve stars Thomas Lewis noted that he had found that both stars were pale yellow. There is considerable

Finder Chart

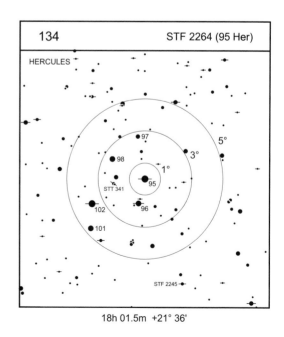

18h 01.5m +21° 36'

difference in the spectral types. Both stars are giants and the primary is more than a whole spectral class earlier than its companion, so some contrast might be expected. Smyth's little book on double stars, usually called *Sidereal Chromatics* for short, explains what his terms for colour hue actually represented. He admitted that he used inexact epithets and then presented a table explaining what these terms meant. The table can be found in the Appendix. By apple green he meant brownish green and cherry red is actually pale red, rather than the deeper hue that modern observers might take it to mean.

The Modern Era

Although the WDS gives a separation for 1777 as the first measure known for this pair, it is an estimate only; William Herschel measured the stars in 1781 and came up with a distance of 6″.1. There has been little change in separation or position angle since that date.

Observing and Neighbourhood

The controversial historical observations of colour in each star make this double a target for the small aperture. Recent estimates show little difference in hue and certainly not the spectacular colours noted by Smyth and Piazzi Smyth. In 1968 with an 8.3-inch reflector RWA found both stars white, whilst John Nanson [535], observing in July 2010 with 3-inch aperture at ×133, noted that both stars were white but 'with just a hint of yellow on closer examination'. With an 8-inch f/5.9 Newtonian at ×240 in 2009, Jeremy Perez [536] noted colours of light blue and light yellow, an impression RWA also received recently with the Cambridge 8-inch. One degree E is STT 341AB, a rapid binary of 20-year period; the orbit is highly inclined and very eccentric (7.8, 8.8, 87°, 0″.20, 2020). The stars will widen to 0″.49 in 2029, which is as far apart as they get; given the difference in magnitude, this pair will test the largest apertures and the seeing. Four degrees S and slightly W is an easier proposition, STF 2245 (7.4, 7.6, 11°, 2″.5, 2017).

Measures

Early measure (STF)	261°.8	6″.06	1829.90
Recent measure (ARY)	258°.0	6″.37	2014.68

135. τ OPH = STF 2262 = WDS J18031−0811AB

Table 9.135 Physical parameters for τ Oph

STF 2262	RA: 18 03 04.92	Dec: −08 10 49.3	WDS: 27(633)	
V magnitudes	A: 5.30	B: 5.94		
(*B − V*) magnitudes	A: +0.43	B: +0.41		
μ(A)	12.09 mas yr^{-1}	± 0.46	−37.65 mas yr^{-1}	± 0.45 (DR2)
μ(B)	31.13 mas yr^{-1}	± 0.68	−30.01 mas yr^{-1}	± 0.44 (DR2)
π(A)	18.62 mas	± 0.34	175 light yr	± 3 (DR2)
π(B)	18.59 mas	± 0.35	175 light yr	± 3 (DR2)
Spectra	A: F2V + ?	B: F5V		
Masses (M$_\odot$)	Aa: 1.93	Ab: 0.34	B: 1.64	
Luminosities (L$_\odot$)	A: 18	B: 10		
Catalogues (A/B)	HR 6734/3	HD 164765/4	SAO	HIP 88404
DS catalogues	H 1 88 (AB)	STF 2262 (AB)		
Radial velocity	−38.39 km s^{-1}	± 0.28		
Galactic coordinates	20°.061	+6°.870		

History

This was first seen by William Herschel on 28 April 1783, and reported in his second double star survey paper in 1784. 'The closest of all my double stars; can only be suspected with 460; but 932 confirms it to be a double star. Pretty unequal. Both p(ale)r(ed) or w(hitish)r(ed).' Star C (magnitude 11.3 at 125°, 100″.8, 2000) was first measured by W. H. Smyth. Although the distance and angle have not changed since 1880, Smyth's distances of 1832 and 1838 were given as 83″.1 and 82″.7, and his angles are about 10° out, too.

The Modern Era

Star A was found to be an SB1 with a period of 184.085 days, but little is known about the companion. Tokovinin in his MSC gives a mass of 0.34 M$_\odot$. Shaya and Olling [270] have shown that the chances of C being physical are almost 100%.

Finder Chart

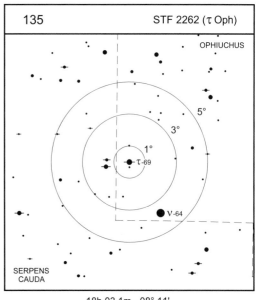

18h 03.1m -08° 11'

Orbit

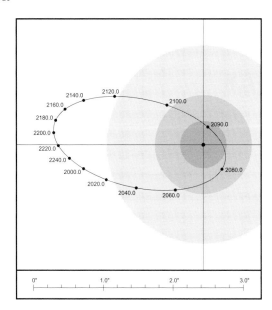

Ephemeris for STF 2262 AB (2015 to 2060)

Orbit by Sod (1999) Period: 257 years, Grade: 3

Year	PA(°)	Sep(")	Year	PA(°)	Sep(")
2015.0	287.0	1.55	2040.0	302.2	1.13
2020.0	289.4	1.47	2045.0	306.8	1.04
2025.0	292.0	1.39	2050.0	312.3	0.94
2030.0	295.0	1.31	2055.0	319.1	0.84
2035.0	298.3	1.22	2060.0	327.7	0.74

The stars are now slowly closing to periastron (0″.22 in 2088 or thereabouts).

Observing and Neighbourhood

The pair can be found 11° S of 70 Oph. It should be readily resolvable in 80-mm providing the star is reasonably high in the sky. From the UK it requires a night of good seeing. Five degrees due E (and just off the chart) is the pair STF 2303 (6.6, 9.3, 240°, 1″.6, 1991), which has halved its separation since 1825 and does not appear to have been observed for almost 20 years. A test for 20-cm.

Measures

Early measure (STF)	192°.9	0″.35
(Orbit	197°.2	0″.38)
Recent measure (ARY)	287°.9	1″.62
(Orbit	286°.5	1″.56)

136. 70 OPH = STF 2272
= WDS J18055+0230AB

Table 9.136 Physical parameters for 70 Oph

STF 2272	RA: 18 05 27.37	Dec: +02 29 59.3	WDS: 2(1700)	
V magnitudes	A: 4.22	B: 6.17		
$(B - V)$	A: +0.94	B: +1.41		
μ	124.16 mas yr^{-1}	± 0.81	−962.82 mas yr^{-1}	± 0.84
π	196.72 mas	± 0.83	16.58 light yr	± 0.07
$\mu(B)$	334.07 mas yr^{-1}	± 0.38	−1069.29 mas yr^{-1}	± 0.31 (DR2)
$\pi(B)$	195.22 mas	± 0.10	16.71 light yr	± 0.01 (DR2)
Spectra	K0V	K4V		
Masses (M$_\odot$)	A: 0.90	± 0.04	B: 0.70	±0.07
Radii (R$_\odot$)	A: 0.83	± 0.003	B: 0.67	± 0.008
Luminosities (L$_\odot$)	A: 0.4	B: 0.1		
Catalogues	HR 6752	HD 165341	SAO 123107	HIP 88601
DS catalogues	H 2 4	STF 2272	BDS 8340	ADS 11046
Radial velocity	−6.87 km s^{-1}	± 0.08		
Galactic coordinates	29°.893	+11°.367		

History

The bright pair AB was discovered by William Herschel in 1779 and called by him 'p (70) Serpentarii'. The elder Herschel noted the speed with which the position angle changed between 1779 and 1804. 'This', he says, 'cannot be owing to the effect of systematic parallax, which could never bring the small star to the preceding side of the large one', which helped Herschel to conclude that what he was seeing was orbital motion under a mutual gravitational attraction. When John Herschel came to consider the system in the 1830s he was worried by the fact that measurements by F. G. W. Struve at Dorpat did not agree with the expected motion from his orbital analysis, leading to speculation about the presence of a third body which was perturbing one or other component of the bright pair. A number of contemporary astronomers seemed to be of the same opinion, including J. H. Mädler.

Finder Chart

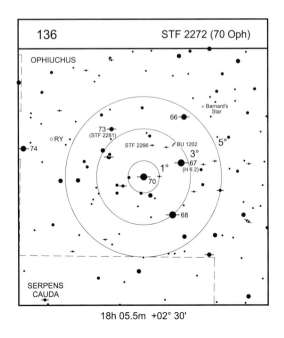

18h 05.5m +02° 30'

Orbit

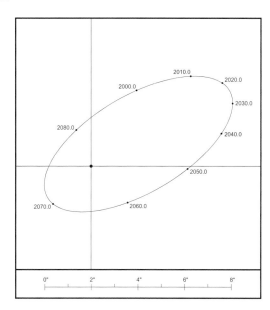

Ephemeris for STF 2272 AB (2018 to 2036)

Orbit by Egn (2008) Period: 88.3770 years, Grade: 1

Year	PA(°)	Sep(")	Year	PA(°)	Sep(")
2018.0	123.5	6.50	2028.0	115.1	6.67
2020.0	121.8	6.61	2030.0	113.4	6.60
2022.0	120.1	6.68	2032.0	111.7	6.49
2024.0	118.4	6.71	2034.0	109.9	6.36
2026.0	116.8	6.71	2036.0	108.0	6.19

W. S. Jacob in 1857 even thought the third body was a planet. Towards the end of the century T. J. J. See, E. Doolittle, and T. Lewis were all convinced that there were clear deviations from Keplerian motion. When R. G. Aitken compiled his double star catalogue in 1932 he noted that more calculations of the orbit of 70 Oph had been made than for any other system. He also included the elements of a contemporary calculation by Pavel which accounted for the a perturbation of A by a body with a period of 6.5 years but suggested that the orbital elements could not be further improved until the next periastron passage in the 1980s. In Baily's *British Catalogue* 70 Oph is designated by 'p' although this does not exist on Baily's map, and Baily rejected it in his edition of *Flamsteed's Catalogue*.

The Modern Era

In 1937 Kai Strand [537] made a thorough analysis of the available material, including some photographic results from Potsdam from 1914 to 1922 and from Johannesburg from 1931 to 1935; he argued that this material is sufficient to

Orbit Measures with 8" Refractor

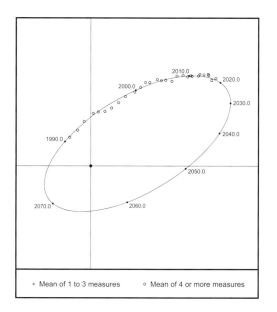

+ Mean of 1 to 3 measures   ○ Mean of 4 or more measures

Year	PA(°)	Sep(")	No.	Year	PA(°)	Sep(")	No.
1990.60	217.4	1.51	5	2005.52	137.7	4.85	8
1991.66	201.1	1.60	5	2006.53	135.1	5.00	7
1992.62	189.0	1.88	6	2007.58	135.1	5.31	8
1993.59	177.2	2.20	5	2008.61	133.4	5.52	8
1994.57	171.5	2.28	7	2009.52	131.5	5.62	5
1995.55	164.9	2.36	6	2010.63	131.0	5.76	5
1996.53	159.7	2.58	6	2011.64	130.4	5.78	6
1997.52	155.8	2.90	6	2012.53	128.8	6.00	6
1998.63	153.0	3.26	5	2013.57	129.0	6.10	6
1999.53	148.5	3.62	6	2014.56	127.8	6.22	6
2000.55	146.7	3.93	8	2015.55	127.0	6.38	7
2001.55	145.7	4.22	8	2016.53	126.5	6.36	6
2002.65	143.6	4.33	7	2017.58	124.3	6.35	6
2003.61	141.2	4.63	8	2018.53	124.0	6.54	8
2004.69	139.2	4.71	6				

prove that large perturbations are impossible. Even so, some six years later Reuyl & Holmberg [538] used photographic material to argue for a third body. In the 1980s radial velocity observations by Batten & Fletcher [539] and astrometric studies by Heintz [540] more or less settled the question of whether there was a third body in the system. The micrometric measurement data show no evidence of another star and Heintz suggested that the earlier observations were affected by the significant difference in brightness between the two stars. He concluded 'Other than systematic errors the hypothesis would remain of a one-time outside perturbation of the binary, a kind of massive interstellar *Nemesis*, which may have altered the orbit some time around 1880. Thus the choice seems to be between an exciting potential discovery and the meek recourse to observational human frailty.' The A star is also in the IBVS as V2391 Oph. SIMBAD allocates it to the

BY Draconis class of variables whilst the AAVSO classify it as belonging to the RS CVn class.

Exoplanet Host?

A long-term radial velocity monitoring programme carried out with the 2.7-metre reflector at McDonald Observatory between 1988 and 2006 was able to rule out any giant planets in the system beyond 8 AU.

Observing and Neighbourhood

A spectacular binary star, one of the finest in the sky. Its brightness and the beautiful colours of its components combined with an ease of resolution make it a top target for the small telescope user. Admiral Smyth noted colours of pale topaz and violet whilst T. W. Webb found yellow and orange in 1850 as did E. J. Hartung about a century later, and the observers of the ASNSW around 1980 made them yellow and orange yellow. Only 17 light years away, this is one of the closest stars to the Sun. It is also one of the most observed. The WDS lists at least 1730 measurements since the pair was discovered. The significant magnitude difference between the two stars amounting to almost two magnitudes does tend to make measurement with a filar micrometer more difficult than for equally bright stars. The pair is currently opening and will reach maximum separation ($6''.7$) in 2025. This is a nearby stellar system, so even at closest separation ($1''.5$) the stars can be seen in a small aperture and can therefore be followed around the whole of the 88-year orbit. In the true orbit, the stars are 12 AU apart at periastron and 35 AU distant at apastron. Two degrees to the NE is 73 Oph (STF 2281 (see Star 138). To the WNW, $1°.5$ is 67 Oph, one of Herschel's class VI pairs, H 6 2 (4.0, 8.0, $142°$, $54''.3$, 2014). The star BU 1202 sits about a degree N of 67 Oph and two distant stars can be seen with 15-cm. These are C (magnitude 10.2) at $28°$, $105''$, 2011, whilst E (8.4) is at $138°$, $90''$, 2011. The pair AB is faint, unequal, and was last seen in 1996 at $0''.6$. Following BU 1202 by about $50'$ is STF 2266 (7.9, 9.6, $188°$, $8''.7$, 2016).

Faint field stars within 5 arcminutes of 70 Oph

Comp	Epochs	PA1	PA2	Sep1	Sep2	Magnitudes	
AB	1777 2014	82	128	7.0	6.2	4.22	6.17
AC	1878 2013	198	326	71.4	93.1		12.05
AD	1905 2009	252	315	70.0	129.3		14.36
AR	1905 2012	73	26	105.0	164.8		12.87
AS	1878 2009	50	10	87.2	202.6		12.45
AT	1905 2013	103	45	120.3	128.4		12.25
AU	1905 2009	329	337	142.4	255.5		13.79
AV	1886 2013	224	276	165.9	144.5		10.83
AY	1924 2009	5	357	158.3	250.5		14.73
AZ	2000 2000	163	163	68.3	68.3		16.04
BC	1899 2000	208	322	50.8	81.2	6.17	12.05
BD	1900 2000	247	311	69.3	120.0		14.36
BR	1908 2000	70	29	104.7	154.3		12.87
BZ	1900 2000	168	163	68.4	70.2		16.04
VT	1905 2009	73	72	247.4	247.5	10.83	12.25
VW	1906 2009	270	270	180.5	180.4		11.77
VX	1906 2012	250	252	16.6	17.4		13.88

Measures

Early measure (STF)	$135°.7$	$5''.31$	1830.84
(Orbit	$134°.4$	$5''.31$)	
Recent measure (ARY)	$125°.2$	$6''.44$	2018.47
(Orbit	$122°.4$	$6''.53$)	

137. HJ 5014 CRA = WDS J18068–4325

Table 9.137 Physical parameters for HJ 5014 CrA

HJ 5014	RA: 18 06 49.91	Dec: −43 25 30.0	WDS: 529(120)	
V magnitudes	A: 5.93	B: 6.02		
(B − V) magnitudes	A: +0.25	B: +0.26		
μ(A)	10.99 mas yr^{-1}	± 0.34	−105.74 mas yr^{-1}	± 0.29 (DR2)
μ(B)	−0.88 mas yr^{-1}	± 0.55	−101.05 mas yr^{-1}	± 0.48 (DR2)
π(A)	22.44 mas	± 0.16	145.3 light yr	± 1.0 (DR2)
π(B)	22.35 mas	± 0.33	145.9 light yr	± 2.2 (DR2)
Spectra	A: A6V	B: A7V		
Luminosities (L$_\odot$)	A: 7	B: 7		
Catalogues	HR 6749	HD 165189	SAO 228708	HIP 88726
DS catalogues	HJ 5014			
Radial velocity	−7.80 km s^{-1}	± 0.4		
Galactic coordinates	349°.447	−10°.861		

History

In his 1847 publication which summarises the results of the observing at the Cape of Good Hope, John Herschel gives two catalogues of measures for this pair. The first is a summary of the 20-foot telescope observations, in which the separation is recorded as 0″.7, and the second includes measures from the 7-foot instrument. The pair HJ 5014 was found in 1836 and no notes are attached. With the 7-foot telescope he observed the pair on four occasions, and whilst he called the system HJ 5013 in error, he gave some background on the pair. He found it during a 'review of large stars', presumably with the smaller telescope, and notes on the last night of four that it 'is excessively close and difficult'.

Finder Chart

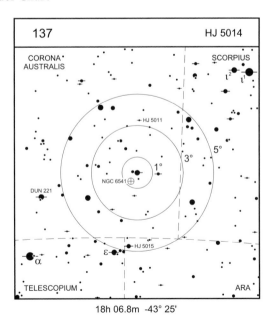

18h 06.8m -43° 25'

Orbit

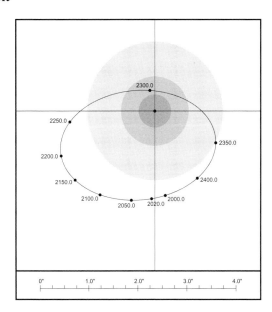

Ephemeris for HJ 5014 (2010 to 2100)

Orbit by Alz (2001a) Period: 450.00 years, Grade: 4

Year	PA(°)	Sep(")	Year	PA(°)	Sep(")
2010.0	2.4	1.72	2060.0	340.9	1.87
2020.0	357.8	1.75	2070.0	337.0	1.91
2030.0	353.4	1.78	2080.0	333.3	1.94
2040.0	349.0	1.81	2090.0	329.7	1.98
2050.0	344.9	1.84	2100.0	326.2	2.01

The Modern Era

This pair is thought to be a part of the β Pictoris Young Moving Group [541]. Such stellar groups are excellent places to find young stars for studies of star formation and stellar evolution. The β Pic group is thought to be only 25 million years old. The proper motion of HJ 5014 will carry it to within 7′ of the cluster NGC 6541 in about 1000 years time.

Observing and Neighbourhood

The pair HJ 5014 can be found by projecting a line through λ and ι^1 Sco by the same length again. It is a beautiful pair for 10-cm and above. The stars are currently at 358°, 1″.8, 2020, and will widen for 150 years when they will be separated by 2″.2. There are a number of John Herschel pairs in the area, the nearest being HJ 5011 (7.6, 8.5, 344°, 28″, 2016), and 2°.5 S and a little W is HJ 5015 (6.2, 9.6, 258°, 4″.3, 2016). Three and a quarter degrees ESE of HJ 5014 can be found DUN 221 (5.2, 10.1, 161°, 74″, 2010), where the separation has increased from the value of 40″ which Dunlop found in 1826. The small values of proper motion assigned to both stars by the WDS seem at odds with the rapid increase in distance, assuming that Dunlop's observation is correct.

Measures

Early measure (COG)	65°.3	1″.67
(Orbit	65°.9	1″.37)
Recent measure (ARY)	2°.7	1″.81
(Orbit	359°.2	1″.74)

138. 73 OPH = STF 2281 = WDS J18096+0400AB

Table 9.138 Physical parameters for 73 Oph

STF 2281	RA: 18 09 33.89	Dec: +03 59 35.8	WDS: 73(406)		
V magnitudes	A: 6.03	B: 7.96	C: 12.6		
$(B - V)$	A: +0.33	B: +0.66			
μ	48.88 mas^{-1}	± 0.65	−2.96 mas yr^{-1}	± 0.50	
π	18.25 mas	± 0.60	179 light yr	± 6	
μ(A)	32.59 mas yr^{-1}	± 1.69	0.01 mas yr^{-1}	± 1.31 (DR2)	
π(A)	10.24 mas	± 1.08	319 light yr	± 34 (DR2)	
Spectra	A: F0V	B: G1V			
Luminosities (L$_\odot$)	A: 30	B: 5			
Catalogues	HR 6795	HD 166233	SAO 123187	HIP 88964	V2666 Oph
DS catalogues	H 1 87	STF 281	BDS 8380	ADS 11111	
Radial velocity (A/B)	−8.80	± 2.8 km s^{-1}			
Galactic coordinates	31°.725	+11°.132			

History

The pair 73 Oph was found as a difficult double star by William Herschel on 27 April 1783. He noted that it was 'A very minute double star. Considerably unequal. L(arge) r(ed). S(mall) r(ed). With 227 not to be suspected unless known to be double, but may be seen wedge-formed, and with long attention I have also perceived a most minute division.' During the nineteenth century the companion began to close slowly and by 1900 it made a close approach to A (to within almost 0″.1) and then around 1930 began to head back towards the third quadrant again.

The Modern Era

In 2003 Fekel & Henry [542] wrote a paper for the *Astrophysical Journal* in which they analysed the light variations of the star V2502 Oph. In their observing they used 73 Oph

Finder Chart

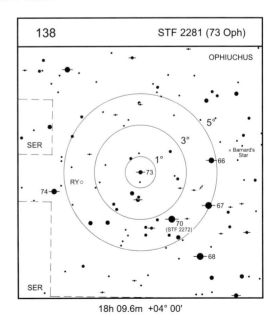

18h 09.6m +04° 00'

Orbit

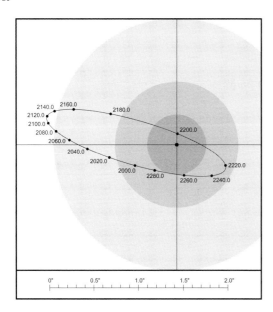

Ephemeris for STF 2281 AB (2015 to 2060)

Orbit by Sod (1999) Period: 294 years, Grade: 2

Year	PA(°)	Sep(")	Year	PA(°)	Sep(")
2015.0	283.5	0.70	2040.0	340.9	1.00
2020.0	280.7	0.77	2045.0	337.0	1.06
2025.0	278.2	0.83	2050.0	333.3	1.11
2030.0	276.1	0.89	2055.0	329.7	1.16
2035.0	274.3	0.95	2060.0	326.2	1.21

as a check star and found that it too also varied in brightness. Further investigation showed that 73 Oph A is a γ Doradûs variable with a period of 0.67 days and an amplitude in the V band of about 0.006 magnitude. The γ Doradûs variables are similar to Cepheids, but the pulsations are far more akin to water waves on the surface of the Earth, and the change in luminosity is much smaller than that in the classical Cepheid variables. The GCVS now includes 73 Oph in the catalogues as V2666 Oph. The current visual orbit shows that the stars will continue to separate for about another century.

Observing and Neighbourhood

A significant factor in resolving this pair at the moment is the question of the magnitude difference. The WDS gives magnitudes for A and B of 5.97 and 7.52, whereas the Hipparcos satellite observed both components in the early 1990s and obtained magnitudes in the Tycho B and V system. These are similar to Johnson's B and V but when the transformation is made the two components have V magnitudes of 6.03 and 7.96 respectively. RWA has made two measures of the pair in the last few years but there is a significant difference between the measured and orbital position angles, which is no doubt due to the difficulty of seeing component B clearly. A faint star of magnitude 12.6, 68″ distant in position angle 194°, is unconnected to the visual binary system. The pair is easily found just over 2° NE of 70 Oph (see Star 136). About 3° slightly N and W of 73 Oph is Barnard's star (see the finding chart).

Measures

Early measure (STF)	259°.7	1″.54	1831.05
(Orbit	257°.2	1″.48)	
Recent measure (TOK)	285°.9	0″.7063	2014.76
(Orbit	283°.6	0″.70)	

139. α LYR = VEGA = H 5 39
= WDS J18369+3846AB

Table 9.139 Physical parameters for α Lyr

H 5 39	RA: 18 36 56.34	Dec: +38 47 01.3	WDS: 900(83)		
V magnitudes	A: 0.09	B: 9.5	C = 11.0	D = 11.0	E = 9.5
(B − V) magnitudes	A: 0.00	B:			
μ (A)	+200.94 mas yr^{-1}	± 0.32	286.23 mas yr^{-1}	± 0.40	
π (A)	130.23 mas	± 0.36	25.04 light yr	± 0.07	
μ (B)	4.06 mas yr^{-1}	± 0.04	0.54 mas yr^{-1}	± 0.04 (DR2)	
π (B)	1.46 mas	± 0.03	2230 light yr	± 45 (DR2)	
Spectra	A0V				
Masses (M$_\odot$)	A: 2.71	± 0.02			
Luminosities (L$_\odot$)	A: 44	B: 60			
Catalogues	3 Lyr	HR 7001	HD 172167	SAO 67174	HIP 92162
DS catalogues	H 5 39 (AB)	STFB 9 (AC/AE)	BDS 8692	ADS 11510	
Radial velocity	−20.60 km s^{-1}	± 0.2			
Galactic coordinates	67°.448	+19°.237			

History

When William Herschel observed Vega on 24 September 1781 he found great difficulty in seeing the magnitude 9.5 field star in the glare of Vega itself. He noted that it was easier to see with no moonlight, and he obtained a separation of 37″. A month later he made a measurement of the diameter of the star and with his micrometer and a power of ×6450 he obtained a value of 0″.3553. Attempting this observation will have necessitated some skill at the telescope since he notes that Vega crossed the field of this high-power eyepiece in less than three seconds. He was of the opinion that with a 6-inch aperture there was enough light from Vega to bear a magnification of ×100,000. Vega was the first star to be photographed (at Harvard in 1850) and in 1872 Henry Draper took the first photographic stellar spectrum – that of Vega.

Finder Chart

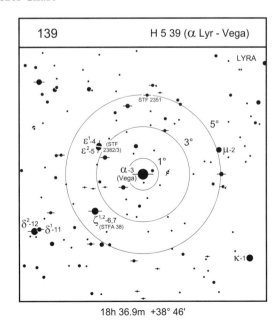

18h 36.9m +38° 46'

The Modern Era

Vega is of considerable interest to modern astronomers because of the disk of material which surrounds it and is confirmed from infrared observations by IRAS and Spitzer, which may presage the formation of a solar system. It is also one of the nearest A stars to our Solar System and appears to be rotating very quickly. Recent measures put the rotational velocity at about 270 km s^{-1}. As a consequence the star is very oblate in shape – the polar radius is 2.36 R$_\odot$ whilst the equatorial radius is 2.82 R$_\odot$ [544]. Dynamically Vega belongs to the Castor moving group, but its membership is uncertain as it is considerably older than the other members. SIMBAD has more than 2300 references to the star. The angular diameter of Vega has been directly measured a number of times and the current value, obtained from observations with the Navy Prototype Optical Interferometer as 2.71 mas ± 0.02 [546], is about 100 times smaller than Herschel's measurement. The NPOI currently has six imaging stations, with baselines between 10 and 97 metres, but four more will be added to extend the baseline to 432 metres. The current limiting magnitude is about $V = 6$.

Observing and Neighbourhood

The brightest star in the northern sky, with the exception of the orange Arcturus, Vega is a member of the Summer Triangle along with Altair in Aquila and Deneb in Cygnus. Unmistakeably, in mid-summer Vega is near the zenith in the UK. The B star is now more than 80″ away from the primary star, making its observation easier than it would have been in Herschel's time. With the Cambridge 8-inch the companion is barely visible when the micrometer field illumination is on. In a dark sky 15-cm should suffice to see it. The group of faint stars listed in the WDS entry on Vega are all unconnected, as indeed is star B. For more information on the field of Vega, see the article by Juerg Schlimmer [543]. The Double-Double (see Star 141) is 1°.5 NNE whilst 1°.5 SE is ζ Lyr (STFA 38) (4.3, 5.6, 150°, 43″.8, 2017) whilst the WDS adds four distant stars with magnitudes between 13 and 16. A neat pair can be found 2°.5 N and slightly E. This is STF 2351 (7.6, 7.6, 160°, 5″.1, 2017).

Measures

Early measure (STF)	137°.8	42″.95	1836.14
Recent measure (SMR)	183°.7	82″.13	2016.37

140. STF 2398 DRA = WDS J18428+5938AB

Table 9.140 Physical parameters for STF 2398 Dra

STF 2398	RA: 18 42 46.69	Dec: +59 37 49.4	WDS 32 (610)	
V magnitudes	A: 9.11	B: 9.96	C: 12.2	D: 13.5
(*B* − *V*) magnitudes	A: +1.79	B: +2.02		
μ(A)	−1311.91 mas yr^{-1}	± 0.18	1792.19 mas yr^{-1}	± 0.22 (DR2)
μ(B)	−1400.02 mas yr^{-1}	± 0.03	1862.41 mas yr^{-1}	± 0.39 (DR2)
π(A)	283.95 mas	± 0.06	11.487 light yr	± 0.002 (DR2)
π(B)	283.86 mas	± 0.11	11.490 light yr	± 0.004 (DR2)
Spectra	A: M4	B: M4.5		
Radii (R$_\odot$)	A: 0.36	± 0.006	B: 0.33	± 0.011
Luminosities (L$_\odot$)	A: 0.002	B: 0.001		
Catalogues (A/B)	HD 173739/40	SAO 31128/9	HIP 91768/72	
DS catalogues	STF 2398	BDS 8798	ADS 11632	
Radial velocity (A)	−1.07 km s^{-1}	± 0.09	B: −1.09 km s^{-1}	± 0.09
Galactic coordinates	89°.288	+24°.231		

History

Found by F. G. W. Struve at Pulkovo, this pair initially encouraged little follow-up interest owing to the faintness of the stars and the relatively large angular separation. By the end of the nineteenth century, however, proper motions had been obtained which were shown to be significantly large. This led to a number of parallax determinations by Kapteyn (0″.35), Flint (0″.32), Lamp (0″.35), and Schlesinger (0″.30)[194], which gave an average value of 0″.32 corresponding to a distance of 10 light years. The magnitude 12.2 star C was first seen by Barnard [553] on 1 August 1904, with the 40-inch at Yerkes at powers of ×700 and ×1300. He was responding to a request, possibly from Burnham, to measure the AB pair. The seeing was reported to be 'often very poor'. The separation between A and C, which was 51″ at the time of Barnard's observation, has now increased to more than 250″.

Finder Chart

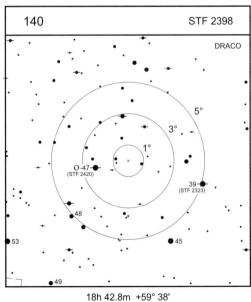

18h 42.8m +59° 38'

Orbit

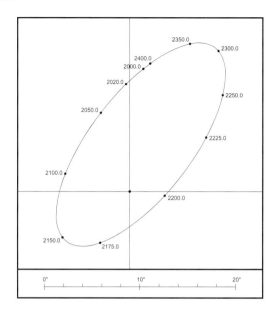

Ephemeris for STF 2398 AB (2015 to 2060)

Orbit by Hei (1987b) Period: 408 years, Grade: 4

Year	PA(°)	Sep(")	Year	PA(°)	Sep(")
2015.0	179.7	11.37	2040.0	193.5	9.33
2020.0	182.0	10.96	2045.0	197.1	8.95
2025.0	184.6	10.54	2050.0	200.9	8.58
2030.0	187.3	10.13	2055.0	205.0	8.24
2035.0	190.3	9.73	2060.0	209.5	7.92

The Modern Era

In 1953 Wieth-Knudsen [548] produced a parabolic orbit for the pair but Heintz [549] re-examined the available data in 1968, and added his own material from Sproul photographs, to state that the orbit was elliptical and that the period appeared to be 453 years. There was, however, considerable speculation about the existence of perturbations to one or both components of the pair. In 1960, Eichhorn & Alden [552], using plates from Yerkes and McCormick observatories and a series of visual measures made by Lamp at Kiel with a 205-mm refractor between 1885 and 1887, suggested that there were two perturbations, with periods of 5.5 and 10.7 years and with amplitudes amounting to about

0″.02. Baize [555], in 1976, had suggested that a third body was perturbing one of the stars but Heintz [550] found no evidence of this when he revisited the system in 1987 and obtained the currently accepted period of 408 years (see the ephemeris above). Popovic [551], in 1988, noted a star of magnitude 13.5 at 91″, which, like the C component, is also being left behind rapidly by the motion of the binary pair.

Exoplanet Host?

Most recently an investigation by the HARPS-N team [554] into the radial velocities of both components to determine the noise induced by observations of late-type stars led to a possible perturbation of component B with period of 2.7 days and amplitude 1.2 m s^{-1}.

Observing and Neighbourhood

One of the more difficult pairs in this volume to find, STF 2398 lies in Draco, situated approximately between the pairs omicron and 39 Dra (STF 2323) (see the finder chart). The stars will continue to close until around 2090, when the separation will have reduced to 6″.9. They then slowly widen to 8″.5 in 2150 before reaching closest separation (2″.8) in 2194. The widest part of the orbit (17″.1) is reached in 2310. A magnitude 12.2 star (C) is being left behind rapidly (158°, 215″, 2000). Nearby are the pairs o and 39 Draconis. Omicron (STF 2420) has stars of magnitudes 4.8 and 8.3 separated by 37″, which is slowly increasing, and the PA is currently 319°. The star 39 Dra (STF 2323, 5.1, 8.1, 347°, 3″.8, 2015) shows slow orbital motion with a preliminary period of almost 4000 years. A third star C of magnitude 8 is 80″ away in PA 19°.

Measures

Early measure (STF)	134°.4	12″.42	1832.17
(Orbit	132°.0	12″.42)	
Recent measure (LOC)	179°.6	11″.57	2015.73
Orbit	180°.0	11″.31)	

141. $\epsilon^{1,2}$ LYR = STF 2382/3 = WDS J18443+3940AB,CD

Table 9.141 Physical parameters for ϵ Lyr

STF 2382 (AB) STF 2383 (CD)	RA: 18 44 20.35 RA: 18 44 22.78	Dec: +39 40 12.5 Dec: +39 36 45.8	WDS: 35 (603) WDS: 28 (617)		
V magnitudes	A: 5.15	B: 6.10	C: 5.25	D: 5.38	
$(B-V)$	A: +0.14	B: +0.23	C: +0.19	D: +0.25	
μ(A)	9.30 mas yr^{-1}	$\pm$ 0.48	75.29 mas yr^{-1}	$\pm$ 0.48 (DR2)	
μ(B)	1.83 mas yr^{-1}	$\pm$ 0.11	49.55 mas yr^{-1}	$\pm$ 0.09 (DR2)	
μ(C)	6.97 mas yr^{-1}	$\pm$ 0.32	49.93 mas yr^{-1}	$\pm$ 0.26 (DR2)	
μ(D)	4.14 mas yr^{-1}	$\pm$ 0.23	66.45 mas yr^{-1}	$\pm$ 0.20 (DR2)	
π(A)	17.97 mas	$\pm$ 0.23	181.5 light yr	$\pm$ 2.3 (DR2)	
π(B)	20.41 mas	$\pm$ 0.05	159.8 light yr	$\pm$ 0.4 (DR2)	
π(C)	20.19 mas	$\pm$ 0.13	161.5 light yr	$\pm$ 1.0 (DR2)	
π(D)	20.06 mas	$\pm$ 0.12	162.6 light yr	$\pm$ 1.0 (DR2)	
Spectra	A: A3V	B: F0V	C: A6Vn	D: A7Vn	
Luminosities (L$_\odot$)	A: 21	B: 7	C: 15	D: 14	
Catalogues (AB/CD)	HR 7051/2	HD 173582/607	SAO 67310/09	HIP 91919/26	
DS catalogues (AB/CD)	Mayer 57/8	H 2 5	STF 2382/3	BDS 8783/5	ADS 11632
Radial velocity (AB)	−31.2 km s^{-1}	$\pm$ 1.7 km s^{-1}			
Radial velocity (CD)	−24.40 km s^{-1}	$\pm$ 1.7 km s^{-1}			
Galactic coordinates (AB)	68°.849	+18°.198			
Galactic coordinates (CD)	68°.795	+18°.171			

History

The unusual nature of ϵ Lyrae was revealed in 1777 when Christian Mayer turned his 2.7-inch aperture telescope onto the system. He noted that each of the two bright components, separated by about 210″ (and therefore visible to the unaided eye) was a double star. Using the most recent orbits allows an estimate of the separations of each pair for 1777. Both ϵ^1 Lyra (also called 4 Lyr, and the slightly brighter of the two) and ϵ^2 were 3″.0 apart. According to Juerg Schlimmer in his discussion of Mayer's catalogue (JDSO), ϵ^2 represents the closest pair in Mayer's list. The system came under the scrutiny of William Herschel on 29 August 1779 and part of his phrase 'A very curious double-double star' has stuck and the Double-Double is now known to most astronomers and often features in public night-observing sessions during the autumn.

Finder Chart

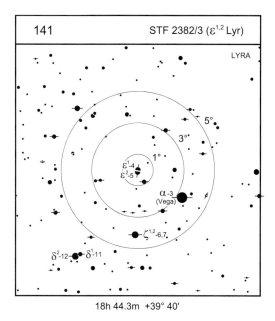

18h 44.3m +39° 40'

The Modern Era

The Double-Double is a true physical quadruple. Stars A and B rotate around their common centre of gravity in around 1725 years whilst C and D perform the same manoeuvre in about 724 years. These values are very uncertain, as is the rotation period of the two pairs, which has been estimated by Andrei Tokovinin *et al.* [177] to be around 340 kiloyears. In 1985 Harold McAlister [556] using the Canada France Hawaii Telescope announced that they had resolved C into two stars at a distance of 0″.18 (CHR 77). Five positive measurements were made between 1985 and 2005 appearing to show considerable orbital motion, but the stars have not been resolved since and no orbit is extant.

Observing and Neighbourhood

The ϵ Lyrae system sits close to Vega and is about 2° N and E of that bright star. The *WDS Catalog* lists ten stars under the systemic names STF 2382 (AB) and 2383 (CD), eleven if the duplicity of C is real. The *Fourth Interferometric Catalog of Double Stars* shows five measurements of the Cc system, several of which are marked as uncertain, but the position angles show a decreasing trend with time, consistent with a period of a few decades. All other components are merely unrelated field stars ranging from $V = 10.1$ to 14.2. Each pair is easy in 10-cm and the two pairs currently have almost the same separation whilst the position angles are 90° different.

Orbit of STF 2382 AB

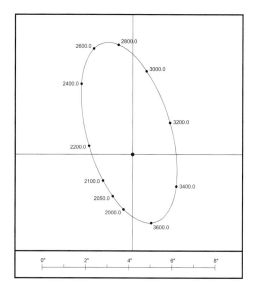

Ephemeris for STF 2382 AB (2000 to 2180)

Orbit by WSI (2004b) Period: 1725 years, Grade: 4

Year	PA(°)	Sep(")	Year	PA(°)	Sep(")
2000.0	350.4	2.47	2100.0	310.6	1.78
2020.0	344.5	2.31	2120.0	299.7	1.74
2040.0	337.7	2.14	2140.0	288.5	1.75
2060.0	329.8	1.97	2160.0	277.8	1.80
2080.0	320.7	1.87	2180.0	267.9	1.91

Orbit of STF 2383 CD

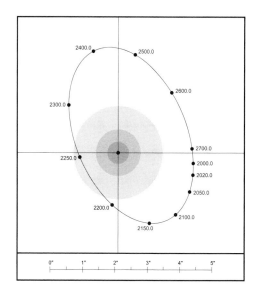

Ephemeris for STF 2383 CD (2010 to 2100)

Orbit by Doc (1984b) Period: 724.307 years, Grade: 4

Year	PA(°)	Sep(")	Year	PA(°)	Sep(")
2010.0	78.2	2.37	2060.0	58.7	2.52
2020.0	74.1	2.40	2070.0	55.0	2.54
2030.0	70.1	2.43	2080.0	51.4	2.55
2040.0	66.2	2.46	2090.0	47.8	2.56
2050.0	62.4	2.49	2100.0	44.2	2.55

The two ϵ stars can be seen with the naked eye. Bob King [558] notes that he was able to resolve this pair and also suggests the following: $\alpha + \beta$ Cap, $\omega^{1,2}$ Sco, and $o^1 + 30$ Cyg. The Struve companion of Vega is a challenging object for the small aperture, although not as hard as it used to be. Herschel found it difficult in moonlight so choose a dark night for the best opportunity. About 2° SE of Vega is ζ Lyrae, a fine, easy pair of white stars of magnitudes 4.3 and 5.6. According to Greg Stone [559] in 2011, with a current separation of 43″ this physically connected pair is within range of hand-held 15×70 binoculars. Also nearby is the wide binocular pair δ^1 and δ^2 Lyrae. These stars, of magnitudes 4.5 and 5.7, are almost 620″ apart but worth finding for the beautiful contrast in colours; δ^1 is an M giant and δ^2 is an early B dwarf.

Measures

AB			
Early measure (STF)	26°.1	3″.03	1831.44
(Orbit	24°.5	3″.19)	
Recent measure (ARY)	344°.8	2″.19	2016.66
(Orbit	345°.4	2″.33)	
CD			
Early measure (STF)	155°.2	2″.57	1831.44
(Orbit	154°.4	2″.33)	
Recent measure (ARY)	76°.5	2″.33	2016.66
(Orbit	75°.4	2″.39)	

142. β LYR = STFA 39 = WDS J18501+3322AB

Table 9.142 Physical parameters for β Lyr

STFA 39	RA: 18 50 04.79	Dec: +33 21 45.6	WDS: 560(115)	
V magnitudes	A: 3.42	B: 7.13	E: 10.2	F: 10.6
$(B - V)$ magnitudes	A: -0.08	B: 0.00		
μ(A)	4.37 mas yr^{-1}	± 0.09	-0.98 mas yr^{-1}	± 0.10 (DR2)
μ(B)	1.57 mas yr^{-1}	± 0.58	-2.52 mas yr^{-1}	± 0.59 (DR2)
μ(E)	1.65 mas yr^{-1}	± 0.05	0.72 mas yr^{-1}	± 0.05 (DR2)
π(A)	3.01 mas	± 0.54	1084 light yr	± 195 (DR2)
π(B)	1.09 mas	± 0.34	2993 light yr	± 933 (DR2)
π(E)	1.62 mas	± 0.34	2013 light yr	± 422 (DR2)
Spectra	A: B6-8II + B0.5	B: B7V		
Masses (M$_\odot$)	Aa1: 2.8	Aa2: 12.7		
Radii (R$_\odot$)	A: 29.4	B:		
Luminosities (L$_\odot$)	A: 4000	B: 1000		
Catalogues	10 Lyr	HR 7106	HD 174638/664	SAO 67451/3	HIP 92420
DS catalogues	Mayer 59 (AB)	H 5 3 (ABEF)	BU 293 (AC,AD)	CIA 3 (Aa1,2)	RBR 11 (AaAb)
	BDS 8868	ADS 11745		
Radial velocity	2.20 km s^{-1}	± 0.7		
Galactic coordinates	63°.188	+14°.784		

History

Found by Christian Mayer, β Lyrae (Sheliak) was observed by William Herschel on 29 August 1779. He noted 'Quadruple. All w(hite). First and second considerably unequal. First and third very unequal. First and fourth very unequal.' On 10 September 1784 John Goodricke [563] noticed that the star varied in brightness and that there were two distinctive brightness minima. The bright pair was later catalogued by Struve, but in his first Appendix catalogue as Σ I 39 (now STFA 39). In 1878, Burnham [219] noted a very faint companion to A at 247° 46″ using the 18.5-inch Dearborn telescope (C), and in the BDS catalogue mentions a second even fainter

(D, magnitude 14) object at 68°, 64″, which was found by Aitken with the 36-inch refractor at Lick.

The Modern Era

The astrophysical interest in this star lies in the nature of the primary. It is the prototype eclipsing binary, whose stars are ellipsoidal in shape owing to rapid rotation. The brightest star is the least massive and is giving mass to the secondary via an accretion disk which surrounds that star. The point on the disk where the stream of material from A to B hits the disk is called the hot spot and is periodically eclipsed by the primary

Finder Chart

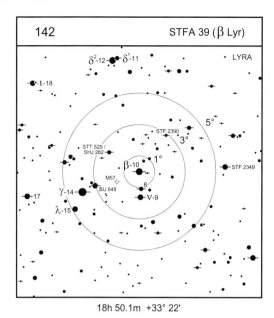

```
142                    STFA 39 (β Lyr)
```

18h 50.1m +33° 22'

star. The close, eclipsing pair has been resolved in the infrared (at 1.6 μm) by the CHARA Interferometric Array – hence the catalogue number of this pair. A paper in 2008 by Zhao *et al.* [561] showed that the orbital plane is almost edge-on to the line of sight and the maximum apparent separation is near 0.9 mas. They were able to confirm that the orbital period of 12.9 days corresponds to the period derived from eclipse photometry. In 2002, Roberts *et al.* [562], using a 3.6-metre telescope on the summit of Haleakala on the Hawaiian island of Maui, detected a close-in star in the near infrared. The

distance was 0″.5 and the new object was of magnitude about 8.2. Despite a number of repeated attempts, only one positive detection of this star remains. All the stars appear to be very distant, but Gaia DR2 measurements still show considerable uncertainty in the parallaxes.

Observing and Neighbourhood

The small aperture sees four stars (A, B, E, and F). Star E is at 317°, 67″, 2014, distance increasing, and F is at 18°, 86″, 2014, distance also increasing. The two Burnham components are very faint and probably need 25- or 30-cm on a dark night to be seen. The star STF 2390 is 1°.5 NNW of β and rather faint and close for the smallest apertures (7.4, 8.6, 155°, 4″.3, 2016). The wide pair SHJ 282 AC is 1° ENE (6.1, 7.6, 350°, 45″.3, 2017) and has a test for 15-cm in star A, which was doubled by Otto Struve at Pulkovo. The star STT 525 is 6.1, 9.1, 129°, 1″.8, 2011. Two point five degrees W is STF 2349, which has a naked-eye primary (5.4, 9.4, 204°, 7″.2, 2016). Just NW of γ is the binary BU 648, a 61.4-year system which was at widest separation in 2018 (5.3, 8.0, 232°, 1″.3, 2020.0)

Measures

Early measure (STF)	149°.8	45″.75	1835.23
Recent measure (RWA)	148°.9	45″.43	2010.53

143. $\theta^{1,2}$ SER = STF 2417 = WDS J18562+0412AB

Table 9.143 Physical parameters for θ Ser

STF 2417	RA: 18 56 13.18	Dec: +04 12 12.9	WDS: 198(224)	
V magnitudes	A: 4.62	B: 4.98		
$(B - V)$ magnitudes	A: +0.17	B: +0.20		
μ(A)	45.85 mas yr^{-1}	$\pm$ 0.46	31.59 mas yr^{-1}	$\pm$ 0.36 (DR2)
μ(B)	50.21 mas yr^{-1}	$\pm$ 0.33	28.51 mas yr^{-1}	$\pm$ 0.28 (DR2)
π(A)	24.95 mas	$\pm$ 0.29	130.7 light yr	$\pm$ 1.5 (DR2)
π(B)	24.80 mas	$\pm$ 0.22	131.5 light yr	$\pm$ 1.2 (DR2)
Spectra	A: A2III	B: A4IV		
Spectra	A5V	A5Vn		
Luminosities (L$_\odot$)	A: 19	B: 13		
Catalogues (A/B)	HR 7141/2	HD 175638/9	SAO 124068/70	HIP 92946/51
DS catalogues	Mayer 60	H 4 6	STF 2417	BDS 8914	ADS 11853
Radial velocity	A: -48.7 km s^{-1}	$\pm$ 0.8	B: -55.7 km s^{-1}	$\pm$ 0.9
Galactic coordinates	37°.219	+0°.854		

History

The pair θ Ser is also known as Alya. It was found first by James Bradley in 1755 and later noted by Christian Mayer in 1777 and by William Herschel, who observed it on 19 October 1779 and thought the stars equally bright and both white.

The Modern Era

The distances to both components, determined by Gaia and announced in the DR2 catalogue, are well defined and equal.

Exoplanet Host

One of the bright, but distant companions (58°, 421″ from A) in a low-power field (HD 175541 – distance 87 light years)

Finder Chart

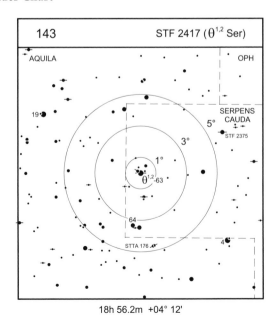

18h 56.2m +04° 12'

hosts an exoplanet (HD 175541b) which has a minimum mass of 0.657 Jupiter masses. It was found by radial velocity variations in 2007 and has a period of 297.3 days. This star forms the very wide pair TOK 620 with another magnitude 8.1 at 270° and 842″. Both have similar proper motions but early Gaia results show that the bright star is about five times closer to us than its companion.

Observing and Neighbourhood

A glorious, brilliant white pair of stars for the small telescope and which also provide a fine sight in binoculars providing that they can be firmly mounted. Alya is probably best found by first acquiring the 3.0 magnitude star ζ Aql, moving 2° NW to the magnitude 4.0 ϵ Aql and then heading due S for a further 15°. The magnitude 6.8 star at 421″ is not related to the pair – this solar-type star is twice as close to us as θ. Two and a half degrees S is the wide pair STTA 176 (7.4, 7.5, 113°, 94″, 2011), whilst 3° WNW is the fine pair STF 2375 (6.3, 6.7, 121°, 2″.4, 2017), which hosts Tweedledum and Tweedledee (see Chapter 7).

Measures

Early measure (STF)	103°.8	21″.65	1830.05
Recent measure (WSI)	103°.2	22″.43	2014.65

144. ζ SGR = HDO 150 = WDS J19026–2953AB

Table 9.144 Physical parameters for ζ Sgr

HDO 150	RA: 19 02 36.72	Dec: −29 52 48.4	WDS: 239(201)		
V magnitudes	A: 3.40	B: 3.58	C: 10.63		
(*B − V*) magnitudes	A: +0.14	B: +0.11			
μ	10.79 mas yr^{-1}	± 1.00	21.11 mas yr^{-1}	± 0.65	
π	36.98 mas	± 0.87	88 light yr	± 2	
Spectra	A: A2III	B: A4IV			
Masses (M$_\odot$)	AB: 5.27	± 0.37 (dyn.)			
Luminosities (L$_\odot$)	A: 27	B: 23			
Catalogues	38 Sgr	HR 7194	HD 176687	SAO 187600	HIP 93506
DS catalogues	H 5 78 (AC)	HDO 150 (AB)	BDS 8965	ADS 11950	
Radial velocity	24.70 km s^{-1}	± 0.7			
Galactic coordinates	6°.840	−15°.354			

History

On 4 August 1782 William Herschel was sweeping in Sagittarius and came across ζ Sgr, noting that it had a faint and distant companion: 'Extremely unequal L(arge). r(ed). S(mall). d(usky). Distance Vth class … A third star. Distance about four times as far as the former…' The third star does not appear in the WDS, probably because it is too distant. Attempts to look at the field taken with the Schmidt telescopes shows that ζ is so overexposed that any near neighbours are obliterated. Between 1866 and 1872 Harvard College Observatory carried out a programme of micrometric measurement of double stars using the East Equatorial (15-inch aperture). The programme was led by Professor Joseph Winlock and he also participated in the observing. Joseph Winlock was born in Kentucky in 1826 and ended up in Cambridge, Massachusetts, as a computer working for the American Ephemeris and Nautical Almanac. He was eventually appointed third director

Finder Chart

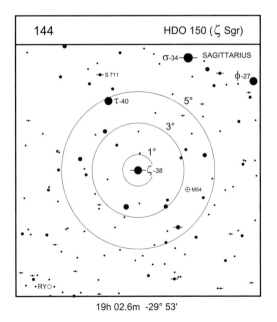

19h 02.6m -29° 53'

Orbit

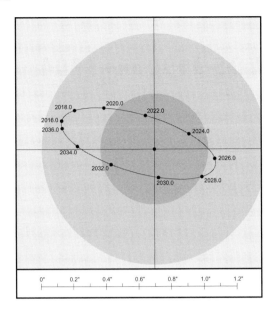

Ephemeris for HDO 150 AB (2019 to 2028)

Orbit by DRs (2012) Period: 21.00 years, Grade: 1

Year	PA(°)	Sep(")	Year	PA(°)	Sep(")
2019.0	239.1	0.48	2024.0	113.0	0.23
2020.0	231.4	0.39	2025.0	93.1	0.31
2021.0	219.0	0.30	2026.0	81.1	0.37
2022.0	195.1	0.21	2027.0	71.3	0.38
2023.0	151.3	0.17	2028.0	60.3	0.34

of the Harvard College Observatory, taking over on the death of G. P. Bond in 1865. On 6 August 1867 Winlock [564] noted that ζ Sgr was double and that the stars were equally bright. His measure gave 257°.7 and 0″.86.

The Modern Era

De Rosa *et al.* [565] included ζ Sgr amongst their target stars when carrying out the VAST survey of A stars. This volume-limited A-star survey was confined to stars within 75 pc of the Sun and was sensitive to companions in the range 30,000 to 45,000 AU. Examining 435 A stars, they found 64 new companions. They recalculated the orbit of ζ Sgr and derived a system mass of 5.26 ± 0.37 M$_\odot$. Comparing this with three theoretical models, each of which gave a mass sum based on the stellar parameters, and all of which gave significantly lower mass sums, allowed them to suggest the ζ Sgr system has a third component for which there is currently no observational evidence.

Observing and Neighbourhood

Easily found, ζ sits at the bottom of the handle of the 'Teapot'. The distant Herschel star can be seen with 15-cm, possibly even 10-cm, whilst the Winlock companion is much more difficult. It was spectacular in the 67-cm refractor at Johannesburg in 2016. At the time of writing it is closing in and probably requires 30-cm even when at its widest (2016 and 2037). By 2023 it will be out of the range of amateur instruments.

Measures

Early measure (BU)	84°.2	0″.42	1878.70
(Orbit	84°.5	0″.36)	
Recent measure (RWA)	251°.8	0″.63	2016.66
(Orbit	250°.0	0″.59)	

145. γ CRA = HJ 5084 = WDS J19064–3704

Table 9.145 Physical parameters for γ CrA

HJ 5084	RA: 19 06 25.14	Dec: −37 03 48.5	WDS: 135(273)	
V magnitudes	A: 5.01	B: 5.05		
(*B − V*) magnitudes	A: +0.51	B: +0.53		
μ	A: 96.74 mas yr^{-1}	± 1.05	B: −281.71 mas yr^{-1}	± 0.58
π	57.79 mas	± 0.75 mas	56.4 light yr	± 0.7
π(A)	57.86 mas	± 0.31	56.4 light yr	± 0.3 (DR2)
π(B)	58.70 mas	± 0.35	55.6 light yr	± 0.3 (DR2)
Spectra	A: F8V	B: F8V		
Masses (M$_\odot$)	A: 1.15	B: 1.14		
Radii (R$_\odot$)	A: 1.47	B: 1.42		
Luminosities (L$_\odot$)	A: 2.5	B: 2.2		
Catalogues (A/B)	HR 7226/7	HD 177474/5	SAO 210928	HIP 93825
DS catalogues	HJ 5084			
Radial velocity	−51.60 km s^{-1}	± 0.3		
Galactic coordinates	0°.159	−18°.736		

History

John Herschel swept up γ CrA on sweep 461 (28 June 1834) and recorded that it was a 'Superb D star'. He also made eight measures of it with the 7-foot equatorial, which showed a steady but small decrease in position angle of 5° over three years, agreeing very well with the current orbit. The first measure was rejected because of 'hitching' (or catching) of the micrometer threads.

The Modern Era

The system has been observed by a number of groups with a view to finding exoplanets but none have yet been seen. The stars are approaching the Sun and around 200,000 years from now will pass by the Solar System at a distance of 24 light years, approximately half the present distance.

Finder Chart

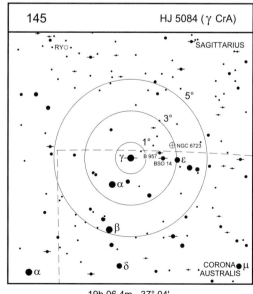

19h 06.4m -37° 04'

Orbit

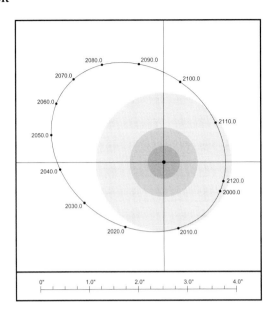

Ephemeris for HJ 5084 (2016 to 2034)

Orbit by Hei (1986b) Period: 121.76 years, Grade: 2

Year	PA(°)	Sep(")	Year	PA(°)	Sep(")
2016.0	345.0	1.42	2026.0	307.9	1.68
2018.0	336.6	1.46	2028.0	301.9	1.74
2020.0	328.7	1.50	2030.0	296.4	1.81
2022.0	321.2	1.56	2032.0	291.3	1.87
2024.0	314.3	1.61	2034.0	286.5	1.94

Observing and Neighbourhood

The star γ CrA can be found about 6° following θ Scorpii. A beautiful and easy pair for the small aperture (both yellowish – ASNSW) with the position angle increasing at about 4° per annum for the next five years or so and then slowing down as the stars move towards their maximum separation of 2″.5 in 2064 before closing to 1″.3 in 2116. One degree preceding γ is BSO 14 – a fine pair of white stars, magnitudes 6.3 and 6.6, at 281°, 13″.4, 2016, moving through space together. With both components spectroscopic binaries, the system is quadruple. The star BSO 14 illuminates the reflection nebula IC 4812 [566]. Fifteen arcminutes N of this position is B 957 AC (7.3, 9.6, 23°, 58″, 2016), but what piqued van den Bos' interest was the 13.4 magnitude companion (B), 4″ distant from A. The main component is also known as TY CrA. Thirty arcminutes away to the NW of BSO 14 is the fine globular NGC 6723 – found by James Dunlop in 1826.

Measures

Early measure (JC)	194°.1	2″.30	1847.32
(Orbit	193°.5	1″.98)	
Recent measure (RWA)	343°.2	1″.56	2016.65
(Orbit	342°.1	1″.43)	

146. GLE 3 PAV = WDS J19172−6640

Table 9.146 Physical parameters for GLE 3 Pav

GLE 3	RA: 19 17 12.23	Dec: −66 39 39.8	WDS: 2483(40)	
V magnitudes	A: 6.12	B: 6.42		
(*B* − *V*) magnitudes	A: +0.17	B: +0.26		
μ	A: 8.10 mas yr^{-1}	± 0.42	−6.72 mas yr^{-1}	± 0.50
π	10.92 mas	± 0.49	299 light yr	± 16
μ(B)	20.93 mas yr^{-1}	± 0.09	5.72 mas yr^{-1}	± 0.07 (DR2)
π(B)	8.14 mas	± 0.04	401 light yr	± 2 (DR2)
Spectra	A: A3/5 IV/V	B:		
Luminosities (L$_\odot$)	A: 25	B: 35		
Catalogues	HR 7278	HD 179366	SAO 254515	HIP 94789
DS catalogues	GLE 3			
Radial velocity	11.50 km s^{-1}	± 1.78		
Galactic coordinates	327°.221	−27°.273		

History

Walter Gale [567] (1865–1945) was an active amateur astronomer who was born in Sydney. He was inspired by the newspaper articles written by John Tebbutt, who also discovered the Great Comet of 1882, and this event encouraged him to take up astronomy and build a 7-inch reflector. Although his main passion was comet hunting he also discovered three double stars and the planetary nebula IC 5148 in Grus. He knew Robert Innes and made him a loan of a 6.25-inch refractor, which Innes used to survey the southern sky for new double stars. The star GLE 1 contains the variable TT Ret and is a close visual binary with a period of 296 years. In 2020.0 the 6.9 and 7.6 magnitude stars are 0″.39 apart in PA 232° and will continue to widen throughout the century. The star GLE 2 is ξ Pavonis, whose magnitudes are 4.5 and 8.2 and the current position is 159°, 3″.7.

Finder Chart

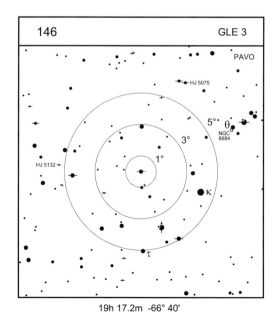

19h 17.2m −66° 40'

Orbit

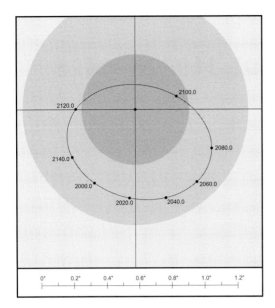

Ephemeris for GLE 3 (2016 to 2052)

Orbit by Doc (2007d) Period: 156.8 years, Grade: 3

Year	PA(°)	Sep(")	Year	PA(°)	Sep(")
2016.0	351.0	0.52	2036.0	15.1	0.55
2020.0	356.0	0.53	2040.0	19.6	0.56
2024.0	0.9	0.54	2044.0	24.0	0.56
2028.0	5.7	0.54	2048.0	28.3	0.57
2032.0	10.4	0.55	2052.0	32.6	0.57

The Modern Era

The current orbit for this pair predicts a period of 156.8 years. After discovery the stars closed slowly, reaching a minimum distance of around $0''.15$ in 1952/3, since when they have opened but are now close to widest separation ($0''.52$) at the time of going to press. The close proximity of the stars suggests that the Gaia astrometry of both will not be issued until DR3.

Observing and Neighbourhood

A testing system for 30-cm, this far southern pair is centrally placed in the rather large constellation of Pavo. It is 4° SE of the magnitude 5.7 star θ Pav, which sits 8′ away from NGC 6684, a bright barred lenticular galaxy. There are two John Herschel pairs within 4° of GLE 3: HJ 5075 lies 3° N and slightly W (7.7, 7.7, 113°, $1''.78$, 2009) whilst HJ 5132 lies 3° E (7.6, 9.7, 309°, $21''.5$, 2016).

Measures

Early measure (I)	39°.5	$0''.82$
(Orbit	41°.0	$0''.57$)
Recent measure (ANT)	349°.6	$0''.51$
(Orbit	347°.9	$0''.52$)

147. β^1 SGR = Δ 226 = WDS J19226–4428

Table 9.147 Physical parameters for β^1 Sgr

DUN 226	RA: 19 22 38.29	Dec: −44 27 32.1	WDS: 3605(31)	
V magnitudes	A: 3.94	B: 7.21		
$(B - V)$ magnitudes	A: −0.10	B: +0.18		
μ(A)	12.48 mas yr^{-1}	± 0.64	−11.93 mas yr^{-1}	± 0.89 (DR2)
μ(B)	13.18 mas yr^{-1}	± 0.08	−14.98 mas yr^{-1}	± 0.70 (DR2)
π(A)	12.48 mas	± 0.64	261 light yr	± 13 (DR2)
π(B)	8.95 mas	± 0.06	364 light yr	± 2 (DR2)
Spectra	A: B8V	B: F0V		
Masses (M$_\odot$)	A: 3.67	± 0.14		
Luminosities (L$_\odot$)	A: 145	B: 14		
Catalogues (A/B)	HR 7337/—	HD 181454/84	SAO 229646/7	HIP 95241/—
DS catalogues	DUN 226			
Radial velocity	−10.7 km s^{-1}	± 2		
Galactic coordinates	353°.613	−23°.928		

History

James Dunlop catalogued this bright, wide pair in 1826 and estimated the position angle as 60° and the separation as 30″. John Herschel observed the pair three times. On 27 July 1834 he noted 'Most dreadfully ill-defined, a violent struggle going forward between the regular S. E. wind and a hot north-wester which is coming on'.

The Modern Era

Although the DR2 parallax of A is not particularly precise, owing to the brightness of the star, it seems likely that there is no physical relationship with the B component even though the proper motions are similar.

Finder Chart

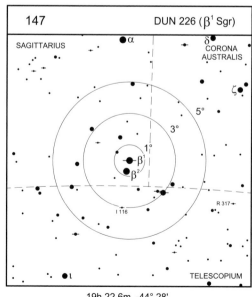

19h 22.6m -44° 28'

Observing and Neighbourhood

The pair β^1 and β^2 Sgr (magnitudes respectively $V = 3.9$ and $V = 4.3$) sit in the extreme SW corner of Sagittarius at declination $-44°$, a full $30°$ from the northern border of the constellation and therefore not visible from northern Europe. They make a naked-eye pair with a separation of about $21'$, but they are not related – β^2 is less than half the distance from the Solar System (135 light years) of its brighter neighbour. This is a fairly empty region of sky for the double star observer. The system R 317, $3°.5$ W and a little S is a triple star. The wide pair (HJ 5078) is 8.0, 8.8, $252°$, $18''.6$, 2016, whilst Russell added a close companion of magnitude 8.8 at $287°$, $1''.7$, 2016, and van den Bos then found a magnitude 12.9 star $268°$ and $6''.2$ from C (B 2873). One point five degrees S and a little E is I 116, which is very similar. It has a close pair, 8.6, 9.4 at $24°$, $2''.7$, 2016, whilst a third star of magnitude 8.6 can be found at $190°$, $16''$, 2016.

Measures

Early measure (JC)	$78°.0$	$28''.26$	1846.23
Recent measure (ANT)	$76°.0$	$28''.7$	2009.71

148. $\beta^{1,2}$ CYG = ALBIREO = STFA 43 = WDS J19307+2758AB

Table 9.148 Physical parameters for β Cyg

STFA 43	RA: 19 30 43.28	Dec: +27 57 34.9	WDS: 133(275)		
V magnitudes	Aa: 3.37	Ac: 5.16	B: 4.68		
$(B - V)$ magnitudes	A: +1.25	B: +0.18			
μ(A)	-7.17 mas yr^{-1}	± 0.25	-6.15 mas yr^{-1}	± 0.33	
μ(B)	-1.90 mas yr^{-1}	± 0.19	-1.02 mas yr^{-1}	± 0.27	
π(A)	7.51 mas	± 0.33	430 light yr	± 19	
π(B)	8.16 mas	± 0.25	400 light yr	± 12	
μ(B)	-0.99 mas yr^{-1}	± 0.26	-0.54 mas yr^{-1}	± 0.28 (DR2)	
π(B)	8.38 mas	± 0.17	389 light yr	± 8 (DR2)	
Spectra	Aa: K3II	Ab: B9.5V	B: B9.5		
Masses	Aa: 14.5	Ab: 3.0 : B: 3.2			
Luminosities (L$_\odot$)	Aa: 580	Ac: 120	B: 155		
Catalogues (A/B)	6 Cyg	HR 7417/8	HD 183912/3	SAO 87301/2	HIP 95947/51
DS catalogues	H 5 5	STFA 43	BDS 9374	ADS 12540	MCA55 (AaAc)
Radial velocity (A/B)	-24.07 km s^{-1}	± 0.12	-18.80 km s^{-1}	± 2.2	
Galactic coordinates	62°.110	+4°.572			

History

The pair β Cygni was first measured by Bradley in 1755 using his meridian circle. He noted that star B transited two seconds later than star A. The pair was then picked up by William Herschel on 12 September 1779. He noted 'L(arge star). pale r(ed)., s(mall star). a beautiful blue'. It was mainly of interest to subsequent generations of astronomers because of the colour contrast. Indeed, W. H. Smyth dedicated 15 pages, in his volume called *Sidereal Chromatics*, summarising the colours of each star as determined by his côterie of astronomical friends and acquaintances, including Sir Rowland Hill, the man who was responsible for the British Post Office, and William Huggins; eventually, even the Astronomer Royal George Airy contributed, via one of the members of staff at the Royal Observatory. Smyth's conclusion was that the colours were yellow and blue. *Sidereal Chromatics* has been re-issued (in black and white) by Cambridge University Press.

The Modern Era

Miss Maury placed the star on the list of composite-spectrum objects in 1897. W. W. Campbell [568] announced in 1918 that a series of spectrographs of the brighter component taken between 1898 and 1918 using the Mills spectrograph on the 36-inch refractor on Mount Hamilton (Lick Observatory) showed a small but definite decrease in radial velocity and noted that this was not due to orbital motion around the B star but rather due to the presence of a much closer star

Finder Chart

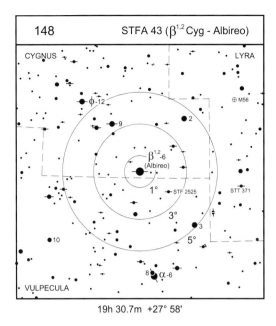

| 148 | STFA 43 (β¹,² Cyg - Albireo) |

19h 30.7m +27° 58'

accompanying A. The duplicity of the brightest star (= MCA 55) was not confirmed until 1976, when McAlister & Hendry [569] made some measures with a speckle camera on the 2.1-metre telescope at McDonald Observatory; it was then measured visually by Charles Worley [570] using the 26-inch refractor at the USNO in Washington. This begs the question, why wasn't it seen by the great observers around the turn of the last century, during the course of the intense surveys? According to the orbit of M. Scardia *et al.* [572], it was certainly somewhat wider at that time (∼0″.6) but perhaps the star was never examined. After Campbell's discovery of the changing radial velocity in the A component, Merrill [571] observed both components with the 100-inch reflector at Mount Wilson and the stellar interferometer and found each 'apparently' single. It may be that Dr Scardia's orbit, which is, at least, a fair reflection of the existing 42 years of observations, needs significant revision and that the closest separation is considerably smaller than previously thought. Professor Roger Griffin has been monitoring the radial velocity of star A for four decades and it is by no means certain that the observed change in radial velocity so far corresponds to the 213-year binary period derived by Dr Scardia and colleagues. Two other close-in components have been announced but the reality of these is not certain. In 1978 Bonneau & Foy [573], using the 1.9-metre reflector at Haute Provence, measured a companion to A at 0″.125 and 160° but this does not seem

to correspond to the McAlister component. In 2000 Prieur *et al.* [574] found a star at 160° and 0″.045, but this does not correspond to MCA 55 either. There is reason to believe that Aac and B are physically linked, albeit with a very long orbital period. An argument to this effect was made by Roger Griffin [577]. The revised Hipparcos parallaxes do not quite agree, even after allowing for the quoted errors, but the duplicity of A will be a factor and Gaia might be expected to resolve this question, although it has not appeared in the DR2 catalogue, whilst the B star has done so. Professor Griffin also argues that for there to be two such bright stars close together is more than coincidental. There is clearly more to be found out about this fascinating system.

Observing and Neighbourhood

Finding β Cygni is very straightforward. It is the bright star at the bottom of the 'Cross' of Cygnus. The smallest apertures, and firmly mounted binoculars, will show this dazzling object. Owners of large apertures (30-cm upwards) should occasionally look at A when the seeing is good. Although the close visual companion to A is about 2 magnitudes fainter, there might be an opportunity to see an elongation. The pair should now start to slowly separate. Christopher Taylor [575], using a 12.5-inch (31-cm) Calver reflector at ×820, saw A divided on 17 October 1996. The glory of the wide pair is the colour contrast. Many people see the stars as yellow and blue but some observers report specific shades. Admiral Smyth in his Bedford Catalogue alludes to topaz yellow and sapphire blue, whilst more recently E. Hartung noted deep yellow and pale bluish. Most recently, John Nanson records gold and blue. Robert Burnham, in his *Handbook*, thinks that one should avoid too high a magnification, or too small or too large an aperture, in order to see the colours of this fine pair at their best. He suggests that with 6-inch (15-cm) aperture, a magnification of ×30 is optimal. Walter Scott Houston [578] reported splitting the wide pair in firmly held 7 × 50 binoculars. The orbital pair STF 2525 can be found 1°.2 SW. This long-period binary (P = 883 years) will widen for another three centuries before reaching 3″.3 (8.2, 8.4, 289°, 2″.2, 2020). Further in the same direction, 3° W of Albireo is the triple star STT 371. The close pair (7.0, 7.6, 159°, 0″.7, 2015) will test a 10-cm aperture but the 9.8 magnitude companion at 271°, 47″, 2007 should be somewhat easier. There is a fourth component at 7°, 47″, 2002, of magnitude 11.5.

Measures

AB			
Early measure (STF)	55°.7	34″.29	1832.18
Recent measure (WSI)	53°.7	34″.28	2014.55
BaBb			
Early measure (MCA)	186°.1	0″.44	1977.48
(Orbit	186°.0	0″.44	
Recent measure (PRI)	100°.8	0″.36	2008.80
(Orbit	101°.7	0″.38	

149. 16 CYG = STFA 46 = WDS J19418+5032AB

Table 9.149 Physical parameters for 16 Cyg

STFA 46	RA: 19 41 48.95	Dec: +50 31 30.2	WDS: 36(583)	
V magnitudes	A: 5.95	B: 6.20		
$(B - V)$ magnitudes	A: +0.72	B: +0.69		
μ(A)	-148.30 mas yr^{-1}	± 0.06	-158.96 mas yr^{-1}	± 0.07 (DR2)
μ(B)	-134.97 mas yr^{-1}	± 0.05	-162.49 mas yr^{-1}	± 0.05 (DR2)
π(A)	47.28 mas	± 0.03	68.99 light yr	± 0.04 (DR2)
π(B)	47.28 mas	± 0.03	68.99 light yr	± 0.04 (DR2)
Spectra	A: G1.5Vb	B: G3V		
Masses (M$_\odot$)	A: 1.07	± 0.05	B: 1.05	± 0.02
Radii (R$_\odot$)	A: 1.22	± 0.02	B: 1.12	± 0.02
Luminosities (L$_\odot$)	A: 1.5	B: 1.2		
Catalogues (A/B)	HR 7503/4	HD 186408/27	SAO 31898/9	HIP 96895/901
DS catalogues	H 5 36	STFA 46	ADS 12815	BDS 9560
Radial velocity (A/B)	-27.61 km s^{-1}	± 0.08	-28.02 km s^{-1}	± 0.08 (DR2)
Radial velocity (A/B)	-27.21 km s^{-1}	± 0.15	-27.73 km s^{-1}	± 0.16 (DR2)
Galactic coordinates	83°.340	+13°.215		

History

This binary was possibly found by Hodierna. His description mentions a pair in the right wing of the Swan. William Herschel picked it up on 5 October 1781. 'Almost equal. Both pale r(ed). Distance 30″.'

The Modern Era

Between 1960 and 2007, a series of photographic plates centred on AB were taken at Pulkovo Observatory using the 26-inch refractor and reported by Kiselev & Romanenko [579]. Astrometric analysis of the plates shows a perturbation in star B in both right ascension and declination with a period of about eight or nine years. The amplitude is about 0″.02. An orbital period of > 20,000 years was given; the orbital semi-major axis was 900 AU and the eccentricity > 0.65. In 1999 Hauser & Marcy [580] found the most likely period of AB to be 135.127 centuries. Adaptive optics observations of the system have revealed very faint images close to both stars. In 2002, Patience et al. [581] reported the existence of a star of magnitude 11.9 in the K band at a distance of 15″.8 from B. They also noted a closer companion to A at 205° and 3″.43. This star had first been found, a few months before, by Turner et al. [582] in 1998 using adaptive optics on the Mount Wilson 100-inch reflector, at 207°and 3″.15. As the large proper motion of A and B would soon leave behind any faint stars which just happened to be in the field of view, this was solid evidence that here was a physical companion

Finder Chart

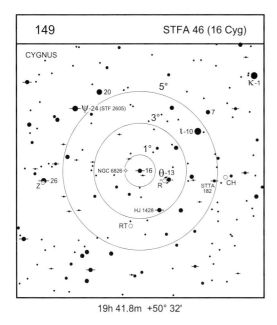

149   STFA 46 (16 Cyg)

CYGNUS

5°

3°

1°

K-1

20

ψ-24 (STF 2605)

7

l-10

NGC 6826   16 θ-13

Z-26   R   STTA 182   CH

HJ 1428

RT

19h 41.8m +50° 32'

to A. Trilling & Brown [583] thought that the companion to A was a distant giant star many times further away than A and hence unrelated.

Exoplanet Host?

The components of 16 Cyg have been investigated both by high-precision radial velocity techniques and by adaptive optics with large telescopes searching for other stellar companions and planets, In 1996 Cochran *et al.* [584]

discovered a body orbiting B with a period of 799 days, a velocity semi-amplitude of 50.6 km s^{-1}, and a likely mass of 2.4 Jupiters.

Observing and Neighbourhood

The pair 16 Cyg is easily found as it is about 6° due N of the magnitude 2.9 star δ Cyg, itself a long-period binary of note and the right-hand star in the 'Cross' of Cygnus. Smyth found both stars to be a pale-fawn colour. Also nearby, 1°.5 to the SW is HJ 1428 (6.6, 10.2, 233°, 34″, distance increasing); another 3° further W brings you to STTA 182 (7.5, 8.6, 297°, 73″, distance increasing), whilst continuing another 3° further one finds the binary star STF 2486 (6.5, 6.7, 202°, 7″.2, 2020, not on the chart), whose projected orbital period of 3100 years means that motion is extremely slow. Our pair 16 Cyg lies just 28′ W of the Blinking Planetary NGC 6826. Three degrees NE is ψ Cyg (STF 2605) (5.0, 7.5, 177°, 2″.8, 2014); in 1996 the primary star was found to be a close pair (YR 2); their separation ranges between 0″.03 and 0″.14 in a period of 55 years.

Measures

Early measure (STF)	136°.3	37″.30	1837.30
(Orbit	136°.0	37″.39)	
Recent measure (WSI)	132°.8	39″.67	2014.68
(Orbit	133°.1	39″.74)	

150. δ CYG = STF 2579 = WDS J19450+4508AB

Table 9.150 Physical parameters for δ Cyg

STF 2579	RA: 19 44 58.44	Dec: +45 07 50.5	WDS: 51(477)		
V magnitudes	A = 2.89	B = 6.27			
$(B - V)$ magnitudes	A: −0.02	B: +0.27			
μ	+44.07 mas yr^{-1}	± 0.46	+48.66 mas yr^{-1}	± 0.48	
π	19.77 mas	± 0.48	165 light yr	± 4	
Spectra	A: B9.5 IV	B: F1V?			
Luminosities (L$_\odot$)	A: 145	B: 6			
Catalogues	18 Cyg	HR 7528	HD 186882	SAO 48796	HIP 97165
DS catalogues	H 1 94	STF 2579	BDS 9605	ADS 12880	
Radial velocity	−23.60 km s^{-1}	± 3			
Galactic coordinates	78°.719	+10°.243			

History

This binary was found by William Herschel on 20 September 1783. He noted that 'the stars were very unequal, with the primary a fine white and the secondary [an] ash colour, tending to red'. In May 1804 Herschel was unable to see B either in his 10-foot reflector or the 20-foot telescope. He assumed that the apparent movement of B was carrying it across A, therefore in agreement with his prediction of the effect of stellar parallax on two stars at noticeably different distances. In 1823 the star was single as observed by South and John Herschel, and in 1825 as observed by South. Others considered that the observed positions were indicative of a short period – perhaps 40 years. However, as the century went on it became clear that the inability of some observers to see B was due to a combination of the magnitude difference and the proximity of the two stars. Dawes was doubtful of any short-period solutions and noted that from 1840 to 1865 his measured distances 'scarcely varied'. Dawes also said that he found no evidence of any brightness change in B. Thomas Lewis

Finder Chart

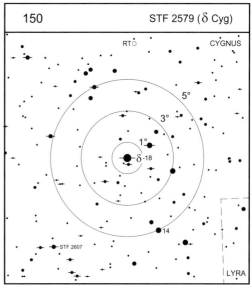

19h 45.0m +45° 08'

Orbit

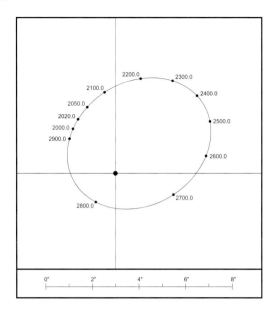

Ephemeris for STF 2579 AB (2010 to 2100)

Orbit by Sca (2012c) Period: 918.1 years, Grade: 4

Year	PA(°)	Sep(")	Year	PA(°)	Sep(")
2010.0	219.7	2.68	2060.0	200.1	3.10
2020.0	215.3	2.77	2070.0	196.8	3.18
2030.0	211.2	2.85	2080.0	193.6	3.26
2040.0	207.3	2.93	2090.0	190.6	3.34
2050.0	203.6	3.02	2100.0	187.8	3.41

in 1906 included a lengthy note on this system and noted particularly the occasional difficulty of seeing the companion and, what he regards as 'unquestionable', the change in this star's perceived colour. Agnes Clerke [586] thought that the B component varied from the seventh to the ninth magnitude, but there is an added factor to consider. During the early nineteenth century the separation of B from A placed it on the first diffraction ring of the considerably brighter primary star, and a number of observers found it easier to measure the two in daylight. Lewis' note on the colour variation of B is as follows: 'STRUVE marked it as ashen grey, 1826–33, but as remarkably red in 1836. DAWES found it blue (1839–41). SECCHI red, 1856 and 1862; blue 1856-7-8; violet 1857 and 1863. DEMBOWSKI grey in 1862–63. ENGELMANN red in 1865. DUNER saw it always red, except on one occasion when it appeared olive. PERROTIN 1883 and 1886 orange. Of late its blue aspect has predominated.' By 1900 significant orbital motion of about 90° retrograde had been observed, but the scatter in separation measures due to the closeness and the brightness difference meant that there was no real prospect of getting even an approximate period for several centuries.

Today, with an extra 120 years of coverage the orbit is still essentially uncertain but the period of 918 years is unlikely to be grossly in error.

The Modern Era

In a paper from 1980, Abt [585] and collaborators noted a range of radial velocity variation in star A, from −55 to −12 km s^{-1} and noted a 'short-term variation'. No confirming observation of a spectroscopic or interferometric component has been made, and it should be noted that the star is a rapid rotator with broad spectral lines. The 12th magnitude star called C, which is 63″ away in PA 66°, appears to be moving with the system. Both components are in the *Suspected Variables Catalogue*, the A star as NSV 12381 and the B star as 12380. The evidence for the variability of A comes from photographic photometry carried out around 1950, which showed a range of 0.03 magnitudes. The B star was also thought by F. G. W. Struve to vary by two magnitudes. The brightness of A will mean that any Gaia result will have to await special processing.

Observing and Neighbourhood

The star is unmistakeable; it is the westernmost bright star in the five that make up the 'Cross' of Cygnus. F. G. W. Struve found the stars to be greenish and ash whilst Smyth recorded pale yellow and sea-green. On a good night the companion is easy to measure in the 20-cm refractor at Cambridge and should be visible in 10-cm aperture. The separation is currently wide enough for there to be no need to examine the star in daylight. The stars will continue to separate for several centuries if the current orbit is correct, reaching 4″.8 around 2400 and closing to 1″.4 in about 2780. Four degrees SE is STF 2607 (6.6, 9.1, 287°, 3″.0, 2012); the primary is a close binary (STT 392), $P = 270$ years, which is only 0″.13 apart in 2020 and is never wider than 0″.36.

Measures

Early measure (STF)	40°.7	1″.91	1826.55
(Orbit	42°.0	1″.87)	
Recent measure (ARY)	218°.1	2″.84	2016.33
(Orbit	216°.7	2″.74)	

151. θ SGE = STF 2637 = WDS 20099+2055AB,C

Table 9.151 Physical parameters for θ Sge

STF 2637	RA: 20 09 56.71	Dec: +20 54 55.6	WDS:459(133)(AB)	803(90) (AC)	
V magnitudes	A: 6.39	B: 8.63	C: 6.95		
$(B-V)$ magnitudes	A: +0.37	B: +0.73	C: +1.15		
μ(A)	58.39 mas yr^{-1}	± 0.05	98.27 mas yr^{-1}	± 0.04 (DR2)	
μ(B)	64.73 mas yr^{-1}	± 0.04	101.80 mas yr^{-1}	± 0.04 (DR2)	
μ(C)	5.40 mas yr^{-1}	± 0.04	−4.78 mas yr^{-1}	± 0.04 (DR2)	
π(A)	22.33 mas	± 0.07	146.1 light yr	± 0.5 (DR2)	
π(B)	22.38 mas	± 0.03	145.7 light yr	± 0.2 (DR2)	
π(C)	3.87 mas	± 0.04	843 light yr	± 9 (DR2)	
Spectra	A: F3V	B: G5V	C: K2III		
Luminosities (L$_\odot$)	A: 5	B: 0.5	C: 100	D:	
Catalogues (AB/C)	17 Sge	HR 7705	HD 191570/1	SAO 88266/70	HIP 93351
DS catalogues	H 3 24	STF 2637	BDS 9955	ADS 13442	
Radial velocity (A/B)	−43.10 km s^{-1}	± 0.6	−5.2 km s^{-1}	± 0.9	
Galactic coordinates	60°.573	-6°.688			

History

Found by William Herschel on 8 August 1780, the close stars were described as extremely unequal and Herschel noted that they were pale red and dusky, with the third and more distant star C given as pale red. He did not measure any angles but gave the distances of AB and AC as 11″ and 67″ respectively. Smyth, writing in the *Bedford Catalogue*, took him unfairly to task by reporting that Herschel's distances were 11″ 4‴ and 59″ 49‴. It is ironic to note that Smyth's own measures were later called in question by no less a person than S. W. Burnham.

Finder Chart

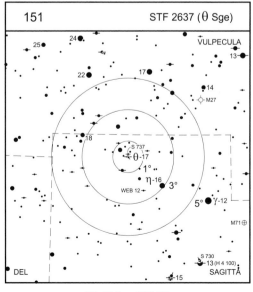

20h 09.9m +20° 55′

The Modern Era

There is some confusion as to the spectral types of the three stars. The WDS does not give consistent values, so the spectral types given above are taken from SIMBAD. The close stars certainly form a physical pair: the Gaia DR2 parallaxes for each component are to all intents and purposes equal. Star C is unrelated and is separating from AB owing to the proper motion of the close couple.

Observing and Neighbourhood

The system θ Sge can be found 3.0° SE of the Dumbbell Nebula (M27) and about 3° NNE of γ Sge, the brightest star in the constellation. It is a superb sight in small telescopes. The close pair are both white or yellowish-white and the distant C component is distinctly orange, as befits a K star. However, the colours of the stars can differ depending on the source of information. W. H. Smyth found the close pair to be pale topaz and grey, whilst C was pearly-yellow; more recently Sissy Haas sees A and B as white, whilst C is grey. Smyth notes a wide double which 'follows A on the parallel'. It is not clear which pair this is but there is a wide South pair (no. 737), just 5′ N with magnitudes of 7.9 and 9.2 at 100″; the primary is a K star. Not on the chart but 2°.5 W of γ and a little S is is the triple star ζ Sge (STF 2585) – a much tougher prospect than θ and which requires apertures in excess of 50-cm to see it as triple. The easier task is to divide AC (magnitudes 5.0 and 9.0 at 8″.4), but, in testing a new 12-inch objective bound for Lick Observatory in 1875, A. G. Clark noted the duplicity of the bright star. This is a short-period highly inclined system with a period of 23.2 years, whose separation will be 0″.19 in 2020 and will increase to a maximum value of 0″.24. One of the pairs attributed to Thomas William Webb can be found 1°.5 SW. WEB 12 has stars of magnitudes 8.4, 8.4 at 77°, 41″, 2015. The stars are fixed. The system 13 Sge (H 4 100), about 4° SW, is a coarse triple (10.1, 10.0, 5.6, 265°, 297°, 24″, 113″) as is S 730, which is only 6′ distant (7.2, 8.5, 10.2, 14°, 338°, 13″, 79″).

Measures

AB			
Early measure (STF)	316°.7	11″.41	1832.82
Recent measure (ARY)	330°.0	11″.74	2014.68
AC			
Early measure (WSI)	226°.6	70″.70	1832.82
Recent measure (SMR)	221°.9	90″.84	2013.73

152. o¹ CYG = STFA 50 = WDS J20136+4644

Table 9.152 Physical parameters for o Cyg

STFA 50	RA: 20 13 37.90	Dec: +46 44 28.8	WDS: 2111 (45)	AC	
V magnitudes (31)	Aa,Ab: 3.80	B:	C: 6.99	D: 4.82	
V magnitudes (32)	Aa,Ab: 3.98	B: 8.32			
$(B-V)$ magnitudes (31)	Aa,Ab: +1.28	B: —	C: −0.15	D: +0.10	
$(B-V)$ magnitudes (32)	Aa,Ab: +1.52	B: +0.30			
μ(A)	0.13 mas yr^{-1}	± 0.07	4.83 mas yr^{-1}	± 0.63 (DR2)	
μ(C)	3.72 mas yr^{-1}	± 0.08	2.05 mas yr^{-1}	± 0.80 (DR2)	
μ(D)	4.38 mas yr^{-1}	± 0.29	−0.50 mas yr^{-1}	± 0.31 (DR2)	
μ(32)	1.89 mas yr^{-1}	± 0.54	2.81 mas yr^{-1}	± 0.59 (DR2)	
π(A)	4.34 mas	± 0.35	752 light yr	± 61 (DR2)	
π(C)	2.90 mas	± 0.04	1125 light yr	± 15 (DR2)	
π(D)	5.06 mas	± 0.18	645 light yr	± 23 (DR2)	
π(32)	2.12 mas	± 0.27	3070 light yr	± 390 (DR2)	
Spectra (31)	A: K4Ib + B4V	C: B9	D: A5III		
Spectra (32)	K7Ib-II + B1V				
Masses (31) (M$_\odot$)	Aa: 11.7	Ab: 7.1			
Masses (32) (M$_\odot$)	Aa: 7.5	Ab: 4.1			
Radii (31) (R$_\odot$)	Aa: 197	Ab: 5			
Radii (32) (R$_\odot$)	Aa: 184	Ab: 3			
Luminosities (31) (L$_\odot$)	A: 1320	C: 155	D: 380		
Luminosities (32) (L$_\odot$)	AaAb: 4700	B: 85			
Catalogues (A/B/D)	31 Cyg/—/—	HR 7735/—/30	HD 192577-8/579/514	SAO 49337/8/2	HIP 99675/—/99639
Catalogues	32 Cyg	HR 7751	HD 192910/9	SAO 49385	HIP 99848
DS catalogues (31)	H 6 10 (AC)	HJ 1485 (AB)	STFA 50 (AB,D)	BU 1495 (AB)	
	WRH 33 (Aa,Ab)	BDS 10036	ADS 13554		
DS catalogues (32)	H 6 33	S 743	BDS 10036	ADS 13554	
Radial velocity (A)	−6.42 km s^{-1}	± 0.03			
Radial velocity (C)	−3.79 km s^{-1}	± 3.7			
Radial velocity (D)	−26.50 km s^{-1}	± 3.7			
Radial velocity (32)	−6.39 km s^{-1}	± 0.03			
Galactic coordinates	82°.676	+6°.777			

History

William Herschel observed 31 Cygni on 2 November 1779. He noted 'Double. Considerably unequal. L. pale r(ed). S. blue. It is the following star of the two o's that are close together. Distance 1′ 39″ 57‴. Position 87° 14′ s(outh). preceding.' Struve noted it at Dorpat but, because of the large separations, put it in his first Appendix catalogue as number 50. John Herschel found a faint companion to A (magnitude 13.4) and in 1878, Burnham added several more stars, of magnitudes 12.3 to 14.2, three of which were measured relative to C and the other to A. In 1897 Antonia Maury [587] published a list of stars with composite spectra which included ζ Aur and 31 Cygni. This explains why 31 Cygni (and indeed, later, in the case of 32 Cyg) the stars had two HD numbers allocated to them. In the 1940s, R. H. Wilson [588], using an interferometer of his own design fitted to the Flower Observatory 18-inch refractor, observed 31 Cyg and made six positive resolutions, which are listed in the WDS observations catalogue but the notes associated with them cast some doubt on their validity. W. H. van den Bos [592] also had reservations. In the modern era only a single instance of positive resolution has been made, by H. A. McAlister [589], in 1985 when the separation was measured at 0″.027. An orbit was calculated by Mason [591] in 2013 with a period of 10.04 years. According to this orbit, the angular separation of the stars varies from 0″.012 to 0″.046.

The Modern Era

The pairs 31 and 32 Cygni, along with ζ Aurigae, form a class of close binaries in which a late-type supergiant primary eclipses a B-type companion. They have been extensively studied, more recently by Professor Roger Griffin [590], who has maintained a monthly watch on the radial velocities of each star for more than 20 years. In addition to the orbital periods, which are 10.36 and 3.14 years respectively, systematic runs of velocity residuals, typically of the order of 300 metres s^{-1}, correspond to periodicities of a little over two years in both 31 and 32 Cyg. Professor Griffin suggests that that these might correspond to large-scale features on the supergiant components and that the periods found reflect the rotation periods of the stars in each case. He resurrects comments made in a paper by Tremblot [594] in 1938 which pointed out that 31 and 32 Cygni 'whose association in space can scarcely be doubted' and notes that the systemic radial velocities are the same to within about 30 metres s^{-1}. Since Griffin's paper the Gaia DR2 astrometry has been issued and is given above. The agreement in proper motion between 31 and 32 Cygni is not as strong as it was and, whilst the parallaxes

Finder Chart

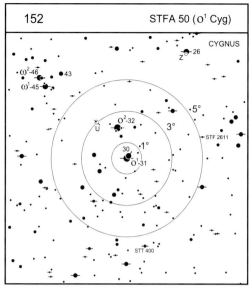

20h 13.6m +46° 44'

overlap when the errors are considered, this is far from conclusive proof that the two stars have a common proper motion. The spatial separation is about 6 or 7 parsecs so it is likely that there is no orbital connection, but the strange similarity between 31 and 32 Cygni is notable. The Gaia DR2 seems to indicate that the stars are not equidistant within the observed errors, and that 32 Cyg is far more distant. The masses and radii of 31 Cyg were determined by Eaton [593].

Observing and Neighbourhood

The nomenclature of this bright, widely spread group of stars is often confused. Here o¹ refers to the three bright stars 31 Cyg, HD 192579, and 30 Cyg. The distant 32 Cyg is o². Three degrees S is STT 400, a binary of 85.6-year period. The stars are magnitudes 7.6 and 9.8 and widest separation (0″.65) occurs in 2019, with the stars closing to 0″.13 by 2050. The system STF 2611 is a neat pair to be found 2°.5 W (8.5, 8.5, 208°, 5″.4, 2010).

Measures

AC			
Early measure (STF)	174°.0	106″.82	1866.18
Recent measure (ARY)	172°.5	106″.51	2013.25
AD			
Early measure (STF)	323°.7	337″.72	1835.95
Recent measure (UC)	322°.7	336″.68	2003.45

153. β DEL = BU 151 = WDS J20375+1436AB

Table 9.153 Physical parameters for β Del

BU 151	RA: 20 37 32.87	Dec: +14 35 42.7	WDS 20(733)		
V magnitudes	A = 4.11	B = 5.02	C = 13.5	D = 11.4	E = 11.63
$(B - V)$ magnitudes	A: +0.43	B: +0.56			
μ	118.09^{-1}	± 0.47	-48.06 mas yr^{-1}	± 0.43	
π	32.33 mas	± 0.47	101.0 light yr	± 1.5	
Spectra	A: F5IV	B: F2V			
Masses	A: 1.75 M$_\odot$	± 0.00	B: 1.47 M$_\odot$	± 0.04	
Luminosities (L$_\odot$)	A: 19	B: 6			
Catalogues	6 Del	HD 196524	HR 7882	SAO 106316	HIP 101769
DS catalogues	BU 151 (AB)	HJ 5545 (AC)	H 4 75 (AD)	STF 2704 (AD)	BDS 10363
	ADS 14073				
Radial velocity	-22.7 km s^{-1}	± 0.9 km s^{-1}			
Galactic coordinates (A)	$58°.881$	$-15°.651$			

History

Sir William Herschel found a faint star almost 26″ distant from β on 1 August 1781. He noted: 'Extremely unequal. hardly visible with 227; pretty strong with 460.' This star is what is now known as component D and the system was included in Struve's *Mensurae Micrometricae* as STF 2704 AD. The WDS gives its V magnitude as 11.4. John Herschel then found a 13.5 magnitude star in 1828. The close pair was discovered by S. W. Burnham in August 1873 using his 6-inch Clark refractor. Burnham had no micrometer at that time and turned to Baron Ercole Dembowski in Italy for help. Dembowski measured the pair in 1874 with his 7-inch refractor, with the result which is given here. By 1903 the star had been observed so assiduously that there were nine orbits extant by that time. Labitzke [597] added another faint companion E of magnitude 11.6 in 1922.

Finder Chart

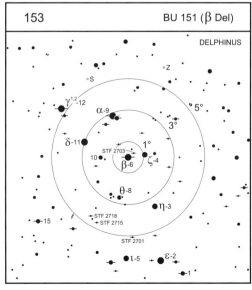

20h 37.5m +14° 36'

Orbit

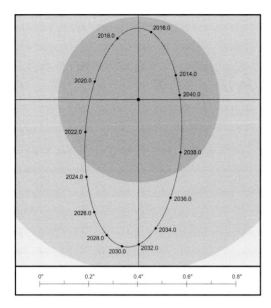

Ephemeris for BU 151 AB (2018 to 2027)

Orbit by Mut (2010e) Period: 26.6831 years, Grade: 1

Year	PA(°)	Sep(")	Year	PA(°)	Sep(")
2018.0	199.4	0.26	2023.0	315.4	0.31
2019.0	220.0	0.22	2024.0	325.2	0.37
2020.0	248.2	0.19	2025.0	332.4	0.43
2021.0	278.2	0.21	2026.0	337.9	0.48
2022.0	300.7	0.25	2027.0	342.4	0.52

The Modern Era

The orbit is now very well known. A series of extremely accurate measurements using the three-mirror ground-based Navy Prototype Optical Interferometer (NPOI) has fixed the period to within 1.4 days. The minimum separation of 0″.19 occurs in 2019 and the maximum distance is 0″.59 in 2030. The John Herschel star has closed significantly with AB and was found at a distance of 12″.1 by M. Fay [595] in 2011. At the time of writing SIMBAD included this star and gave its distance from A as 0″.9. This is incorrect. The results from Gaia have to await the DR3 catalogue.

Exoplanet Host?

During their observations with the NPOI, Muterspaugh *et al.* [620] subtracted the visual binary orbit from their astrometric observations of A and found a residual periodic perturbation, which they attributed to a giant planet of period 435 days and mass 9 ± 6 Jupiters. The NPOI consists of three 40-cm mirrors spaced by baselines of 87 metres, 87 metres, and 110 metres and which operate at wavelengths of 1.2, 1.6, and 2.2 microns.

Observing and Neighbourhood

Even at its widest separation β Del needs 20-cm to resolve it but it may be seen elongated at times other than this. The brightness of the stars allows significant magnifications to be employed in examining this star, but the most important factor is likely to be the state of the atmosphere. A night on which there is a slight haze, which usually also conspires to coincide with little or no wind, is best. It will not be until about 2023 that the stars will appear divided in 30-cm but it will get significantly easier within a few years after that. At magnitude 13.5 the John Herschel companion will be increasingly difficult to see in the glare of AB. In 2011, M. Fay [595] measured D (318°, 46″.7) and E (271°, 112″.2). Both are being left behind by the proper motion of AB. Just 15′ NW is STF 2703 (8.3, 8.5, 290°, 25″, 2016), whilst three further STF pairs can be found in an arc to the S and SE: STF 2701 is 2°.5 S (8.3, 8.6, 221°, 2″.1, 2016); 2° SE is STF 2715 (7.8, 10.2, 3°, 12″, 2013); and lastly STF 2718 is a further 20′ NE of 2715. The stars here are magnitudes 8.3 and 8.4; they are at 58°, 8″.6, 2016.

Measures

Early measure (D)	15°.5	0″.65	1874.66
(Orbit	11°.9	0″.48)	
Recent measure (RAO)	73°.5	0″.19	2012.72
(Orbit	73°.1	0″.18)	

154. $\gamma^{1,2}$ DEL = STF 2727 = WDS J20467+1607AB

Table 9.154 Physical parameters for γ Del

STF 2727	RA: 20 46 39.50	Dec: +16 07 27.4	WDS: 42(524)		
V magnitudes	A: 4.36	B: 5.03			
$(B - V)$ magnitudes	A: +1.17	B: +0.62			
μ(A)	-29.85 mas yr^{-1}	± 0.64	-196.06 mas yr^{-1}	± 0.83 (DR2)	
μ(B)	-11.15 mas yr^{-1}	± 0.27	-203.34 mas yr^{-1}	± 0.19 (DR2)	
π(A)	28.52 mas	± 0.48	114.4 light yr	± 2 (DR2)	
π(B)	28.41 mas	± 0.15	114.8 light yr	± 0.6 (DR2)	
Spectra	A: K1IV	B: F7V			
Luminosities (L$_\odot$)	A: 18	B: 10			
Catalogues (A/B)	12 Del	HR 7948/7	HD 197964/3	SAO 106476/5	HIP 102532/1
DS catalogues	Mayer 67	H 3 10	STF 2727	BDS 10609	ADS 14279
Radial velocity (A/B)	-6.15 km s^{-1}	± 0.15	-7.70 km s^{-1}	± 0.3	
Galactic coordinates	$61°.494$	$-16°.578$			

History

According to Thomas Lewis in 1906, both components were observed for the first time by James Bradley in 1755. Christian Mayer then included the pair in his catalogue; William Herschel also observed them on 27 September 1779 and then subsequently measured them in the following three months, noting that he suspected one of the stars of moving. Herschel also noted the two stars as white.

The Modern Era

The pair γ Del is a member of the Wolf 630 group, a spatially unassociated group of stars moving through the solar neighbourhood with common kinematics. The group as a whole does not have a similar element abundance, but there are certain subsamples of the group which do have similar abundances. In 1994 Hale [638] published a paper in which

Finder Chart

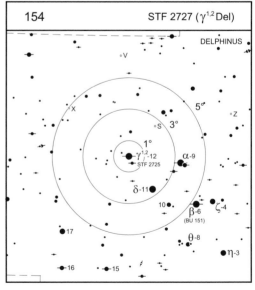

20h 46.7m +16° 07'

Orbit

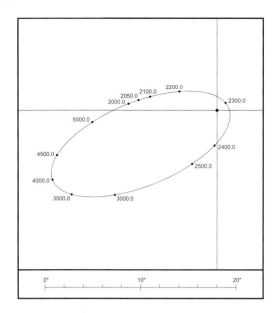

Ephemeris for STF 2727 AB (2000 to 2180)

Orbit by Hle (1994) Period: 3249 years, Grade: 4

Year	PA(°)	Sep(")	Year	PA(°)	Sep(")
2000.0	265.9	9.27	2100.0	258.8	7.13
2020.0	264.7	8.88	2120.0	256.9	6.64
2040.0	263.5	8.47	2140.0	254.5	6.12
2060.0	262.1	8.04	2160.0	251.8	5.58
2080.0	260.6	7.60	2180.0	248.4	5.00

he gave an orbit for AB with a period of 3249 years, but this is clearly a very preliminary result. Irwin & VandenBurg [601] obtained theoretical masses of 1.72 and 1.57 $M_\odot$ and radii of 6.43 and 2.21 $R_\odot$ for A and B respectively. DR2 moves the stars significantly closer to the Solar System compared with the distances found by Hipparcos.

Exoplanet Host?

In 1999 Larson *et al.* [599] announced the result of the radial velocity study of the star γ^2. They found a velocity amplitude of 11.6 ± 2.2 metres s^{-1} with a period of 1.44 years. This corresponds to a 0.7 Jupiter-mass object, which could be stable given the large separation between γ^2 and γ^1. Some years later both bright components were the subject of another intensive radial velocity study by Toyota *et al.* [600]. Over a 4.5-year period they used a high-resolution echelle spectrograph with a precision of 10 metres s^{-1} to measure the radial velocities on a monthly basis. They concluded that no objects with more than 1.8 Jupiter masses were attached to either of the bright stars.

Observing and neighbourhood

The pair γ Del is the easternmost star in the compact kite-shaped asterism which constitutes the brightest stars in Delphinus: α, β, γ, and δ Del. It is a wonderfully attractive pair for any aperture but needs sufficient power to move the stars well apart for the best view. RWA found colours of golden-yellow and blue-green using a 8.3-inch (21-cm) mirror. Smyth found yellow and light emerald. In the same low-power field is the pretty pair STF 2725 (7.5, 8.2, 12°, 6″.2, 2020 – an orbital object with a period of 2945 years). A real test of resolution for larger apertures can be found in β.

Measures

	PA	Sep	Epoch
Early measure (STF)	273°.8	11″.90	1830.89
(Orbit	272°.9	12″.06)	
Recent measure (ARY)	266°.0	8″.92	2015.80
(Orbit	264°.9	8″.96)	

155. λ CYG = STT 413 = WDS J20474+3629AB

Table 9.155 Physical parameters for λ Cyg

STT 413	RA: 20 47 24.54	Dec: +36 29 26.6	WDS: 74(405)	
V magnitudes	Aa: 5.4	Ab: 5.8	B: 6.26	
(*B* − *V*) magnitudes	Aab: −0.12	B: −0.03		
μ	+14.71 mas yr^{-1}	±0.32	−8.96 mas yr^{-1}	± 0.49
π	4.24 ± 0.43 mas	769 light yr	± 78 light yr	
μ(A)	18.58 mas yr^{-1}	± 0.88	−7.56 mas yr^{-1}	± 0.60 (DR2)
π(A)	4.99 mas	± 0.34	654 light yr	± 45 (DR2)
Spectra	Aa:	Ab:	B:	C:
Luminosities (L$_\odot$)	Aa: 230	Ab: 160	B: 100	
Catalogues	HR 7963	HD 198183	SAO 70505	HIP 102589
DS catalogues	H 6 32 (AC)	STT 413 (AaAb-B)	MCA 63 (AaAb)	BDS 10533	ADS 14296
Radial velocity (A/B)	−23.20 km s^{-1}	± 1.1 km s^{-1}		
Galactic coordinates	78°.084	−4°.338		

History

Sir William Herschel first noted the ninth magnitude star C at a distance of 'about 1 min.' from λ on 20 September 1781. The close pair, forming AB, was discovered by Otto Struve, using the 15-inch refractor at Pulkovo in 1842. The slow initial motion of this pair led Burnham to doubt whether it was a physical pair, although an orbit with period 93 years was published when only 50° of the apparent orbit had been traversed and the motion in separation looked linear.

The Modern Era

H. McAlister [602] and colleagues resolved the primary component in 1978 using the 4-metre Mayall reflector at Kitt Peak. The two components form a binary with a period of 11 years and a maximum separation of 0″.05. More recently,

Finder Chart

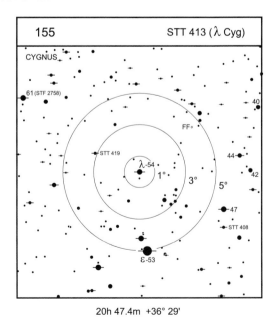

20h 47.4m +36° 29′

Orbit

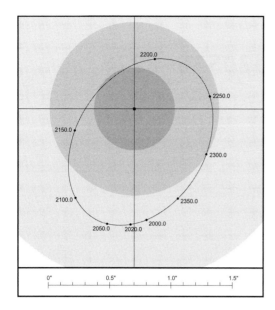

Ephemeris for STT 413 AB (2015 to 2060)

Orbit by Rab (1948b) Period: 391.30 years, Grade: 4

Year	PA(°)	Sep(")	Year	PA(°)	Sep(")
2015.0	0.2	0.92	2040.0	350.3	0.94
2020.0	358.1	0.92	2045.0	348.3	0.94
2025.0	356.1	0.93	2050.0	346.4	0.95
2030.0	354.1	0.93	2055.0	344.5	0.94
2035.0	352.2	0.94	2060.0	342.5	0.94

Marcel Fay [660] added three distant stars of $V \sim 14$ which are not connected to the multiple star. In 2015 Xia & Fu [606] published a study of the star which extended the period of the wide binary to 536 years. They derived a mass for star B of 7.1 $M_\odot$, which they were inclined to believe indicated that this star was itself a close pair. Some

corroborative evidence had been obtained in 2013 when a paper by Touhaimi *et al.* [604] suggested that either Aa or B was a single-lined spectroscopic binary. The DR2 results are not spectacularly precise, partly owing perhaps to the brightness of the primary star and the presence of the close companion to A.

Observing and Neighbourhood

The pair λ Cyg is easy to find. Start with ε Cyg in the Cross and move N 3° in declination. The companion has moved in retrograde to PA 5° and separation near 1 arcsecond. This makes it a feasible target for 15-cm with high magnification on a steady night. Star B is easily seen, but not so easily measured, with the 20-cm refractor at Cambridge and a measure made recently put the companion about 5° behind the ephemeris, which is derived from the 391-year orbit of Rabe, the current entry for this star in the USNO *Sixth Orbit Catalog*. This orbit dates from 1948 and does not take into account the duplicity of the primary star. Two further Otto Struve discoveries are in the neighbourhood: STT 408 is 3°.5 SSW (6.8, 9.9, 193°, 1″.6, fixed); STT 419 is 1°.5 ENE. Similarly to STT 408, the stars are slightly fainter: 7.2, 10.0 at 23°, 1″.6, 2012.

Measures

Early measure (STT)	116°.9	0″.69	1842.66
(Orbit	114°.5	0″.59)	
Recent measure (ARY)	3°.6	1″.05	2016.89
(Orbit	358°.8	0″.92)	

156. RMK 26 PAV = WDS J20516−6226

Table 9.156 Physical parameters for RMK 26 Pav

RMK 26	RA: 20 51 38.47	Dec: −62 25 45.6	WDS: 1322(64)	
V magnitudes	A: 6.27	B: 6.44		
$(B − V)$ magnitudes	A: +0.22	B: +0.29		
μ(A)	82.70 mas yr^{-1}	± 0.50	−46.56 mas yr^{-1}	± 0.07 (DR2)
μ(B)	82.08 mas yr^{-1}	± 0.06	−42.34 mas yr^{-1}	± 0.08 (DR2)
π(A)	13.51 mas	± 0.05	241 light yr	± 1 (DR2)
π(B)	13.53 mas	± 0.06	241 light yr	± 1 (DR2)
Spectra	A: A2III	B: A3III		
Luminosities (L$_\odot$)	A: 14	B: 12		
Catalogues (A/B)	HR 7959/60	HD 198160/1	SAO 254883/4	HIP 102962
DS catalogues	RMK 26			
Radial velocity (A/B)	−16.00 km s^{-1}	± 7.4	−10 km s^{-1}	± 5
Galactic coordinates	333°.318	−37.624		

History

This pair was found by Rümker in 1826. Christian Carl Ludwig Rümker was born at Stargard, Meklinburg-Strelitz, the son of a court councillor [607]. Having trained as a builder he was more interested in mathematics and taught the subject for two years in Hamburg. He came to England in 1809 and having been press-ganged into the Navy in 1813 he subsequently taught cadets on three Royal Navy ships. Whilst in the Mediterranean he met Franz Xavier von Zach, who induced him to pursue astronomy. He left the Navy in 1814 and became private astronomer to Sir Thomas Brisbane, who was appointed Governor of Australia, and on arrival Brisbane set up a private observatory at Parramatta, which Rümker used. Rümker's relations with Brisbane were fractious on occasion and he left Parramatta, only to return in 1826 on the appointment of a new Governor. He started the observations which led to his small catalogue of bright and

Finder Chart

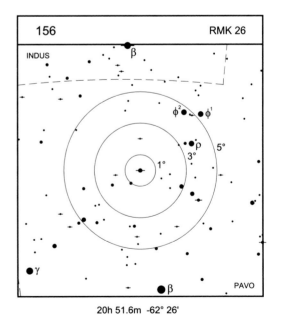

20h 51.6m -62° 26'

wide double stars and also discovered a comet in September of that year. He was awarded the RAS Gold Medal in 1854 and retired to Lisbon in 1857, where he died five years later.

The Modern Era

The scatter on the radial velocities of the two components, and the rather poorly defined parallax from Hipparcos, suggests that one or both stars may be composite; however, the DR2 results show excellent agreement in parallax value. The stars clearly form a long-period binary system.

Observing and Neighbourhood

The nearest double star of note is the Peacock star, α Pavonis (HJ 5193), which is about 6°.5 NNW (1.9, 9.1, 9.7, 80°, 249″, 77°, 245″) and therefore not on the finder chart. The B and C stars form a 16″ pair about 4′ from the $V = 1.9$ primary. In 2010 RWA observed this group with the Johannesburg refractor. The colours appeared white, reddish, and blue respectively. Four degrees SSW of the Peacock star is an isosceles triangle of fourth and fifth magnitude stars. The star RMK 26 can be found 3° SW of this group. 'Superb D *'. 'Superb'. 'Fine'. These are some of the comments from John Herschel during his Cape observations of this star. Ross Gould using 175-mm at ×100 reports 'neat and close, pale yellow pair … gemlike at ×180. The field is sparse and faint.'

Measures

Early measure (R)	94°.0	3″.38	1873.59
Recent measure (ARY)	80°.8	2″.49	2011.12

157. ε EQU = STF 2737
= WDS J20591+0418AB,C

Table 9.157 Physical parameters for ε Equ

STF 2737	RA: 20 59 04.54	Dec: +04 17 37.8	WDS(AB): 50(479)		
			WDS(AB,C): 116(300)		
V magnitudes	A: 5.96	B: 6.31	C: 7.05		
(B − V) magnitudes	A: +0.51	B: +0.47	C: +0.52		
μ(A)	−115.75 mas yr^{-1}	−1.52 mas yr^{-1}	−151.70	± 0.70	
π(A)	18.49 mas	± 1.35	176 light yr	± 13	
μ(C)	−116.44 mas yr^{-1}	−1.41 mas yr^{-1}	−152.18	± 0.71	
π(C)	18.16 mas	± 0.24	180 light yr	± 3	
μ(C)	−115.76 mas yr^{-1}	± 0.07	−143.02 mas yr^{-1}	± 0.07 (DR2)	
π(C)	18.08 mas	± 0.04	180.3 light yr	± 0.4 (DR2)	
Spectra	AB: F6IV	C: dF4			
Luminosities (L$_\odot$)	A: 10	B: 7	C: 4		
Catalogues (AB/C)	1 Equ	HD 199796	SAO 126428/9	HIP 103569/71	
DS catalogues	H 3 21 (BC)	STF 2737 (AB,C)	HJL 288 (BC)	BDS 10643	ADS 14499
Radial velocity (AB)	8.20 km s^{-1}	± 0.2			
Radial velocity (C)	8.00 km s^{-1}	± 0.3			
Galactic coordinates	52°.864	−25°.783			

History

William Herschel noted the star now known as C on 2 August 1780. His notes say 'Double. Considerably unequal. L.(arge) w.(hite)'; S.(mall) much inclining to r.(ed)... A third small star follows.' The WDS gives magnitude 13.1 for this object. It was next noted by Burnham [632] in 1912, when he used it to measure the proper motion of the close binary system.

The Modern Era

A study by Zeller in 1965 [608] concluded that the C star was in a hyperbolic orbit around AB and gave predicted position angles and separations up to 2042. Interpolating between the points for 2002.4 and 2042.9 gives a position for 2015.70 (see below) of 10″.78 and 65°.4. Gaia DR2 only gives details of the C component but they seem to agree with the Hipparcos distance and proper motion derived for the close pair.

Observing and Neighbourhood

The close pair is now beyond most amateur apertures for the next 17 years or so. The apparent orbit is a highly elongated and highly eccentric ellipse in which the companion spends about 20 years in a close approach to the primary star; this

Finder Chart

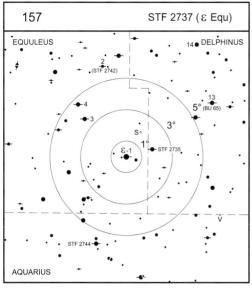

157   STF 2737 (ε Equ)

EQUULEUS   14 ● DELPHINUS

2
(STF 2742)

● 4   13
● 3   5°   (BU 65)

3°
S°
1°
ε-1   ● STF 2735

V°

STF 2744 ●

AQUARIUS

20h 59.1m +04° 18'

Orbit

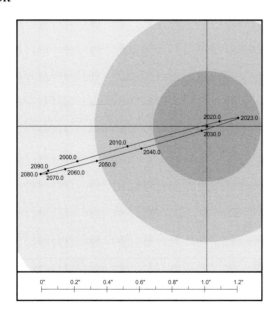

0"   0.2"   0.4"   0.6"   0.8"   1.0"   1.2"

Ephemeris for STF 2737 AB (2020 to 2047)

Orbit by Zel (1965) Period: 101.485 years, Grade: 2

Year	PA(°)	Sep(")	Year	PA(°)	Sep(")
2020.0	110.6	0.08	2035.0	290.3	0.24
2023.0	104.7	0.20	2038.0	288.9	0.35
2026.0	100.6	0.13	2041.0	288.2	0.46
2029.0	35.5	0.02	2044.0	287.7	0.55
2032.0	294.3	0.12	2047.0	287.3	0.63

takes it within 20 mas before swinging around and heading out again towards PA 285°. RWA was just able to see it as double in 2012 using the Cambridge 20-cm refractor when the separation was 0″.44. Components AB, C remain easy in small telescopes and, for a stiffer test, the faint star mentioned by William Herschel is magnitude 13.1 at 69″; the distance is slowly decreasing. One degree WNW is the fine pair STF 2735 (6.5, 7.5, 284°, 2″.3, 2011), which T. W. Webb found to be orange and purple in 1833 and Dembowski found to be white and blue in 1854. The binary STF 2744, an easy target for 15-cm, lies 3°.5 S. The stars, of magnitudes 6.8 and 7.3, orbit each other every 1532 years and are never closer than 0″.9; at 2020.0 they will be found at 100°, 1″.2. A similar distance to the N is 2 Equ (STF 2742) (7.4, 7.6, 215°, 2″.9, 2015). The binary 13 Del was found by Burnham using his 6-inch Clark (BU 65). The magnitudes, 5.6 and 8.2, make it a stiff test for 15-cm. The latest position is 199°, 1″.5, 2009.

Measures

AB			
Early measure (STF)	294°.0	0″.35	1835.67
(Orbit	289°.1	0″.38)	
Recent measure (RAO)	283°.7	0″.43	2012.76
(Orbit	283°.0	0″.37)	
A-C			
Early measure (STF)	75°.8	10″.85	1831.45
Recent measure (WSI)	66°.5	10″.29	2015.70

158. 61 CYG = STF 2758 = WDS J21069+3845AB

Table 9.158 Physical parameters for 61 Cyg

STF 2758	RA: 21 06 53.95	Dec: +38 44 57.9	WDS 1(1711)		
V magnitudes	A: 5.20	B: 6.05			
$(B - V)$ magnitudes	A: +1.27	B: +1.54			
μ(A)	4164.17 mas yr^{-1}	± 0.19	3249.99 mas yr^{-1}	± 0.25 (DR2)	
μ(B)	4105.79 mas yr^{-1}	± 0.09	3155.76 mas yr^{-1}	± 0.10 (DR2)	
π(A)	286.15 mas	± 0.06	11.398 light yr	± 0.002 (DR2)	
π(B)	285.95 mas	± 0.10	11.406 light yr	± 0.004 (DR2)	
Spectra	A: K5V	B: K7V			
Masses (M$_\odot$)	A: 0.70		B: 0.63		
Radii (R$_\odot$)	A: 0.665	± 0.005	B: 0.595	± 0.008	
Luminosities (L$_\odot$)	A: 0.08	B: 0.04			
Catalogues (A/B)	61 Cyg	HR 8085/6	HD 201091/2	SAO 70919	HIP 104214/7
DS catalogues	Mayer 70	H 4 18	STF 2758	BDS 10732	ADS 14636
Radial velocity (A/B)	−65.74 km s^{-1}	± 0.09	−64.07 km s^{-1}	± 0.08	
Galactic coordinates	82°.320	−5°.818			

History

R. G. Aitken [1] noted that Bradley had been the first to observe both components, in 1753, and quoted the relative positions of the stars as noted by John Herschel. Mayer included 61 Cyg in his catalogue and the pair came to the attention of Sir William Herschel on 20 September 1780. He recorded: 'Pretty unequal. L(arge). pale r(ed).; s(mall) r.; or L(arge). r(ed).; s(mall). garnet'. About a decade later Piazzi observed them and when he compared his positions with those of Bradley, 40 years earlier, he noted the significant differences and as a result called 61 Cygni 'The flying star'. In 1838, F. W. Bessel [612] announced that his researches using the Konigsberg heliometer had indicated that 61 Cygni had a measurable parallax and the value was 0″.3136, the first time that the distance to a star had been geometrically determined and preceding similar work on α Centauri, by Henderson, by a few months. Bessel's value is remarkably accurate;

Finder Chart

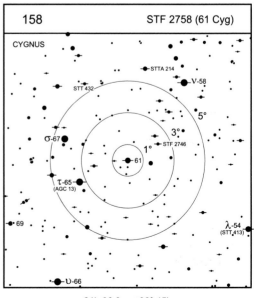

21h 06.9m +38° 45'

Orbit

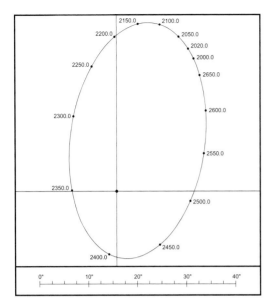

Ephemeris for STF 2758 AB (2000 to 2180)

Orbit by PkO (2006b) Period: 678 years, Grade: 4

Year	PA(°)	Sep(")	Year	PA(°)	Sep(")
2000.0	149.2	30.63	2100.0	165.1	34.26
2020.0	152.7	31.84	2120.0	168.0	34.19
2040.0	156.0	32.82	2140.0	171.0	33.82
2060.0	159.1	33.56	2160.0	174.0	33.12
2080.0	162.1	34.04	2180.0	177.2	32.08

the current value as determined by Gaia places the star a further 10% away.

The Modern Era

Recent deep-imaging surveys in the infrared have failed to reveal anything which is physically connected to either star. Heinze *et al.* [648] found faint background stars in the L-band. One is 11″.24 distant in PA 227°.5 whilst another lies 7″.78 away in PA 83°.2. In 1960 Deutsch [613] found a perturbation in the A star with a period of 4.9 years, agreeing with Strand's earlier value. The Hipparcos satellite determined the proper motion and parallax of A but with very large quoted errors, possibly an indication of multiplicity. The perturbation in A is still in the USNO Sixth Catalogue of binary star orbits. DR2 confirms the Hipparcos results but to much greater precision.

Exoplanet host?

In 1942, K. A. Strand [650] announced a perturbation in the system but did not specify which star was affected. Using refractor plates taken at Potsdam, Lick, and Sproul, a period of 4.9 years was found. In 1977 astronomers at Pulkovo Observatory [614] announced that component A was accompanied by two planetary bodies and B had one. In the following year W. D. Heintz [549] dismissed the idea that there were any perturbing influences in the 61 Cygni system.

Observing and Neighbourhood

The stars are now widening and can be seen with firmly mounted binoculars or the smallest telescopic apertures. They should reach a maximum distance of about 34″.3 in 2105, and will then close, reaching the minimum distance of 9″ in around 2360. The late spectral type of each component leaves the observer in little doubt about the colours. The rapid motion of AB through the surrounding starfield is shown by the two measures (below) of star H with respect to star A, measured over a period of about 13 years. A fine test for 10-cm is STT 432 (7.9, 8.0, 115°, 1″.4, 2016), 3° NE. One degree WNW is STF 2746 (7.9, 8.7, 323°, 1″.2, 2016). The PA has increased 50° since discovery and the stars are now slightly wider. The star system STTA 214 is a binocular pair 3° N (6.4, 8.6, 185°, 57″, 2017).

Measures

AB

	PA	Sep	Date
Early measure (STF)	91°.2	15″.63	1831.70
(Orbit	90°.0	15″.70)	
Recent measure (ARY)	152°.7	31″.41	2015.89
(Orbit	151°.5	31″.61)	

AH

	PA	Sep	Date
Earlier measure (ARY)	307°.9	70″.68	2003.03
Recent measure (ARY)	271°.3	108″.00	2015.89

159. H 1 48 CEP = WDS J21137+6424

Table 9.159 Physical parameters for H 1 48 Cep

H 1 48	RA: 21 13 42.46	Dec: +64 24 15.1	WDS: 387(148)	
V magnitudes	A: 7.21	B: 7.33		
$(B - V)$ magnitudes	A: +0.77			
μ	+16.89 mas yr^{-1}	±0.51	−105.13 mas yr^{-1}	± 0.43
π	23.39 mas	± 0.42	42.8 light yr	± 0.8
Spectra	G2IV	G2IV		
Luminosities (L$_\odot$)	A: 2	B: 1.8		
Catalogues	HR 8133	HD 202582	SAO 19257	HIP 104788
DS catalogues	H 1 48	AC 19	BDS 10863	ADS 14783
Radial velocity	30.30 km s^{-1}	± 0.2 km s^{-1}		
Galactic coordinates	101°.969	+10°.812		

History

Sir William Herschel called it 'A minute and beautiful double star' in his log of 27 September 1782. He goes on to say 'A little unequal. Both pr (pink-rose colour). Almost in contact with 460, with 625 better divided with 657 still better.' After that the star appears to have been completely ignored by observers. It closed up to below 0″.5 in 1825 and remained below that value until the late 1840s. In 1859 Alvan Clark rediscovered it when it was separated by 0″.9. It was subsequently noted to be a known pair, but even in Burnham's *General Catalogue* it was dismissed with the phrase 'change doubtful'. Burnham did concede that 'the later measures appear to show a little direct motion'. In the period 1910–1920 the star underwent another unobserved close approach and Philip Fox measured it in 1925 at a separation of 0″.5. In 1932, Robert Aitken, in his *General Catalogue*, commented that 'A complete set of measures of this pair is greatly to be desired'. In fact it would be almost another century before a representative orbit would be calculated.

Finder Chart

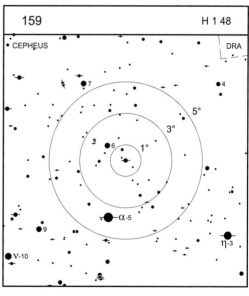

21h 13.7m +64° 24'

Orbit

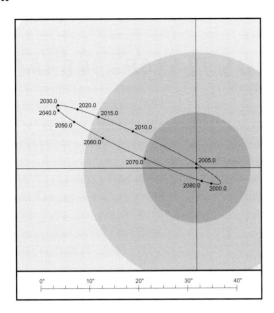

Ephemeris for H 1 48 (2017 to 2044)

Orbit by Hrt (2014b) Period: 81.64 years, Grade: 3

Year	PA(°)	Sep(")	Year	PA(°)	Sep(")
2017.0	243.4	0.73	2032.0	246.3	0.93
2020.0	244.1	0.81	2035.0	246.8	0.93
2023.0	244.7	0.86	2038.0	247.3	0.92
2026.0	245.3	0.90	2041.0	247.9	0.91
2029.0	245.8	0.92	2044.0	248.4	0.88

The Modern Era

The apparent orbit of H 1 48 is an elongated ellipse with the relatively short period of 81.64 years [615]. In reality, the plane of the true orbit is inclined to the line of sight by 82° and the eccentricity is 0.78. As a result, the apparent motion is largely in separation with rapid motion through periastron. In consequence, through sheer bad luck, the pair was never observed at this point in the orbit, which would have alerted the observers at once. We now know that between 1910 and 1915 the stars were less than 0″.1 apart and, as Burnham had essentially downplayed the importance of the system, no one attempted to observe the pair. Another close approach occurred around 2004, for which again there are few precise observations, and at present the stars are separating quickly to reach a maximum separation of 0″.93 by about 2030.

Observing and Neighbourhood

It is extremely rare for such a relatively short-period binary to retain its original Herschel number. Most of that observer's discoveries were subsumed into the Dorpat catalogue compiled by F. G. W. Struve. RWA has only been able to make a positive measure of the pair in 2016 as it has become considerably easier. An aperture of 20-cm will certainly show the stars separated. The system H 1 48 lies about 2°.5 N and slightly W of Alderamin (α Cep). The same distance S of that star lies the fine white pair STF 2780 (6.1, 6.8, 213°, 1″.0, 2015, not on the chart), which has moved just 12° retrograde since it was first measured in 1828.

Measures

Early measure (DA)	246°.2	0″.88	1859.73
(Orbit	245°.9	0″.86)	
Recent measure (ARY)	248°.1	0″.63	2016.06
(Orbit	243°.0	0″.70)	

160. τ CYG = AGC 13 = WDS J21148+3803AB

Table 9.160 Physical parameters for τ Cyg

AGC 13	RA: 21 14 47.49	Dec: +38 02 43.1	WDS: 93(346)		
V magnitudes	A: 3.87	B: 6.65	F: 11.95	I: 16.02	
(*B* − *V*) magnitudes	A: +0.41	B: +0.83			
μ	+196.99 mas yr^{-1}	±0.30	+410.28 mas yr^{-1}	± 0.30	
π	49.16 mas	± 0.40	65.9 light yr	± 0.5	
μ(A)	136.61 mas yr^{-1}	± 0.71	451.92 mas yr^{-1}	± 0.80 (DR2)	
π(A)	49.58 mas	± 0.46	65.8 light yr	± 0.6 (DR2)	
Spectra	F3V	F7V			
Masses (M$_\odot$)	A: 1.0	B: 0.75			
Luminosities (L$_\odot$)	A: 9	B: 0.7			
Catalogues	65 Cyg	HR 8130	HD 202444	SAO 71121	HIP 104887
DS catalogues	AGC13 (AB)	BDS 10846	ADS 14787	JOD 20 (FaFb)	
Radial velocity	−20.90 km s^{-1}	± 0.8			
Galactic coordinates	82°.854	−7°.432			

History

The star was first resolved by Alvan G. Clark, who was testing a 26-inch objective bound for the McCormick Observatory in October 1874. It was first measured during the same year by Simon Newcomb [622] using the 26.5-inch refractor at Washington. In 1908, spectrograms of τ Cyg taken by Barrett [616] showed that the star had a variable radial velocity and a period of only 3.5 hours, a record at the time. In 1921 Paraskévopoulos [621] found a radial velocity variation with a period of 0.1425 days but also argued that it could be due to radial pulsation.

The Modern Era

In 1960 Abt [617] re-observed τ Cyg spectroscopically and found no sign of a variation in radial velocity. He also argued that combining a period of 0.143 days with a parallax of 0″.47

Finder Chart

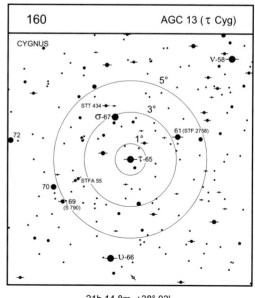

21h 14.8m +38° 03′

Orbit

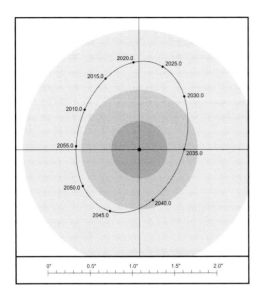

Ephemeris for AGC 13 AB (2018 to 2036)

Orbit by Mut (2010e) Period: 49.6257 years, Grade: 1

Year	PA(°)	Sep(")	Year	PA(°)	Sep(")
2018.0	192.1	0.98	2028.0	150.1	0.91
2020.0	183.8	1.01	2030.0	138.9	0.81
2022.0	175.8	1.02	2032.0	124.4	0.69
2024.0	167.8	1.01	2034.0	103.7	0.58
2026.0	159.4	0.97	2036.0	74.6	0.50

would yield a rotational velocity at the equator of 750 km s^{-1}, which is not viable. Heintz [618], ten years later, also dismissed the idea that the 0.143-day period could be due to a close companion of the primary star. A number of independent reports indicated that photometrically τ Cyg was shown to be variable with an amplitude of about 0.02 magnitudes, but the most recent observations, by Bartolini & Dapergolas [619] in 1980, indicate no period. The best interpretation of the radial velocity and light variations is that the star is a bright member of the δ Scuti class of pulsating variables. The WDS catalogue lists a number of faint stars within 10 arcminutes of the A component but only two of them (F and I) share the same considerable proper motion. Imaging of star F with the 2.2-metre reflector at Almeria, Spain, by Jodar et al. [623] showed that F is a close pair with a slightly fainter companion 0″.4 distant in PA 18°.

Figure 9.4 τ Cygni, observed with 12.5-inch Calver reflector in 1972 (left) and 1996 (right) showing orbital motion over almost half an orbital period (49 years) (C. Taylor)

Exoplanet Host?

The most recent observational investigation of τ Cyg was done by Muterspaugh et al. [620], who used the extremely precise ground-based NPOI interferometer to yield an orbit with a period of 49.626 years. They speculated on the existence of a substellar companion (possibly having a period of 810 days) but suggested that this needed confirmation.

Observing and Neighbourhood

The ephemeris shows that for the next ten years or so the stars are near the widest separation that they can achieve. A 20-cm aperture certainly will resolve them and a good 15-cm should also show them clearly. This system is one of the few in which orbital motion can be seen over a few years. Christopher Taylor, who operates a 12.5-inch Calver reflector in Oxfordshire, has made drawings of a number of bright binaries and his observations of τ Cyg, shown above, clearly show the stars on opposite sides of their apparent orbit. Close to τ (1°.6 NW) is the famous nearby binary 61 Cygni. The binocular pair STFA 55 (6.6, 6.6, 304°, 364″) is 2°.0 ESE of τ. For the telescopic observer there are four faint stars between magnitudes 11 and 14 in the field: S 790 (= 69 Cyg, 5.9, 10.2, 99°, 53″, 2012, and another 11.7 at 34″, 2012) is 1°.0 SSE of STFA 55. Marcel Fay adds a magnitude 13.9 (FYM 124) at 19″, 84° from C (2014). Two degrees N and slightly E is STT 434 (6.7, 9.9, 122°, 24″, 2006). A third star (magnitude 11) is at 316°, 98″.

Measures

Early measure (HAL)	160°.2	1″.04	1876.90
(Orbit	160°.9	0″.98)	
Recent measure (RAO)	213°.4	0″.86	2013.62
(Orbit	213°.0	0″.88)	

161. θ IND = HJ 5258 = WDS J21199–5327

Table 9.161 Physical parameters for θ Ind

HJ 5258	RA: 21 19 51.89	Dec: −53 26 57.4	WDS: 986(78)	
V magnitudes	A: 4.50	B: 6.93		
(B − V) magnitudes	A: +0.18	B:		
μ(A)	106.23 mas yr⁻¹	± 0.54	−67.03 mas yr⁻¹	± 0.45 (DR2)
μ(B)	92.93 mas yr⁻¹	± 0.77	−76.52 mas yr⁻¹	± 0.06 (DR2)
π(A)	33.93 mas	± 0.27	96.1 light yr	± 0.8 (DR2)
π(B)	33.46 mas	± 0.04	97.5 light yr	± 0.1 (DR2)
Spectra	A: A5IV-V	B: G0V		
Luminosities (L⊙)	A: 11	B: 1		
Catalogues (A/B)	HR 8140	HD 202730	SAO 246965/4	HIP 105319
DS catalogues	HJ 5258 (AB)	MRN 3 (AaAb)		
Radial velocity	−14.5 km s⁻¹	± 2		
Radial velocity (B)	−6.49 km s⁻¹	± 0.16		(DR2)
Galactic coordinates	343°.686	−43.320		

History

This pair was found by John Herschel on 8 July 1834. He notes it as 'beautiful' in the section marked 'Reduced observations of double stars' in the 1847 catalogue. However, on three subsequent nights out of four in which he made micrometric measures, he noted the magnitudes of A and B to be 5 and 10 and says that in less than ideal conditions the companion could be 'difficult'. The modern visual magnitude for B is 6.9.

The Modern Era

The primary is a fast-rotating star with equatorial rotational velocity measured at 210 km s⁻¹. As a relatively nearby star it has been examined [624] for the presence of exo-zodiacal dust using an infrared photometer fitted to the four outrigger telescopes of the VLT, giving high-resolution information

Finder Chart

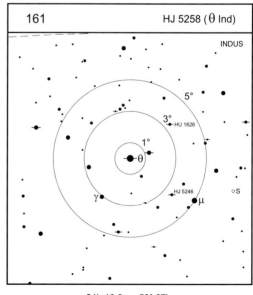

21h 19.9m -53° 27'

on the immediate stellar surroundings. When the star was observed it was found that it had a companion, close in, of almost equal brightness in the H band, suggesting that this star was also of spectral type A5. The apparent separation of the stars was $0''.062$ on 2012.563. Several years before, Legrange *et al.* [625] had measured the radial velocity and found little or no variation, which implies that the orbital plane of the new binary is almost face-on and the period must be around 1.3 years.

Observing and Neighbourhood

This is a beautiful pair. There is a faint pair of stars $4'$ due S, Ary 79 (11, 11, 98°, 29") noted by Hartung, which RWA chanced upon during an observation of the bright pair at Johannesburg in 2010. For θ itself Hartung gives the colours pale yellow and distinctly reddish, which might explain the apparently large difference in magnitude estimates for the companion although this does not agree with the spectral type, which is given as G0V. RWA made the colours yellowish and lilac with the Johannesburg telescope. Hartung [626] considered that the large and similar proper motions constitute a physical system whilst Hartkopf [627] has computed a linear ephemeris for the motion of B with respect to A. Gaia DR2 shows that the bright stars are at the same distance and more likely to be physically connected. One point five degrees NW is the binary pair HU 1626 (7.3, 8.8, 287°, $1''.0$, 2020) with a period of 148 years, whilst $1°.5$ SW is HJ 5246 (7.8, 8.0, 131°, $4''.1$, 2001).

Measures

Early measure (JC)	300°.5	3".53	1850.65
(Linear ephemeris	294°.2	3".61)	
Recent measure (RWA)	268°.2	7".26	2010.62
(Linear ephemeris	269°.9	7".18)	

162. β CEP = STF 2806 = WDS J21287+7034AB

Table 9.162 Physical parameters for β Cep

STF 2806	RA: 21 28 39.60	Dec: +70 33 38.6	WDS: 668(102)		
V magnitudes	Aab: 3.17	B: 8.63			
(*B* − *V*) magnitudes	A: −0.23	B:			
μ(A)	14.77 mas yr^{-1}	± 1.66	7.94 mas yr^{-1}	± 1.75 (DR2)	
μ(B)	9.98 mas yr^{-1}	± 0.06	6.67 mas yr^{-1}	± 0.06 (DR2)	
π(A)	9.67 mas	± 0.99	337 light yr	± 35(DR2)	
π(B)	4.72 mas	± 0.03	691 light yr	± 4 (DR2)	
Spectra	Aa: B2III+?	B: A1V			
Masses (M$_\odot$)	Aab: 12	± 1.0			
Radii (R$_\odot$)	Aab: 6.5	± 1.2			
Luminosities (L$_\odot$)	A: 460	B: 13			
Catalogues	8 Cep	HR 8238	HD 205021	SAO 10057	HIP 106032
DS catalogues	H 3 6 (AB)	STF 2806 (AB)	BDS 11046	ADS 15032	LAB 6 (Aa)
Radial velocity	−8.2 km s^{-1}	± 2			
Galactic coordinates	107°.539	+14°.026			

History

William Herschel observed β Cep on 31 August 1779, with the following notes 'Double. Very unequal. L(arge). blueish w(hite).; S(mall). garnet. Distance 13″.125, Position 15° 28′ preceding.' Smyth found the stars white and blue and noted that the pair had been seen by Piazzi, who estimated the companion at magnitude 11. F. G. W. Struve notes greenish-white and blue whilst, more recently, Sissy Haas records a colour of green for the secondary in 60-mm aperture at ×25.

The Modern Era

The pioneer of speckle interferometry, Antoine Labeyrie, first discovered [628] the duplicity of the bright star during observations to measure stellar diameters made with the 200-inch (508-cm) reflector on Palomar Mountain in 1971. Subsequently Paul Couteau tried to resolve the star visually using

Finder Chart

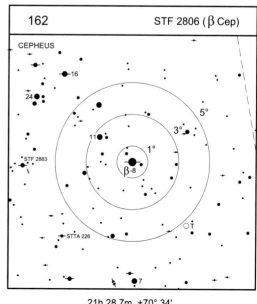

21h 28.7m +70° 34'

the large refractors in the south of France but was unable to do so, confirming that the difference in magnitude was at least 3 in *V*. The pair β Cep has a magnetic field that might be responsible for the spherical shell or ring-like structure found around the star [630] whilst high-resolution data taken with the VEGA instrument on the CHARA long baseline array was being analysed.

Observing and Neighbourhood

The Aa system has a period of 83 years and is very highly inclined to the line of sight, so that it is possible that an eclipse might take place close to the next periastron in the 2060s.

The separation is currently near its maximum of $0''.3$ and will reduce to under $0''.01$ in 2076. It now seems clear from the DR2 results that the Struve component is not physical. Four degrees E is STF 2883 (5.6, 8.6, 253°, $14''.2$, 2016) whilst 3° SE is a pair marked STTA 266 on *CDSA2* but which is actually STTA 226.

Measures

Early measure (STF)	250°.0	13″.57	1832.26
Recent measure (FYM)	248°.3	14″.1	2013.74

163. μ CYG = STF 2822 = WDS J21441+2845AB

Table 9.163 Physical parameters for μ Cyg

STF 2822	RA: 21 44 08.57	Dec: +28 44 33.4	WDS: 21(718)		
V magnitudes	A: 4.75	BaBb: 6.18			
$(B-V)$ magnitudes	A: +0.47	B: +0.61			
μ	+260.72 mas yr^{-1}	$\pm$ 0.36	-243.21 mas yr^{-1}	$\pm$ 0.21	
π	44.97 mas	$\pm$ 0.43	72.5 light yr	$\pm$ 0.7	
μ(A)	257.01 mas yr^{-1}	$\pm$ 0.35	-239.01 mas yr^{-1}	$\pm$ 0.39 (DR2)	
π(A)	45.22 mas	$\pm$ 0.24	72.1 light yr	$\pm$ 0.4 (DR2)	
Spectra	A: F5IV	Ba: F6IV	Bb: K5V		
Masses	A: 1.55	Ba: 1.66	Bb: 0.82		
Luminosities (L$_\odot$)	A: 5	Ba: 1	Bb: —		
Catalogues (A/B)	78 Cyg	HR 8309/10	HD 206826/7	SAO 89940/39	HIP 107310
DS catalogues	Mayer 72	H 3 15	STF 2822	BDS 11214	ADS 15270
Radial velocity	16.95 km s^{-1}	$\pm$ 0.16			
Galactic coordinates	80°.585	-18°.341			

History

Discovered by Christian Mayer on 7 September 1777. According to J. Schlimmer the separation at this time was 11″.2, but the current orbit when interpolated back gives a distance of 6″.7 for that epoch. Significant errors in observed distance are not uncommon in Mayer's catalogue and are understandable, considering the instrument he had at his disposal. The stars closed through the nineteenth century but in a rectilinear fashion, causing Burnham to opine that μ Cyg was a member of the 61 Cyg class of double stars i.e. travelling through space together but with no obvious signs of orbital motion. Thomas Lewis noted that it was a binary and by the time that Aitken's catalogue had been published (1932) the stars were close to minimum separation (0″.55 in 1938). A number of faint and distant companions have been added by Burnham and Espin amongst others but all are field stars.

Finder Chart

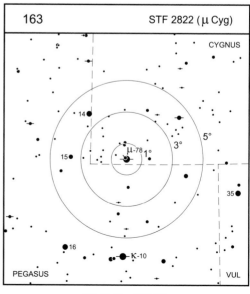

21h 44.1m +28° 45′

Orbit

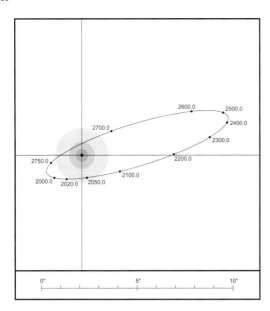

Ephemeris for STF 2822 AB (2010 to 2100)

Orbit by Hei (1995) Period: 789 years, Grade: 4

Year	PA(°)	Sep(")	Year	PA(°)	Sep(")
2010.0	316.4	1.66	2060.0	29.8	1.26
2020.0	326.5	1.46	2070.0	43.5	1.43
2030.0	339.4	1.29	2080.0	53.9	1.64
2040.0	355.5	1.19	2090.0	61.7	1.89
2050.0	13.2	1.17	2100.0	67.7	2.16

The Modern Era

The current orbit by Heintz [631], which has a period of 789 years, and recent micrometric observations by ARY (author RWA) and J.-F. Courtot (CTT) have shown a consistent residual of 2° or 3° in PA, in the sense of observations lagging the orbital prediction, possibly indicating that the period is too short. The stars will now continue to close very slowly to 1″.17 in 2047 and then will widen for several centuries, with maximum separation occurring in 2440 at 7″.8. The companion is an SB1.

Observing and Neighbourhood

Smyth records the colours as white and blue. At that time the stars were over 5″ apart. A beautiful sight in 20-cm but resolvable in 10-cm. The WDS gives a list of five stars, all unconnected with μ Cyg, in the field, four of which have V = 12 to 13. The fifth, D has magnitude 7.0 and can be found at 43° and 197″. The distance is decreasing owing to the approach of the binary pair.

Measures

Early measure (STF)	114°.5	5″.56	1831.63
(Orbit	113°.4	5″.56)	
Recent measure (ARY)	320°.8	1″.81	2013.62
(Orbit	322°.9	1″.52)	

164. ξ CEP = STF 2863 = WDS J22038+6438AB

Table 9.164 Physical parameters for ξ Cep

STF 2863	RA: 22 03 47.45	Dec: +64 37 40.7	WDS: 129(277)	
V magnitudes	Aa: 4.8	Ab: 4.6	B: 6.3	C: 12.6
$(B-V)$	A: +0.36	B: +0.52		
$\mu(A)$	222.36 mas yr^{-1}	± 1.00	93.22 mas yr^{-1}	± 0.97 (DR2)
$\mu(B)$	202.02 mas yr^{-1}	± 0.04	81.99 mas yr^{-1}	± 0.04 (DR2)
$\pi(A)$	32.17 mas	± 0.62	101.4 light yr	± 1.9 (DR2)
$\pi(B)$	32.12 mas	± 0.02	101.5 light yr	± 0.1 (DR2)
Spectra	Aab: A7? + F2-F5 ?	B: F8V		
Masses (M$_\odot$)	Aab: 1.045	± 0.032	B: 0.409 M$_\odot$	± 0.066
Luminosities (L$_\odot$)	A: (10 + 11)	B: 2.5		
Catalogues (A/B)	17 Cep	HR 8417	HD 209790/1	SAO 19827/6	HIP 108917
DS catalogues (AB)	H 2 16	STF 2863	BDS 11483	ADS 15600	MCA 69 (Aa/Ab)
Radial velocity	−10.74 km s^{-1}	± 0.34		
Galactic coordinates	106°.157	+7°.366		

History

First observed by Sir William Herschel on 7 November 1779: 'A fine double star. Considerably unequal. L(arge) w(hite) inclining to r(ed). S(mall). dusky grey.' Burnham classified this system as a '61 Cygni' type – i.e. a pair of stars with similar proper motions but indiscernible orbital motion. In 1912, Burnham [632] noted a faint star of magnitude 12.6 (D) at 95″.3. The distance has since increased to 110″, confirming the optical nature of this star. Hynek [633] in 1938 included ξ Cep A in his list of stars with composite spectra and Abt [637] noted a slow but significant decrease in the velocity of the bright star over several years, which he attributed to the A star being a single-lined spectroscopic binary.

Finder Chart

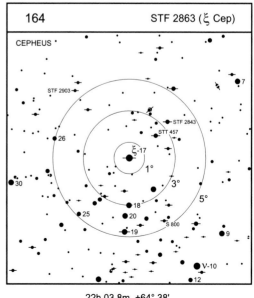

22h 03.8m +64° 38'

Orbit

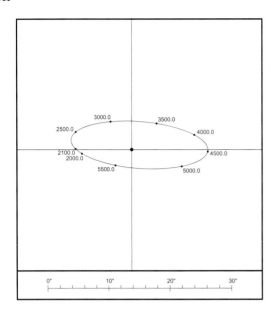

Ephemeris for STF 2863 AB (2000 to 2225)

Orbit by Zel (1965) Period: 3800 years, Grade: 5

Year	PA(°)	Sep(")	Year	PA(°)	Sep(")
2000.0	274.6	8.22	2125.0	268.2	9.43
2025.0	273.1	8.51	2150.0	267.2	9.60
2050.0	271.8	8.78	2175.0	266.1	9.75
2075.0	270.6	9.02	2200.0	265.1	9.87
2100.0	269.4	9.24	2225.0	264.1	9.96

The Modern Era

Vickers & Scarfe [634] in 1976 derived an orbit for the A component with a period of 810 days and, as the paper was in press, news came in that Harold McAlister [635], using a speckle camera on the 4-metre Mayall reflector at Kitt Peak Observatory, had resolved the A component into two stars and found the separation to be 0″.055. Vickers and Scarfe obtained masses of 2.2 ± 1.3 M$_\odot$ for the Aa star and 0.8 ± 0.5 M$_\odot$ for the Ab star. In 2014 Farrington *et al.* [636] discussed the AaAb system with the benefit of a number of very precise relative positions obtained with ground-based interferometers. Using both spectroscopic and astrometric measurements they derived the masses given in the table

above for Aab and B. Vickers and Scarfe considered that the spectral types of Aa and Ab were A7 and F2-F5 III-IV, but these do not fit the masses found by Farrington. The more distant companion, C, is a F5 dwarf according to SIMBAD.

Observing and Neighbourhood

ξ Cep is located in a rich area of Cepheus, approximately half-way between the magnitudes of α and ι. This is an attractive pair for the small aperture. Smyth thought both components were bluish but in 1854 Dembowski noted that they were white and violet. In 1910 Storey called them yellow and blue. In 1968 RWA noted white and blue in a 21-cm reflector at ×96. Star C (12.6 at 206°, 110″, 2000) was not seen in the Cambridge telescope with the field illumination on, but should be visible in 15-cm under normal conditions. Recent measurements of the wide pair all agree that the actual separation is around 0″.3 below that predicted by Zeller's orbit, which is, in any case, speculative. The star system STT 457 (6.0, 8.2, 245°, 1″.3) is 1° NW. This pair is now at half the separation that it was at discovery and needs 15-cm to pick up the relatively faint and close secondary. Another 0°.7 in the same direction brings you to STF 2843, at a similar separation to STT 457 but the stars are almost equal (7.0, 7.3, 151°, 0″.3, 2016). Another star, of magnitude 11, can be found at 277° and 55″. To the NE distant 2°.5 is STF 2903, a neat pair which is probably a long-period binary (7.1, 7.8, 97°, 4″.1, 2017). Two degrees SSW is the open cluster NGC 7160, in which can be found the wide binocular pair S 800 (7.0, 7.9, 145°. 63″, 2016). The primary star is EM Cep, a β-Lyrae variable which is also a close pair (0″.1).

Measures

Early measure (STF)	288°.9	5″.60	1831.77
(Orbit	289°.1	5″.76)	
Recent measure (ARY)	275°.5	8″.02	2015.74
(Orbit	273°.2	8″.41)	

165. 53 AQR = SHJ 345 = WDS J22266–1645AB

Table 9.165 Physical parameters for 53 Aqr

SHJ 345	RA: 22 26 34.30	Dec: −16 44 31.9	WDS: 163(250)	
V magnitudes	A: 6.24	B: 6.39		
$(B − V)$ magnitudes	A: +0.68	B: +0.70		
μ	A: 200.59 mas yr^{-1}	± 1.76	B: −14.51 mas yr^{-1}	± 0.89
π	49.50 mas	± 1.23 mas	66 light yr	± 2
$\mu(A)$	254.95 mas yr^{-1}	± 0.43	−62.14 mas yr^{-1}	± 0.21 (DR2)
$\pi(A)$	51.51 mas	± 0.13	63.3 light yr	± 0.2 (DR2)
Spectra	A: G2V	B: G3V + ?		
Masses (M$_\odot$)	A: 1.06	B: 1.03		
Radii (R$_\odot$)	A: 1.02	B: 0.98		
Luminosities (L$_\odot$)	A: 1.1	B: 0.9		
Catalogues (A/B)	HR 8545/4	HD 212698/7	SAO 165078/7	HIP 110778
DS catalogues	SHJ 345	BDS 11715	ADS 15934	
Radial velocity	2.28 km s^{-1}	± 0.15		
Galactic coordinates	42°.553	−54°.980		

History

Herschel first noted the duplicity of 53 Aqr on a cloudy night (20 September 1786). He revisited the system on 6 September 1793 and found the stars to be in the second or third class of separation, rather nearer the latter, so one might assume from this a separation of about 10″. Not long afterwards Piazzi also noted both components and in his catalogue of 1814 gives the differences in RA of 12″.7 and in declination of 4″.3, which yield a position angle 288° and a separation of 13″.4 for the epoch 1800. Both stars were assigned the magnitude 6.7. In 1901 Burnham noted star C of magnitude 12.8 at a distance of 46″.7 and PA 339° and used this to check on the proper motions of both components; because of their mutually large size and direction, this led him to classify 53 Aqr as being in the 61 Cygni class of binary stars. He later found D, at magnitude 13.8, just 1″.8 from C, but this has only ever been measured twice and there is not enough data

Finder Chart

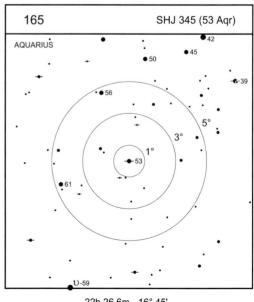

22h 26.6m -16° 45'

Orbit

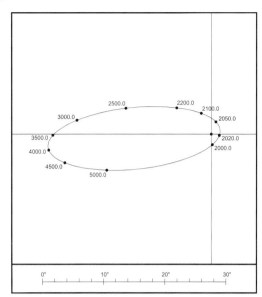

Ephemeris for SHJ 345 AB (2010 to 2100)

Orbit by Hle (1994) Period: 3500 years, Grade: 4

Year	PA(°)	Sep(")	Year	PA(°)	Sep(")
2010.0	38.9	1.31	2060.0	173.1	2.36
2020.0	81.7	1.33	2070.0	184.5	2.67
2030.0	115.6	1.59	2080.0	193.4	2.99
2040.0	139.9	1.84	2090.0	200.5	3.33
2050.0	158.6	2.08	2100.0	206.4	3.68

to decide whether it is a physical part of the system in its own right. Components C and D are certainly field stars.

The Modern Era

The stars have continued to close through the twentieth century and at the time of writing should just be starting to open up, according to Hale's 1994 orbit [638], having reached a minimum separation of 1″.27 in 2014. It is now believed that the B component is a single-line SB. Its nature was deduced from CORAVEL observations of radial velocity by Cutispoto *et al.* [639], who point out that it should be a very late-type star. Star B does not appear in the DR2 catalogue, which suggests that DR2 sees an image which is complex. More recently, Willmarth *et al.* [640] have calculated an orbit with a period of 257 days. The masses and radii come from Fuhrmann *et al.* [641].

Observing and Neighbourhood

RWA has been observing this interesting system since 1990 and has made annual micrometric measurements. Mean measures made in 2015 and 2016, in particular, appear to show that the companion star is well ahead of the ephemeris by between 5° and 10° in angle. Andrei Tokovinin, using the

Measures with 8″ Refractor

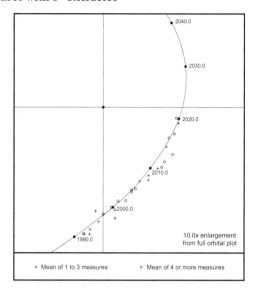

+ Mean of 1 to 3 measures	○ Mean of 4 or more measures

Year	PA(°)	Sep(")	No.	Year	PA(°)	Sep(")	No.
1991.91	350.8	2.16	9	2005.780	22.0	1.48	5
1992.84	351.3	2.16	3	2006.75	26.9	1.40	3
1993.82	353.7	2.15	3	2007.845	32.6	1.46	3
1994.81	353.7	2.07	4	2008.775	34.5	1.40	3
1995.82	355.8	1.75	3	2009.890	42.5	1.41	3
1996.73	357.8	1.93	3	2010.890	45.1	1.44	4
1997.72	0.1	1.80	4	2011.868	49.1	1.41	5
1998.93	6.3	1.88	4	2012.788	56.4	1.45	4
1999.87	4.8	1.69	5	2013.818	58.5	1.29	2
2000.80	6.7	1.74	6	2014.753	65.4	1.26	4
2001.785	9.2	1.60	5	2015.793	69.8	1.32	4
2002.831	13.9	1.62	6	2016.886	78.0	1.34	3
2003.876	19.0	1.51	6	2017.812	81.4	1.29	4
2004.825	19.4	1.48	5				

4-metre SOAR telescope in Chile in 2017, found the companion 5°.5 ahead of the ephemeris, confirming that the present orbit will need some adjustment, but right now accurate measurements are vital as the companion heads off towards its widest separation of almost 27″ in 3890 or so. It seems likely that minimum separation has now been reached and that the position angle is increasing by about 4° per year. This bright pair is resolvable in 10-cm but from northern latitudes the problems in seeing B are more likely to be atmospheric owing to its low declination. From Cambridge few nights are really good enough to allow acceptable measures. Both stars appear white; 53 Aqr is in a relatively sparse area of the sky but is 16° due S of ζ Aquarii.

Measures

Early measure (HJ)	302°.5	12″.0	1830.57
(Orbit	301°.7	9″.67)	
Recent measure (TOK)	77°.3	1″.20	2017.60
(Orbit	71°.8	1″.29)	

166. KRÜGER 60 CEP = WDS J22280+5742AB

Table 9.166 Physical parameters for KR 60 Cep

KR 60	RA: 22 27 59.47	Dec: +57 41 45.23	WDS: 125(289)	
V magnitudes	A: 9.93	B: 11.41		
$(B - V)$ magnitudes	A: +0.41	B: +2.2		
$\mu(A)$	-725.23 mas yr^{-1}	± 0.54	-223.46 mas yr^{-1}	± 0.35 (DR2)
$\mu(B)$	-934.10 mas yr^{-1}	± 1.32	-686.24 mas yr^{-1}	± 1.41 (DR2)
$\pi(A)$	249.39 mas	± 0.17	13.078 light yr	± 0.009 (DR2)
$\pi(B)$	249.97 mas	± 0.24	13.048 light yr	± 0.012 (DR2)
Spectra	A = M2.0V	B = M4.0V		
Masses (M$_\odot$)	A: 0.271		B: 0.176	
Radii (R$_\odot$)	A: 0.35		B: 0.24	
Luminosities (L$_\odot$)	A: 0.001	B: 0.0004		
Catalogues (A/B)	DO Cep	HD 239960	HIP 110893	
DS catalogues (AB)	BDS 11761	ADS 15972		
Radial velocity	-33.94 km s^{-1}	± 0.1		
Galactic coordinates	$104°.687$	$-0°.003$		

History

The star is named after A. Krüger [643] at Helsinki, who noticed, on plates taken for the Astronomische Gesellschaft for the Helsingfors/Gotha zone, that there was a fainter star (C) 12″ away from the ninth magnitude primary. In 1890, Burnham [644] observed this system and noted that there was a closer companion (B) about 2″.3 distant, nearly due S. In 1898, when it was next observed, by Doolittle [645], 'a very decided change had taken place …' and it was subsequently followed with interest, particularly by Barnard [642], who observed it annually from 1903 until the year of his death (1921) and measured it on 28 nights in 1916.

Finder Chart

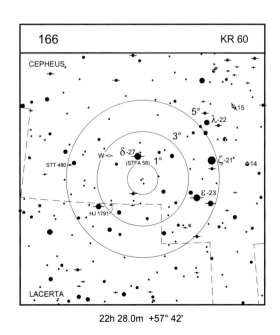

22h 28.0m +57° 42'

Orbit

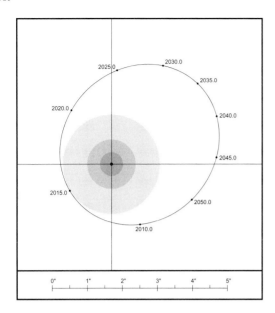

Ephemeris for KR 60 AB (2018 to 2036)

Orbit by Hei (1986b) Period: 44.67 years, Grade: 2

Year	PA(°)	Sep(")	Year	PA(°)	Sep(")
2018.0	244.5	1.60	2028.0	160.6	2.94
2020.0	217.2	1.89	2030.0	151.7	3.10
2022.0	197.6	2.20	2032.0	143.6	3.22
2024.0	182.7	2.48	2034.0	136.0	3.30
2026.0	170.7	2.73	2036.0	128.6	3.34

The Modern Era

Starting in 1931, observers at Sproul Observatory took a long series of plates with the intention of obtaining a visual orbit and mass sum for the two components. In a short note to PASP in 1951, P. van de Kamp & Lippincott [646] noted that on one occasion in 1939 the two stars, normally 1.5 magnitudes apart in the visual band, appeared to be equally bright and concluded that star B was a flare star, one of only six known at the time. The component KR 60 has been surveyed a number of times in the past decade in an attempt to find substellar companions or exoplanets. Heinze *et al.* [648] found an object of $K = 16$ some $7''$ from A (which is 4.7 in the same infrared band), but it turned out to be a background object. The WDS catalogue contains 16 companions to A, from magnitudes 8 to 16, all of which, with the exception of B, are background objects whose positions with respect to AB are changing rapidly owing to the large proper motion of the binary. The trigonometric parallax found by Hipparcos is not as precisely defined as some single stars observed by the satellite, because of the proximity of the fainter star, but, at a distance of 13 light years, KR 60 is one of the most nearby stars. It is also getting nearer to the Solar System and will approach to within 6.3 light years in around 89,000 years' time.

Observing and Neighbourhood

One of the more difficult pairs to observe, because of the faintness and closeness of the stars, this object is nevertheless worth the attempt to see, providing you have 30-cm. It is certainly beyond the range of the 8-inch at Cambridge at the present time, although Robert Burnham suggested that a good 6-inch (15-cm) at high power will suffice to show the pair (which was separated by $2''$ when he made his comments in 1976). His *Celestial Handbook*, Volume 2, p. 601, shows the flare in star B captured at Sproul in 1939. Having reached periastron in 2015, the stars are now separating and the annual angular motion is between $10°$ and $15°$ for the near future, so even without a measuring device the movement in position angle ought to be obvious. The system KR 60 is close to δ Cephei (see Star 168), being about $45'$ S, and is shown in the finder chart. There is a comprehensive paper on this star by Wilfried Knapp and John Nanson, which is freely available in JDSO [649]. *CDSA2* shows two stars within $2°$ or so. One point five degrees ESE is HJ 1791 (7.7, 9.7, $59°$, $17''$, 2016). The galactic cluster NGC 7380 is $2°$ E and on the SW edge is STT 480 (7.7, 8.6, $116°$, $30''.7$, 2016).

Measures

Early measure (BU)	$126°.5$	$3''.36$	1902.81
(Orbit	$126°.0$	$3''.34$)	
Recent measure (WSI)	$326°.1$	$1''.47$	2013.73
(Orbit	$327°.7$	$1''.44$)	

167. ζ AQR = STF 2909 = WDS J22288−0001AB

Table 9.167 Physical parameters for ζ Aqr

STF 2909	RA: 22 28 49.81	Dec: −00 01 12.2	WDS: 8(1169)		
V magnitudes	Aa: 4.79	Ab: 11.3	B: 4.89		
$(B-V)$ magnitudes	A: +0.45	B: +0.40			
μ(A)	180.83 mas yr^{-1}	± 0.83	33.07 mas yr^{-1}	± 0.75 (DR2)	
μ(B)	224.30 mas yr^{-1}	± 0.52	25.40 mas yr^{-1}	± 0.48 (DR2)	
π(A)	34.51 mas	± 0.47	94.5 light yr	± 1.3 (DR2)	
π(B)	34.45 mas	± 0.31	94.7 light yr	± 0.9 (DR2)	
Spectra	A: F3V	B: F6IV			
Masses (M$_\odot$)	Aa: 1.4 ± 0.14	Ab: 0.6 ± 0.06	B: 1.4 ± 0.14		
Luminosities (L$_\odot$)	Aa: 8	Ab: 0.2	B: 9		
Catalogues (A/B)	55 Aqr	HR 8559/8	HD 213052/1	SAO 146108/7	HIP 110960
DS catalogues	H 2 7 (AB)	STF 2909 (AB)	EBE 1 (AaAb)		
Radial velocity	25.30 km s^{-1}	± 0.5			
Galactic coordinates	65°.355	−46°.331			

History

Lewis, in 1906, noted that ζ Aqr was first seen as double by Christian Mayer in 1777, but the star does not appear in Schlimmer's catalogue. It is listed in Bode's Astronomisches Jahrbuch for 1784 but the name of the discoverer is not known. It was then found by William Herschel on 12 September 1779. Herschel's description was 'Equal, or the preceding rather the largest. Both w(hite). With 227 $1\frac{1}{4}$ diameter; with 449, $1\frac{1}{3}$ diameter; with 460, 2 diameters; with 910 near 2 diameters; with 932 $2\frac{1}{2}$ diameters; with 2010, pretty distinct; but too tremulous to estimate.' Lewis noted that the motion between 1777 and 1906 was virtually a straight line but the common proper motion between the two stars confirmed that it was a binary.

Finder Chart

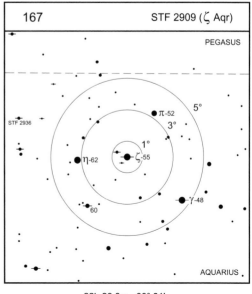

22h 28.8m -00° 01'

Orbit

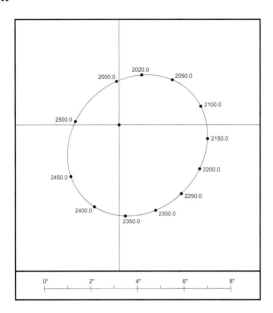

Ephemeris for STF 2909 AB (2015 to 2060)

Orbit by Tok (2016f) Period: 540 years, Grade: 3

Year	PA(°)	Sep(")	Year	PA(°)	Sep(")
2015.0	161.0	2.16	2040.0	136.7	2.75
2020.0	155.2	2.29	2045.0	133.0	2.85
2025.0	149.9	2.41	2050.0	129.5	2.94
2030.0	145.1	2.53	2055.0	126.3	3.03
2035.0	140.7	2.64	2060.0	123.2	3.11

The Modern Era

Strand [650] wrote a paper in which he presented a visual orbit for stars AB with particular emphasis on a series of precise photographic positions taken between 1914 and 1941. He announced that he had found 'deviations from Keplerian motion that were entirely too large and regular to be explained as systematic [or] accidental errors of observation'. He attributed these to the effects of an invisible third body orbiting the B component. It was not until 1979 that an experimental camera, being used with the 3.6-metre reflector at ESO in Chile [651], made the first direct detection of this body.

In 1982, McCarthy *et al.* [652], also using infrared detectors, claimed to have found an object close to the A component. Two years later, Wulff Heintz at Sproul Observatory discussed the astrometry of the whole system with the benefit of several more decades of photographic material to help him. He concluded that the third body does revolve around the A star and that there was no evidence for anything connected with A. Heintz [654] also cast doubt on the reported detections by Ebersberger and McCarthy. The latest word comes from Andrei Tokovinin [653] who analysed the motions of all three stars and confirmed that the infrared object revolves around A. He finds a period of 25.95 years and notes that A and B have a rotation period of 540 years. He also states that his observations of ζ Aqr A with the 4.1-metre SOAR telescope unambiguously show the Ab component for the first time, and consequently the close pair should bear the designation TOK 201.

Observing and Neighbourhood

A bright pair of yellow stars, which is a fine sight in 7.5-cm and above, ζ Aqr is in the top ten of the most measured visual binaries. F. G. W. Struve notes that both components are greenish whilst Smyth thought that they were very white and white. The combination of brightness and relative ease of resolution has attracted observers to the pair for more than two centuries. The closest approach was 1″.6 in 1981 but the stars are now widening and will reach a maximum separation of 3″.9 about two centuries hence. The system STF 2936 is 3° to the ENE. The stars are magnitudes 7.0 and 9.6, at 52°, 4″.2, 2013.

Measures

Early measure (STF)	356°.6	3″.50	1830.79
(Orbit	355°.8	3″.73)	
Recent measure (ARY)	162°.4	2″.39	2016.88
(Orbit	158°.7	2″.21)	

168. δ CEP = STFA 58 = WDS J22292+5825AB

Table 9.168 Physical parameters for δ Cep

STFA 58	RA: 22 29 10.25	Dec: +58 24 54.7	WDS: 789(91)		
V magnitudes	Aab: 4.21v	B: 13.0	Cab: 6.11		
(*B* − *V*) magnitudes	A: +0.89	C: +0.07			
μ(A)	15.35 mas yr^{-1}	± 0.22	3.52 mas yr^{-1}	± 0.18	
μ(B)	16.19 mas yr^{-1}	± 0.59	4.28 mas yr^{-1}	± 0.50	
π(A)	3.66 mas	± 0.15	891 light yr	± 37	
π(B)	3.65 mas	± 0.15	894 light yr	± 37	
μ(C)	14.10 mas yr^{-1}	± 0.09	3.93 mas yr^{-1}	± 0.09 (DR2)	
π(C)	3.36 mas	± 0.05	971 light yr	± 14 (DR2)	
Spectra	A: F5Iab+?	C: B7+F5V?			
Masses	Aab: 5.1 M$_\odot$		Cab: 4.0 M$_\odot$		
Luminosities (L$_\odot$)	Aab: 1290	Cab: 270			
Catalogues (A/C)	27 Cep	HD 213306/7	HR 8571(A)	HIP 110991/8	SAO 34508/6
DS catalogues	H 5 4 (AC)	STFA 58 (AC)	BU 702 (AB)	BDS 11772	ADS 15987
Radial velocity (A/B)	−24.00 km s^{-1}	± 0.2	−21.60 km s^{-1}	± 0.9	

History

Noted by William Herschel on 31 August 1779: 'Double. Considerably unequal. L(arge). reddish w(hite).; S(mall). blueish w.' The pair δ Cephei has been compared with Albireo with respect to the brightness and colours of the stars and the separation. Coincidentally, Herschel noted Albireo 13 days after he observed δ and catalogued it as H 5 4. The star δ Cephei is most noted for its regular brightness variability. In 1784, a talented young astronomer called John Goodricke [655] noticed δ and started to carry out a series of visual estimations of the star's brightness. In 1898, Burnham [678] found a 13th magnitude star 20″ away in PA 284°, which is now known as B. The astronomer Bélopolsky [657], using the 30-inch refractor at Pulkovo in 1895, took spectra of δ Cep and announced that it was a spectroscopic binary with

Finder Chart

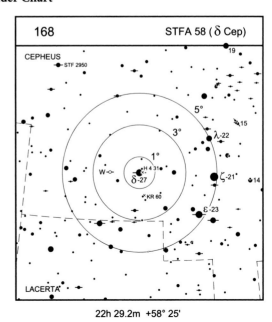

22h 29.2m +58° 25'

a period of five days and nine hours. In 1914, however, Harlow Shapley [658] discussed the problems associated with linking the observed radial velocity variations with binarity. He argued that the shape of the curve was non-Keplerian, there was never a secondary spectrum visible, and the gradual change of spectral type with time could not be argued by the presence of two stars. He proposed that the variations in radial velocity were due to the pulsation of the star's outer layers rather than the presence of a companion star. In 1912 Leavitt & Pickering [659] found that the luminosity of a Cepheid was directly proportional to its period of pulsation, and this can be used to form the basis of distance measurement in stellar astronomy. Unfortunately Cepheids are, on the whole, rare and distant objects and even the nearest one, Polaris, is about 430 light years away. Even using the Hipparcos satellite, the distance is only determined to an accuracy of about 4% for δ Cephei. The advent of Gaia promises a revolution in the definition of the distance scale, with hundreds of Cepheids becoming accessible. In the case of δ Cep, DR2 contains the astrometry only of the wide Herschel star, but it does appear to be at a very similar distance to the Cepheid.

The Modern Era

Marcel Fay [660] added four faint (13–15 magnitude) and distant companions to A, whilst Jim Daley [664] measured a 13.9 magnitude star at 109″, which has a close partner of magnitude 14.0 at 23° and 1″.4. The system was imaged by Benedict *et al.* [661] with the HST and they found that A and C have the same parallaxes (see above), which put the stars slightly further away than the Hipparcos result, π = 3.77 mas (865 light years). Star C is known to be a variable and also to be a binary. A preliminary period of 390 days was found

from astrometry carried out with the HST in 2002. In 2015 Anderson *et al.* [662] published a paper in which they found that δ itself was also a spectroscopic binary, with a period of 2201 days. The new component was found with a high-resolution spectrograph, called Hermes, fitted to the Flemish 1.2-metre Mercator telescope on La Palma. Once the known pulsation period of the Cepheid was allowed for, the residuals showed the effect of the companion star, which they suggest is a young main sequence object. They derived a trigonometric parallax which puts δ Cephei at a distance of 778 light years. A subsequent search for this new companion using the CHARA array [663] in the infrared was unsuccessful.

Observing and Neighbourhood

The two bright stars form a splendid sight in a small telescope and show a colour contrast. Admiral Smyth recorded orange tint and fine blue, whilst John Nanson in 2010 found 'the primary is a rich yellow with a tinge of red to it and 'C' is pronounced blue leaning a bit towards white'. Move 5′.8 W and you'll reach the coarse triple star H 4 31 (8.5, 10.5, 4°, 25″, and a magnitude 9.5 at 321°, 78″). About 45′ S is the nearby red dwarf binary Krueger 60. The object STF 2950, which is 4°.5 NE, is an orbital pair with period 804 years. The discovery separation was 3″ but the pair are now expected to be at 272°, 1″.1 in 2020. The stars are 6.1 and 7.1.

Measures

Early measure (STF)	192°.0	40″.86	1835.15
Recent measure (WSI)	190°.9	40″.94	2013.55

169. 8 LAC = STF 2922 = WDS J22359+3938AB

Table 9.169 Physical parameters for 8 Lac

STF 2922	RA: 22 35 52.28	Dec: +39 38 03.6	WDS: 340(163)AB	
V magnitudes	Aab: 5.73	B: 5.67	C: 10.49	D: 9.09
(*B − V*) magnitudes	Aab: −0.15	B: −0.15	C: +0.76	D: −0.12
μ(A)	−1.82 mas yr^{-1}	± 0.10	−4.68 mas yr^{-1}	± 0.14 (DR2)
μ(B)	−1.22 mas yr^{-1}	± 0.10	−4.40 mas yr^{-1}	± 0.11 (DR2)
μ(C)	−4.32 mas yr^{-1}	± 0.26	−5.21 mas yr^{-1}	± 0.34 (DR2)
μ(D)	13.02 mas yr^{-1}	± 0.04	7.26 mas yr^{-1}	± 0.05 (DR2)
π(A)	1.82 mas	± 0.10	1792 light yr	± 98 (DR2)
π(B)	1.67 mas	± 0.07	1953 light yr	± 82 (DR2)
π(C)	2.03 mas	± 0.20	1607 light yr	± 158 (DR2)
π(D)	7.36 mas	± 0.03	443 light yr	± 2 (DR2)
Spectra	A: B2Ve	B: B5		D: F0
Luminosities (L$_\odot$)	A: 1300	B: 1550	C: 13	D: 60
Catalogues (A/B)	HR 8603	HD 214167/8	SAO 72509/8	HIP 111546/4
DS catalogues	H 4 86	STF 2922	A1469	BDS 11839	ADS 16095
	CHR 112(AaAb)			
Radial velocity (A/B)	−11.00 km s^{-1}	± 0.9	−11.70	± 0.9
Galactic coordinates	96°.365	−16°.146		

History

William Herschel (4 October 1782): 'Quadruple. The two largest and nearest a little unequal. Distance 17″ 14‴. Position 84° 30′ s(outh). preceding. The next two very unequal, of the fourth class. The two remaining considerably unequal, of the fifth class. They form an arch.'

The Modern Era

Although there are multiple components in the group there is little evidence to support the fact that they are at the same distance. A recent Gaia observation of star A shows that it is certainly remote whilst D, which is HIP 111505, is four times as close to the Sun. The A component was found to be a close pair by McAlister *et al.* [665]. Orbital motion appears established but there is not enough coverage to determine an orbital period, which is probably a few decades. The maximum separation appears to be about 0″.05. A number of fainter stars in the group have been seen and imaged since Herschel's observation. Aitken found a 14th magnitude star just 1″.3 from C, which has been observed only by him and not since then. (An attempt by Gili & Prieur [666] using the 76-cm refractor at Nice and a CCD camera did not resolve the pair.)

Finder Chart

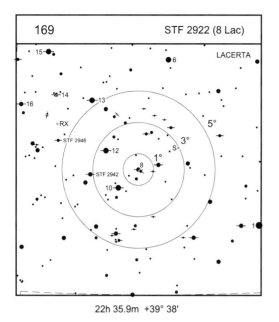

169   STF 2922 (8 Lac)

LACERTA

22h 35.9m +39° 38'

Observing and Neighbourhood

One point five degrees E is STF 2942 (6.2, 8.9, 277°, 3″.2, 2011). Burnham added a 11.7 magnitude star to C at 247°, 9″ (1998). Two point five degrees ENE is STF 2946 (7.3, 8.6, 5°, 2″.6, 2015).

Measures

AB			
Early measure (STF)	185°.7	22″.43	1831.61
Recent measure (WSI)	185°.1	22″.36	2014.68
AC			
Early measure (ENH)	168°.1	49″.01	1887.65
Recent measure (WSI)	167°.9	48″.88	2014.68
AD			
Early measure (DA)	144°.3	82″.07	1858.33
Recent measure (WSI)	143°.8	81″.56	2014.68

170. STF 2944 AQR = WDS J22478−0414AB,C

Table 9.170 Physical parameters for STF 2944 Aqr

STF 2944	RA: 22 47 50.19	Dec: −04 13 44.5	WDS: 75(405) (AB)	
			WDS: 523(121) (AC)	
V magnitudes	A: 7.30	B: 7.68	C: 8.52	
(B − V) magnitudes	A: +0.70	B: +0.75	C: +0.47	
μ(A)	−215.09 mas yr^{-1}	± 0.15	−313.43 mas yr^{-1}	± 0.12 (DR2)
μ(B)	−191.22 mas yr^{-1}	± 0.13	−300.45 mas yr^{-1}	± 0.11 (DR2)
μ(C)	6.25 mas yr^{-1}	± 0.42	−8.17 mas yr^{-1}	± 0.45 (DR2)
π(A)	30.11 mas	± 0.11	108.3 light yr	± 0.4 (DR2)
π(B)	29.91 mas	± 0.06	109.0 light yr	± 0.2 (DR2)
π(C)	9.16 mas	± 0.20	356 light yr	± 8 (DR2)
Spectra	A: G2V	B: G4	C: F0V	
Luminosities (L$_\odot$)	A: 1.0	B: 0.8	C: 4	
Catalogues (A/B/C)		HD 215812/—/3	SAO 146315/5/7	HIP 112559
DS catalogues	H 2 57	STF 2944	BDS 11968	ADS 16270
Radial velocity (A/B)	−16.70 km s^{-1}	± 0.1	−25 km s^{-1}	± 5
Galactic coordinates	65°.296	−52°.675		

History

'Treble. About $2\frac{1}{2}$ degrees following κ, in a line parallel to α and η Aquarii ... Both r ... With 460 $2\frac{1}{2}$ diameters of L(arger star).' So ran the report of William Herschel after his observation of 27 September 1782. Smyth notes 'yellowish, flushed white and flushed white'.

The Modern Era

Gaia has observed all three components. It confirms that the determined parallax of C shows that it is unassociated with AB. The linear ephemeris fits the observed positions of C with respect to A pretty well. The current orbit of AB is based on 60° of prograde motion around the apparent orbit although the separation has decreased from more than 4″ at discovery

Finder Chart

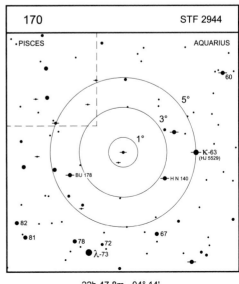

22h 47.8m -04° 14'

Orbit

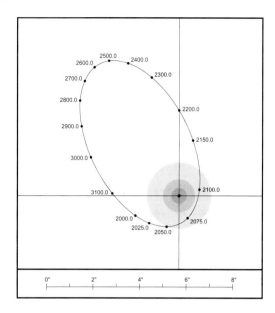

Ephemeris for STF 2944 AB (2010 to 2100)

Orbit by Zir (2007) Period: 1160.28 years, Grade: 4

Year	PA(°)	Sep(")	Year	PA(°)	Sep(")
2010.0	300.2	1.91	2060.0	351.9	1.25
2020.0	307.6	1.78	2070.0	10.1	1.09
2030.0	316.1	1.65	2080.0	35.6	0.90
2040.0	326.0	1.52	2090.0	71.0	0.80
2050.0	337.7	1.40	2100.0	106.4	0.91

to its present value of below 2″. According to Zirm's [667] orbit, with period 1160 years, the stars will close slowly to 0″.8 around 2090 and then reach maximum separation of 6″.5 in about 2600.

Observing and Neighbourhood

The close pair should be well seen in 15-cm and the distant star C is easy in 10-cm. The group is located in a rather sparse area of sky about 5° SSE of η Aqr. It is 2°.5 due E of κ Aqr – a John Herschel pair (5529) in which the stars are magnitudes 5.2 and 12.2 at 256° and 87″(2016), with separation decreasing quite rapidly owing to the significantly different proper motions of A and B. One point five degrees WSW of STF 2944 is H N 140 (6.7, 10.9, 267°, 62″, 2010), with a distant 9.7 magnitude star at 186°, 160″, 2012. Two degrees ESE is BU 178, a difficult binary, $P = 96.4$ years, which will test 40-cm because of the magnitude difference. The stars are 6.0, 7.8 at 326°, 0″.34, 2020.

Measures

AB			
Early measure (STF)	247°.0	4″.12	1832.95
(Orbit	246°.4	4″.18)	
Recent measure (TOK)	305°.8	1″.85	2017.64
(Orbit	303°.8	1″.81)	
AB-C			
Early measure (STF)	157°.0	55″.40	1834.25
(Linear	156°.8	55″.46)	
Recent measure (RWA)	85°.4	60″.07	2014.80
(Linear	85°.6	60″.79)	

171. θ GRU = JC 20 = WDS J23069–4331AB

Table 9.171 Physical parameters for θ Gru

JC 20	RA: 23 06 52.77	Dec: −43 31 13.2	WDS: 1326(64)	
V magnitudes	A: 4.45	B: 6.60	C: 7.78	
(B − V) magnitudes	A: +0.37	B: +0.54	C: +0.64	
μ(AB)	−47.17 mas yr⁻¹	± 0.41	B: −13.49 mas yr⁻¹	± 0.41
π(AB)	24.73 mas	± 0.45	132 light yr	± 3
μ(A)	−49.04 mas yr⁻¹	± 0.62	−13.45 mas yr⁻¹	± 0.65 (DR2)
π(A)	27.74 mas	± 0.49	118 light yr	± 2 (DR2)
μ(C)	−47.92 mas yr⁻¹	± 0.09	B: −22.15 mas yr⁻¹	± 0.10 (DR2)
π(C)	24.85 mas	± 0.09	131.3 light yr	± 0.5 (DR2)
Spectra	A: F5m	B:	C: G2V	
Luminosities (L☉)	A: 17	B: 3	C: 1	D:
Catalogues (A/C)	HR 8787	HD 218227/05	SAO 231444/3	HIP 114131/12
DS catalogues	JC 20 (AB,C)	SHY 366 (AB,C)		
Radial velocity (A/C)	9.6 km s⁻¹	± 0.9	15.10 km s⁻¹	± 0.1
Galactic coordinates	348°.242	−63°.297		

History

This was found by Captain Jacob [668] whilst stationed at Pune in India. Using a 5-foot Dollond telescope equatorially mounted, he sent in a series of micrometric measurements of double stars to the RAS which appeared in *RAS Memoirs*, Volumes 16 and 17. On 1845.88 he found that the primary star of θ Gru was a close pair and estimated that the stars were magnitudes 4.5 and 8.5 and separated by 2″.5. In addition he made the following comments about the colours of the stars in the group: 'A: yellow; B: orange; C: orange and D: blue?' Jacob also wrote 'Is D not a nebula or a close double star?' It is interesting to speculate on why Dunlop or Rümker did not see the AC pair. In the first few decades after discovery, many observers gave an estimated magnitude difference of three or four magnitudes, but modern values, as determined by Tycho and presented here, show that there is only a 2.2 magnitude difference.

Finder Chart

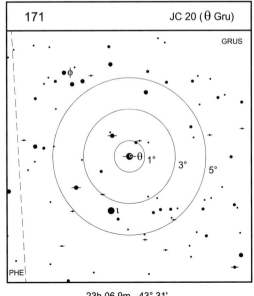

23h 06.9m -43° 31'

The Modern Era

The separation of the close pair has been diminishing since discovery and the current separation is half what it was. More than 90° of position angle have been traced out but it is still too early to say what sort of orbital period the two stars have. Tokovinin in his MSC suggests it could be optical. The data from the Hipparcos satellite shows that the proper motions of A and C are very similar and both stars are at the same distance from us, but DR2 indicates that A and B are at significantly different distances. Tokovinin *et al.* [669] also studied the radial velocity of C and say that it is 'similar' to that of A.

Observing and Neighbourhood

Observers will need at least 10-cm and preferably 15-cm to get a good view of the close unequal pair that forms AB. Both these apertures should suffice to show the magnitude 11.2 companion to C at 355° and 101″. Binoculars will suffice to see the distant C.

Measures

AB			
Early measure (JC)	10°.6	2″.91	1851.85
Recent measure (ANT)	114°.9	1″.53	2013.72

AC			
Early measure (JC)	292°.8	160″.2	1851.86
Recent measure (ANT)	291°.5	158″.9	2002.69

172. π CEP = STT 489 = WDS J23079+7523AB

Table 9.172 Physical parameters for π Cep

STT 489	RA: 23 07 58.34	Dec: +75 23 15.3	WDS: 157(358)		
V magnitudes	A: 5.57	B: 7.28	C: 12.2		
$(B - V)$ magnitudes	A: +0.96	B: +0.48			
μ	6.81 mas yr^{-1}	$\pm$ 1.05	-34.06 mas yr^{-1}	$\pm$ 0.88	
π	13.22 mas	$\pm$ 0.95	247 light yr	$\pm$ 18	
μ(A)	7.32 mas yr^{-1}	$\pm$ 1.23	-31.63 mas yr^{-1}	$\pm$ 1.06 (DR2)	
π(A)	13.73 mas	$\pm$ 0.53	238 light yr	$\pm$ 9 (DR2)	
π	13.8 mas	$\pm$ 0.41 mas	236 light yr	$\pm$ 7 (dyn.)	
Spectra	Aa: K0III	Ab: K0IV	B: A8V		
Luminosities (L$_\odot$)	A: 25	B:5			
Catalogues	33 Cep	HR 8819	HD 218658	SAO 10629	HIP 114222
DS catalogues	STT 489 (AB)	HJ 1852 (AC)			
Radial velocity	-18.6 km s^{-1}	$\pm$ 0.9			
Galactic coordinates	116.419	+13.841			

History

In 1843 Otto Struve [670], using the 15-inch refractor at Pulkovo, found a close companion to π Cephei, and when the news reached W. H. Smyth he examined the star with his 5.9-inch refractor and estimated the companion to be at 330° and 1″.8, whilst Struve himself found 350°.1 and 1″.16 when he measured it in 1846. In August 1899 Campbell [671], who had been taking plates of bright stars with the Mills spectrograph on the 36-inch refractor at Lick, noted that the star exhibited variable radial velocity. In 1925 Harper [672] at Dominion Astrophysical Observatory calculated an orbit for the stars; this was improved by Scarfe *et al.* [673] in 1983, who found a single-lined system with a period of 557 days. After Smyth's observation the visual pair began to close and between 1925 and 1931 only George van Biesbroeck was observing them as they reached a minimum separation of less than 0″.3. Between 1916 and 1924 a number of observers noted the star,

Finder Chart

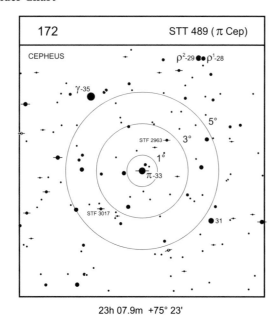

23h 07.9m +75° 23'

Orbit

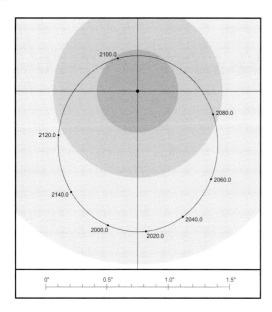

Ephemeris for STT 489 AB (2016 to 2052)

Orbit by Sca (2009a) Period: 162.8 years, Grade: 3

Year	PA(°)	Sep(")	Year	PA(°)	Sep(")
2016.0	0.3	1.12	2036.0	16.8	1.08
2020.0	3.5	1.12	2040.0	20.3	1.06
2024.0	6.8	1.11	2044.0	23.9	1.04
2028.0	10.0	1.11	2048.0	27.7	1.02
2032.0	13.4	1.10	2052.0	31.7	0.99

as single leading to some speculation that the B component may be variable.

The Modern Era

The large mass function derived from the spectroscopic orbit drew some attention from investigators, as they thought this suggested that the unseen spectroscopic companion was rather massive for an object which did not show up in the spectra. Trimble & Thorne [674], who were on the look-out for collapsed stars or neutron stars, had it on a list of SB1s with large mass functions. Gatewood *et al.* [675] used a series of Allegheny Observatory photographic plates of π Cep from 1925, extending about 80 years, to look at the astrometric history of the system. They found a 'wobble' with a period of 1.524 years, which coincided with the period of the spectroscopic binary AaAb. This most probably explains why the DR2 parallax barely improves on the Hipparcos value and will require further processing.

Observing and Neighbourhood

Smyth noted colours of deep yellow and purple for the two stars. The Pulkovo pair is now at widest separation and the stars were seen just separated by Bart Fried [676] in New York State in early October 2017 using a 115-mm f/15 Brashear refractor at ×429. The right-hand star at the apex of the pentangle that forms the main shape of Cepheus is γ Cephei;. π can be found 3° S preceding. The John Herschel component C is currently 244° and 58″ from A and appears to have changed little since discovery. About 10° away is β Cephei, the next star around the pentangle in a clockwise fashion (see Star 162). There are two STF pairs in the field, both of which need at least 15-cm owing to their faintness and closeness: STF 2963 (8.0, 8.5, 3°, 1″.9, 2016) is 1°.5 NNW of π whilst STF 3017 (7.6, 8.5, 19°, 1″.2, 2003) is 1°.5 SE. This pair has closed in from 2″.8 at discovery. A third star of magnitude 13.8 is at 145°, 88″, 2003.

Measures

Early measure (STT)	350°.1	1″.16	1846.48
(Orbit	355°.7	1″.12)	
Recent measure (DRU)	355°.3	1″.11	2010.68
(Orbit	356°.0	1″.12)	

173. 72 PEG = BU 720 = WDS J23340+3120

Table 9.173 Physical parameters for 72 Peg

BU 720	RA: 23 33 57.19	Dec: +31 19 31.0	WDS: 62(433)		
V magnitudes	A: 5.67	B: 6.11			
$(B - V)$	A: +1.67	+1.60			
μ	+50.81 mas yr^{-1}	± 0.49	-17.46 mas yr^{-1}	± 0.24	
π	5.94 mas	± 0.45	549 light yr	± 42	
Spectra	A: K4III	B: K5III?			
Luminosities (L$_\odot$)	A: 120	B: 80			
Catalogues (A/B)	72 Peg	HR 8943	HD 221673	SAO 73341	HIP 116310
DS catalogues	BU 720	BDS 12432	ADS 16836		
Radial velocity	-24.71 km s^{-1}	± 0.22			
Galactic coordinates	$104°.087$	$-28°.695$			

History

The pair 72 Peg yielded to the resolving power of the Dearborn 18.5-inch refractor in the hands of S. W. Burnham [677] in early September 1878 on the same night that the more difficult pairs 64 Peg and 85 Peg were also found.

The Modern Era

Baize & Petit [144] included 72 Peg in their catalogue of double stars with a variable component based on visual estimates of the difference in brightness between the two stars. However, the Hipparcos satellite shows only a variation of 0.006 magnitude over the length of that mission. Muterspaugh *et al.* [620] observed 72 Peg intensively with the PHASES (Palomar High-Precision Search for Exoplanet Systems) instrument using a 110-metre baseline at 2.1 μm, achieving accuracies of considerably better than 1 mas in separation. Superimposed on the measured separation was a short-term sinusoidal variation with a period of 1539 days, which they ascribed to a brown dwarf with a mass of 35 ± 4 Jupiter masses.

Finder Chart

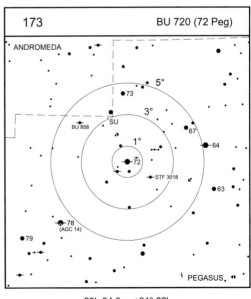

23h 34.0m +31° 20'

Orbit

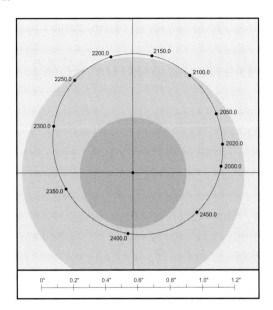

Ephemeris for BU 720 (2010 to 2100)

Orbit by Mut (2010e) Period: 492.31 years, Grade: 4

Year	PA(°)	Sep(")	Year	PA(°)	Sep(")
2010.0	100.7	0.56	2060.0	129.5	0.64
2020.0	107.1	0.58	2070.0	134.6	0.65
2030.0	113.1	0.60	2080.0	139.4	0.66
2040.0	118.8	0.61	2090.0	144.2	0.67
2050.0	124.3	0.62	2100.0	148.7	0.68

Observing and Neighbourhood

This pair is a textbook test for telescopic resolution. The stars are almost equally bright, around $V = 6$, and the separation is nicely placed to test a 20-cm aperture in terms of objective quality and steadiness of the atmosphere. The separation, although increasing, is doing so extremely slowly so that the star will remain a test for some decades to come. The star will be at 107°, 0″.58 in 2020. There are two other visual binaries nearby. Neither is easy. The pair STF 3018 can be found 1° WSW of 72 Peg. It consists of 7.4, 9.8 at 202°, 19″.2, 2012, but Burnham divided A into a very close pair (BU 1266), which turns out to have a period of only 48.4 years. At 2020 they will be at 129°, 0″.19 and will widen to only 0″.26. Star D is 13.8 at 355°, 29″, 2015. Two point five degrees SE is one of the pairs found by Alvan G. Clark: 78 Peg is AGC 14, the stars are 5.1, 8.1, and the period is 620 years. By 2020 they will be at 286°, 0″.91. Two degrees NE is BU 858 (7.9, 8.8, 22°, 0″.9, 2015) perhaps the easiest of the group of pairs in this area.

Measures

Early measure (STF)	307°.7	0″.40	1878.74
(Orbit	310°.2	0″.38)	
Recent measure (SCA)	101°.7	0″.57	2011.90
(Orbit	102°.0	0″.57)	

174. 107 AQR = H 2 24 = WDS J23460−1841

Table 9.174 Physical parameters for 107 Aqr

H 2 24	RA: 23 46 00.85	Dec: −18 40 42.1	WDS: 474(130)		
V magnitudes	A: 5.65	B: 6.46			
(*B − V*) magnitudes	A: +0.28	B: +0.35			
μ(A)	132.91 mas yr^{-1}	± 0.36	16.01 mas yr^{-1}	± 0.35 (DR2)	
μ(B)	139.26 mas yr^{-1}	± 0.64	10.05 mas yr^{-1}	± 0.62 (DR2)	
π(A)	16.14 mas	± 0.22	202 light yr	± 3 (DR2)	
π(B)	20.02 mas	± 0.40	163 light yr	± 3 (DR2)	
Spectra	A: A9IV	B: F2V			
Luminosities (L$_\odot$)	A: 17	B: 5	C:	D:	
Catalogues (A/B)	i^2 Aqr	HR 9002	HD 223024	SAO 165867/8	HIP 117218
DS catalogues	H 2 24	SHJ 356	BDS 12543	ADS 16979	
Radial velocity	−0.7 km s^{-1}	± 0.5			
Galactic coordinates	58°.820	−72°.751			

History

William Herschel picked up 107 Aqr on 23 August 1780 and made the following notes, having referred to the star as 108 Aqr: 'Unequal. With 227, 2 diameters, with 460, almost 3 diameters. In Harris's maps it is marked i.' Joseph Harris (1704–1764) produced two maps, one of each hemisphere. They are based on the stars in Flamsteed's British catalogue, which Halley issued without permission. It seems likely that Harris was not actually responsible for the northern map, according to Wolfgang Steinicke [679], who concludes that John Senex, an experienced London cartographer, engraved and published it, so the use of the term 'Harris's maps' is strictly incorrect. The stars 106, 107, and 108 Aqr are respectively labelled by Bayer as i^1, i^2, and i^3 Aqr. W. H. Smyth and John Herschel included the pair in their catalogue of 1823, where it is known as SHJ 356. Smyth recorded the colours as bright white and blue in the *Bedford Catalogue*.

Finder Chart

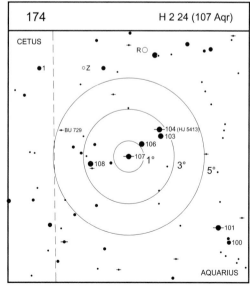

23h 46.0m -18° 41'

The Modern Era

Gaia DR2 not only obtained significantly more accurate values of parallax compared with those of Hipparcos, which only looked at the system as a whole, but also indicates that the two components are at different distances, although the quoted errors are large compared with what is usually obtained for stars of this brightness.

Observing and Neighbourhood

The pair 107 Aqr is in a group of fifth and sixth magnitude stars 15° due W of Deneb Kaitos (β Cet, V = 2.01).

SIMBAD calls the star Diphda. One point five degrees NW is 104 Aqr, a very wide John Herschel pair (HJ 5413) (4.9, 8.6, 4°, 118″, 2012). Star C is only magnitude 12.5, at 75°, 93″ from B. Two point five degrees ENE of 107 is BU 729, a rather nondescript object (7.8, 11.7, 346°, 11″.2, 2012).

Measures

Early measure (JC)	140°.5	5″.78	1845.89
Recent measure (WSI)	135°.6	6″.94	2015.74

175. STF 3050 AND = WDS J23595+3343AB

Table 9.175 Physical parameters for STF 3050 And

STF 3050	RA: 23 59 29.33	Dec: +33 43 26.9	WDS: 62(433)		
V magnitudes	A: 6.46	B: 6.72			
$(B - V)$ magnitudes	A: +0.56	B: +0.59			
μ(A)	-62.46 mas yr^{-1}	$\pm$ 0.06	-109.74 mas yr^{-1}	$\pm$ 0.04 (DR2)	
μ(B)	-47.49 mas yr^{-1}	$\pm$ 0.06	-78.63 mas yr^{-1}	$\pm$ 0.04 (DR2)	
π(A)	33.71 mas	$\pm$ 0.04	96.8 light yr	$\pm$ 0.1 (DR2)	
π(B)	33.92 mas	$\pm$ 0.04	96.2 light yr	$\pm$ 0.1 (DR2)	
Spectra	A: F8V	B:			
Luminosities (L$_\odot$)	A: 1.8	B: 1.4			
Catalogues	37 And	HR 9074	HD 224635	SAO 73656	HIP 118281
DS catalogues	Mayer 80	H N 58	STF 3050	BDS 12675	ADS 17149
Radial velocity	-7.90 km s^{-1}	$\pm$ 0.2			
Galactic coordinates	110°.720	-27°.913			

History

This star was found to be double by Christian Mayer in 1777 and later swept up by William Herschel in his double star survey in 1792 and announced as H N 58. Thomas Lewis notes in his 1906 volume that, as an easy pair, the measured separations should not have been quite so scattered, and he mentioned that one of the stars might be double. The mean positions plotted on his apparent orbit chart do show a sinusoidal variation with a possible period of 60 years, but no modern confirmation of higher multiplicity has been forthcoming. Lewis was convinced that the star was a binary but Burnham in his catalogue said 'apparently rectilinear motion', a view that pervaded the entry for the star in Espin's 1917 revision of *Celestial Objects for Common Telescopes*.

The Modern Era

The current orbit, whilst only graded 4, does a good job of representing the observed motion over the last 180 years

Finder Chart

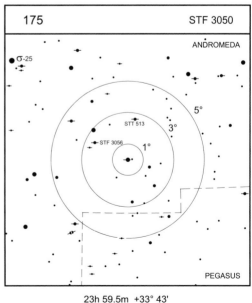

23h 59.5m +33° 43'

Orbit

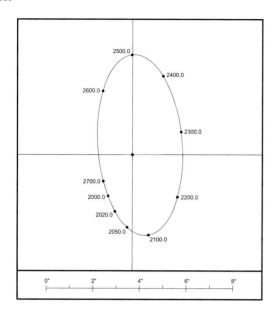

Ephemeris for STF 3050 AB (2010 to 2100)

Orbit by Hrt (2011a) Period: 717 years, Grade: 4

Year	PA(°)	Sep(")	Year	PA(°)	Sep(")
2010.0	336.3	2.26	2060.0	359.4	3.19
2020.0	342.4	2.48	2070.0	2.7	3.29
2030.0	347.5	2.69	2080.0	5.8	3.37
2040.0	352.0	2.88	2090.0	8.8	3.41
2050.0	355.9	3.05	2100.0	11.7	3.43

but it does not cover either end of the major axis of the apparent orbit. Curiously, Christopher Taylor [680], without prior knowledge of Lewis' comment, thought that he could see an elongation in one of the stars in late 2016. DR2 gives very precise values for the parallax of each component with no obvious sign of higher multiplicity. The stars are visible at all times in their orbit with 10-cm aperture and they will continue to widen for the next 80 years or so.

Observing and Neighbourhood

This is a beautiful binary, both stars white or yellowish and easily resolvable in small apertures, which is now widening. Measures made with the Cambridge 20-cm since 1990 show that the angle has increased by 20°, during which the separation has slowly increased. There is a magnitude 12.8 star at 298° and 80″ (2001). Just over 1° ENE is STF 3056 (7.7, 8.1, 143°, 0″.8, 2016), which appears to be slowly widening. One point five degrees N is STT 513 (6.8, 9.3, 17°, 3″.7, 2014).

Measures

Early measure (STF)	191°.0	3″.78	1832.65
(Orbit	190°.7	3″.83)	
Recent measures (ARY)	343°.3	2″.47	2018.11
(Orbit	341°.3	2″.44)	

APPENDIX

Double Star Nomenclature

The normally accepted nomenclature for a double star is its WDS name, typically WDS J00003+1430 where the J denotes Julian and the standard epoch and equinox of 2000 defines the position. Many observers prefer to refer to stars by the original discoverer's names and numbers, as given in the WDS. Note that this may not necessarily refer to the astronomer who first found the star. Many of William Herschel's discovery names have been replaced by others – most notably the STF designation of F. G. W. Struve, who in 1827 produced the results of the first systematic search for new double stars in the Dorpat catalogue.

Later in this appendix there is a list of the most common abbreviations. The original nomenclature for some astronomers such as F. G. W. Struve, Burnham, or Finsen was originally a greek letter, i.e. Σ, β, or ϕ. These have now been romanized into STF, BU, and FIN. The list is not comprehensive but does cover all the discovery designations to be found in this volume. A full list can be found on the USNO website http:ad.usno.mil/wds.

William Herschel's Double Star Classes

When Herschel published his first list of double star discoveries in 1782 he used a system of classification which allocated a class type to a particular range of angular separations. The closest discoveries went into Class I and so on. His second list, in 1784, also used the same system whilst his last lists, in 1800 and 1822, just used the classification N for New.

Class	Range	Example
I	$0''$–$2''$*	H I 16 = η CrB = STF 1937
II	$2''$–$5''$†	H II 13 = μ Dra = STF 2130
III	$5''$–$15''$	H III 16 = γ Del = STF 2727
IV	$15''$–$30''$	H IV 17 = α CVn = STF 1692
V	$30''$–$60''$	H V 55 = ν Dra = STFA 35
VI	$60''$–$120''$	H VI 40 = γ Lep
N	all	H N 23 = STF 427

* Herschel noted 'requiring indeed a very superior telescope, the utmost clearness of air, etc.'.
† Herschel noted 'those suitable for very delicate measures of the micrometer'.

Resolution as a Function of Aperture

This is a subject on which there is no definitive view, but there are a number of relationships which can act as a reasonable guide to the limit of double star separation that might be expected for a given aperture. If a telescope aperture is not large enough then no amount of magnification will make any difference. The star images need to be separated in the focal plane by the objective or primary mirror. The most well-known criterion is Dawe's simple relation, which is $4''.56/D$ (inches) or $11''.6/D$ (cm). Dawes [127] worked this out on the basis of many observations of double stars with a range of apertures D. It is a good guide, provided that the aperture is relatively small and the stars are about magnitude 6 and of similar brightness. Once one of these conditions is relaxed then the numerator in the equation starts to increase. In 1914, Thomas Lewis [2] repeated Dawes' work but for a much larger range of aperture and separation, and he found that for equally bright stars (typically magnitudes 5.7 and 6.4) the relation was $4''.8/D$ (inches) whilst for equally faint stars (8.5 and 9.1), it increased to $8''.5/D$ (inches). R. G. Aitken [1] later discussed these figures and found that for the 36-inch Lick refractor he could do significantly better than Lewis – in the cases just mentioned, $4''.3/D$ (inches) and $6''.1/D$ (inches) respectively, no doubt due to his keen eyesight and the superior seeing. The theoretical relationship, worked out by Airy from the wave theory of light, is called the Rayleigh limit. It is slightly more pessimistic than Dawes and comes in at $4''.8/D$ (the value also depends on the wavelength of light, a factor which Dawes did not include). Modern telescopes working in the infrared have a slight advantage. All this assumes that the two disks are fully separated, i.e. in tangential contact. Once a double star is so close that the image appears like a figure '8' or an oval then the criteria are not reliable. From years of experience RWA is aware that the Cambridge 20-cm refractor gives separate star images at around $0''.6$ separation. In 2005

RWA observed the periastron passage of γ Virginis when the stars closed to just under $0''.4$ and he could still see an elongation in the image. It seems likely that S. W. Burnham, with a 15-cm aperture, could see pairs of stars closer than that, albeit still elongated, which demonstrates again that there is a personal factor involved in this business.

Decimals of a Year for Each Month

When double star measures are made for the purpose of orbit determination, the date of observation is always converted into a year and its fractional component. A Julian year is 365.2422 days long and so each day is 0.0027 of a year. The two look-up tables on this page allow a conversion to be made from a calendar date to a decimal date. The first table gives the fractional part of the year on the first of the month, for each of the four years following a leap year. The second table gives the additional fraction for each day of the month after the first day. The following example is given: what is the decimal date for 0 hours on 2014 July 27? From the tables, since 2014 is an even year, 27 July 2014 = 2014 + 0.4956 + 0.0712 = 2014.5678. For most purposes three places of decimals will be sufficient. (The tables were prepared by Michael Greaney.)

Month	Leap yr +1	Even yr	Leap yr −1	Leap yr
Jan	0.0007	0.0000	−0.0007	−0.0014
Feb	0.0856	0.0849	0.0842	0.0835
Mar	0.1622	0.1615	0.1608	0.1629
Apr	0.2471	0.2464	0.2457	0.2478
May	0.3292	0.3285	0.3278	0.3299
Jun	0.4141	0.4134	0.4127	0.4148
Jul	0.4963	0.4956	0.4949	0.4970
Aug	0.5811	0.5803	0.5797	0.5818
Sep	0.6660	0.6653	0.6646	0.6667
Oct	0.7481	0.7474	0.7467	0.7488
Nov	0.8330	0.8323	0.8316	0.8337
Dec	0.9151	0.9144	0.9137	0.9158

Day	Fraction	Day	Fraction	Day	Fraction
2	0.0027	12	0.0300	22	0.0575
3	0.0055	13	0.0328	23	0.0602
4	0.0082	14	0.0356	24	0.0630
5	0.0109	15	0.0383	25	0.0657
6	0.0137	16	0.0411	26	0.0684
7	0.0164	17	0.0438	27	0.0712
8	0.0192	18	0.0465	28	0.0739
9	0.0219	19	0.0493	29	0.0766
10	0.0246	20	0.0520	30	0.0794
11	0.0274	21	0.0547	31	0.0821

The Top Visual Observers

The WDS website lists the 25 most prolific observers or groups (the first author or group gets all the credit). 'N means' refers to an accumulation of individual measures into a mean value to increase the weight of the measurement. In the case of visual observers this number needs to be multiplied by between two and three, approximately, to get the number of individual measures. W. H. van den Bos is the most prolific visual observer in history and has made more than 73,000 individual measures (24,574 means).

The second table lists the number of discoveries (Nsys) in descending order. W. J. Luyten did his work using photography so the greatest visual discoverer in recent history is John Herschel, followed by Robert Rossiter and F. G. W. Struve.

These lists were supplied by Dr B. D. Mason and are correct as at March 2018. They are to be found on the following page.

Observers

The following table ranks observers by N means.

		Observer	N means
1	TMA	Two Mass Cat.	239,547
2	WIS	WISE IR Explorer	165,696
3	TYC	Tycho	89,566
4	AAV	AAVSO Phot. Survey	87,285
5	WFC	Washington. Fund. Cat.	76,992
6	UC	USNO CCD Cat.	75,577
7	DEN	DENIS IR survey	35,740
8	WSI	Wash. Speckle Int.	29,130
9	B	W. H. van den Bos	24,574
10	HIP	Hipparcos	22,702
11	HEI	W. D. Heintz	20,015
12	SKF	B. A. Skiff	18,571
13	DAM	F. Damm	16,278
14	VBS	G. van Biesbroeck	14,727
15	A	R. G. Aitken	14,567
16	BU	S. W. Burnham	13,249
17	RST	R. A. Rossiter	13,212
18	WOR	C. E. Worley	12,799
19	TOK	A. A. Tokovinin	11,876
20	MCA	H. McAlister	11,547
21	USN	US Naval Obs.	11,480
22	COU	R. Jonckheere	11,312
23	HOR	E. P. Horch	10,781
24	DHT	E. Horch	10,137
25	HRT	W. I. Hartkopf	10,036

The next table ranks observers by Nsys.

		Observer	Nsys
1	TDS	Tycho	9893
2	LDS	W. J. Luyten	6118
3	HJ	J. F. W. Herschel	5968
4	POU	A. Pourteau	5707

contd

		Observer	Nsys
5	RST	R. Rossiter	5606
6	UC	USNO CCD Cat.	5271
7	STF	F. G. W. Struve	4428
8	TOK	A. A. Tokovinin	4277
9	A	R. G. Aitken	3507
10	HDS	Hipparcos	3382
11	J	R. Jonckheere	3291
12	BRT	S. G. Barton	3255
13	GWP	Garraf Wide Pairs	3202
14	ES	T. H. Espin	3187
15	B	W. H. van den Bos	3113
16	SKF	B. Skiff	2865
17	BPM	Bruce Proper Motion	2773
18	COU	P. Couteau	2742
19	KOI	Kepler mission	2640
20	BAL	J. Baillaud	2320
21	BU	S. W. Burnham	2293
22	STI	J. Stein	2154
23	CVR	J. Chivers	1740
24	HU	W. J. Hussey	1672
25	I	R. T. A. Innes	1671

Smyth's Colours

In an attempt to standardize the description of stellar colours Smyth produced a coloured sheet with six main categories of colour and four sub-divisions of shades for each colour. The main colours were red, orange, yellow, green, blue and purple. Using this scheme he designated the colours of α Herculis A and B as yellow 2 and blue 3, whilst he had originally classified them as orange, and emerald or bluish green.

The following table is reproduced from Smyth's *Sidereal Chromatics* (1864).

Amethyst	Purple
Apple green	Brownish green
Ash colour	Pale dull grey
Beet hue	Crimson

contd

Cinereous	Wood-ash tint
Cherry-colour	Pale red
Creamy	Pale white
Cobalt	Bluish white
Crocus	Deep yellow
Damson	Dark purple
Dusky	Brownish hue
Emerald	Lucid green
Dawn-coloured	Whitey-brown
Flushed	Reddened
Garnet	Red of various shades
Golden hue	Bright yellow
Grape red	A variety of purple
Jacinth	Pellucid orange tint
Lemon-coloured	Bright but pale yellow
Lilac	Light purple
Livid	Lead colour
Melon tint	Greenish yellow
Orpiment	Bright yellow
Pale	Deficient in hue
Pearl colour	Shining white
Plum colour	Pale purple
Radish tint	Dull purple
Rose tint	Flushed crimson
Ruby colour	Pellucid red
Ruddy	Flesh coloured
Sapphire	Blue tint
Sardonyx	Reddish yellow
Sea green	Faint cold green
Silvery	Mild white lustre
Smalt	Fine deep blue
Topaz	Lucid yellow
Vanilla tint	Dark brown or chocolate

F. G. W. Struve's Colours

F. G. W. Struve used the following terms; obscurissima, obscura, pallida, livida, alba, sub-flava, flava, sub-caerulea, caerulea, rubicunda, and rubra.

Journals

To save space, references to astronomical literature in the catalogue have been shortened. Below is a guide to the symbols used:

A&A	Astronomy and Astrophysics
A&AS	Astronomy and Astrophysics Supplement Series
A+ARv	Astronomy and Astrophysics Reviews
AJ	Astronomical Journal
AN	Astronomische Nachrichten
ApJ	Astrophysical Journal
ApJS	Astrophysical Journal Supplements
ApJL	Astrophysical Journal Letters
AP+SS	Astrophysics and Space Science
Ast. Rep.	Astronomical Reports
BAAS	Bulletin of the American Astronomical Society
IAUDS	Information Circular of IAU Double Star Commission
IBVS	International Bulletin of Variable Stars
JBAA	Journal of the British Astronomical Association
JDSO	Journal of Double Star Observers
J. Hist. Astron.	Journal for the History of Astronomy
JRASC	Journal of the Royal Astronomical Society of Canada
Mems. RAS	Memoirs of the Royal Astronomical Society
MN	Monthly Notices of the RAS
MNASSA	Monthly Notices of the Astronomical Society of Southern Africa
Phil. Trans.	Philosophical Transactions of the Royal Society
PASP	Publications of the Astronomical Society of the Pacific
PDAO	Publications of the Dominion Astrophysical Observatory
SvaL	Soviet Astronomy Letters
VSSC	Variable Star Section Circulars, BAA.

Observer Key

A	R. G. Aitken
AC	A. Clark
AGC	A. G. Clark
ALM	Atacama Large Millimetre Array
ANT	R. Anton
ARN	D. Arnold
ARY	R. W. Argyle
B	W. H. van den Bos
BLL	R. S. Ball
BND	G. P. Bond
BSO	Brisbane Observatory
BU	S. W. Burnham (also β)
CHR	CHARA consortium
COG	W. Cogshall
COO	Cordoba Observatory
COP	R. Copeland
CTT	J.-F. Courtot
D	E. Dembowski
DA	W. R. Dawes
DAW	B. H. Dawson (also δ)
DOO	E. Doolittle
DRS	R. J. de Rosa
DRU	J. Drummond
DUN	J. Dunlop (also Δ)
ENH	B. von Engelhardt
FIN	W. S. Finsen
FYM	M. Fay
GAT	G. Gatewood
GLE	W. Gale
GLI	J. M. Gilliss
GNT	J. W. Grant
GRB	S. Groombridge
H	W. Herschel
HAL	A. Hall
HDO	Harvard Observatory
HJ	J. F. W. Herschel
HOR	E. P. Horch
HSW	R. Harshaw

HU	W. J. Hussey
HUB	S. Hubrig
HWE	H. A. Howe
I	R. T. A. Innes
JC	W. S. Jacob
JOD	E. Jodar
JOY	A. H. Joy
KR	A. Krüger
LBO	L. Boss
LOC	G. Locatelli
LV	F. P. Leavenworth
MCA	H. A. McAlister
MLO	Melbourne Observatory
MRN	L. Marion
MTL	O. M. Mitchel
NPI	Navy Prototype Optical Interferometer
PRU	J.- L. Prieur
PWL	E. B. Powell
PZ	C. Piazzi
R	H. C. Russell
RAO	RoboAO (Palomar)
RHD	J. Richaud
RMK	C. Rümker
RST	R. A. Rossiter
SCA	M. Scardia
SEE	T. J. J. See (also λ)
SER	J. Serro
SHJ	J. South/J. Herschel
SLR	R. P. Sellors
SMR	J. Schlimmer
SP	G. V. Schiaparelli
STF	F. G. W. Struve (also Σ)
STT	O. Struve (also $O\Sigma$)
TEB	J. Tebbutt
TMA	Two Mass Catalogue
TOK	A. A. Tokovinin
UC	USNO UCAC4
USNO	26-inch
WSI	Washington Speckle Interferometer

contd

REFERENCES

[1] R. G. Aitken, *The Binary Stars* (New York: Dover, 1964).

[2] T. Lewis, On the Class of Double Stars which can be Observed with Refractors. *Observatory* 37 (1914), 372–378.

[3] C. J. R. Lord, www.brayebrookobservatory.org/ BrayObsWebSite/BOOKS/TELESCOPIC% 20RESOLUTION.pdf.

[4] M. Beech, Cigarette- and Trade Card-Astronomy: C. 1900 – C. 2000. A Journey from Engaged Imagination to Passive Data Consumption. *Observatory* 137 (2017), 288–295.

[5] J. Daley, CCD Imaging of STF 93C and D. *JDSO* 2 (2006), 51–53.

[6] B. D. Mason & W. I. Hartkopf, *Sixth Catalog of Visual Double Star Orbits*, https://ad.usno.navy.mil/wds/ orb6.html USNO, 2019.

[7] B. MacEvoy & W. Tirion, *Cambridge Double Star Atlas* (Cambridge: Cambridge University Press, 2015).

[8] T. W. Webb, Celestial Objects for Common Telescopes (Cambridge: Cambridge University Press, 2010).

[9] R. Burnham, *Burnham's Celestial Handbook*, three volumes (New York: Dover, 1978).

[10] W. H. Smyth, *The Bedford Catalogue* (Richmond: Willmann-Bell, 1986).

[11] W. T. Olcott & E. W. Putnam, *The Field Book of the Skies* (New York: Putnam, 1929).

[12] J. Dunlop, Approximate Places of Double Stars in the Southern Hemisphere, Observed at Paramatta in New South Wales. *Mems. RAS* 3 (1829), 257–275.

[13] S. Haas, *Double Stars for Small Telescopes* (Cambridge: Sky Publishing, 2006).

[14] T. Lewis, On the Magnifying Powers Used by Double Star Observers. *Observatory* 36 (1913), 423–428.

[15] W. S. Finsen (private communication), 13 October, 1973.

[16] W. H. van den Bos (private communication), 1973.

[17] J.- A. Docobo (private communication), 2016.

[18] A. Tokovinin (private communication), 2016.

[19] M. Scardia (private communication), 2018.

[20] T. Hockey *et al.* (eds.), *Biographical Encyclopaedia of Astronomers* (New York: Springer, 2014).

[21] F. Losse (private communication), 2017.

[22] T. E. R. Phillips, Obituary of T. H. E. C. Espin. *MN* 95 (1934), 319–322.

[23] J. Jackson, Obituary of S. W. Burnham. *MN* 82 (1922), 258–263.

[24] W. Knapp (private communication), 2018.

[25] B. D. Mason (private communication), 2017.

[26] W. I. Hartkopf (private communication), 2017.

[27] Lowell Observatory website, https://lowell.edu /staff-member/brian-a-skiff/.

[28] J. - F. Courtot (private communication), 2018.

[29] J. Schlimmer, Christian Mayer's Double Star Catalogue. *JDSO* 3, 4 (2007), 151–158.

[30] T. R. Williams, *Biographical Encyclopaedia of Astronomers* Volume 2, p. 753 (Heidelberg: Springer, 2007).

[31] H. J. Augener and E. H. Geyer, https://aas.org/ obituaries/wulff-dieter-heintz-1930-2006.

[32] R. C. Tanguay, Observing Double Stars for Fun and Science. *Sky & Telescope* 97 (1999), 116–120.

[33] W. Herschel, Catalogue of Double Stars. *Phil. Trans.* 72 (1782), 112–162.

[34] W. Herschel, Catalogue of Double Stars. *Phil. Trans.* 75 (1785), 40–126.

[35] W. Herschel, On the Places of 145 New Double Stars, *Mems. RAS* 1 (1822), 166–181.

[36] R. G. Aitken, Measures of Double Stars in the Years 1923–1926. *Lick Obs. Bull.* 12 (1927), 173–182.

[37] A. H. Joy, *Publ. AAS* (1922), 29th Meeting, 14.

[38] L. B. Sokoloski & L. Bilsen, Evidence for the White Dwarf Nature of Mira B. *ApJ* 723 (2010), 1188–1194.

[39] W. Vlemmings, S. Ramstedt, E. O'Gorman, *et al.*, Resolving the Stellar Activity of the Mira AB Binary with ALMA. *A&A* 577 (2015), L4, 1–5.

[40] A. Winnecke, Schreiben des Herrn Professors Dr Winnecke, Directors des Sternwarte im Strassbrug und den Herausgeben. *AN* 87 (1876), 160.

[41] R. E. Nather & D. S. Evans, Discovery and Measurement of Double Stars by Lunar Occultations. *Ap+SS* 11 (1971), 28.

[42] H. A. Abt, R. C. Barnes, E. S. Biggs & P. S. Osmer, The Frequency of Spectroscopic Binaries in the Pleiades. *ApJ* 142 (1965), 1604–1615.

[43] J. T. McGraw, D. W. Dunham, D. S. Evans & T. J. Moffett, Occultation of the Pleiades. Photoelectric Observations at Tonantzintla with a Discussion of the Duplicity of Atlas. *AJ* 79 (1974), 1299–1303.

[44] P. Bartholdi, Photoelectric Observations of Occultations in the Pleiades and the Incidence of Duplicity in the Cluster. *AJ* 80 (1975), 445–448.

[45] N. Zwahlen, P. North, L. Debernardi, L. Eyer, F. Galland, M. A. T. Groenewagen *et al.*, A Purely Geometric Distance to the Binary Star Atlas, a Member of the Pleiades, *A&A* 425 (2004), L45–L48.

[46] R. G. Aitken, Eighty-Five New Double Stars. Twenty-Fifth List. *Lick Obs. Bull.* 12 (1927), 170–172.

[47] A. A. Tokovinin, Speckle Interferometry and Orbits of 'Fast' Binaries. *AJ* 144 (2012), 56, 1–11.

[48] R. Snow, *MN* 2 (1832), 126.

[49] Anon., *MN* 15 (1855), 117.

[50] Cited by Lewis (1906) [194].

[51] Cited by Lewis (1906) [194].

[52] S. W. Burnham, Double Star Observations Made in 1877–8 in Chicago with the 18.5-inch Refractor of the Dearborn Observatory Comprising I. A Catalogue of 251 New Double Stars with Measures; II. Micrometrical Measures of 500 Double Stars. *Mems. RAS* 44 (1878), 141–305.

[53] F. Holden, Double Star Measures at Lick Observatory, Mount Hamilton, California. *PASP* 86 (1974), 902–906.

[54] G. van Biesbroeck, Micrometric Measurements of Double Stars. *ApJS* 28 (1974), 414–448.

[55] B. D. Mason, The High Angular Resolution Multiplicity of Massive Stars. *AJ* 137 (2009), 3358–3378.

[56] R. F. Sanford, The Spectrographic Orbit of the Companion to Rigel. *ApJ* 95 (1942), 421–424.

[57] W. W. Campbell, The Spectroscopic Binary Capella. *ApJ* 10 (1899), 177.

[58] H. F. Newall, Variable Velocities of Stars in the Line of Sight. *Observatory* 22 (1899), 436–437.

[59] H. F. Reese, A Determination of the Orbit of Capella. *Lick Obs. Bull.* 1 (1901), 32–35.

[60] J. Ashbrook, Capella as a Close Visual Double Star. *Sky & Telescope* 51 (1976), 322–323.

[61] J. A. Anderson, Application of Michelson's Interferometer Method to the Measurement of Close Double Stars. *ApJ* 51 (1920), 263–275.

[62] G. Torres, A. Claret, K. Pavlovski & A. Dotter, Capella (α Aurigae) Revisited: New Binary Orbit, Physical Properties, and Evolutionary State. *ApJ* 807 (2015), 26.

[63] R. Furuhjhelm, Ein Schwacher Begleiter zu Capella. *AN* 197 (1914), 181–184.

[64] C. Stearns, Note on the Duplicity of Capella H. *AJ* 45 (1936), 120.

[65] P. Fox, Measures of Miscellaneous Double Stars. *Ann. Dearborn Obs.* 2 (1925), 80–219.

[66] W. S. Finsen, Measures of Double Stars. *Union Obs. Circ.* 80 (1928), 85–98.

[67] D. Benest & J. L. Duvent, Is Sirius a Triple Star? *A&A* 299 (1995), 621–628.

[68] F. W. Bessel, On the Variation of the Proper Motions of Procyon and Sirius. *MN* 6 (1844), 136–141.

[69] H. E. Bond, R. L. Gillilano, G. H. Schaefer, P. Demarque, T. M. Girard, J. B. Holberg *et al.*, HST Astrometry of the Procyon System. *ApJ* 813 (2015), 106, 1–19.

[70] C. E. Worley, Measures of 2589 Double Stars. *Publ. USNO* 25, Pt. III (1989), 1–211.

[71] W. H. van den Bos, Double Stars That Vex the Observer. *PASP Leaflets* 7 (1958), 347.

[72] R. Shobbrook & J. Robertson, Velocity Variations in Beta Centauri. *Pub. Australian Ast. Soc.* 1 (1968), 82.

[73] A. Pigulski, H. Cugler, A Popowicz, R. Kuschnig, A. F. J. Moffat, S. M. Rucinski *et al.*, Massive Pulsating Stars Observed by BRITE-Constellation. The Triple System β Centauri (Agena). *A&A* 588 (2016), A55, 1–17.

[74] J. G. Voûte, Measures of Double Stars. 5th Series November 1932 – May 1937. *Ann. Bosccha Obs.* 6 (1947), D71.

[75] R. Hanbury Brown, The Angular Diameter of 32 Stars. *MN* 167 (1974), 121.

[76] J. Davis, A. Mendez, E. B. Seneta, W. J. Tango, A. J. Booth, J. W. O'Byrne *et al.*, Orbital Parameters, Masses, and Distances to β Centauri Determined with the Sydney University Stellar Interferometer and High Resolution Spectroscopy. *MN* 356 (2005), 1362–1370.

[77] R. E. Wilson, *Lick Obs Bull.*, 8 (1915), 124–127.

[78] A. Tokovinin, *MSC Catalogue*: www.ctio.noao.edu/~atokovinin/stars/intro.html.

[79] T. J. J. See, Discovery of a Companion to θ Scorpii. *AN* 142 (1896), 43.

[80] M. Kerr, D. J. Frew & R. Jaworski, The Companion of θ Scorpii. *Webb Society Deep-Sky Observer* 141 (2008), 8–9.

[81] W. S. Finsen, New Double Stars. VI. A Case of Tweedledum and Tweedledee. *MNASSA* 12 (1953), 86.

[82] N. Voronov, New Orbits of Two Double Stars. *Tashkent Ast. Obs. Circ.* 27 (1934).

[83] W. H. van den Bos, Is This Orbit Really Necessary. *PASP* 74 (1962), 297–301.

[84] W. Beavers & J. J. Eitter, E. W. Fick Observatory Stellar Radial Velocity Measurements. *ApJS* 62 (1986), 147–228.

[85] R. W. Argyle, Filar Micrometer. In *Observing and Measuring Visual Double Stars*, R. W. Argyle, ed. (New York: Springer, 2012), pp. 169–181.

[86] B. MacEvoy & W. Tirion, *Cambridge Double Star Atlas*, 2nd edition (Cambridge: Cambridge University Press, 2015).

[87] J. Perez, Double Star Sketching. In *Observing and Measuring Visual Double Stars*, R. W. Argyle, ed. (New York: Springer, 2012), pp. 33–52.

[88] R. Anton, Lucky Imaging. In *Observing and Measuring Visual Double Stars*, R. W. Argyle, ed. (New York: Springer, 2012), pp. 231–252.

[89] J. Perez, www.perezmedia.net/beltofvenus.

[90] R. C. Brooks, The Development of Micrometers. *J. Hist. Astron* 22 (1991), 127–173.

[91] F. Losse, www.astrosurf.com/hfosaf.

[92] J. Lodigruss, www.astropix.com/html/observing/20_fun_naked_eye_doubles.html.

[93] https://optcorp.com/blogs/astronomy/double-stars-for-binoculars/.

[94] https://occultations.org.

[95] T. Teague, Simple Techniques of Measurement. In *Observing and Measuring Visual Double Stars*, R. W. Argyle, ed. (New York: Springer, 2012), pp. 149–167.

[96] N. Webster, Double Star Measurements Made with a Meade 12 mm Astrometric Eyepiece in 2017. *Webb Society Double Star Circulars* 26 (2018), 32–42.

[97] R. Anton, Double Star Measurements at the Southern Sky with a 50-cm reflector in 2016. *JDSO* 13 (2017), 495.

[98] J. Daley, A Method of Measuring High Δm Doubles. *JDSO* 3 (2007), 159–164.

[99] D. Arnold, Divinus Lux Observatory Bulletin: Report 27. *JDSO* 9 (2013), 10–18.

[100] R. Benavides J. González & E. Masa, *Observacion de Estrellas Dobles (Observation of Visual Double Stars)* (Barcelona: Marcombo, 2017).

[101] W. Herschel, Catalogue of Double Stars. *Phil. Trans.* 72 (1782), 112–162.

[102] W. Herschel, Catalogue of Double Stars. *Phil. Trans.* 75 (1785), 40–126.

[103] W. Herschel, On the Places of 145 New Double Stars. *Mems. RAS* 1 (1822), 166–181.

[104] W. Herschel, *Phil. Trans*, 1800.

[105] K. Schwarzschild & W. Villiger, Ueber Messung von Doppelstern durch Interferenzen. *AN* 139 (1896), 353–360.

[106] The International Occultation Timing Association, https://occultations.org.

[107] G. Popovic, *Bull. Obs. Ast. Belgrade* 155 (1990), 97.

[108] L. Kiyaeva, A. A. Kiselev, E. V. Polyakov & V. B. Rafal'skij, An Astrometric Study of the Triple Star ADS 48. *Soviet Ast. Letters* 27 (2001), 456–463.

[109] E. J. Shaya & R. Olling, Very Wide Binaries and Other Comoving Stellar Companions: A Bayesian Analysis of the Hipparcos Catalog. *ApJS* 192 (2011), 2, 1–17.

[110] W. E. Harper, The Radial Velocities of 125 Stars. PDAO 2 (1923), 189–203.

[111] R. F. Griffin, Spectroscopic Binary Orbits from Photoelectric Radial Velocities Paper 144: HR5B. *Observatory* 119 (1999), 27–45.

[112] O. Brettmann, D. S. Hall, C. H. Poe, R. E. Fried, W. Duvall & J. S. Shaw, HR 5 = ADS 61: A New Variable Star. *IBVS* 2389 (1983), 1–3.

[113] S. W. Burnham, Seventh Catalogue of New Double Stars. *AN* 88 (1875), 2103, 225–230.

[114] G. van Belle *et al.*, Interferometric Observations of Rapidly Rotating Stars. *Ast. & Ap. Rev.* 20 (2012), 51, 1–49.

[115] W. Luyten, University of Minnesota PM survey no. 25 (1970).

[116] S. W. Burnham, Double Star Observations Made in 1879 and 1880 with the 18.5-inch Refractor of the Dearborn Observatory: I. Catalogue of 151 New Double Stars with Micrometric Measures. II. Micrometric Measures of 770 Double Stars. *Mems. RAS* 47 (1881), 167–324.

[117] S. L. Lippincott, Parallax and Orbital Motion of the Two Nearby Long-Period Visual Binaries Groombridge 34 and ADS 9090. *AJ* 77 (1972), 165–168.

[118] A. Howard, G. W. Marcy, D. A. Fischer, H. Isaacson, P. S. Muirhead, G. W. Henry *et al.*, The NASA-UC-UH Eta-Earth Program. IV. A Low Mass Planet Orbiting an M-Dwarf 3.6 Pc from Earth. *ApJ* 794 (2014), 51 1–9.

[119] A. M. Tanner, C. M. Gelino & N. M. Law, A HIGH Contrast Imaging Survey of SimLITE Planet Search Targets. *PASP* 122 (2010), 1195–1206.

[120] T. Trifonov, M. Kurster, M. Zechmeister, L. Tal-Or, J. A. Caballero, A Quirrenbach *et al.*, The CARMENES Search for Exoplanets around M-Dwarfs. First Visual-Channel Radial Velocity Measures and Orbital Parameter Update of 7 M-dwarf Planetary Systems. *A&A* 607 (2018), A117, 1–24.

[121] M. G. Edmunds, Founder of the RAS: Stephen Groombridge. *A& G* 59 (2018), 2–15.

[122] R. T. A. Innes, Fifth List of New Double Stars. *AN* 116 (1898), 369–372.

[123] S. Bailey, Catalogue of Southern Double Stars. *Ann. Harvard Obs.* 32 (1900), 2, 296–308.

[124] P. van de Kamp & E. Flathers, Parallax and Mass-Ratio of eta Cassiopeiae from Photographs Taken with the Sproul 24-inch Refractor. *AJ* 60 (1955), 448.

[125] D. Fischer, G. W. Marcy & J. F. Spronk, The Twenty Five Year Lick Planet Search. *ApJS* 210 (2014), 5, 1–11.

[126] Star Splitters website: https:bestdoubles.word press.com.

[127] W. R. Dawes, Micrometrical Measurements of the Positions and Distances of 121 Double Stars, Taken at Ormskirk During the Years 1830, 1831, 1832 and 1833. *Mems. RAS* 8 (1835), 61–94.

[128] A. A. Tokovinin & S. Lépine, Wide Companions to Hipparcos Stars within 67 pc of the Sun. *AJ* 144 (2012), 102, 1–12.

[129] P. Fox, Measures of Miscellaneous Double Stars. *Ann. Dearborn Obs.* 2 (1925), 80–219.

[130] S. W. Burnham, Measures of Double Stars. *AJ* 31 (1918), 141–144.

[131] R. P. Sellors, Observations of Double Stars Made at Sydney Observatory. *MN* 54 (1893), 123.

[132] R. W. Argyle, A. Alzner & E. Horch, Orbits for Five Southern Visual Binaries. *A&A* 384 (2002), 171–179.

[133] B. Warner, The Double Star Discoveries of Fearon Fallows. *MNASSA* 36 (1977), 134–135.

[134] R. E. Wilson, Five Stars Whose Radial Velocities Are Variable. *Lick Obs. Bull.* 257 (1914), 80–81.

[135] A. Colacevich, Provisional Elements of the Spectroscopic Binary δ Phoenicis. *PASP* 47 (1935), 84–86.

[136] A. R. Hogg, The Eclipsing System ζ Phoenicis. *MN* 111 (1951), 315–324.

[137] R. A. Rossiter, New Southern Double Stars (First List) Found at the Lamont-Hussey Observatory of the University of Michigan, Bloemfontein. *Mems. RAS* 65 (1933), 73–182.

[138] D. Malin & D. J. Frew, *Hartung's Astronomical Objects for Southern Telescopes* (Melbourne: Melbourne University Press, 1995).

[139] J. Anderson, Spectroscopic Observations of Eclipsing Binaries. *A&A* 118 (1983), 255–261.

[140] J.-L. Halbwachs, Common Proper Motion Stars in the AGK3. *A&AS* 66 (1986), 131–148.

[141] C. Worley, Micrometer Measures of 1164 Double Stars, Pt 6. *Publ. USNO* 18 (1967), 1–133.

[142] J. F. W. Herschel, Results of Astronomical Observations at the Cape of Good Hope (London: Stewart & Murray, 1847).

[143] R. T. A. Innes, A New Quadruple Stellar System. *MN* 57 (1897), 456–457.

[144] P. Baize & M. Petit, Étoiles Doubles Orbitales a Composantes Variables. *A&AS* 77 (1989), 497–511.

[145] W. S. Jacob, On the Orbit of p Eridani. *MN* 10 (1850), 171.

[146] G. van Albada, The Orbit of p Eridani. *Contr. Bosscha Obs.* 5 (1957), 11–38.

[147] M. Scardia, *IAUDS* 186 (2015).

[148] S. W. Burnham, Double Star Observations Made in 1877–8 in Chicago with the 18.5-inch Refractor of the Dearborn Observatory Comprising I. A Catalogue of 251 New Double Stars with Measures; II. Micrometrical Measures of 500 Double Stars. *Mems. RAS* 44 (1878), 141–305.

[149] E. J. Powell, Observations of Double Stars Taken at Madras in 1853, 4, 5 and the Beginning of 1856. *Mems. RAS* 25 (1857), 55–97.

[150] D. Pourbaix, *Ninth Spectroscopic Binary Catalogue*, sb9 .astro.ulb.ac.be.

[151] W. D. Heintz, Observations of Double Stars and New Pairs XVII. *ApJS* 105 (1996), 475–480.

[152] C. Mayer, Verzeichniss aller Dopplesterne. In *Astronomisches Jahrbuch fur das Jahr 1784*, by J. E. Bode (1784), p. 183.

[153] F. G. W. Struve – cited by Bishop (*Bishop's Observatory Astronomical Observations*), 1852.

[154] S. W. Burnham, Double Star Observations Made in 1877–8 in Chicago with the 18.5-inch refractor of the Dearborn Observatory Comprising I. A Catalogue of 251 New Double Stars with Measures; II. Micrometrical Measures of 500 Double Stars. *Mems. RAS* (1878), 141–305.

[155] R. G. Aitken, Measures of Double Stars in the Years 1923–1926. *Lick Obs. Bull.* 12 (1927), 173–182.

[156] W. Vlemmings, S. Ramstedt, E. O'Gorman, E. M. Humphrys, M. Wittkowski, A. Baudry *et al.*, Resolving the Stellar Activity of the Mira AB Binary with ALMA. *A&A* 577 (2015), L4, 1–5.

[157] J. L. Sokoloski & L. Bilsen, Evidence for the White Dwarf Nature of Mira B. *ApJ* 723 (2010), 1188–1194.

[158] H. A. McAlister, W. I. Hartkopf, D. J. Hutter & O. Franz, ICCD Speckle Observations of Binary Stars. II – Measurements During 1982–1985 from the Kitt Peak 4-m telescope. *AJ* 93 (1987), 688–723.

[159] J. Drummond, S. Mister, P. Ryan & L. C. Roberts, ι Cassiopeiae: Orbit, Masses and Photometry from Adaptive Optics Imaging in the I and H Bands. *ApJ* 585 (2003), 1007–1014.

[160] W. D. Heintz, A Study of Multiple Star Systems. *AJ* 111 (1996), 408–411.

[161] L. Seidel, IIK1. *D. K. Akad. Wiss. München* 6 (1852), 564.

[162] J. Schmidt, Ueber Verandliche Sterne. *AN* 46 (1856), 293–298.

[163] W. W. Campbell, On the Variable Velocity of Polaris in the Line of Sight. *PASP* 11 (1899), 195–203.

[164] W. W. Campbell, Concerning the Radial Velocity of Polaris. *PASP* 22 (1910), 5–35.

[165] R. H. Wilson, 82 Geminorum. Polaris Observed Double with the Interferometer. *PASP* 49 (1937), 202.

[166] K. Kamper, Polaris Today. *JRASC* 90 (1996), 140–157.

[167] T. E. Nordgren, J. T. Armstrong, M. E. Germain, R. B. Hindsley, A. R. Hajian, J. T. Sudol *et al.*, Astrophysical Quantities of Cepheid Variables Measured with the NPOI. *ApJ* 543 (2000), 972–978.

[168] N. R. Evans, G. H. Schaefer, H. E. Bond, G. Bono, M. Karovska, E. Nelan *et al.*, Direct Detection of the Close Companion of Polaris with the HST. *AJ* 136 (2008), 1137–1146.

[169] B. P. Gerasimovič, The System of Polaris. *ApJ* 84 (1936), 229–234.

[170] J. Daley, CCD Imaging of STF 93 C and D. *JDSO* 2 (2006), 51–53.

[171] H. E. Bond, E. Nelan, N. R. Evans, G. H. Schaefer & D. L. Harmer, HST Trigonometrical Parallax of Polaris B, Companion of the Nearest Cepheid. *ApJ* 853 (2018), 55.

[172] H. Alden & P. van de Kamp, Faint Stars of Appreciable Proper Motion. *AJ* 35 (1924), 167–166.

[173] W. Hartkopf, Binary Stars Unresolved by Speckle Interferometry III. *PASP* 96 (1985), 105–116.

[174] K. Katarzyński, M. Gawroński & K. Góździewski, Search for Exoplanets and Brown Dwarfs using VLBI. *MN* 461 (2016), 929–938.

[175] K. Fuhrmann, R. Chini, L. Katerhandt & Z. Chen, Multiplicity Among Solar-Type Stars. *ApJ* 846 (2017), 139, 1–23.

[176] E. C. Pickering, Spectra of Bright Southern Stars. *ApJ* 6 (1897), 349–351.

[177] A. Tokovinin, *MSC Catalogue*: www.ctio.noao.edu/ atokovinin/stars/intro.html.

[178] P. Hurly & B. Warner, Area Scanner Observations of Close Visual Double Stars. II – Results for 153 Southern Stars. *MN* 202 (1983), 761–765.

[179] W. H. Smyth, A Cycle of Celestial Objects (Richmond: Willmann Bell, 1986).

[180] W. H. Smyth, *Sidereal Chromatics* (Cambridge: Cambridge University Press, 2010).

[181] F. Rica, Orbital Elements for BU 741 AB, STF 333 AB, BU 920 and R 207. *JDSO* 8 (2012), 127–139.

[182] K. Fuhrmann, R. Chini, L Kaderhandt, Z. Chen & R. Lachaume, Evidence for Very Near Hidden White Dwarfs. *MN* 459 (2016), 1682–1686.

[183] M. Zechmeister, M. Kürster, M. Endl, G. lo Curto, H. Hartman, H. Nilsson *et al.*, The Planet Search Programme at the ESO CES and HARPS. IV. The Search for Jupiter Analogues Around Solar-Like Stars. *A&A* 552 (2013), A78, 1–62.

[184] W. H. van den Bos, Measures of Double Stars Made with the 26.5- and 9-inch refractors of the Union Observatory at Johannesburg. *Annal. Leiden Sterren* 14 (1928), 1–143.

[185] W. D. Heintz, The Triple Star HJ 3556. *PASP* 91 (1979), 356–357.

[186] R. P. Sellors, Measures of Double Stars. *AN* 186 (1910), 65–72.

[187] R. T. A. Innes, New Double Stars Found at the Cape Observatory in 1906. *MN* 57 (1897), 533–541.

[188] S. Bailey, Catalogue of Southern Double Stars. *Ann. Harvard Obs.* 32 (1900), 2, 296–308.

[189] O. J. Eggen, Masses, Luminosities, Colours and Space Motion of 228 Visual Binaries. *AJ* 70 (1965), 19–93.

[190] O. J. Eggen, Stellar Groups. V – Luminosities, Motions and Masses of the Late-Type Subgiants. *MN* 120 (1960), 430–443.

[191] R. G. Aitken, A Double Star Problem. *PASP* 18 (1906), 70–73.

[192] W. J. Luyten, NLTT Cat. 1–4 (1979).

[193] G. M. Popovic, *Bull. Obs. Ast. Belgrade* 156 (1997), 189.

[194] T. Lewis, Measures of the Double Stars Contained in the *Mensurae Micrometricae* of F. G. W. Struve. *Mems. RAS* 56 (1906).

[195] R. Genet, H. Zirm, F. Rica *et al.*, Two New Triple Systems with Detectable Inner Orbital Motion and Speckle Interferometry of 40 Other Double Stars. *JDSO* 11 (2015), 200–213.

[196] H. Zirm & F. Rica Romero, *IAUDS* 183 (2014).

[197] L. C. Roberts, A. A. Tokovinin, W. I. Hartkopf & R. L. Riddle, Observation of Hierarchical Solar-Type Multiple Star Systems. *AJ* 150 (2015), 130, 1–7.

[198] B. Baillaud, Observation d'étoiles doubles. *Ann. Toulouse Obs.* 3 (1899), 77–141.

[199] R. P. Butler, S. S. Vogt, G. Laughlin, J. A. Burt, E. J. Rivera, M. Tuomi *et al.*, The LCES HIRES/Keck Precision Radial Velocity Exoplanet Survey. *AJ* 153 (2017), 208, 1–19.

[200] T. Boyarjian, K. von Braun, G. van Belle, H. A. McAlister, Th. ten Brummelaar, S. R. Kane *et al.*, Stellar Diameters and Temperatures II. Main Sequence K and M Stars. *ApJ* 757 (2012), 112, 1–31.

[201] B. D. Mason, W. I. Hartkopf & K. N. Miles, Binary Star Orbits. V – The Nearby White Dwarf/Red Dwarf Pair 40 Eri BC. *AJ* 154 (2017), 200, 1–9.

[202] W. Hartkopf & H. A. McAlister, Binary Stars Unresolved by Speckle Interferometry III. *PASP* 96 (1984), 105–116.

[203] E. Dembowski, Misure Micrometriche Stelle Doppie e Multiple Fatte negli Anni 1852–1878, Roma (1883).

[204] J. Percy, *BAAS* 7 (1975), 39.

[205] W. D. Heintz, A Study of Multiple Star Systems. *AJ* 111 (1996), 408–411.

[206] J. W. Davidson Jr., B. J. Baptista, E. P. Horch, O. Franz & W. van Altena, A Photometric Analysis of Seventeen Binary Stars Using Speckle Imaging. *AJ* 138 (2009), 1354–1365.

[207] O. Eggen, Three Color Photometry of the Components in 228 Wide Double and Multiple Systems. *AJ* 68 (1963), 483–514.

[208] J. S. Plaskett, The Spectroscopic Binary ψ Orionis. *ApJ* 28 (1909), 266–273.

[209] T. W. Webb, Celestial Objects for Common Telescopes, 6th edition (New York: Dover, 1962).

[210] A. Blazit, D. Bonneau, L. Koechlin & A. Labeyrie, The Digital Speckle Interferometer: Preliminary Results on 59 Stars. *ApJ* 214 (1977), L79–L84.

[211] R. F. Sanford, The Spectrographic Orbit of the Companion to Rigel. *ApJ* 95 (1942), 421–424.

[212] F. Galland, A.-M. Lagrange, S. Udry, A Chelli, F. Pepe, J.-L. Beuzit *et al.*, Extrasolar Planets and Brown Dwarfs Around A-F Type Stars II. A Planet Found with Elodie Around the F6V Star HD 33564. *A&A* 444 (2005), L21–L24.

[213] W. R. Dawes, Letter from Mr Dawes Announcing the Discovery of Two New Double Stars. *MN* 8 (1848), 53.

[214] W. Hartkopf, A. A. Tokovinin & B. D. Mason, Speckle Interferometry at SOAR in 2010 and 2011. *AJ* 143 (2012), 42, 1–19.

[215] W. Adams, The Orbit of the Spectroscopic Binary η Orionis. *ApJ* 17 (1903), 68–71.

[216] A. Veramendi & J. P. González, Spectroscopic Study of Early-Type Multiple Systems. I – Orbits of Spectroscopic Binary Subsystems. *A&A* 563 (2014), A138, 1–15.

[217] A. Alzner, *IAUDS* 170 (2010).

[218] R. T. A. Innes, Micrometrical Measurements of Double Stars 1849–1868 and 1899–1903. *Ann. Cape Obs.* 2 (1905), 4, 1–98.

[219] S. W. Burnham, Double Star Observations Made in 1879 and 1880 with the 18.5-inch Refractor of the Dearborn Observatory: I. Catalogue of 151 New Double Stars with Micrometric Measures. II. Micrometric Measures of 770 Double Stars. *Mems. RAS* 47 (1881), 167–324.

[220] J. Hartmann, Investigations on the Spectrum and Orbit of δ Orionis. *ApJ* 19 (1904), 268–286.

[221] H. Deslandres, *Comptes Rendus* 130 (1900), 379.

[222] W. D. Heintz, Micrometer Observations of Double Stars and New Pairs X. *ApJS* 44 (1980), 111–136.

[223] T. Shenar, L. Oskinova, W. R. Hammann, M. F. Corcoran, A. F. J. Moffat, H. Pablo *et al.*, A Coordinated X-ray and Optical Campaign of the Nearest Massive Eclipsing Binary δ Orionis. IV – A Multiwavelength Non-LTE Spectroscopic Analysis. *ApJ* 809 (2015), 135, 1–20.

[224] N. Richardson, A. D. Moffat, T. R. Gull, D. J. Lindler, D. R. Gies *et al.*, HST/STIS Ultraviolet Spectroscopy of the Components of the Massive Triple Star δ Orionis A. *ApJ* 808 (2015), 88, 1–6.

[225] J. Hartmann, Investigations on the Spectrum and Orbit of δ Orionis. *ApJ* 19 (1904), 268–286.

[226] C. Hummel, N. M. White, N . M. Elias II, A. R. Hajian & T E. Nordgren, ζ Orionis is a Double Star. *ApJ* 540 (2000), L91–L93.

[227] C. Hummel, Th. Rovinius, M.-F. Nieva, O. Stahl, G. van Belle, R. T. Zavala *et al.*, Dynamical Mass of the O-Type Supergiant in ζ Orionis A. *A&A* 554 (2013), A52, 1–7.

[228] M. Bakich, *Constellations* (Cambridge: Cambridge University Press, 1995).

[229] J. Krtička, Z. Mikulasek, T. Lufteye & M. Jagelka, Visual and Ultraviolet Flux Variability of the Bright CP Star θ Aurigae. *A&A* 576 (2015), A82, 1–11.

[230] J. Winzer, The Photometric Variability of the Peculiar A Stars. Unpublished PhD thesis, University of Toronto (1974).

[231] J. M. Gilliss, Auszug aus einem Briefe des Herrn Lieutenant J. M. Gilliss USN and den Prof. Gering in Marburg. *AN* 34 (1852), 337–340.

[232] J. B. Tatlock, *Sid. Mess.* 4 (1884), 18.

[233] www.aavso.org.

[234] A. Richichi & G. Calamai, Infrared High Angular Resolution Measures of Stellar Sources. VI – Accurate Angular Diameters of X Cnc, U Ori & η Geminorum. *A&A* 399 (2006), 275–278.

[235] D. B. McLaughlin & S. E. van Dyke, The Spectrographic Orbit and Light Variations of η Geminorum. *ApJ* 100 (1944), 63–68.

[236] R. Wasson, Speckle Interferometry with the OCA 22-inch Telescope. *JDSO* 14 (2018), 223–241.

[237] M. Scardia, J.-L. Prieur, L. Pansecchi, R. W. Argyle, A. Zanutta & E. Aristidi, Speckle Observations with PISCO in Merate (Italy) XVI: Astrometric Measurements on Visual Binaries in 2015 and New Orbits for DUN 5, ADS 5958, 6276, 7294, 8211 and 13169, *AN* 339 (2018), 571–585.

[238] S. W. Burnham, Measures of Proper Motion Stars Made with the 40-inch Refractor of the Yerkes Observatory in the Years 1907–1912. *Carnegie Inst. Washington* 168 (1913).

[239] R. S. Ball, In Search of Stars with an Annual Parallax. *Publ. Dunsink Obs.* 5 (1884), 1–157.

[240] H. A. McAlister, W. I. Hartkopf, J. R. Sowell, E. G. Dembrowski & O. G. Franz, ICCD Speckle Observations of Binary Stars. IV – Measurements During 1986–1988 from the Kitt Peak 4m Telescope. *AJ* 97 (1989), 510–531.

[241] C. Fabricius & V. Makarov, Two-colour Photometry for 9473 Components of Close Hipparcos Double and Multiple Stars. *A&A* 356 (2000), 141–145.

[242] E. A. Pluzhnik, Differential Photometry of Speckle-Interferometric Binary and Multiple Stars, *A&A* 431 (2005), 587–596.

[243] H. Zirm, *IAUDS* 179 (2013).

[244] S. W. Burnham, Measures of Proper Motion Stars Made with the 40-inch Refractor of the Yerkes Observatory in the Years 1907–1912. *Carnegie Inst. Washington* 168 (1913).

[245] S. W. Burnham, Double Star Observations Made in 1877–8 in Chicago with the 18.5-inch refractor of the Dearborn Observatory Comprising I. A Catalogue of 251 New Double Stars with Measures; II. Micrometrical Measures of 500 Double Stars. *Mems. RAS* 44 (1878), 141–305.

[246] H. A. McAlister, B. D. Mason, W. I. Hartkopf & M. M. Shara, ICCD Speckle Observations of Binary Stars. X. A Further Survey for Duplicity Among the Bright Stars. *AJ* 106 (1993), 1639–1655.

[247] H. C. Russell, Sydney Observatory Results 1871–81 (1882).

[248] S. Bailey, Catalogue of Southern Double Stars. *Ann. Harvard Obs.* 32 (1900), 2, 296–308.

[249] D. Trilling, J. Stansbury & K. R. Stapelfeldt, Debris Disks in Main-Sequence Binary Systems. *ApJ* 658 (2007), 1289–1311.

[250] J. J. Tobin, L. Hartmann, G. Fürucz & M. Mako, Kinematic and Spatial Substructure in NGC 2264. *AJ* 149 (2015), 119, 1–11.

[251] D. Gies, B. D. Mason, W. G. Bagnuolo Jr. *et al.*, The O-Type Binary 15 Monocerotis Near Periastron. *ApJ* 475 (1997), L49–L52.

[252] H. Sung, M. S. Bessell & S.-W. Lee, UBVRIHα Photometry of the Open Young Cluster NGC 2264. *AJ* 114 (1997), 2644–2657.

[253] E. J. Baxter, K. R. Covey, A. A. Muench, G. Furücz, L. Rebull, A. Szentgyorgi *et al.*, The Distance to NGC 2264. *AJ* 138 (2009), 963–975.

[254] H. A. McAlister, B. D. Mason, W. I. Hartkopf & M. M. Shara, ICCD Speckle Observations of Binary Stars. X. A Further Survey for Duplicity Among the Bright Stars. *AJ* 106 (1993), 1639–1655.

[255] G. P. Bond, On the Companion of Sirius. *AN* 57 (1862), 131–134.

[256] H. E. Bond, G. H. Schaefer, R. C. Gilliland, The Sirius System and its Astrophysical Puzzles: HST and Ground-Based Astrometry. *ApJ* 840 (2017), 70.

[257] F. W. Bessel, On the Variation of the Proper Motions of Procyon and Sirius. *MN* 6 (1844), 136–141.

[258] W. Huggins, Further Observations on the Spectra of Some of the Stars and Nebulae, with an Attempt to Determine Therefore Whether These Bodies are Moving Towards or From the Earth. *Phil Trans.* 158 (1868), 529–569.

[259] R. S. Ball, In Search of Stars with an Annual Parallax. *Publ. Dunsink Obs.* 5 (1884), 1–157.

[260] B. D. Mason, W. I. Hartkopf, G. L. Wycoff & E. R. Holdenreid, Speckle Interferometry at the US Naval Observatory – XII. *AJ* 132 (2006), 2219–2231.

[261] A. A. Tokovinin & S. Lépine, Wide Companions to Hipparcos Stars. *AJ* 144 (2012), 102, 1–12.

[262] H. A. McAlister, B. D. Mason, W. I. Hartkopf & M. M. Shara, ICCD Speckle Observations of Binary Stars. X. A Further Survey for Duplicity Among the Bright Stars. *AJ* 106 (1993), 1639–1665.

[263] P. de Cat & C. Aerts, A Study of Bright Southern Slowly Pulsating B Stars. II – The Intrinsic Frequencies. *A&A* 393 (2002), 965–981.

[264] R. Gould, *Southern Astronomy* 52, March/April 1994.

[265] A. V. Kazarovets, N. N. Samus, O. V. Durlevich, M. S. Frolov, S. V. Antipin, N. N. Kireeva *et al.* The 74th Special Name-List of Variable Stars. *IBVS* 4659 (1999), 1–27.

[266] R. Jaworski, www.skyandtelescope.com/observing /celestial-objects-to-watch/southern-double-star-gems/, 2006.

[267] S. Haas, *Double Stars for Small Telescopes* (Cambridge: Sky Publishing Corporation, 2006).

[268] G. Stone, Star Splitters website: https:bestdoubles .wordpress.com.

[269] W. Hartkopf, Catalogue of rectilinear elements: www.usno .navy.mil/USBO/astrometry/optical-IT-prod/wds/lin1.

[270] E. J. Shaya & R. Olling, Very Wide Binaries and Other Comoving Stellar Companions: A Bayesian Analysis of the *Hipparcos Catalog*. *ApJS* 192 (2011), 2, 1–17.

[271] C. Bailer-Jones, Close Encounters of the Stellar Kind. *A&A* 575 (2015), A35, 1–13.

[272] H. A. Abt, The Frequency of Binaries Among Normal A-Type Stars. *ApJS* 11 (1965), 429–460.

[273] S. D. Tremaine, E. J. Groth & M. R. Nelson, Observation of a Lunar Occultation of δ Geminorum. *AJ* 79 (1974), 649–650.

[274] The Hipparcos and Tycho catalogues (ESA SP-1200) (ESA Noordwijk, 1997).

[275] L. E. Pasinelli-Fracassini, L. Pastori, S. Covino & A. Pozzi, Catalogue of Apparent Diameters and Absolute Radii of Stars (CADARS), 3rd edition, Comments and Statistics. *A&A* 367 (2001), 521–524.

[276] J. Nanson, Star Splitters website: `https:bestdoubles.wordpress.com`.

[277] B. H. Dawson, Medidas Micrométricas de Estrellas Dobles. *Publ. La Plata* 6 (1937), 85–127.

[278] W. H. van den Bos, Measures of Double Stars – 10th Series. *Union Obs. Circ.* 80 (1929), 59.

[279] F. van Leeuwen, Parallaxes and Proper Motions for 20 Open Clusters as Based on the New Hipparcos Catalogue. *A&A* 497 (2009), 209–242.

[280] G. Torres & I. Ribes, Absolute Dimensions of the M-Type Eclipsing Binary YY Geminorum (Castor C): A Challenge to Evolutionary Models in the Lower Main Sequence. *AJ* 567 (2002), 1140–1165.

[281] J. Butler, N. Erhan, E. Budding, J. G. Doyle, B. Foing, G. E. Bromage *et al.*, A Multi-Wavelength Study of the M Dwarf Binary YY Geminorum. *MN* 446 (2015), 4205–4219.

[282] A. Bélopolsky, On the Spectroscopic Binary α^1 Geminorum. *ApJ* 5 (1897), 1–7.

[283] H. D. Curtis, The System of Castor. *ApJ* 23 (1906), 351–369.

[284] A. Richichi & I. Percheron, First Results From the ESO VLTI Calibration Programme. *A&A* 434 (2005), 1201–1209.

[285] W. S. Adams & A. Joy, The Spectrum of the Companion to Castor and of W.B. 16^h 908. *PASP* 32 (1920), 158.

[286] A. Joy & R. F. Sanford, The Dwarf Companion to Castor as a Spectroscopic Binary and an Eclipsing Variable. *ApJ* 64 (1926), 250–257.

[287] J. Herschel, Micrometrical Observations of 364 Double Stars with a 7-Foot Equatorial Achromatic Telescope, Taken at Slough in the Years 1828, 1829 and 1830. *Mems. RAS* 5 (1833), 13–90.

[288] H. Zirm, *IAUDS* 185 (2015).

[289] B. Gary, `brucegary.net/yygem`.

[290] H. E. Bond, R. L. Gilliland, G. H. Schaefer, P. Demarque, T. M. Girard, J. B. Holbert *et al.*, Hubble Space Telescope Astrometry of the Procyon System. *ApJ* 813 (2015), 106, 1–19.

[291] G. Gatewood & I. Han, An Astrometric Study of Procyon. *AJ* 131 (2016), 1015–1021.

[292] C. Worley, Micrometer Measures of 2589 Double Stars, *USNO Publ.* 25 (1986), 3, 1–211.

[293] J. Schaeberle, Discovery of the Companion to Procyon. *PASP* 8 (1896), 314.

[294] D. L. Nidever, G. W. Marcy, R. P. Butler, D. A. Fischer & S. S. Vogt, Radial Velocities for 889 Late-Type Stars. *ApJS* 141 (2002), 503–522.

[295] J. H. Moore, Twenty Three Stars Whose Radial Velocities Vary. *Lick Obs. Bull.* 6 (1911), 150–152.

[296] C. A. Hernández & J. Sahade, The Spectroscopic Binary γ_1 Velorum. *PASP* 92 (1980), 819–824.

[297] A. Tokovinin, A. Chalabaev, N. I. Shatsky & J. L. Beuzit, A Near IR Adaptive Optics Search for Faint Companions to Early-Type Multiple Stars. *A&A* 346 (1999), 481–486.

[298] F. Millour, R. G. Petrov, O. Chesnau, D. Bonneau, L. Dessart, C. Bechet *et al.*, Direct Constraints on the Distance of γ^2 Velorum from AMBER/VLTI Observations. *A&A* 464 (2007), 107–118.

[299] J. R. North, P. G. Tuthill, W. J. Tango & J. Davis, γ^2 Velorum: Orbital Solution and Fundamental Parameters Determined with SUSI. *MN* 377 (2007), 415–424.

[300] R. D. Jeffries, T. Naylor, F. M. Walter, M. P. Bozzo & C. R. Devey, The Stellar Association Around Gamma Velorum and its Relationship with Vela OB2. *MN* 393 (2009), 538.

[301] B. D. Mason, ICCD Speckle Observations of Binary Stars. XIX. An Astrometric and Spectroscopic Survey of O Stars. *AJ* 115 (1998) 821–847.

[302] R. T. A. Innes, *Southern Double Star Catalogue* (Johannesburg: Union Observatory 1927).

[303] R. A. Rossiter, Catalogue of Southern Double Stars, *Publ. Univ. Michigan Obs.*, 11 (1955), 1.

[304] S. W. Burnham, Invisible Double Stars. *MN* 51 (1891), 388–395.

[305] O. Struve, *Comptes Rendus* 79 (1874), 1463.

[306] R. F. Griffin, Spectroscopic Binary Orbits from Photoelectric Radial Velocities Paper 150: ζ Cnc C. *Observatory* 120 (2000), 1–47.

[307] A. Richichi, An Investigation of the Multiple Star ζ Cnc by a Lunar Occultation. *A&A* 364 (2000), 225.

[308] J. B. Hutchings & R. F. Griffin, *PASP* 112 (2000), 833.

[309] W. D. Heintz, A Study of Multiple Star Systems. *AJ* 111 (1996), 408–411.

[310] R. W. Argyle, A. Alzner & E. Horch, Orbits for Five Southern Visual Binaries. *A&A* 384 (2002), 171–179.

[311] W. J. Tango, J. Davis, R. J. Thompson & R. Hanbury Brown, A "Narrabri" Binary Star Resolved by Speckle Interferometry. *Proc. Astron. Soc. Australia* 3 (1979), 323–324.

[312] S. Otero, P. Fieseler & C. Lloyd, Delta Velorum is an Eclipsing Binary. *IBVS* 4999 (2000), 1–4.

[313] R. T. A. Innes, A List of Probably New Double Stars. *MN* 55 (1895), 312–315.

[314] P. Kervella, T. Mérand, M. G. Petr-Gotzens, T. Pribulla & F. Thevenin, The Nearby Eclipsing Stellar System δ Velorum. IV – Differential Astrometry with VLT/NACO at the 100 Microarcsecond Level. *A&A* 552 (2013), A18.

[315] A. Massarotti, D. W. Latham & R. P. Stefanik, Rotational and Radial Velocities for a Sample of 761 Hipparcos Giants and the Role of Binarity. *AJ* 135 (2008), 209–231.

[316] J. Gledhill, Double Stars for April. *Observatory* 2 (1879), 415–416.

[317] A. Underhill, Radial Velocity Observations of Eight Short-Period Visual Binaries. *PDAO* 12 (1967), 159–171.

[318] J. D. Drummond, Binary Stars Observed with Adaptive Optics at the Starfire Optical Range. *AJ* 147 (2014), 65, 1–10.

[319] R. Sanford, On the Orbits of Four Spectroscopic Binaries. *ApJ* 64 (1926), 172–193.

[320] E. Crossley, J. Gledhill & J. M. Wilson, *A Handbook of Double Stars* (London: Macmillan, 1879).

[321] H. A. Abt & S. G. Levy, Visual Multiples. I. ADS 7251. *AJ* 78 (1978), 1093–1095.

[322] C. Morbey & R. F. Griffin, On the Reality of Certain Spectroscopic Orbits. *ApJ* 317 (1987), 343–352.

[323] K. Ward-Duong, J. Patience, R. J. de Rosa, J. Bulger, A. Rajan, S. P. Goodwin *et al.*, The M Dwarfs in Multiples (MINMS) Survey. I – Stellar Multiplicity Among Low-Mass Stars within 15 Parsecs. *MN* 449 (2015), 2618–2637.

[324] J. Gagné, P. Plavchan, P. Gao, G. Anglaa-Escudé, E. Furlan, C. Davison *et al.*, A High Precision Near-Infrared Survey for Radial Velocity Low-Mass Stars using CSHELL and a Methane Gas Cell. *ApJ* 822 (2016), 1–24.

[325] K. Chang, Parallax, Proper Motion and Orbital Motion of the Visual Binary Σ1321. *AJ* 77 (1972), 759–761.

[326] Cited by Burnham (1913) [238].

[327] T. Espin, Micrometric Measures of Double Stars (Fifth Series). *MN* 68 (1908), 202–209.

[328] B. A. Skiff & G. Gatewood, The Variability of Solar-Type Stars. V. The Standard Stars 10 and 11 Leo Minoris. *PASP* 98 (1986), 338.

[329] W. D. Heintz, Orbits of Twenty Visual Binaries. *A&AS* 72 (1988), 543–549.

[330] R. J. de Rosa, J. Bulger, J. Patience, B. Leland, B. Macintish, A. Schneider *et al.*, The Volume Limited A-Star (VAST) Survey. I – Companions and the Unexpected X-Ray Detection of B6–A7 Stars. *MN* 415 (2011), 854–866.

[331] H. A. McAlister, B. D. Mason, W. I. Hartkopf & M. M. Shara, ICCD Speckle Observations of Binary Stars. X. A Further Survey for Duplicity Among the Bright Stars. *AJ* 106 (1993), 1639–1655.

[332] W. I. Hartkopf, B. D. Mason & Th. J. Rafferty, Speckle Interferometry at the USNO Flagstaff Station: Observations obtained in 2003–2004 and 17 New Orbits. *AJ* 135 (2008), 1334–1342.

[333] R. S. Ball, In Search of Stars with an Annual Parallax. *Publ. Dunsink Obs.* 5 (1884), 1–157

[334] S. Arend, Measures Micrométriques d'Étoiles Doubles Visuelles Effectuées a l'Equatorial de 45 cm. *Ann. R. Obs. Belgique, Series 3* (1963), 9, 93–103.

[335] M. Muterspaugh, W. I. Hartkopf, B. F. Lane, J. O'Connell, M. Willamson, S. Kulkarni *et al.* The PHASES Differential Astrometry Archive II. Updated Binary Star Orbits and a Long-Period Eclipsing Binary. *AJ* 140 (2010), 1623–1630.

[336] R. Copeland, An Account of Some Recent Astronomical Experiments at High Elevation in the Andes. *Copernicus* 3 (1884), 193–231.

[337] K. Fuhrmann & R. Chini, Multiplicity Among F-Type Stars. *ApJS* 203 (2012), 20, 1–20.

[338] L. E. Pasinelli-Fracassini, L. Pastori, S. Covino & A. Pozzi, Catalogue of Apparent Diameters and Absolute Radii of Stars (CADARS), 3rd edition, Comments and Statistics. *A&A* 367 (2001), 521–524.

[339] C. Mandrini & V. Niemela, On the Visual Binary υ Carinae. *MN* 223 (1986), 79–85.

[340] B. Warner, The Double Star Discoveries of Fearon Fallows. *MNASSA* 36 (1977), 134–135.

[341] R. Letchford, G. L. White & A. D. Ernest, The Southern Double Stars of Carl Rumker I: History, Identification, Accuracy. *JDSO* 13 (2017), 220–232.

[342] W. R. Dawes, New Double Stars Discovered by Mr. Alvan Clark, Boston, US; with Appended Remarks. *MN* 17 (1857), 257–259.

[343] J. A. Zaera, *IAUDS* 93 (1984).

[344] I. Han, B. C. Lee, K. M. Kim, D. E. Mrkrtichian, A. P. Hatzes, G. Valyavin *et al.*, Detection of a Planetary Companion Around the Giant Star γ^1 Leonis, *A&A* 509 (2010), A24, 1–5.

[345] D. Reuyl, The Variable Proper Motion of the Star Cincinnati 1244. *ApJ* 97 (1943), 186–189.

[346] S. L. Lippincott, Astrometric Studies of +20° 2465 from Photographs Taken with the 24-inch Sproul Refractor. *AJ* 74 (1969), 224–228.

[347] Y. Y. Balega, D. Bonneau & R. Foy, Speckle Interferometric Measurements of Binary Stars. II. *A&AS* 57 (1984), 31–36.

[348] L. Romanenko & A. Kisselev, *Astr. Rep.* 58 (2014), 30.

[349] M. Veramendi & J. F. Gonzalez, Spectroscopic Study of Early-Type Multiple Stellar Systems – II. New Binary Subsystems. *A&A* 567 (2014), A35, 1–10.

[350] R. Sanford, The Orbit of the Spectroscopic Binary ϵ Volantis. *Lick Obs. Bull.*, 8 (1914), 127–130.

[351] R. Chini, A Spectroscopic Survey on the Multiplicity of High Mass Stars. *MN* 424 (2012), 1925–1929.

[352] M. M. Kiminki, M. Reiter & N. Smith, Ancient Eruption of η Carinae: A Tale Written in Proper Motions. *MN* 463 (2016), 845–857.

[353] G. Weigelt & J. Ebersberger, Eta Carinae Resolved by Speckle Interferometry. *A&A* 163 (1986), L5–L6.

[354] T. R. Ayres, R. A. Osten & A. Brown, The Rise and Fall of μ Velorum: A Remarkable Flare on a Yellow Giant Star Observed with Extreme Ultraviolet Explorer. *ApJ* 526 (1999), 445–450.

[355] R. R. Shobbrook, UBV(RIC)$_c$ Observations for 13 Bright Cepheids. *MN* 255 (1992), 486.

[356] J. C. Becker, J. A. Johnson, A. Vanderburg & L. T. Morton, Extracting Radial Velocities of A- and B-Type Stars from Echelle Spectrograph Calibration Data. *ApJS* 217 (2015), 29, 1–13.

[357] A. Slettebak, The Spectra and Axial Rotational Velocities of the Components of 116 Visual Double Star Systems. *ApJ* 138 (1963), 118–139.

[358] R. F. Griffin, Spectroscopic Binary Orbits from Photoelectric Radial Velocities. Paper 142: ξ Ursae Majoris. *Observatory* 118 (1998), 273–298.

[359] N. Norlund, Determination de l'orbite de ξ Ursae Majoris. *AN* 170 (1905), 117–132.

[360] W. Campbell & W. Wright, A List of Nine Stars Where Velocities in the Line of Sight are Variable. *ApJ* 12 (1900), 254–257.

[361] B. D. Mason, H. McAlister, W. I. Hartkopf & M. M. Shara, Binary Star Orbits from Speckle Interferometry. VII – The Multiple System ξ Ursae Majoris. *AJ* 109 (1995), 332–340.

[362] E. L. Wright, M. F. Skrutskie, J. D. Kirkpatrick, C. R. Gelino, R. L. Griffith, K. A. Marsh *et al.*, A T8.5 Brown Dwarf Member of the ξ Ursae Majoris System. *AJ* 145 (2013), 84, 1–8.

[363] W. Herschel, Account of the Changes That Have Happened, During the Last Twenty-Five Years, in the Relative Situation of Double Stars, With an Investigation of the Cause to Which They Are Owing. *Phil. Trans.* 93 (1803), 339–382.

[364] F. Savary, *Conn. du Temps pour l'an 1830* (1830).

[365] D. Bonneau, *Observations y Travaux*, 52 (2000), 8.

[366] W. H. van den Bos & D. Kgl. Danske. *Naturvidensk. og Math Afd. ser 8* (1928), 12, 293.

[367] W. Hartkopf, *IAUDS* 181 (2013), 1.

[368] J.-L. Prieur, M. Scardia, L. Pansecchi, R. W. Argyle & M. Sala, Speckle Observations with PISCO in Merate: XI Astrometric Measurements of Visual Binaries in 2010. *MN* 422 (2012), 1057.

[369] H. A. Abt & S. G. Levy, Multiplicity Among Solar-Type Stars. *A&AS* 30 (1976), 273–306.

[370] S. Söderhjelm, Visual Binary Orbits and Masses Post-Hipparcos. *A&AS* 341 (1999), 121–140.

[371] T. Pribulla, D. Chochol, H. Rovithis-Livanion & P. Rovithis, The Contact Binary AW UMa as a Member of a Multiple System. *A&A* 345 (1999), 137.

[372] A. Gimenez & J. A. Quesada, Photoelectric Photometry of 65 UMa (HR 4568). *IBVS* 1648 (1979), 1–4.

[373] P. Zasche, R. Uhlárr, M. Slechta, M. Wolf, P. Harmanec, J. A. Nemravová *et al.*, Unique Sextuple System: 65 Ursae Majoris. *A&A* 542 (2012), A78, 1–6.

[374] Y. Y. Balega, V. V. Dyachenko, A. F. Maksimov, E. V. Malogolovets & D. A. Rasteyev, Speckle Interferometry of Magnetic Stars with the BTA. 1. First Results. *Astr. Bull.* 67 (2012), 44–56.

[375] P. P. Eggleton & A. A. Tokovinin, A Catalogue of Multiplicity Among Bright Stellar Systems. *MN* 389 (2008), 869–879.

[376] A. Giménez & J. A. Quesada, The Period of DN UMa (HR 4560): A Bright Eclipsing Binary. *IBVS* 2068 (1982), 1–3.

[377] R. Letchford, G. L. White & A. D. Ernest, The Southern Double Stars of Carl Rumker I: History, Identification, Accuracy. *JDSO* 13 (2017), 220–232.

[378] R. Jaworski, www.skyandtelescope.com/observing/celestial-objects-to-watch/southern-double-star-gems/, 2006.

[379] C. A. Guerrero, V. G. Orlov, M. A. Monroy-Rodriguez & M. Borges Fernandes, Stellar Multiplicity of the Open Cluster Melotte 111. *AJ* 150 (2015), 16, 1–9.

[380] D. Olevic, *IAUDS* 141 (2000).

[381] N. Shatsky & A. Tokovinin, The Mass-Ratio Distribution of B-Type Visual Binaries in the Sco–Cen Association. *A&A* 382 (2002), 92–103.

[382] R. Anton, Double Star Measurements at the International Amateur Sternwarte (IAS) in Namibia in 2007. *JDSO* 4 (2008), 40–51.

[383] A. D. Thackeray & P. Hill, The System of α Crucis. *MN* 165 (1974), 55–59.

[384] F. J. Neubauer, The Radial Velocities of Nineteen Stars. *Lick Obs. Bull.* 15 (1932), 190–193.

[385] J. H. Moore, Six Stars Where Radial Velocities Vary. *Lick Obs. Bull.* 9 (1916), 37–38.

[386] J. Kaler, stars.astro.illinois.edu/sow/sowlist.html.

[387] C. Hernández & E. B. de Hernández, The Orbital Elements of 25 G Crucis (HD 108250). *Revista Mexicana* 4 (1979), 297–300.

[388] K. Murdoch & M. Clark, The Radial Velocity Variability of Gamma Crucis, *MN* 254 (1992), 27–29.

[389] M. Petit, Catalogue des Etoiles Variables ou Suspectes dans le Voisinage du Soleil. *A&AS* 85 (1990), 971–987.

[390] J. Kaler, stars.astro.illinois.edu/sow/sowlist.html.

[391] W. Adams, Seven Spectroscopic Binaries. *PASP* 27 (1915), 132.

[392] R. M. Petrie, The Spectrographic Orbits of HD 109510. *PDAO* 6 (1937), 365–368.

[393] A. Richichi, I Percheron & M. Khristoforova, CHARM2: An Updated Catalog of High Angular Resolution Measurements. *A&A* 431 (2005), 773–777.

[394] R. L. Riddle, A. A. Tokovinin, B. D. Mason, W. I. Hartkopf, L. C. Roberts, C. Baranec, *et al.*, A Survey of the High Order Multiplicity of Nearby Solar-Type Binary Stars with ROBO-AO. *ApJ* 799 (2015), 4, 1–21.

[395] E. Horch, D. Falta, L. M. Anderson, M. D. DeSousa, C. M. Minter, T. Ahmed *et al.*, CCD Speckle Observations of Binary Stars with the WIYN Telescope. VI. Measures During 2007–2008. *AJ* 139 (2010), 205–215.

[396] J. Lequeux (private communication), 2012.

[397] S. Richaud, *L'Histoire de l'Academie Royale des Sciences* (Paris, 1733).

[398] N. Law (private communication), 2008.

[399] M. Scardia, R. W. Argyle , J.-L. Prieur, L. Pansecchi, S. Basso, N. M. Law *et al.*, The Orbit of the Visual Binary ADS 8630 (γ Vir). *AN* 328 (2007), 146–153.

[400] A. Rizzutto, M. J. Ireland, J. G. Robertson, Y. Kok, P. G. Tuthill, B. A. Warrington *et al.*, Long Baseline Interferometric Multiplicity Survey of the Sco–Cen OB Association. *MN* 436 (2013), 1694–1707.

[401] J. L. Halbwachs, M. Mayor & S. Udry, Double Stars with Wide Separations in the AGK3 – I. Components That Are Themselves Spectroscopic Binaries. *MN* 422 (2012), 14–24.

[402] J. Le Beau, *Observations y Travaux* 22 (1990), 39.

[403] J. D. Drummond, Binary Stars Observed with Adaptive Optics at the Starfire Optical Range. *AJ* 147 (2014), 65, 1–10.

[404] J. Perez, www.perezmedia.net/beltofvenus.

[405] S. Jakate, A Search for Beta Cephei Stars. III – Photometric Studies of Southern B-Type Stars. *AJ* 84 (1979), 552–558.

[406] A. A. Tokovinin & S. Lépine, Wide Companions to Hipparcos Stars within 67 pc of the Sun. *AJ* 144 (2012), 102, 1–12.

[407] M. Stupar, Q. A. Parker & M. D. Filopovic, The Optical Emission Nebulae in the Vicinity of WR48 (θ Muscae): The Wolf–Rayet Ejecta and Unconnected Supernova Remnant. *MN* 401 (2010), 1760–1769.

[408] R. Sota, T. Maíz-Appelániz, N. I. Morrell, R. H. Barbá, N. R. Walborn, R. C. Gamen *et al.*, The Galactic O-Star Spectroscopic Survey (GOSSS). II. Bright Southern Stars. *ApJS* 211 (2014), 10, 1–84.

[409] R. Chini, V. H. Hoffmeister, A. Nasseri, O. Stahl & H. Zinnecker, A Spectroscopic Survey on the Multiplicity of High Mass Stars. *MN* 424 (2012), 1925–1929.

[410] B. D. Mason, D. R. Gies, W. I. Hartkopf, W. J. Bagnuolo, Th. ten Brummelaar, H. A. McAlister *et al.*, ICCD Speckle Observations of Binary Stars. XIX An Astrometric/Spectroscopic Survey of O stars. *AJ* 115 (1998), 821–847.

[411] W. I. Hartkopf, B. D. Mason, D. R. Gies, Th. ten Brummelaar, H. McAlister, A. F. J. Moffat *et al.*, ICCD Speckle Observations of Binary Stars. XXII. A Survey of Wolf–Rayet Stars for Close Visual Companions. *AJ* 118 (1999), 509–514.

[412] A. F. J. Moffatt & W. Seggewiss, The Wolf–Rayet Binary Theta Muscae. *A&A* 54 (1977), 607–616.

[413] S. Zhekov, T. Tomov, M. P. Gawrónski, L. N. Georgiev, J. Borissova, R. Kurtev *et al.*, A Multi-Wavelength View on the Dusty Wolf–Rayet Star WR 48a. *MN* 445 (2014), 1663–1678.

[414] H. Sana, J.-P. Le Bouguin, S. Lacour, J.-P. Berger, G. Duvert, L. Gauchet *et al.*, Southern Massive Stars at High Angular Resolution: Observational Campaign and Companion Detection. *ApJS* 215 (2014), 15, 1–35.

[415] W. I. Hartkopf & H. A. McAlister, Binary Star Orbits from Speckle Interferometry II. Combined Visual/Spectroscopic Orbits of 28 Close Systems. *AJ* 98 (1989), 1014.

[416] M. Muterspaugh, M. Wijngaarden, H. F. Henrichs, B. F. Lane, W. I. Hartkopf, G. W. Henry *et al.*, Predicting the α Comae Berenices Time of Eclipse: How 3 Ambiguous Measurements out of 609 Caused a 26 Year Binary's Eclipse to be Missed. *AJ* 150 (2015), 140, 1–4.

[417] N. Shatsky & A. Tokovinin, The Mass-Ratio Distribution of B-type Visual Binaries in the Sco OB2 Association. *A&A* 382 (2002), 92–103.

[418] M. Veramendi & J. F. González, Spectroscopic Study of Early-Type Multiple Stellar Systems – II. New Binary Subsystems. *A&A* 567 (2014), A35, 1–10.

[419] A. Tokovinin, B. D. Mason & W. I. Hartkopf, Speckle Interferometry at the Blanco and SOAR Telescopes in 2008 and 2009. *AJ* 139 (2010), 743–756.

[420] A. Tokovinin, B. D. Mason & W. I. Hartkopf, Speckle Interferometry at SOAR in 2012 and 2013. AJ 147 (2014), 123, 1–12.

[421] W. Finsen, Measures of Double Stars – 21st Series. *Union Obs. Circ.* 96 (1936), 263–296.

[422] L. Ondra, A New View of Mizar. *Sky & Telescope* 108 (2004), 72–75.

[423] E. C. Pickering, On the Spectrum of ζ UMa. *Observatory* 13 (1890), 80–81.

[424] F. G. Pease, Interferometer Notes III: Measurement of the Spectroscopic Binary Mizar. *PASP* 37 (1925), 155.

[425] F. Gutmann, The Spectroscopic Orbit of ζ^1 Ursae Majoris, *PDAO* 12 (1965), 361–375.

[426] C. Hummel, J. T. Armstrong, D. F. Buscher *et al.*, Orbits of Small Angular Scale Binaries Resolved with the Mark III Interferometer. *AJ* 110 (1995), 376–390.

[427] E. Mamajek, M. A. Kenworthy, P. M. Hinz & M. R. Meyer, Discovery of a Faint Companion to Alcor Using MMT/AO 5μm Imaging. *AJ* 139 (2010), 919–925.

[428] N. Zimmermann, B. Oppenheimer, S. Hinkley, D. Brenner, I. R. Parry, A. Sivaramakrishnan *et al.*, Parallactic Motion for Companion Discovery: An M-Dwarf Orbiting Alcor. *ApJ* 709 (2010), 733–740.

[429] G. Gontcharev & O. Kiyaeva, Photocentric Orbits from a Direct Combination of Ground-Based Astrometry with Hipparcos. II. Preliminary Orbits for Six Astrometric Binaries. *New Astronomy* 15 (2010), 324–331.

[430] R. J. de Rosa, J. R. Patience, P. A. Wilson, A. Schneider, S. J. Wiktorowicz, A. Vigan *et al.*, The VAST Survey. III – The Multiplicity of A-Type Stars within 75 Pc. *MN* 437 (2014), 1216–1240.

[431] J. Plaskett, W. E. Harper, R. K. Young & H. H. Plaskett, The Radial Velocities of 594 Stars. *PDAO* 2 (1921), 3–128.

[432] G. Bakos, Spectroscopic Orbital Elements of κ^2 Bootis. *AJ* 91 (1986), 1416–1417.

[433] O. Kiyaeva, Astrometric Study of the Triple Star ADS 9173, *SvaL* 32 (2006), 836–844.

[434] H. A. Abt, Catalogue of Individual Radial Velocities, 0^h–12^h, Measured by Astronomers of the Mount Wilson Observatory. *ApJS* 19 (1970), 387–505.

[435] H. A. Abt, Catalogue of Individual Radial Velocities, 12^h–24^h, Measured by Astronomers of the Mount Wilson Observatory. *ApJS* 23 (1973), 365–489.

[436] A. B. Schultz, H. M. Hart, J. M. Hershey, F. C. Hamilton, M. Kochte, F. C. Bruhweiler *et al.*, A Possible Companion to Proxima Centauri. *AJ* 115 (1998), 345–350.

[437] P. Kervella, Close Stellar Conjunction of α Centauri A and B Until 2050. An M_k = 7.8 Star May Enter the Einstein Ring of α Centauri A in 2028. *A&A* 594 (2016), A107, 1–15.

[438] H. L. Alden, Alpha and Proxima Centauri. *AJ* 39 (1928), 20–23.

[439] G. Benedict, B. McArthur, D. W. Chappell, E. Nelan, W. H. Jeffreys, W. van Altena *et al.*, Interferometric Astrometry of Proxima Centauri and Barnard's Star Using Hubble Space Telescope Fine Guidance Sensor: 3. Detection limits for Substellar Companions. *AJ* 118 (1999), 1086–1100.

[440] G. Anglada-Escudé, P. J. Amado, J. Barnes, Z. M. Berdiñas, R. P. Butler, G. A. L. Coleman *et al.*, A Terrestrial Planet Candidate in a Temperate Orbit Around Proxima Centauri. *Nature* 536 (2016), 437.

[441] M. Beech, *Alpha Centauri: Unveiling the Secrets of Our Nearest Stellar Neighbor* (Springer: New York, 2015).

[442] I. Glass, *Proxima: The Nearest Star (Other Than the Sun!)* (Cape Town: Mons Mensa, 2008).

[443] R. T. A. Innes, A Faint Star of Large Proper Motion. *Union Obs. Circ.* 30 (1915), 235–236.

[444] P. Kervella, F. Thévenin, & C. Loris, Proxima's Orbit Around Alpha Centauri. *A&A* 598 (2017), L7, 1–7.

[445] J. South & J. F. W. Herschel, Observations of the Apparent Distance and Position of 380 Double and Triple Stars, Made in the Years 1821, 1822 and 1823, and compared with Those of Other Astronomers... *Phil. Trans.* 114 (1824), 1–412.

[446] M. Muterspaugh, W. I. Hartkopf, B. F. Lane, J. O'Connell, M. Willamson, S. Kulkarni *et al.*, The PHASES Differential Astrometry Archive II. Updated Binary Star Orbits and a Long-Period Eclipsing Binary. *AJ* 140 (2010), 1623–1630.

[447] M. Scardia, *IAUDS* 163 (2007).

[448] J. Herschel, *Mems. RAS* 5 (1826), 46.

[449] B. Wood & J. L. Linset, Resolving the ξ Bootis Binary with Chandra, and Revealing the Spectral Type Dependence of the Coronal 'FIP Effect'. *ApJ* 717 (2010), 1279–1290.

[450] K. Strand, The Triple System μ Draconis. *PASP* 55 (1943), 26–27.

[451] A. Howard & B. J. Fulton, Limits on Planetary Companions from Doppler Surveys of Nearby Stars. *PASP* 128 (2016), 114401–114401.

[452] A. Clerke, An Historical and Descriptive List of Some Double Stars Suspected to Vary in Light. *Nature* 39 (1888), 55–58.

[453] J. Percy, *Understanding Variable Stars* (Cambridge: Cambridge University Press, 2007).

[454] W. S. Adams, A. H. Joy, G Strömberg & C. G. Burwell, The Parallaxes of 1646 Stars Derived by the Spectroscopic Method. *ApJ* 53 (1921), 13–94.

[455] J. Schilt, Two New Variable Stars of the Type of W Ursae Majoris. *ApJ* 64 (1926), 215.

[456] J. M. Saxton, Recent Observations of 44 Bootis. *BAA Variable Star Section Circ.* 91 (1997), 11–16.

[457] I. Pustylnik, P. Kalv, T. Aas, V. Harvig & M. Mars, Light-Time Effects in Selected Semi-Detached and Contact Binaries with an Observed Third Component. In *The Light-Time Effect in Astrophysics. Causes and Cures of the O-C Diagram*, ed. C. Sterken, ASP Conference Series 335 (San Francisco: PASP, 2005), pp. 321–332.

[458] C. Nitschelm, Discovery and Confirmation of Some Double-Lined Spectroscopic Binaries in the Sco–Cen Complex. In *Spectroscopically and Spatially Resolving the Components of Close Binary Stars*, eds. R. W. Hilditch, H. Hensberge & K. Pavlovski, ASP Conference Series 318 (San Francisco: PASP, 2004), pp. 291–293.

[459] C. Nitschelm & M. David, An Investigation on Close Binaries in the Sco–Cen complex. In *Evolution of Compact Binaries*, eds. L. Schmidtobreick, M. R. Schreiber & C. Tappert, ASP Conference Series 447 (San Francisco: PASP, 2011), pp. 75–80.

[460] W. Buscombe & Stockley, *AJ* 37 (1975), 197.

[461] S. Hubrig, D. Le Mignant, P. North & J. Krautter, Search for Low-Mass Prestellar Companions around X-Ray Selected Late B-Stars. *A&A* 372 (2001), 152–164.

[462] M. Veramendi & J. F. Gonzalez, Spectroscopic Study of Early-Type Multiple Stellar Systems – II. New Binary Subsystems. *A&A* 567 (2014), A35, 1–10.

[463] A. Rizzutto, M. J. Ireland, J. G. Robertson, Y. Kok, P. G. Tuthill, B. A. Warrington *et al.*, Long Baseline Interferometric Multiplicity Survey of the Sco–Cen OB Association. *MN* 436 (2013), 1694–1707.

[464] A. D. Thackeray, The Double-Lined Spectroscopic Binary ϵ Lupi (HD 136504). *MN* 149 (1970), 75–80.

[465] K. Uytterhoeven, P. Harmanec, J. H. Telting & C. Aerts, The Orbit of the Close Spectroscopic Binary ϵ Lupi and the Intrinsic Variability of Its Early B-type Components. *A&A* 440 (2005), 249–260.

[466] J. D. Kirkpatrick, C. C. Dahn, D. G. Monet, I. N. Reid, J. E. Gizis, J. Liebert *et al.*, Brown Dwarf Companions to G0-Type Stars. I – Gliese 417B and Gliese 584C. *AJ* 121 (2001), 3235–3253.

[467] S. Parsons, New and Confirmed Triple Systems with Common Cool Primaries and Hot Companions. *AJ* 127 (2004), 2915–2930.

[468] W. Hartkopf, *IAUDS* 170 (2010).

[469] H. Abt, The Frequency of Binaries Among Normal A-Stars, *ApJS* 11 (1965), 429–460.

[470] R. Niehaus & C. Scarfe, Radial Velocity Observations of μ^1 Bootis. *PASP* 82 (1970), 1111–1118.

[471] H. A. McAlister, B. D. Mason, W. I. Hartkopf & M. M. Shara, ICCD Speckle Observations of Binary Stars. X. A Further Survey for Duplicity Among the Bright Stars. *AJ* 106 (1993), 1639–1655.

[472] M. Muterspaugh, W. I. Hartkopf, B. F. Lane, J. O'Connell, M. Willamson, S. Kulkarni *et al.*, The PHASES Differential Astrometry Archive II. Updated Binary Star Orbits and a Long-Period Eclipsing Binary. *AJ* 140 (2010), 1623–1630.

[473] O. V. Kiyaeva, A. A. Kiselev, L. G. Romanenko & O. A. Kalinichenko, Accurate Relative Positions and Motions of Poorly Studied Binary Stars. *Astr. Reports* 56 (2012), 952–965.

[474] H. Levato, S. Malaroda, W. Morrell & G. Solivella, Stellar Multiplicity in the Scorpius–Centaurus Association. *ApJS* 64 (1987), 487–503.

[475] A. Rizzutto, M. J. Ireland, J. G. Robertson, Y. Kok, P. G. Tuthill, B. A. Warrington *et al.*, Long Baseline Interferometric Multiplicity Survey of the Sco–Cen OB Association. *MN* 436 (2013), 1694–1707.

[476] J. Plaskett, The Orbits of Two Double-Lined B-Type Binaries. *PDAO* 3 (1927), 179–188.

[477] K. Abhyankar & M. Sarma, A Study of the Spectroscopic Binary ζ^2 CrB A. *MN* 133 (1966), 437–445.

[478] K. D. Gordon & C. L. Mulliss, ζ^2 Coronae Borealis, a Spectroscopic Triple System Including an Asynchronous Close Binary. *PASP* 109 (1997), 221–225.

[479] R. Lake, Photoelectric Magnitudes and Colours for Bright Southern Stars (Sixth List). *MNASSA* 24 (1965), 41–45.

[480] W. Jacob, Note from Captain Jacob to the Editor, Relative to the Ternary Star 57 Librae. *MN* 18 (1858), 317.

[481] R. G. Aitken, *Lick Obs. Bull.* 80 (1905), 147–149.

[482] H. McAlister, B. D. Mason, W. I. Hartkopf & M. M. Shara, ICCD Speckle Observations of Binary Stars. X – A Further Survey for Duplicity Among the Bright Stars. *AJ* 106 (1993), 1639–1655.

[483] R. Grellmann, Th. Ratzica, R. Köhler, Th. Preibisch & P. Mucciarelli, New Constraints on the Multiplicity of Massive Young Stars in Upper Scorpius. *A&A* 578 (2015), A84, 1–11.

[484] E. P. Horch, L. A. P. Bahi, J. R. Gaulin, S. B. Howell, W. H. Sherry, R. B. Gallé *et al.*, Speckle Observations of Binary Stars with the WIYN Telescope. VII – Measures During 2008–2009. *AJ* 143 (2012), 10, 1–10.

[485] W. D. Heintz, Photometric Astrometry of Binary and Proper-Motion Stars. V. *AJ* 99 (1990), 420–423.

[486] S. French, Deep-Sky Wonders. *Sky & Telescope* 135 (2018), 54–56.

[487] O. M. Mitchel, *Sidereal Messenger* 1 (1846) 7 + 49.

[488] E. Gledhill, J. Crossley & J. Wilson, *A Handbook of Double Stars* (London: Macmillan, 1879).

[489] W. H. Wright, The Variable Radial Velocity of Antares, *ApJ* 25 (1907), 58.

[490] B. L. Morgan, D. R. Beddoes, R. J. Scaddan & J. C. Dainty, Observations of Binary Stars by Speckle Interferometry. *MN* 183 (1978), 701–710.

[491] Reported by C. Rümker in *MN* 2 (1823), 25.

[492] E. B. Frost, S. B. Barrett & O. Struve, Radial Velocities of 500 Stars of Spectral Class A. *Publ. Yerkes Obs.* 7 (1929), 1–79.

[493] W. D. Heintz & C. Strom, The Visual Binary λ Ophiuchi. *PASP* 105 (1993), 293–293.

[494] X. Pan, M. Shao, M. M. Colavita, Sub-Millarcsecond Binary Star Astrometry. In *Complementary Approaches to Double and Multiple Star Research*, eds. H. A. McAlister & W. I. Hartkopf, ASP Conference Series 32 (San Francisco: PASP, 1992), pp. 502–509.

[495] P. R. Allen, A. J. Burgasser, J. K. Faherty & J. D. Kirkpatrick, Low Mass Tertiary Companions to Spectroscopic Binaries. I – Common Proper Motion Survey for Wide Companions using 2MASS. *AJ* 144 (2012), 62, 1–12.

[496] D. Hutter, R. T. Zavala, C. Tycner, J. A. Benson, C. A. Hummel, J. Sanbora *et al.*, Surveying the Bright Stars by Optical Interferometry. A Search for Multiplicity Among the Stars of Spectral Type F-K. *ApJS* 227 (2016), 4, 1–34.

[497] P. Baize, Éléments Orbitaux de Dix-Huit Étoiles Doubles Visuelles. *A&AS* 26 (1976), 177–193.

[498] D. McCarthy, Near Infrared Imaging of Unseen Companions to Nearby Stars. *IAU Colloq.* 76 (1983), 107–112.

[499] G. Comstock, Observations of Double Stars. *Publ. Washburn Obs.* 10 (1916), 4, 1–167.

[500] W. D. Heintz, *Veroff. Sternw. Munchen* (1966), 7, 19.

[501] W. D. Heintz, The Visual Binary μ Draconis. *PASP* 93 (1981), 90.

[502] K. Strand, The Triple System μ Draconis. *PASP* 55 (1943), 26–28.

[503] G. Ishida, On the Radial Velocity Variation of Two Visual Binaries: HR 404 and μ Draconis. *PASP* 78 (1966), 273–278.

[504] J.- L. Prieur, M. Scardia, L. Pansecchi, R. W. Argyle & M. Sala, Speckle Observations with PISCO in Merate: XI. Astrometric Measurements of Visual Binaries in 2010. *MN* 422 (2012), 1057–1070.

[505] R. F. Sanford, The Orbit of Seven Spectroscopic Binaries. *ApJ* 53 (1921), 201–223.

[506] I. Theiring & D. Reimers, Ultraviolet Observations of the Circumstellar Envelope of α^1 Her in the Line of Sight of α^2 Her. *A&A* 274 (1993), 838–846.

[507] E. Moravveji, Investigating the Semi-Regular Light Variation of the Bright M5 Supergiant α Herculis. *Ap+SS* 328 (2010), 113–117.

[508] E. Moravveji, E. F. Guinan, H. Khosroshahi & R. Wasatonic, The Age and Mass of the Alpha Herculis Triple-Star System from a MESA Grid of Rotating Stars with 1.3 $M_\odot \leq M \leq 8.0$ $M_\odot$. *AJ* 146 (2013), 148, 1–13.

[509] H. McAlister, W. I. Hartkopf, D. J. Hutter, M. M. Shara & O. G. Franz, ICCD Speckle Observations of Binary Stars I. A Search for Duplicity Among the Bright Stars. *AJ* 92 (1987), 183–194.

[510] P. Merrill, Interferometric Observations of Double Stars. *ApJ* 56 (1922), 40–52. 1921.

[511] G. Abetti, Misure Micrometriche di Coppie di Stella Eseguite nell'Anno 1921. *Pub. R. Obs. Astrof. Arcetri* 39 (1922), 1–23.

[512] D. Bonneau & R. Foy, Speckle Interferometry Observations of Binary Systems at the Haute Provence 1.93 m Telescope. *A&A* 86 (1980), 295–298.

[513] R. Ismailov, Interferometer Measures of Double Stars in 1986–1990. *A&AS* 96 (1992), 375–377.

[514] R. J. de Rosa, J. R. Patience, P. A. Wilson, A. Schneider, S. J. Wiktorowicz, A. Vigan *et al.*, The VAST Survey. III – The Multiplicity of A-Type Stars within 75 Pc. *MN* 437 (2014), 1216–1240.

[515] L. E. Pasinelli-Fracassini, L. Pastori, S. Covino & A. Pozzi, Catalogue of Apparent Diameters and Absolute Radii of Stars (CADARS), 3rd edition, Comments and Statistics. *A&A* 367 (2001), 521–524.

[516] G. Stone, Star Splitters website: `https:bestdoubles.wordpress.com`, 2011.

[517] M. Malagnini & C. Morossi, Accurate Absolute Luminosities, Effective Temperatures, Radii, Masses and Surface Gravities for a Selection of Field Stars. *A&AS* 85 (1990), 1015–1020.

[518] J. South & J. F. W. Herschel, Observations of the Apparent Distance and Position of 380 Double and Triple Stars, Made in the Years 1821, 1822 and 1823, and Compared with Those of Other Astronomers. *Phil. Trans.* 114 (1824), 1–412.

[519] R. Wittenmyer, M. Endl, W. D. Cochran, A. P. Hatzes, G. A. H. Walker, S. L. S. Yang *et al.*, Detection Limits from the McDonald Observatory Planet Search Program. *AJ* 132 (2006), 177–188.

[520] A. Irwin, S. L. Yang & G. H. Walker, 36 Ophiuchi AB: Incompatibility of the Orbit and Precise Radial Velocities. *PASP* 108 (1996), 580.

[521] W. J. Luyten, *Publ. Univ. Minnesota Obs.* 3 (1941), 3.

[522] G. Anglada-Escudé, M. Tuomi, E. Gerlach, R. Barnes, R. Heller, J. S. Jenkins. A Dynamically Packed Planetary System Around GJ 667C with Three Super-Earths in Its Habitable Zone. *A&A* 556 (2013), A126, 1–24.

[523] S. W. Burnham, Seventh Catalogue of New Double Stars. *AN* 88 (1876), 225–230.

[524] H. Howe, Micrometrical Measurements of 517 Double Stars Observed with the 11-inch Refractor During the Years 1877, 1878. *Pub. Cincinnatidare* 4, 1–74.

[525] H. C. Russell, Sydney Observatory Results of Double Star Measurements Made at the Sydney Observatory, New South Wales 1871–1881 under the Direction of H. C. Russell (1882).

[526] *Melbourne Observations*, 1869.

[527] R. T. A. Innes, Micrometrical Measurements of Double Stars 1849–1868 and 1899–1903. *Ann. Cape Obs.* 2 (1905), 4, 1–98.

[528] M. Zechmeister, M. Kürster, M. Endl, G. Lo Curto, H. Hartman, H. Nilsson *et al.*, The Planet Search Programme at the ESO CES and HARPS. IV. The Search for Jupiter Analogues Around Solar-Like Stars. *A&A*, 552 (2013), A78, 1–62.

[529] H. A. McAlister, W. I. Hartkopf, B. J. Gaston & E. M. Hendry, Speckle Interferometric Measures of Binary Stars IX. *ApJS* 54 (1984), 251–257.

[530] M. Scardia, J.-L. Prieur, L. Pansecchi *et al.*, Speckle Observations with PISCO in Merate. II – Astrometric Measurements of Visual Binaries in 2004. *MN* 367 (2006), 1170.

[531] B. D. Mason, W. I. Hartkopf, D. R. Gies, T. J. Henry & T. W. Heisel, The High Angular Resolution Multiplicity of Massive Stars. *AJ* 137 (2009), 3358–3377.

[532] A. Veramendi & J. P. González, Spectroscopic Study of Early-Type Multiple Systems. I – Orbits of Spectroscopic Binary Subsystems. *A&A* 563 (2014), A138, 1–15.

[533] G. B. Hodierna, *De Systemate Orbis Cometici Deque Admirandis Coeli Characteribus Opuscula Duo in Quorum Primo*, 1654.

[534] G. Seronik, *Binocular Highlights* (Sky Publishing, 2006).

[535] J. Nanson, Star Splitters website: https:bestdoubles .wordpress.com.

[536] J. Perez www.perezmedia/beltofvenus/archives.

[537] K. Strand, Photographic Measurements of the Six Double Stars η Cassiopeiae, γ Virginis, ξ Bootis, 44i Bootis, σ Coronae Borealis, 70 Ophiuchi and the Computation of Their Orbits with Special Attention to Those Measurements. *Ann. Leiden Obs.* 18 (1937), 2, 1–138.

[538] D. Reuyl & E. Holmberg, On the Existence of a Third Component in the System 70 Ophiuchi. *AJ* 97 (1943), 41.

[539] A. H. Batten & J. M. Fletcher, On the Orbital Period of 70 Oph. *PASP* 103 (1991), 546–555.

[540] W. Heintz, The Binary Star 70 Ophiuchi Revisited. *JRASC* 82 (1981), 140–145.

[541] B. Zuckermann, R. A. Webb, M. Schwarz & E. E. Becklin, The TW Hya Association: Discovery of T Tau Star Members near HR 4796. *ApJ* 549 (2001), L233–L236.

[542] F. Fekel & G. W. Henry, The Orbit and Pulsation Period of the γ Doradus Variable HR 6844 (V2502 Ophiuchi). *ApJ* 125 (2003), 2156–2162.

[543] J. Schlimmer, An Investigation of the Relative Proper Motion of Some Optical Double Stars. *JDSO* 5 (2009), 1, 10–17.

[544] J. Yoon, D. M. Peterson, R. L. Kurucz & R. J. Zagarello, A New View of Vega's Composition, Mass and Age. *ApJ* 708 (2010), 71–79.

[545] G. van Belle *et al.*, Interferometric Observations of Rapidly Rotating Stars. *Ast.+Ap. Rev.* 20 (2012), 51, 1–49.

[546] E. Baines, J. T. Armstrong, H. R. Schmitt, R. T. Zavala, J. T. Benson, D. J. Hutter *et al.*, Fundamental Parameters of 87 Stars from the Navy Precision Optical Interferometer. *AJ* 155 (2018), 30.

[547] E. Lamp, Ueber die Parallaxe von Σ2398. *AN* 117 (1887), 361–382.

[548] N. Wieth-Knudsen, Adaptation of Thiele's Method of Computing the Orbit of a Visual Binary in the Parabolic and Quasi-Parabolic Case. The Orbit of Σ2398 = ADS 11632. *Ann. Obs. Lund* 12 (1953), A18–A39.

[549] W. D. Heintz, Re-examination of Suspected Resolved Binaries. *ApJ* 220 (1978), 931–934.

[550] W. D. Heintz, The Red Dwarf Binary Σ2398. *PASP* 99 (1987), 1084–1088.

[551] G. Popovic, *Bull. Obs. Astr. Belgrade* 140 (1989), 83.

[552] H. Eichhorn & H. L. Alden, Parallax, Proper Motion and Mass Ratio of Σ2398 (ADS 11632). *AJ* 65 (1960), 148–153.

[553] E. Barnard, Observations of the Double Stars Castor, Σ2398 and μ^1 Herculis. *AN* 172 (1906), 383–384.

[554] Z. Berdiñas, P. J. Amado, G. Anglada-Escudé, C. Rodríguez-López & J. Barnes, High-Cadence Spectroscopy of M Dwarfs. I – Analysis of Systematic Effects in HARPS-N Line Profile Measurements on the Bright Binary GJ 725A + B. *MN* 459 (2016), 3551–3564.

[555] P. Baize, Éléments Orbitaux de Dix-Huit Étoiles Doubles Visuelles. *A&AS* 26 (1976), 177–193.

[556] H. McAlister, W. I. Hartkopf, D. J. Hutter, M. M. Shara & O. G. Franz, ICCD Speckle Observations of Binary Stars I. A Search for Duplicity Among the Bright Stars. *AJ* 92 (1987), 183–194.

[557] A. Tokovinin, Comparative Statistics and Origin of Double and Triple Stars. *MN* 389 (2008), 925–938.

[558] R. King, www.skyandtelescope.com/astronomy-news/observing-news/see-summers-best-naked-eye-double-stars-07092014/, 2014.

[559] G. Stone, Star Splitters website: https:bestdoubles .wordpress.com, 2011.

[560] S. W. Burnham, Double Star Observations Made in 1877–8 in Chicago with the 18.5-inch refractor of the Dearborn Observatory Comprising I. A Catalogue of 251 New Double Stars with Measures; II. Micrometrical Measures of 500 Double Stars. *Mems. RAS* (1878), 141–305.

[561] M. Zhao, D. R. Gies, J. D. Monnier, N. Thureau, E. Pedretti, F. Baron *et al.*, First Resolved Images of the Eclipsing and Interacting Binary β Lyrae. *ApJ* 684 (2008), L95–L98.

[562] L. Roberts, N. H. Turner & Th. ten Brummelaar, Adaptive Optics Photometry and Astrometry of Binary Stars. A Multiplicity Survey of B Stars. *AJ* 133 (2007), 545–552.

[563] J. Goodricke, Observations of a New Variable Star. *Phil. Trans* 75 (1785), 153–164.

[564] J. Winlock, Double Stars. *Harvard Ann.* 13 (1882), 17–61.

[565] R. J. de Rosa, J. Patience, A. Vigen, P. A. Wilson, A. Schneider, N. J. O'Connell *et al.*, The Volume-Limited A-Star (VAST)

Survey – II. Orbital Motion Monitoring of A-type Star Multiples. *MN* 422 (2012), 2765–2785.

[566] W. Keller, *Webb Society Deep Sky Observer* (2018), 179, 30.

[567] W. Orchiston, *John Tebbutt* (Heidelberg: Springer, 2017), 402–403.

[568] W. Campbell, The Variable Velocity of β Cygni. *PASP* 31 (1918), 38.

[569] H. McAlister & E. Hendry, Speckle Interferometric Measurements of Binary Stars VI. *ApJS* 48 (1982), 273–278.

[570] C. Worley, Albireo as a Triple Star. *Sky & Telescope* 59 (1980), 210.

[571] P. Merrill, Interferometric Observations of Double Stars. *ApJ* 56 (1922), 40–52.

[572] M. Scardia, J.-L. Prieur, L. Pansecchi, R. W. Argyle, M. Sala, S. Basso *et al.*, Speckle Observations with Pisco in Merate. IV: Astrometric Measures of Visual Binaries in 2005. *AN* 329 (2008), 54–68.

[573] D. Bonneau & R. Foy, Speckle Interferometric Observations of Binary Systems with the Haute-Provence 1.93 metre telescope. *A&A* 86 (1980), 295–298.

[574] J.-L. Prieur, L. Koechlin, N. Ginestet, J. M. Carquillat, E. Aristidi, M. Scardia *et al.*, Speckle Observations of Composite Spectrum Stars with PISCO in 1993–1998. *ApJS* 142 (2002), 95–104.

[575] C. Taylor (private communication).

[576] A. C. Maury & E. C. Pickering, Spectra of Bright Stars. *Harvard Ann.* 28 (1897), 1–128.

[577] R. F. Griffin, In Defence of Albireo. *JRASC* 93 (1999), 208–209.

[578] W. S. Houston, Deep-Sky Wonders. *Sky & Telescope* 60 (1980), 344–345.

[579] A. Kiselev & L. G. Romanenko, Dynamical Study of the Wide Visual Binary ADS 12815 (16 Cygni). *Astr. Rep.* 55 (2011), 6, 487–496.

[580] H. M. Hauser & G. W. Marcy, The Orbit of 16 Cygni AB. *PASP* 111 (1999), 321–334.

[581] J. Patience, R. J. White, A. M. Ghez, C. McCabe, I. S. McLean, J. E. Larkin *et al.*, Stellar Companions to Stars with Planets. *ApJ* 582 (2004), 654–665.

[582] N. Turner, Th. ten Brummelaar, H. A. McAlister, B. D. Mason, W. I. Hartkopf, L. C. Roberts *et al.*, Search for Faint Companions to Nearby Solar-Like Stars Using the Adaptive Optics System at Mount Wilson Observatory. *AJ* 121 (2001), 3254–3258.

[583] D. Trilling & R. H. Brown, A Circumstellar Disk Around a Star with a Known Planetary Companion. *Nature* 395 (1998), 775–777.

[584] W. D. Cochran, A. P. Hatzes, R. P. Butler & G. W. Marcy, The Discovery of a Planetary Companion to 16 Cygni B. *ApJ* 483 (1997), 457–463.

[585] H. Abt, Visual Multiples V. Radial Velocities of 160 Systems. *ApJS* 43 (1980), 549–575.

[586] A. Clerke, *The System of the Stars* (London: Longmans, 1890), p. 181.

[587] A. Maury, *Harvard Ann.* 28 (1897), 1, 93.

[588] R. H. Wilson, Observations of Double Stars. *AJ* 55 (1950), 153–159.

[589] H. A. McAlister, W. I. Hartkopf, D. J. Hutter & O. G. Franz, ICCD, Speckle Observations of Binary Stars II. Measurements during 1982–1985 from the Kitt Peak 4m Telescope. *AJ* 93 (1987), 688–727.

[590] R. F. Griffin, Spectroscopic Binary Orbits from Photoelectric Radial Velocities. Paper 202: 31 and 32 Cygni. *Observatory* 128 (2008), 362–406.

[591] B. D. Mason, *IAUDS* 179 (2013).

[592] W. H. van den Bos, Notes on Some Series of Interferometric Measures of Double Stars. *J. Obs.* 34 (1951), 85–89.

[593] J. A. Eaton, 31 Cygni: The B Star and the Wind. *AJ* 106 (1993), 2081–2095.

[594] R. Tremblot, Contribution à l'Étude des Étoiles a Spectre Composite. *Bulletin Astron. Ser.* 2, 11 (1938), 377–407.

[595] M. Fay, Una Premier: Medidas CCD de 210 Estrellas Dobles visuales en Remoto con un Telescopio Go-To. *El Observador de Estrellas Dobles* 10 (2013), 38–41.

[596] M. Muterspaugh, W. I. Hartkopf, B. F. Lane, J. O'Connell, M. Willamson, S. Kulkarni *et al.*, The PHASES Differential Astrometry Archive II. Updated Binary Star Orbits and a Long-Period Eclipsing Binary. *AJ* 140 (2010), 1623–1630.

[597] P. Labitzke, Dopplesternmessungen. *AN* 224 (1925), 197–206.

[598] A. Hale, Orbital Coplanarity in Solar-Type Binary Systems: Implications for Planetary Formation and Detection. *AJ* 107 (1994), 306–332.

[599] A. M. Larson, S. L. S. Yang & G. A. H. Walker, The Puzzling K and Early-M Giants: A Summary of Precise Radial Velocity Results for 15 Stars. In *Precise Stellar Radial Velocities*, eds. J. B. Hearnsahw & C. D. Scarfe, ASP Conference Series 185 (San Francisco: PASP, 1999), pp. 193–202.

[600] E. Toyota, Y. Itoh, S. Ishiguma, S. Urakawa, D. Murata, Y. Oasa *et al.*, Radial Velocity Search for Extrasolar Planets in Visual Binary Systems. *PASJ* 61 (2009), 19–28.

[601] A. Irwin & D. VandenBurg, A program for the Analysis of Long-Period Binaries: The Case of γ Delphini. In *Precise Stellar Radial Velocities*, eds. J. B. Hearnshaw & C. D. Scarfe, ASP Conference Series 185 (San Francisco: PASP, 1999), pp. 297–307.

[602] V. H. A. McAlister, Speckle Interferometric Observations of Binary Stars. *ApJS* 43 (1980), 327–337.

[603] O. Struve, *Pulkovo Obs.* 9 (1878).

[604] Y. Touhaimi, D. R. Gies, G. R. Schaefer, H. A. McAlister, S. T. Ridgway, N. B. Richardson *et al.*, A CHARA Array Survey of Circumstellar Disks around Nearby Be-type Stars. *ApJ* 768 (2013), 128, 1–24.

[605] M. Fay, Misure di 110 Stelle Doppie Eseguite con un Telescopio Go-To in Controllo Remoto. *Il Bollettino Delle Stelle Doppie* 6 (2013), 5–9.

[606] F. Xia & Y. Fu, The Dynamical State and Long Term Stability of HIP 102589. *ApJ* 814 (2015), 64, 106.

[607] G. Bergman, *Australian Dictionary of Biography*, Volume 2 (Melbourne: Melbourne University Press, 1967).

[608] G. Zeller, Die Systemkonstanten der Dopple und Mehrfachsterne ξ Cephei (ADS 15600), ϵ Equ (ADS 14499) und ϵ Hya (ADS 6993). *Ann. Sternw. Wien* 26 (1965), 111–126.

[609] S. W. Burnham, Measures of Proper Motion Stars Made with the 40-inch Refractor of the Yerkes Observatory in the Years 1907–1912. *Carnegie Inst. Washington* 168 (1913).

[610] A. N. Heinze, P. M. Hinz, M. Kenworthy *et al.*, Constraints on Long Period Planets from an L- and M-band Survey of Nearby Sun-Like Stars. Modelling Results. *ApJ* 714 (2010), 1570–1581.

[611] K. A. Strand, 61 Cygni as a Triple System. *PASP* 55 (1943), 29–31.

[612] F. Bessel, Bestimmung der Entfernung des 61sten des Schwans. *AN* 16 (1839), 65–96.

[613] A. N. Deutsch, An Investigation of the Motion of the Dark Companion of 61 Cygni. *Ist. Glav. Astr. Obs. Pulkova* 22 (1960), 166, 150.

[614] A. H. Deich & O. N. Orlova, Invisible Companion of the NBinary Star 61 Cygni. *Soviet Astronomy* 21 (1977), 6, 715–719.

[615] W. Hartkopf & B. Mason, *IAUDS* 184 (2014).

[616] E. B. Frost, On Certain Spectroscopic Binaries. *AN* 177 (1908), 171–174.

[617] H. A. Abt, Observations of the Variable Star τ Cygni. *ApJ* 134 (1961), 1013–1015.

[618] W. D. Heintz, The Triple System Tau Cygni. *AJ* 75 (1970), 848–850.

[619] C. Bartolini & A. Dapergolas, Tau Cygni. *IBVS* 1884 (1980).

[620] M. Muterspaugh, W. I. Hartkopf, B. F. Lane, J. O'Connell, M. Williamson, S. R. Kulkarni *et al.*, The PHASES Differential Astrometry Data Archive. II - Updated Binary Star Orbits and a Long-Period Eclipsing Binary. *AJ* 140 (2010), 1623–1631.

[621] J. Paraskévopoulos, The Orbit of the Short Period Spectroscopic Binary 65 τ Cygni. *ApJ* 53 (1921), 144–149.

[622] S. Newcomb, Observations Made with the XXVI-inch Refractor. *Washington Obs.* (1874), 280.

[623] E. Jodar, A. Pérez-Garrido & A. Díiaz-Sánchez, New Companions to Nearby Low-Mass Stars. *MN* 429 (2013), 859–867.

[624] L. Marion, O. Absil, S. Erbel, J.-B. Le Bouquin, J.-C. Augereau, N. Blind *et al.*, Searching for Faint Companions with VLTI/Pionier. II – 92 Main Sequence Stars from the Exozodl Survey. *A&A* 570 (2014), A127, 1–12.

[625] A.-M. Legrange, M. Desort, F. Galland, S. Udry & M. Mayor, Extrasolar Planets and Brown Dwarfs Around A-F Type Stars. VI. High Precision RV Survey of Early-Type Dwarfs with HARPS. *A&A* 495 (2009), 335–352.

[626] E. J. Hartung, *Astronomical Objects for Southern Telescopes* (Cambridge: Cambridge University Press, 1968).

[627] W. I. Hartkopf, ad.usno.navy.mil/lin1/lelements .html.

[628] D. Gezari, A. Labeyrie & R. V. Stachnik, Speckle Interferometry: Diffraction Limited Measurement of Nine Stars with the 200-Inch Telescope. *ApJL* 173 (1972), L1–L5.

[629] J.-F. Donati, G. A. Wade, J. Babel, H. F. Henrichs, J. A. de Jong & T. J. Harris, The Magnetic Field and Wind Confinement of β Cephei. New Clues for Interpreting the Phenomenon. *MN* 326 (2001), 1265–1278.

[630] N. Nardetto, D. Howard & I. Tallon-Bose, An Investigation of the Close Environment of β Cephei with the VEGA/CHARA Interferometer. *A&A* 525 (2011), A67, 1–6.

[631] W. D. Heintz, Observations of Double Stars and New Pairs. XVI. Results in 1992–1994. *ApJS* 99 (1995), 693–700.

[632] S. W. Burnham, cited by Aitken in *ADS Catalogue* (1932).

[633] J. A. Hynek, A Survey of Stars with Composite Spectra. *Contrib. Perkins Obs.* 1 (1938), 10, 1–72.

[634] C. R. Vickers & C. D. Scarfe, A Spectroscopic Study of the Triple System ξ Cephei. *PASP* 88 (1976), 944–948.

[635] H. A. McAlister, Speckle Interferometric Measurements of Binary Stars. I. *ApJ* 215 (1977), 159–165.

[636] C. Farrington, Th. ten Brummelaar, B. D. Mason, W. I. Hartkopf, D. Mourard, E. Moravveji *et al.*, Separated Fringe Packet Observations with the CHARA Array. *AJ* 148 (2014), 48, 1–8.

[637] H. A. Abt, The Frequency of Binaries Among Metallic-Line Stars. *ApJS* 6 (1961), 37–74.

[638] A. Hale, Orbital Co-planarity in Solar-Type Binary Systems: Implications for Planet System Formation and Detection. *AJ* 107 (1994), 306–332.

[639] G. Cutispoto, L. Pastori, L. Pasquini, J. R. de Medeiros, G. Tagliaferri, J. Anderson *et al.*, Fast Rotating Nearby Solar-Type Stars. I – Spectral classification $v \sin i$, Li Abundances and X-Ray Luminosities. *A&A* 384 (2002), 491–593.

[640] D. W. Willmarth, F. C. Fekel, H. A. Abt & D. Pourbaix, Spectroscopic Orbits of 15 Late-Type Stars. *AJ* 152 (2016), 46, 1–13.

[641] K. Fuhrmann, R. Chini, L. Kaderhandt & Z. Chen, Multiplicity Among Solar-Type Stars. *ApJ* 836 (2017), 139, 1–23.

[642] E. E. Barnard, Observations of the Double Star Krueger 60. *MN* 76 (1916), 592–606.

[643] A. Krueger, *Catalog von 14680 Sternen Zwischen 54° 55′ und 65° 10′ Nördlicher Declination 1855 für das Aequinoctium 1875* (Leipzig: Astronomische Gesellschaft, 1875).

[644] S. W. Burnham, Measures of the Stars Noted as Double in Krueger's Catalogue of the Astronomische Gesellschaft, Zone +55° to +65°. *AN* 127 (1891), 289.

[645] E. Doolittle, On the Probable Motion in the Stellar System, 'Krueger 60'. *AJ* 21 (1901), 47–48.

[646] P. van de Kamp & S. L. Lippincott, Flare-Up of Krueger 60 B. *PASP* 63 (1951), 141.

[647] W. D. Heintz, Orbits of 20 Visual Binaries. *A&AS* 65 (1986), 411–417.

[648] A. N. Heinze, P. M. Hinz, S. Sivanadram, M. Kenworthy, M. Meyer, D. Miller *et al.*, Constraints on Long Period Planets from an L- and M-Band Survey of Nearby Sun-Like Stars. Observations. *ApJ* 714 (2010), 1551–1569.

[649] W. Knapp & J. Nanson, Kruger 60. *JDSO* 14 (2018), 3–21.

[650] K. A. Strand, The Orbital Motion of the Triple System Zeta Aquarii. *AJ* 49 (1942), 165–172.

[651] J. Ebersberger & G. Wiegelt, Speckle Interferometry and Speckle Holography with the 1.5m and 3.6m ESO Telescopes. *ESO Messenger* 18 (1979), 24–29.

[652] D. McCarthy, F. J. Low, F. J. Kleinmann *et al.*, Infrared Detection of the Low Mass Companion to Zeta Aquarii B. *ApJL* 259 (1983), L75–L78.

[653] A. A. Tokovinin, The Triple System Zeta Aquarii. *ApJ* 853 (2016), 151, 1–6.

[654] W. D. Heintz, The Triple Star Zeta Aquarii. *ApJ* 284 (1984), 806–809.

[655] J. Goodricke, A Series of Observations and a Discovery of the Period of the Variability of the Light of the Star Marked δ by Bayer, near the Head of Cepheus. *Phil. Trans.* 76 (1786), 48–61.

[656] S. W. Burnham, Double Star Observations Made in 1877–8 in Chicago with the 18.5-inch refractor of the Dearborn Observatory Comprising I. A Catalogue of 251 New Double Stars with Measures; II. Micrometrical Measures of 500 Double Stars. *Mems. RAS* (1878), 141–305.

[657] A. Bélopolsky, The Spectrum of δ Cephei. *ApJ* 1 (1895), 160–161.

[658] H. Shapley, On the Nature and Cause of Cepheid Variation. *ApJ* 40 (1914), 448–465.

[659] H. Leavitt & E. C. Pickering, *Harvard Coll. Obs. Circs.* (1912), 173, 1.

[660] M. Fay, Decouvertes et Résultats de Measures de 224 Étoiles Doubles Visuelles avec un Télescope GoTo piloté à distance. Campagne d'Observation 2011–2012. *Il Bolettino Delle Stelle Doppie* 6 (2013), 46–56.

[661] F. Benedict, B. E. McArthur, L. W. Fredericks *et al.*, Astrometry with the HST: A Parallax of the Fundamental Distance Calibrator δ Cephei. *AJ* 124 (2002), 1695–1705.

[662] R. Anderson, J. Sahlmann, B. Holl, L. Eyer, L. Palaversa, N. Mowlavi *et al.*, Revealing δ Cephei's Secret Companion and Intriguing Past. *ApJ* 804 (2015), 144, 1–11.

[663] A. Galenne, A. Meŕand, P. Kervella. J. D. Monnier, G. Schaefer, R. M. Roettenbacher *et al.*, Multiplicity of Galactic Cepheids from Long Baseline Interferometry. III Sub-Percent Limits on the Relative Brightness of a Close Companion of δ Cephei. *MN* 461 (2016), 1451–1456.

[664] J. Daley, Ludwig Schupmann Observatory Measures of Large Δm Pairs – Part Three. *JDSO* 5 (2009), 149–155.

[665] H. A. McAlister, W. I. Hartkopf, D. J. Hutter, M. M. Shara & O. G. Franz, ICCD Speckle Observations of Binary Stars I. Survey For Duplicity Among the Bright Stars. *AJ* 92 (1987), 183–194.

[666] R. Gili & J. -L. Prieur, Relative Astrometric and Photometric Measurements of Visual Binaries Made with the Nice 76-cm Refractor. *AN* 333 (2012), 727–735.

[667] H. Zirm, *IAUDS* 161 (2007).

[668] W. S. Jacob, Catalogue of Double Stars, Deduced from Observations at Poona from November 1845 to February 1848. *Mems. RAS* 17 (1849), 79–91.

[669] A. A. Tokovinin, T. Pribulla & D. Fischer, Radial Velocities of Southern Visual Multiple Stars. *AJ* 149 (2015), 8, 1–9.

[670] O. Struve, Catalogue de 514 Etoiles Double et Multiples Decouvertes sur l'Hemisphère Celeste Boréal par la Grande Lunette de l'Observatoire Centrale de Poulkova et Catalogue de 256 Etoiles Doubles Principale où la Distance des Composantes est de 32 Secondes à 2 Minutes et qui se trouvent sur l'Hemisphere Boreál. St Petersburg: *Imperial Acad. Sci.* (1843), 41.

[671] W. Campbell, A List of Six Stars Whose Velocities in the Line of Sight are Variable. *ApJ* 14 (1901), 138–140.

[672] W. Harper, The Orbits of Five Spectroscopic Binaries. *PDAO* 3 (1925), 189–297.

[673] C. D. Scarfe, J, Regan & D. Barlow, Revised Orbits for 105 Herculis and π Cephei and a Model for the π Cephei System. *MN* 203 (1983), 103–116.

[674] V. Trimble & K. Thorne, Spectroscopic Binaries and Collapsed Stars. *ApJ* 156 (1965), 1013–1019.

[675] G. Gatewood, I. Han, J. K. de Jonge, C. T. Reiland & D. Pourbaix, Hipparcos and MAP Studies of the Triple Star π Cephei. *ApJ* 549 (2001), 1145–1150.

[676] B. Fried (private communication), 2017.

[677] S. W. Burnham, Double Star Observations Made in 1877–8 in Chicago with the 18.5-inch Refractor of the Dearborn Observatory Comprising I. A Catalogue of 251 New Double Stars with Measures; II. Micrometrical Measures of 500 Double Stars. *Mems. RAS* (1878), 141–305.

[678] P. Baize & M. Petit, Étoiles doubles orbitales a composantes variables. *A&AS* 77 (1989), 497–511.

[679] W. Steinicke, & William Herschel, Flamsteed Numbers and Harris's Star Maps. *J. Hist. Astron.* 45 (2014), 289–307.

[680] C. Taylor (private communication), 2016.

[681] Gaia Collaboration, Description of the Gaia Mission (Spacecraft, Instruments, Survey and Measurement Principles, and Operations) (2016).

[682] Gaia Collaboration, Summary of the Contents and Survey Properties (2018).

[683] D. Wright, Thomas Lewis: A Lifetime of Double Stars, *JBAA* 102 (1992), 95–101.

[684] https://www.webbdeepsky.com/double-star/mizarA-orbital-motion-animation.

[685] http://www.handprint.com/ASTRO/bineye5.html.

[686] C. Worley, Is This Orbit Really Necessary (II)? In *Error, Bias, and Uncertainties in Astronomy*, C. Jascheck & F. Murtagh, eds. (Cambridge: Cambridge University Press, 1990), 419–422.

OBJECT INDEX